Introduction to tropical fish stock assessment

Part I - Manual

by
Per Sparre
Marine Resources Service
Fishery Resources and Environment Division
FAO Fisheries Department
Rome, Italy

Siebren C. Venema
Project Manager
GCP/INT/392/DEN
Rome, Italy

FAO
FISHERIES
TECHNICAL
PAPER

306/1

Rev. 1

DANIDA

FOOD
AND
AGRICULTURE
ORGANIZATION
OF THE
UNITED NATIONS
Rome, 1992

The designations employed and the presentation of material in this publication do not imply the expression of any opinion whatsoever on the part of the Food and Agriculture Organization of the United Nations concerning the legal status of any country, territory, city or area or of its authorities, or concerning the delimitation of its frontiers or boundaries.

M-43
ISBN 92-5-103272-6

All rights reserved. No part of this publication may be reproduced, stored in a retrieval system, or transmitted in any form or by any means, electronic, mechanical, photocopying or otherwise, without the prior permission of the copyright owner. Applications for such permission, with a statement of the purpose and extent of the reproduction, should be addressed to the Director, Publications Division, Food and Agriculture Organization of the United Nations, Viale delle Terme di Caracalla, 00100 Rome, Italy.

© FAO 1992

PREPARATION OF THIS DOCUMENT

The first edition of the manual "Introduction to tropical fish stock assessment" was prepared by the FAO/DANIDA project "Training in fish stock assessment and fisheries research planning" (GCP/INT/392/DEN) for use in a series of regional and national training courses on fish stock assessment.

In 1984 the author, Per Sparre, then employed by the Danish Institute for Fisheries and Marine Research (DIFMAR), Charlottenlund, Denmark was asked to write it on the basis of lecture notes and case studies prepared by the team of lecturers engaged in the courses.

The first edition was printed in July 1985 in Manila, the Philippines, and distributed by the project through the Network of Tropical Fisheries Scientists of the International Center for Living Aquatic Resources Management (ICLARM) and training courses.

In 1989 the manual underwent a thorough revision by Mr. P. Sparre, Dr. E. Ursin, former Chief Scientist and Director of DIFMAR, and Mr. S.C. Venema, Project Manager, GCP/INT/392/DEN. For practical reasons the exercises and their solutions were placed in a separate volume. This version was published in 1989 as FAO Fisheries Technical Paper 306.1 and 306.2. It was widely distributed and used extensively in training courses.

Unfortunately, it soon became evident that there were a lot of errors and imperfections in text, tables and figures. This is particularly annoying when the material is used for training purposes. In the meantime the proces of translation into Spanish, Portuguese and Indonesian had started. The translators, lecturers, course participants and numerous other users signalled errors and imperfections. Early in 1991 when the English version was out of print, it was decided to undertake another thorough revision mainly for didactical purposes, and to base the translations on this revised text.

The original manual was closely connected to the LFSA (Length Frequency Stock Assessment) program produced by Mr. Sparre, and also to the COMPLEAT ELEFAN programs produced by ICLARM. These two programs will be replaced by one program called FiSAT (FAO/ICLARM Stock Assessment Tools). The references in the text to computer programs have been adapted to the new program.

The number of scientists who have provided corrections and comments is too large to list by name. We gratefully acknowledge their contributions.

The figures were partly revised in Chile by Dr. P. Arana and Mr. Alvara Nuñez. Typing and word processing was taken care of by Ms. J. Ugilt in Denmark.

Versions in Spanish, Portuguese and Indonesian will appear in 1992, while translation into French will also be initiated.

Sparre, P.; Venema, S.C.
Introduction to tropical fish stock assessment. Part 1.
Manual. FAO Fisheries Technical Paper No. 306.1, Rev. 1.
Rome, FAO. 1992, 376 p.

ABSTRACT

In Part 1, Manual, a selection of methods on fish stock assessment is described in detail, with examples of calculations. Special emphasis is placed on methods based on the analysis of length-frequencies. After a short introduction to statistics, it covers the estimation of growth parameters and mortality rates, virtual population methods, including age-based and length-based cohort analysis, gear selectivity, sampling, prediction models, including Beverton and Holt's yield per recruit model and Thompson and Bell's model, surplus production models, multispecies and multifleet problems, the assessment of migratory stocks, a discussion on stock/recruitment relationships and demersal trawl surveys, including the swept area method. The manual is completed with a review of stock assessment, where an indication is given of methods to be applied at different levels of availability of input data, a review of relevant computer programs produced by or in cooperation with FAO, and a list of references, including material for further reading.

In Part 2, Exercises, a number of exercises is given with solutions. The exercises are directly related to the various chapters and sections of the manual.

Distribution:

DANIDA
Participants at courses on
    Fish Stock Assessment organized by
    project GCP/INT/392/DEN
Members of ICLARM's Network of Tropical
    Fisheries Scientists
Institutes specialised in Tropical
    Fish Stock Assessment
Institutes of Fisheries Education
Interested National and Inter-
    national Organizations
Marine and Inland Selectors
FAO Fisheries Department

CONTENTS

	Page
LIST OF EXAMPLES	x
LIST OF SYMBOLS	xi

1 INTRODUCTION   1

    1.1 THE PRIMARY OBJECTIVE OF FISH STOCK ASSESSMENT   1

    1.2 THE STOCK CONCEPT   2

    1.3 MODELS   5

         1.3.1 Analytical models   7
         1.3.2 Holistic models   9

    1.4 ASSESSMENT OF STOCKS IN TROPICAL WATERS   9

    1.5 DEFINITIONS OF BODY LENGTH   11

    1.6 AGE AND RECRUITMENT   13

    1.7 THE UNDERLYING ASSUMPTION OF RANDOM SAMPLES   14

    1.8 THE ORGANIZATION OF THE MANUAL   15

    1.9 FURTHER READING   19

2 BIOSTATISTICS   20

    2.1 MEAN VALUE AND VARIANCE   20

    2.2 THE NORMAL DISTRIBUTION   23

    2.3 CONFIDENCE LIMITS   26

    2.4 ORDINARY LINEAR REGRESSION ANALYSIS   28

    2.5 THE CORRELATION COEFFICIENT AND FUNCTIONAL REGRESSION   33

    2.6 LINEAR TRANSFORMATIONS   37

3 ESTIMATION OF GROWTH PARAMETERS   44

    3.1 THE VON BERTALANFFY GROWTH EQUATION   44

         3.1.1 Variability and applicability of growth parameters   48
         3.1.2 The weight-based von Bertalanffy growth equation   50

    3.2 INPUT DATA FOR THE VON BERTALANFFY GROWTH EQUATION   50

         3.2.1 Data from age readings and length measurements   51
         3.2.2 Length composition data (without age compositions)   57
         3.2.3 Data from commercial catches   59

    3.3 METHODS FOR ESTIMATION OF GROWTH PARAMETERS FROM LENGTH-AT-AGE DATA   59

         3.3.1 The Gulland and Holt plot   59
         3.3.2 The Ford-Walford plot and Chapman's method   62
         3.3.3 The von Bertalanffy plot   64
         3.3.4 The least squares method   66

		Page
3.4	ESTIMATION OF AGE COMPOSITION FROM LENGTH-FREQUENCIES	67
	3.4.1 Bhattacharya's method	74
	3.4.2 Modal progression analysis	88
	3.4.3 The probability paper and parabola methods	96
3.5	FITTING GROWTH CURVES BY MEANS OF COMPUTER PROGRAMS	98
	3.5.1 ELEFAN I	99
	3.5.2 The seasonalized von Bertalanffy growth equation	103
	3.5.3 Maximum likelihood methods	104
	3.5.4 Limitations of length-frequency analysis	108

4 ESTIMATION OF MORTALITY RATES — 111

4.1 THE CONCEPT OF A COHORT AND SOME BASIC NOTATION — 111

4.2 THE DYNAMICS OF A COHORT, THE EXPONENTIAL DECAY MODEL — 113

4.3 ESTIMATION OF Z FROM CATCH PER UNIT OF EFFORT DATA AND THE CONCEPT OF THE CATCHABILITY COEFFICIENT — 120

    4.3.1 Heincke's method — 124
    4.3.2 Robson and Chapman's method — 125

4.4 ESTIMATION OF Z FROM A LINEARIZED CATCH CURVE — 126

    4.4.1 The constant parameter system — 126
    4.4.2 The linearized catch curve equation — 128
    4.4.3 The linearized catch curve based on age composition data — 128
    4.4.4 The linearized catch curve based on age compositions with variable time intervals — 131
    4.4.5 The linearized catch curve based on length composition data — 132
    4.4.6 The cumulated catch curve based on length composition data. (The Jones and van Zalinge method) — 137
    4.4.7 Summary of the linearized catch curve methods — 140

4.5 BEVERTON AND HOLT'S Z-EQUATIONS — 140

    4.5.1 Beverton and Holt's Z-equation based on length data — 142
    4.5.2 Beverton and Holt's Z-equation based on age data — 143
    4.5.3 Beverton and Holt's Z-equation based on length-at-first-capture — 144
    4.5.4 The Powell-Wetherall method — 145

4.6 A PLOT OF Z ON EFFORT FOR SEPARATE ESTIMATES OF F AND M — 147

4.7 NATURAL MORTALITY — 149

    4.7.1 Natural mortality and longevity — 150
    4.7.2 Pauly's empirical formula — 151
    4.7.3 Rikhter and Efanov's formula — 152

5 VIRTUAL POPULATION METHODS — 153

5.1 VIRTUAL POPULATION ANALYSIS (VPA) — 153

5.2 AGE-BASED COHORT ANALYSIS (POPE'S COHORT ANALYSIS) — 160

5.3 JONES' LENGTH-BASED COHORT ANALYSIS — 165

	Page

6 GEAR SELECTIVITY — 172

    6.1   ESTIMATION OF TRAWL NET SELECTION — 172

    6.2   ESTIMATION OF GILL NET SELECTION — 175

        6.2.1   Symmetrical selection curves — 175
        6.2.2   The product of two logistic curves — 181

    6.3   DISCUSSION OF SELECTION BY OTHER GEARS — 185

    6.4   OTHER ASPECTS OF GEAR SELECTIVITY — 186

        6.4.1   Knife-edge selection — 186
        6.4.2   Recruitment and selectivity — 187
        6.4.3   Selectivity as a function of age — 188

    6.5   ESTIMATION OF THE SELECTION OGIVE FROM A CATCH CURVE — 190

    6.6   GEAR SELECTIVITY AND VPA METHODS — 194

        6.6.1   Gear selectivity and fishing mortality — 194
        6.6.2   Estimation of selection curves from cohort analysis — 196

    6.7   USING A SELECTION CURVE TO ADJUST LENGTH-FREQUENCY SAMPLES — 196

7 SAMPLING — 200

    7.1   SIMPLE RANDOM SAMPLING — 200

    7.2   STRATIFIED RANDOM SAMPLING — 203

    7.3   PROPORTIONAL SAMPLING — 208

    7.4   SAMPLING COMMERCIAL CATCHES — 209

    7.5   ESTIMATION OF THE TOTAL CATCH IN WEIGHT OF SPECIES S — 212

    7.6   ESTIMATION OF THE LENGTH COMPOSITION OF SPECIES S IN THE TOTAL CATCH — 213

8 PREDICTION MODELS — 222

    8.1   ASSUMPTIONS AND MODELS UNDERLYING THE YIELD PER RECRUIT MODEL OF BEVERTON AND HOLT — 224

    8.2   BEVERTON AND HOLT'S YIELD PER RECRUIT MODEL — 225

    8.3   BEVERTON AND HOLT'S BIOMASS PER RECRUIT MODEL — 231

    8.4   BEVERTON AND HOLT'S RELATIVE YIELD PER RECRUIT MODEL — 233

    8.5   YIELD PER RECRUIT FROM LENGTH DATA — 235

    8.6   AGE-BASED THOMPSON AND BELL MODEL — 235

    8.7   LENGTH-BASED THOMPSON AND BELL MODEL — 244

    8.8   PREDICTION OF THE EFFECTS OF CHANGES OF MESH SIZES USING THE THOMPSON AND BELL METHOD — 251

	Page
9 ESTIMATION OF MAXIMUM SUSTAINABLE YIELD USING SURPLUS PRODUCTION MODELS	253
9.1 THE SCHAEFER AND FOX MODELS	253
9.2 GULLAND'S FORMULA	260
9.3 CADIMA'S FORMULA	262
9.4 MSY ESTIMATORS BASED ON THE SURPLUS PRODUCTION MODEL	263
9.4.1 Validation of estimates of MSY based on empirical formulas	264
9.5 MUNRO AND THOMPSON PLOT	265
9.6 STANDARDIZATION OF EFFORT	267
9.7 THE DERISO/SCHNUTE DELAY DIFFERENCE MODEL	271
10 MULTISPECIES/MULTIFLEET PROBLEMS	272
10.1 SURPLUS PRODUCTION MODELS APPLIED TO MULTISPECIES/MULTIFLEET SYSTEMS	272
10.2 BIOLOGICAL INTERACTION	274
10.3 ECONOMIC INTERACTION	275
10.4 TECHNICAL INTERACTION	275
10.4.1 A yield per recruit model for mixed fisheries	275
10.4.2 Assessment of mixed fisheries based on length-frequency data	276
10.4.3 Multifleet mixed fisheries	278
11 ASSESSMENT OF MIGRATORY STOCKS	282
11.1 THE CONCEPT AND STUDY OF MIGRATION	282
11.2 BIAS CAUSED BY MIGRATION	285
11.3 THE ANNUAL-RETURN MATCHED SAMPLES METHOD	291
11.3.1 Estimation of growth parameters by the annual-return matched samples method	291
11.4 THE GENERAL MATCHED SAMPLES METHOD	295
11.5 ASSESSMENT BASED ON TAGGING DATA	297
11.6 ESTIMATION OF THE GROWTH PARAMETERS OF A MIGRATORY STOCK: THE ATLANTIC MACKEREL	298
12 THE STOCK/RECRUITMENT RELATIONSHIP	301
12.1 CLASSICAL S/R CONSIDERATIONS	302
12.2 THE STABILITY OF RECRUITMENT	305
12.3 TOWARDS MODELLING RECRUITMENT	306

		Page
13	**DEMERSAL TRAWL SURVEYS**	307
13.1	THE BOTTOM TRAWL	307
13.2	PLANNING A DEMERSAL TRAWL SURVEY	308
13.3	DATA RECORDING	309
13.4	DECK SAMPLING AND CATCH RECORDING PROCEDURES	310
13.5	THE SWEPT AREA	311
13.6	BIOMASS ESTIMATION BY THE SWEPT AREA METHOD	313
13.7	PRECISION OF THE ESTIMATE OF BIOMASS	314
13.8	ESTIMATION OF MAXIMUM SUSTAINABLE YIELD	316
14	**SUMMARY OF FISH STOCK ASSESSMENT**	317
14.1	GENERAL ASPECTS OF FISH STOCK ASSESSMENT	318
14.2	REVIEW OF METHODS TO BE USED ACCORDING TO THE TYPE OF DATA AVAILABLE	321
15	**MICROCOMPUTER PROGRAM PACKAGES**	330
15.1	THE LFSA PACKAGE	330
	15.1.1 Length-frequency (LF) programs	330
	15.1.2 Age/length analysis: estimation of growth parameters from age/length data	333
	15.1.3 Miscellaneous programs	334
15.2	THE COMPLEAT ELEFAN PACKAGE	334
15.3	THE FiSAT PACKAGE	335
15.4	OTHER FISH STOCK ASSESSMENT PROGRAMS PRODUCED BY FAO	337
	15.4.1 The ANACO package	337
	15.4.2 The ANALEN package	337
	15.4.3 BEAM 1 and BEAM 2	337
	15.4.4 BEAM 3	338
	15.4.5 BEAM 4	338
	15.4.6 The NAN-SIS package	338
	15.4.7 CLIMPROD	338
16	**REFERENCES**	339
	SUBJECT INDEX	363
	APPENDIX TABLES	369
A1	LIST OF IMPORTANT FORMULAS	370
A2	METHODS BASED ON LINEAR TRANSFORMATIONS AND ORDINARY LINEAR REGRESSION ANALYSIS: $y = a + b*x$	374
A3	IMPORTANT DATES EXPRESSED AS FRACTIONS OF A YEAR FROM 1 JANUARY	376
A4	FRACTILES OF THE t-DISTRIBUTION (STUDENT'S DISTRIBUTION)	376

LIST OF EXAMPLES

		Page
1:	Length-weight relationship	37
2:	Linearization of a normal distribution	40
3:	Age/length composition data from a single survey	51
4:	Age/length composition data from multiple surveys	54
5:	The use of age/length keys	55
6:	Estimating K and L_∞ with the Gulland and Holt plot	60
7:	Estimating K and t_o with the von Bertalanffy plot	65
8:	Estimating the age of species from temperate waters	67
9:	Estimating the age of coral trout, a tropical species	68
10:	A Bhattacharya analysis of a constructed data set	75
11:	Modal progression analysis, based on the data of Example 4	88
12:	The application of ELEFAN I to the coral trout data (Ex. 9)	99
13:	Catch curve with constant time intervals, North Sea whiting	129
14:	Catch curve based on length composition data, <u>Upeneus vittatus</u>	134
15:	The Jones and van Zalinge method, <u>Upeneus vittatus</u>	138
16:	The Powell-Wetherall method	145
17:	Estimation of M and q of a tropical fish	147
18:	Virtual population analysis (VPA), North Sea whiting	154
19:	Pope's cohort analysis, North Sea whiting	162
20:	Jones' length-based cohort analysis, hake, Senegal	165
21:	Covered codend experiment, <u>Nemipterus japonicus</u>, South China Sea	172
22:	Estimation of gill net selection curves, Tilapia, Lake Victoria	177
23:	Estimation of the resultant selection ogive from a catch curve, hypothetical data	191
24:	Using a selection curve to adjust the length-frequency sample of Table 6.5.1	199
25:	Stratified random sampling	205
26:	Stratified random sampling, considering costs	206
27:	Sampling scheme for a tropical demersal fishery	210
28:	Y/R as a function of F, for a tropical species	228
29:	Age-based Thompson and Bell analysis, tropical shrimp	237
30:	Length-based Thompson and Bell analysis, hake, Senegal	247
31:	Schaefer and Fox models, demersal fish, Java Sea	255
32:	Summation of effort for different effort units	269

LIST OF SYMBOLS

Section

A. Symbols used in formulas for fish stock assessment

A	attrition rate	11.5
a	swept area (effective path swept by a trawl)	13.5
ASP	available sum of peaks (ELEFAN)	3.5
b	constant in length-weight relationship $W = q*L^b$	2.6
B	biomass	8.6
Bv	virgin (unexploited) biomass	8.3, 9.1
B/R	biomass per recruit	8.2
C	catch in numbers (VPA)	5.0
$C(t,\infty)$	cumulated catch (from age t to maximum age)	4.4
C	amplitude (0-1) (ELEFAN)	3.5
C_o	fixed costs of a sampling programme	7.2
CPUA	catch per unit of area	13.6
CPUE	catch per unit of effort	4.3, 9.0, 9.5
D	number of natural deaths (VPA)	5.0
D50%	deselection, length at which 50% is <u>not</u> caught	6.2
dL	interval size of length	2.1
E	fishing effort	7.4
E	exploitation rate (F/Z)	8.4
ESP	explained sum of peaks (ELEFAN)	3.5
f	fishing effort	4.3
F	fishing mortality coefficient or instantaneous rate (per time unit)	4.2
Fm	maximum fishing mortality	6.6
F-array	array of F-at-age, fishing pattern	5.1
F-factor	multiplication factor of F (Thompson and Bell), X	8.6
G	natural mortality factor in Pope's cohort analysis	5.2
H	natural mortality factor in Jones' length-based cohort analysis	5.3
I	separation index	3.5
K	curvature parameter	3.1
L	length	general
L1-L2	length class	general
L1,L2	from length L1 to length L2	general
L_∞ or $L\infty$	L infinity, asymptotic length (mean length of very old fish)	3.1
L'	some length for which all fish of that length and larger are under full exploitation (lower limit of corresponding length interval)	4.5
$\overline{Lc}$	average length of the entire catch	4.5
Lc or L50%	length at which 50% of the fish is retained by the gear and 50% escape	4.5
L75% or L75	length at which 75% of the fish is retained in the gear	6.1
Lm	optimum length for being caught	6.2
m	= K/Z	8.4
M	natural mortality coefficient or instantaneous rate of natural mortality or natural mortality rate (per time unit)	4.1, 4.7
MSE	Maximum Sustainable Economic Yield	8.7
MSY	Maximum Sustainable Yield	1.1, 4.5, 8.2, 9.1-9.7, 13.7
N	number of survivors (VPA)	4.1, 5.0
N(t)	number of survivors of a cohort attaining age t	4.1
<u>N</u>(Tr)	number of recruits to the fishery	4.1
$\overline{N}$	average numbers of survivors of a cohort	4.2
ϕ'	(phi prime), $\ln K + 2*\ln L_\infty$	3.4

		Section
q	condition factor, constant in length-weight relationship	2.6, 3.1
q	catchability coefficient	4.3, 4.6, 9.2
R	recruitment, number of recruits, N(Tr)	4.1
S	survival rate	4.2
S_L or S(L)	logistic curve (length-based gear selectivity)	6.1
S_t or S(t)	logistic curve (age-based gear selectivity)	6.4
S1 and S2	constants in the formula for the length-based logistic curve	6.1
SF	selection factor	6.1
SR	reversed logistic curve	6.2
S/R	stock recruitment relationship	12.0
t	time (usually in years)	general
t'	some age for which all fish of that age and older are under full exploitation	4.5
$\bar{t}$	mean age of all fish of age t' and older	4.5
T	ambient temperature in °C	4.7
Tc	age-at-first-capture (start of exploited phase)	4.1
Tm	longevity (maximum age)	4.7
Tm50%	age of massive maturation (50% of population mature)	4.7
t_o	t-zero, initial condition parameter (in years)	3.1
Tr	age-at-recruitment to the fishery	4.1
ts	summerpoint (0-1) (ELEFAN)	3.5
tw	winterpoint (0-1) (ELEFAN)	3.5
t50%	age at which 50% of the fish is retained in the gear (Thompson and Bell)	6.4
T1 and T2	constants in the formula for the age-based logistic curve	6.4
U	$1 - L_c/L_\infty$	8.4
$\bar{v}$	average price (Thompson and Bell)	8.6
V	value (Thompson and Bell)	8.6
VPA	Virtual Population Analysis	5.0
w	weight (usually of one specimen)	general
W_∞ or W∞	weight infinity, asymptotic weight (W infinity, mean weight of very old fish)	3.1
X	multiplication factor of F (Thompson and Bell)	8.6
y	year (usually as an index)	8.6
Y/R	yield per recruit (Beverton and Holt)	8.2
(Y/R)'	relative yield per recruit (Beverton and Holt)	8.4
Z	total mortality coefficient, instantaneous rate of total mortality or total mortality rate (per time unit)	4.2

B: Mathematical notation (general)

*	multiplication sign
/	division sign
ln	natural logarithm (base e = 2.7182818)
log	10 based logarithm
exp(x) or e^x	exponential function, exp(x) = e^x
$\sum_{i=1}^{i=n} X(i)$	sum of all values of X(i), for i from 1 to n; the sum X(1)+X(2)+...+X(n)
$\sqrt{}$ or $\sqrt{}$	square root

∞	infinity
Δx	delta x, a small increment of the variable x
$\mathrm{MAX}_j\{X(j)\}$	maximum value among the elements in the set $\{X(j)\} = \{X(1), X(2), \ldots X(j), \ldots\}$
$\bar{x}$	mean value of x
x(i,j)	i,j indices of x (usually printed as $x_{i,j}$)
π	pi = 3.14159
a < b	a smaller than b
a > b	a greater than b
a => b	a greater than or equal to b
tanh	hyperbolic tangent

C. Statistical notation

ϵ	(epsilon) maximum relative error
f	degrees of freedom
F	observed frequency
Fc	calculated or theoretical frequency
n	number of observation
s	standard deviation
s^2	variance
$s/\sqrt{n}$	standard error
$s/\bar{x}$	relative standard deviation or coefficient of variation
$\bar{x}$	mean value of x
y = a+b*x	linear regression
a	intercept of ordinary regression
a'	intercept of functional regression
b	slope of ordinary regression
b'	slope of functional regression
r	correlation coefficient
sa	standard deviation of the intercept (a)
sa^2	variance of the intercept (a)
sb	standard deviation of the slope (b)
sb^2	variance of the slope (b)
sx	standard deviation of the independent variable (x)
sx^2	variance of the independent variable (x)
sxy	covariance
sy	standard deviation of the dependent variable (y)
sy^2	variance deviation of the dependent variable (y)
x	independent variable
y	dependent variable

1 INTRODUCTION

Several excellent textbooks and manuals dealing with fish stock assessment explain the theory behind the various models and methods, including the mathematical derivation of formulas (see for example Gulland, 1969 and 1983 and Csirke, 1980a). The problem is that from such manuals the novice fishery scientists can usually not derive the precise instructions needed to perform each analysis. Where experienced scientists are available such instructions can easily be provided on-the-job and through participation in working group meetings. However, there are still many countries where such a transfer is not possible. This manual is an attempt to document that part of the instructions that is usually transferred through on-the-job training. It concentrates on the application of methods, while less attention is given to detailed explanations of the theory behind them.

With the help of this manual the fishery scientist should be able to make a start with data analyses and building up the necessary skills and insight in solving stock assessment problems. After this initial step it should be easier to access more complex textbooks.

Stock assessment of tropical resources has developed rapidly in the last decade in particular through the work of Pauly (1979, 1980, 1984), Saila and Roedel (1980), Pauly and David (1981), Garcia and Le Reste (1981) and Munro (1983), but also because of the rapid development of microcomputer hard- and software. This manual is intended to further contribute to this development and for that reason emphasis has been placed on those methods which are particularly useful in tropical areas, while most of the examples given are based on tropical stocks.

The rapid introduction of special software for fish stock assessment, in particular that based on length-frequency data, such as the packages FiSAT (in press), COMPLEAT ELEFAN (Gayanilo, Soriano and Pauly, 1988) and LFSA (Sparre, 1987) also means that inexperienced fishery scientists may be placed in a position where they are using methods and models without fully realising the limitations of each method. The present manual should provide the necessary background knowledge on the methods to the users of the software mentioned above. This does not mean that this manual is directly related to computers. On the contrary, every method and exercise can be applied with the help of a good, programmable scientific pocket calculator. Further details on the use of this manual in training courses can be found in Venema, Christensen and Pauly (1988a).

1.1 THE PRIMARY OBJECTIVE OF FISH STOCK ASSESSMENT

The basic purpose of fish stock assessment is to provide advice on the optimum exploitation of aquatic living resources such as fish and shrimp. Living resources are limited but renewable, and fish stock assessment may be described as the search for the exploitation level which in the long run gives the maximum yield in weight from the fishery.

Fig. 1.1.1 illustrates this basic objective of fish stock assessment. On the horizontal axis is the fishing effort measured, for example, in number of boat days fished. On the other axis is the yield, i.e. the landings in weight. (If the landings consist of different groups of animals, for example shrimp, finfish and squid, it may be more appropriate to express the yield in terms of value.) It shows that up to a certain level we gain by increasing the fishing effort, but after that level the renewal of the resource (the reproduction and the body growth) cannot keep pace with the removal caused by fishing, and a further increase in exploitation level leads to a reduction in yield.

The fishing effort level which in the long term gives the highest yield is indicated by F_{MSY} and the corresponding yield is indicated by "MSY", which stands for "Maximum Sustainable Yield". The phrase "in the long term" is used because one may achieve a high yield in one year by suddenly increasing the effort, but then meager years will follow, because the resource has been fished down. Normally, we are not aiming at such single years with maximum yield, but at a fishing strategy which gives the highest steady yield year after year.

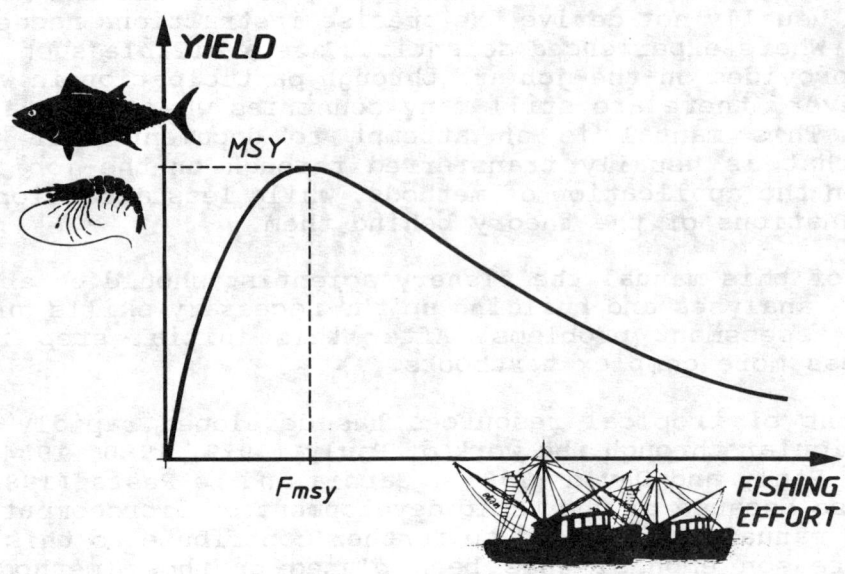

Fig. 1.1.1 The basic objective of fish stock assessment

1.2 THE STOCK CONCEPT

When describing the dynamics of an exploited aquatic resource, a fundamental concept is that of the "stock".

A stock is a sub-set of a "species", which is generally considered as the basic taxonomic unit. A prerequisite for the identification of stocks is the ability to distinguish between different species. Because of the great number of different, but often similar, species observed in tropical fisheries their identification can be problematic. The fishery scientist, however, must master the techniques of species identification if any meaningful fish stock assessment is to come out of the data collected. An aid to solve problems in species identification is provided by the "FAO species identification sheets for fishery purposes" (Fischer, 1978; Fischer and Bianchi, 1984; Fischer, Bianchi and Scott, 1981; Fischer and Hureau, 1985; Fischer, Schneider and Bauchot, 1987; Fischer and Whitehead, 1974) and in the "FAO species catalogues" (Allen, 1985; Carpenter, 1988; Carpenter and Allen, 1989; Cohen et al., 1990; Colette and Nauen, 1983; Compagno, 1984 and 1984a; Holthuis, 1980 and 1990; Márquez, 1990; Nakamura, 1985; Roper, Sweeney and Nauen, 1984; Russell, 1990; Whitehead, 1985, Whitehead, Nelson and Wongratana, 1988).

By a "stock" we mean a sub-set of one species having the same growth and mortality parameters, and inhabiting a particular geographical area.

To this definition we can add that stocks are discrete groups of animals which show little mixing with the adjacent groups. One essential feature is that the growth and mortality parameters remain constant over the distribution area of a stock, so that we can use them for making assessments.

This definition may be too superficial for the taste of many biologists, and in the following paragraphs a few more aspects of the stock concept are mentioned.

A group of animals for which the geographical limits can be defined may be considered a "stock" in terms of fish stock assessment. Such a group of animals should belong to the same race within the species, i.e., share a common gene pool. For species showing little migratory behaviour (mainly demersal species) it is easier to identify a stock than for highly migratory species, such as tunas.

A definition of the term "stock" acceptable to everyone with an interest in intraspecific grouping may be unattainable. For reviews of the stock concept see Booke (1981), Ihssen et al. (1981) and MacLean and Evans (1981).

Cushing (1968) defines a fish stock as one that has a single spawning ground to which the adults return year after year. Larkin (1972) defines a stock as "a population of organisms which, sharing a common gene pool, is sufficiently discrete to warrant consideration as a self-perpetuating system which can be managed", while Ihssen et al. (1981) define a stock as "an intraspecific group of randomly mating individuals with temporal or spatial integrity".

Ricker (1975) defines a fish stock as "the part of a fish population which is under consideration from the point of view of actual or potential utilization". This definition reflects a completely different approach to the stock concept. In this manual we will not follow this definition at all, but will adhere to the biological approach given above.

Perhaps the most suitable definition in the context of fish stock assessment was given by Gulland (1983) who stated that for fisheries management purposes the definition of a "unit stock" is an operational matter, i.e., a subgroup of a species can be treated as a stock if possible differences within the group and interchanges with other groups can be ignored without making the conclusions reached invalid.

This means that it is preferable to start by making stock assessments over the entire area of distribution of a species, as long as there are no indications that separate unit stocks exist in that area. If it becomes clear that the growth and mortality parameters differ significantly in various parts of the area of distribution of the species, then it will be necessary to assess the species on a stock by stock basis. The identification of separate stocks is a complex matter, which usually requires many years of data collection and analysis.

Fish stock assessment should be made for each stock separately. The results may (or may not) subsequently be pooled into an assessment of a multispecies fishery. Therefore, the input data must be available for each stock of each species considered. The stock concept is closely related to the concepts of growth and mortality parameters. The "growth parameters" are numerical values in an equation by which we can predict the body size of a fish when it reaches a certain age. The "mortality parameters" reflect the rate at which the animals die, i.e., the number of deaths per time unit. The mortality parameters considered in this manual are the "fishing mortality", which reflects the deaths created by fishing and the "natural mortality", which accounts for all other causes of death (predation, disease, etc.).

An essential characteristic of a stock is that its growth and mortality parameters remain constant throughout its area of distribution. Let us, as an example, partition that area into two parts, sub-areas A and B:

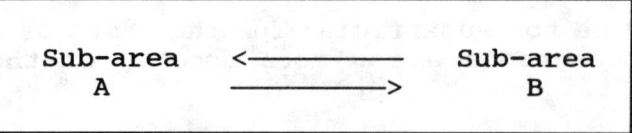

The growth and mortality parameters must be the same in sub-areas A and B, or in other words:

1) The animals in sub-area A must have the same body growth rate as the animals in sub-area B

2) The animals in sub-area A must have the same probability of death as the animals in sub-area B

If fishing takes place only in sub-area A, it is assumed that each individual fish in the stock has the same probability of being encountered in sub-area A and thereby also that it has the same probability of being caught. The individuals are supposed to move freely between the two sub-areas.

In order to determine whether a species forms one or more distinct stocks, we should examine its spawning areas, growth and mortality parameters and morphological and genetic characteristics. We should also compare the fishing patterns in various areas and carry out tagging studies. The process is complicated, and often it is not possible with the knowledge in hand to determine whether there are several stocks of that species or not. There are two main reasons for failing to define a stock properly:

1) The full distribution area of the stock is not covered, so that only a part of the stock is considered, or the opposite

2) Several independent stocks are lumped together, for example because their areas of distribution overlap

Several countries may exploit the same stock. This is the case for many migratory stocks, e.g. tunas. It sometimes happens that a country assesses such a "<u>shared stock</u>" as if it were a national stock only exploited by that country. On the other hand, a fishery of a single country may exploit several independent stocks. Coral reef fish stocks may fall in this category.

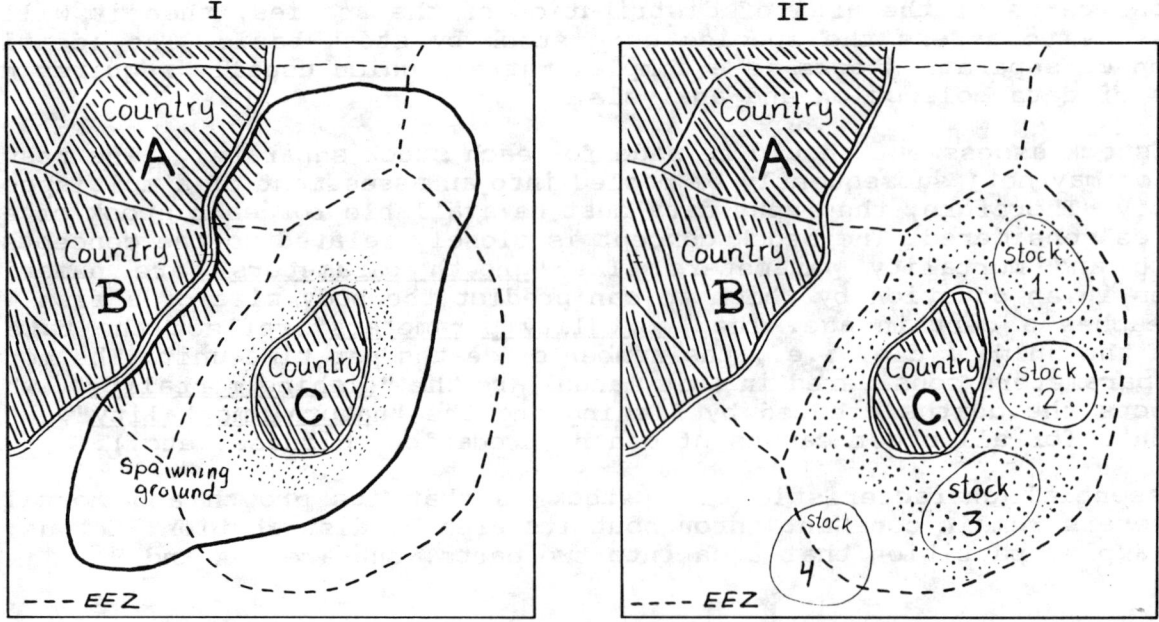

Fig. 1.2.1 **Distributions of stocks related to management problems**

Fig. 1.2.1 illustrates those two cases. In part I, we consider a fish stock, of which the distribution area is indicated by a full line. It is exploited by three countries, A, B and C, and we look at the stock definition from the point of view of the island country C. The broken lines show the EEZ (Exclusive Economic Zone) of each country, i.e. the national jurisdiction over fisheries. The dotted area indicates the fishing area of country C and the hatched areas those of countries A and B. From this it can be seen that if country C would base its assessment on the assumption that the unit stock is limited to its own fishing area, thus ignoring the fisheries of countries A and B, it is likely to draw wrong conclusions. If, for example, countries A and B have intensive fisheries on the stock in question so that it is over-exploited (i.e. a reduc-tion of the fishing intensity would increase the yield) there is little country C can do on its own to improve the situation. From the assessment based on the assumption of a stock limited to country C's waters, country C may conclude that the stock is over-exploited and it may introduce management measures to reduce fishing. However, the expected effect of the management measures will not materialize, if countries A and B do not follow country C.

Part II of Fig. 1.2.1 illustrates the case where one fishery exploits several stocks. In this case the assessment becomes that of the average stock, since it will be impossible to separate the catches by stocks. If the fishing effort expended is similar for each stock, the result of the assessment should come out correctly. However, there may also be difficulties in this case. Suppose that the three stocks currently fished (1, 2 and 3) are heavily overfished, and that the fishery is expanded to include the unexploited stock 4. In that case the average catch rate will increase and this might lead to wrong conclusions regarding the status of stocks 1, 2 and 3.

Nearly all exploited marine organisms undertake migrations, for example to their spawning grounds. A basic key to an understanding of stock structures is the knowledge of migration routes. This can be obtained from tagging experiments, but also from data and information provided by the commercial fisheries. Often the fishermen know where the spawning grounds are and they know where the high concentrations of fish are found at different times of the year.

Some general conclusions may be drawn from the above. Firstly it is usually safer to assume that species in neighbouring fishing areas form one unit stock than to consider each separate fishery to exploit its own unit stock. Further, it is evident that proper assessments can only be carried out when the biology of the species, including its migrations, spawning habits etc., is fully understood. Fish stocks are not bound by human geographical limits and this means that proper assessments can only be made when such limits can be ignored through interstate or international cooperation.

1.3 MODELS

A description of a fishery consists of three basic elements:

1) the <u>input</u> (the fishing effort, e.g. the number of fishing days)

2) the <u>output</u> (the fish landed) and

3) the <u>processes</u> which link input and output (the biological processes and the fishing operations)

Fish stock assessment aims at describing those processes, the link between input and output and the tools used for that are called "<u>models</u>". A model is a simplified description of the links between input data and output data. It consists of a series of instructions on how to perform calculations and it is constructed on the basis of what we can observe or measure, such as for example fishing effort and landings.

The actual processes which go from a certain number of days fishing with a certain number of boats to a certain number of fish being landed are extremely complicated. However, the basic principles are usually well understood, so that by processing the input data by aid of models we can predict the output.

```
INPUT  ———>  PROCESSES  ———>  OUTPUT
observation    model         observation
```

A model is a good one if it can predict the output with a reasonable precision. However, since it is a simplification of reality it will rarely (and only by chance) be exact.

The instructions for the calculations that make up the model are given in the form of mathematical equations. These are composed of three elements: "<u>variables</u>", "<u>parameters</u>" and "<u>operators</u>". For example, the mathematical equation:

$$y = 2.5 + 3*x$$

has the variables y and x, the parameters 2.5 and 3 and the operators "+" and "*" The equation is used to predict the value of y for some value of x.

GENERAL PROCEDURE OF FISH STOCK ASSESSMENT

```
INPUT    :  FISHERIES DATA (+ ASSUMPTIONS)
                    ⇓
PROCESS:    Analyses of historical data
                    ⇓
OUTPUT   :  ESTIMATES OF GROWTH AND MORTALITY PARAMETERS
INPUT    :
                    ⇓
PROCESS:    Predictions of yield for a range of
            alternative exploitation levels
                    ⇓
OUTPUT   :  OPTIMUM FISHING LEVEL
            MAXIMUM SUSTAINABLE YIELD
```

Fig. 1.3.0.1 General flow-chart for fish stock assessment

Fish stock assessment involves five basic steps as illustrated in Fig. 1.3.0.1. The first step is to collect data on the fishery, the INPUT to the assessment, which often have to be supplemented by assumptions or qualified guesses. Then we process the data by applying a model to estimate the growth and mortality parameters, the OUTPUT from the processing of "the historical data". (The term "historical" is used to distinguish it from the subsequent process, the prediction of future yield.) This prediction is based on the previous OUTPUT (= INPUT) and on a model, and the prediction is repeated for a series of alternative options. (Such options could be, for example, a fishing effort reduction of 10%, 20% and 30%, no change in fishing effort or a fishing effort increase of 10%, 20% and 30%.) Among the alternative assumptions the best one is eventually selected as the final OUTPUT. The original INPUT data may be research survey data, data from samples drawn from the commercial fisheries or a combination of both.

Two main groups of fish stock assessment models are covered in this manual: "holistic models" and "analytical models". The simple holistic models use fewer population parameters than the analytical models. They consider a fish stock as a homogeneous biomass and do not take into account, for example, the length or age-structure of the stock. The analytical models are based on a more detailed description of the stock and they are more demanding in terms of quality and quantity of the input data. On the other hand, as a compensation, they are also believed to give more reliable predictions.

The type of model to be used depends on the quality and quantity of input data. If data are available for an advanced analytical model then such a model should be used, while the simple models should be reserved for situations when data are limited. We are often in the situation where a complete set of input data for an analytical approach is not available, but where the available data exceed the demand of the simple models. As an alternative to using simple models in this case, the lacking input data can be replaced by assumptions or qualified guesswork. Often, the lacking parameter for a particular stock can be replaced by known parameters from another, similar stock.

1.3.1 Analytical models

A basic feature of analytical models as developed by, among others, Baranov (1914), Thompson and Bell (1934) and Beverton and Holt (1956), is that they require the age composition of catches to be known. For example, the number of one year old fish caught, the number of two year old fish caught, etc. may form the input data.

The basic ideas behind the analytical models may be expressed as follows:

1) If there are "too few old fish" the stock is overfished and the fishing pressure on the stock should be reduced

2) If there are "very many old fish" the stock is underfished and more fish should be caught in order to maximize the yield

(Some suggestions for more exact definitions of the term "overfishing" are given in Chapter 8).

The analytical models are "age-structured models" working with concepts such as mortality rates and individual body growth rates.

The basic concept in age-structured models is that of a "cohort". To put it simply, a "cohort" of fish is a group of fish all of the same age belonging to the same stock. (We shall further elaborate on the definition of a cohort in Chapter 4.) For example, a cohort of the threadfin bream, (Nemipterus marginatus) could be all the fish of that species that hatched from June to August in 1976 near Tanjung Pinang in the South China Sea. Suppose there were one million specimens in that cohort. After August 1976 the original one million fish would gradually decrease in number because of deaths due to natural causes (predation, diseases, etc.) or fishing. However, while the number of survivors of the cohort decreases with time the average individual body length and body weight increase.

Fig. 1.3.1.1 shows an (hypothetical) example of the dynamics of a cohort, in the form of plots against age of the number of survivors (A), body length (B), body weight (C) and total biomass (D). Curve A shows the decay in the number of survivors as a function of the age of the cohort. Curve B shows how the average body length increases as the cohort grows older. Curve C shows the corresponding body weight, while curve D is a plot of the total biomass of the cohort, i.e. the number of survivors times the average body weight against the age of the cohort.

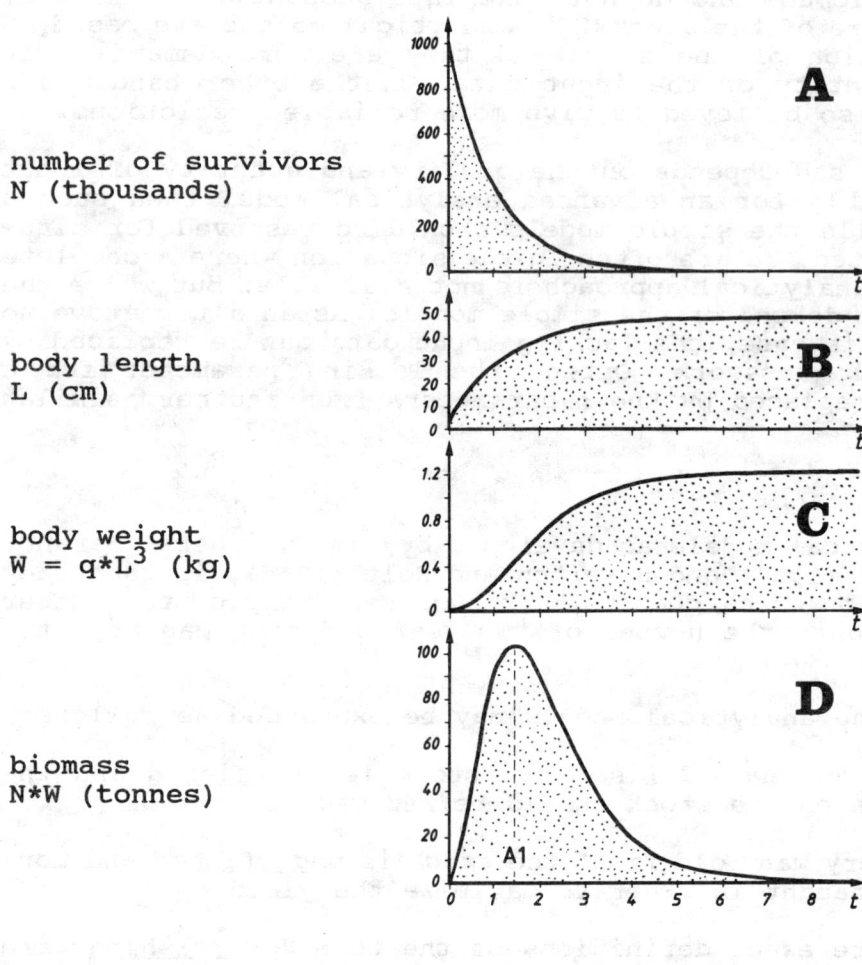

Fig. 1.3.1.1 The dynamics of a cohort

Note that curve D has a maximum (at age A1). Thus, to get the (hypothetical) maximum yield in weight from that cohort all fish should be caught exactly when the cohort has reached age A1. This, of course, is not possible in practice. However, you may say that the goal of fish stock assessment is to manage fisheries in such a way that catches come as close as possible to this theoretical maximum.

The implication is that the fish should neither be caught too young nor too old. If the fish are caught too young there is "growth overfishing" of the stock.

There are thus two major elements in describing the dynamics of a cohort:

1) The average body growth in length and weight

2) The death process

Both elements will be dealt with in greater detail in Chapters 3 and 4 respectively.

1.3.2 Holistic models

In situations where data are limited, for example, when starting up the exploitation of an hitherto unexploited resource, or in cases of limited capability of sampling, one may not have input data of the quality and in the quantity required for an analytical model. One solution would be to start up the collection of the data types required for the analytical approach and then wait until a sufficient amount is available. This approach is, of course, recommendable, because it solves the problem in the long run, but that may take years, while often advice on an exploitation or development strategy may be needed now. The approach taken in this manual is that no matter which type of data you have, there is always some information to be extracted from it, and that advice based on an analysis of a limited data set is usually better than complete guesswork.

In order to cover such data-limited situations, some simple holistic, less data demanding methods have been included in the manual. These methods disregard many of the details of the analytical models. They do not use age or length structures in the description of the stocks, but consider a stock as a homogeneous biomass.

Two types of simple methods are presented, namely the "swept area method" (in Chapter 13) and the "surplus production model" (in Chapter 9).

The swept area method is based on research trawl survey catches per unit of area. From the densities of fish observed (the weight of the fish caught in the area swept by the trawl) we obtain an estimate of the biomass in the sea from which an estimate of the MSY is obtained. This method is rather imprecise and it predicts only the order of magnitude of MSY.

The surplus production methods use catch per unit of effort (for example kg of fish caught per hour trawling) as input. The data usually represent a time series of years and usually stem from sampling the commercial fishery. The models are based on the assumption that the biomass of fish in the sea is proportional to the catch per unit of effort as shown in Fig. 1.3.2.1. An estimate of the yield is obtained by multiplying effort by catch per unit of effort.

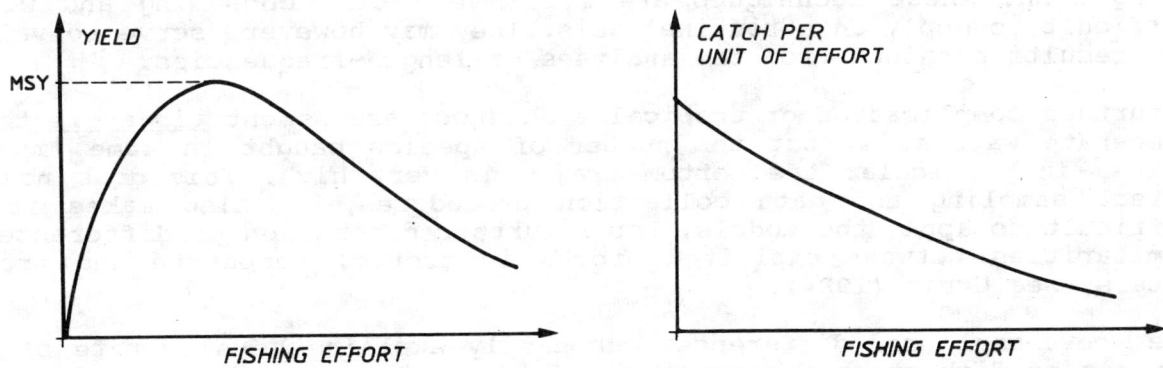

Fig. 1.3.2.1 Surplus production model

1.4 ASSESSMENT OF STOCKS IN TROPICAL WATERS

The literature on fishery biology dealing with species in temperate zones is extensive compared to that on tropical fisheries. The major part of the literature on tropical fish stock assessment was published recently. As will appear in the following chapters of the manual, this can partly be attributed to the fact that tropical resources are somewhat more complex than those of temperate waters.

The present manual has the word "tropical" in its title. Although the methods described in the manual resemble those used in temperate waters, there are special features which justify the use of the word "tropical". Perhaps the most conspicuous difference between fish stock assessment in tropical waters and temperate waters lies in the nature of the basic input data rather than in the models.

For the analytical models we need the number of fish caught of each age group as input. In temperate waters stock assessment methods used are heavily dependent on the fortunate fact that ages of fish can be readily determined by "ageing" them. Ageing is most often done by counting rings in hard parts of the fish body, such as ear-bones (otoliths) or scales. The so-called year-rings are formed through a daily addition (daily-ring) to the size of the scale or otolith. The chemical composition and thereby the transparency of the addition depends (among other things) on the amount of food available and is therefore seasonal. The difference in deposits made in the winter and in the summer can be detected and one year-ring, composed of a summer and a winter part, can be distinguished from the next. Moreover, temperate fish species usually spawn once per year in a relatively short time-span, which makes it easy to distinguish year-classes or cohorts.

Also in tropical fish material is added daily to hard parts, which can be distinguished as daily growth rings. However, the lack of a strong seasonality makes the distinction of seasonal rings and therefore also of year-rings problematic for many tropical species. Moreover, the same absence of strong seasons results in less distinct spawning periods for most species. Many tropical species spawn at least twice per year and often over long periods. Fortunately, due to periodic changes in winds (monsoons) and shifts in oceanographic conditions (upwelling) in many tropical areas, a certain level of seasonality can still be detected. This seasonality may be reflected in the spawning patterns and growth of tropical fish species albeit less pronounced and much more difficult to detect than in temperate waters. These seasonal differences make it possible to detect also in tropical species the existence of different cohorts (often two per year), through the analyses of length-frequency samples.

In recent years, techniques have been developed to read daily rings in the otoliths of many fish species. This has enabled the development of age reading on tropical species, in particular of fish with short life spans, or young fish. These techniques are still very time consuming and will be difficult to apply on a routine basis. They may however, serve to validate the results obtained from the analyses of length-frequencies.

A further complication of tropical fish stock assessment vis-à-vis that in temperate waters is that the number of species caught in some important gears, in particular the bottom trawl, is very high. This does not only affect sampling and data collection procedures, it also makes it more difficult to apply the models. For a further discussion of differences and similarities between exploited stocks in arctic, temperate and tropical waters, see Ursin (1984).

The above-mentioned differences can easily explain the slow rate of development of fish stock assessment in the tropics compared to that in temperate areas. The present manual works with methods which are the length-based parallels to the traditional age-based methods of temperate waters.

Clearly, there is a relationship between age and length, and if the relationship is known we can convert length-frequencies into age frequencies. Fig. 1.4.1 shows a resolution of a length-frequency sample into age groups (cohorts). There are several techniques available for the separation of length groups and conversion into age groups, most of which are computerized. Several of these are discussed in the manual and one of them, the Bhattacharya method, is illustrated by examples and exercises. This method, although applicable in several computerized versions, can also be performed by using simply paper, pencil and a (scientific) pocket calculator.

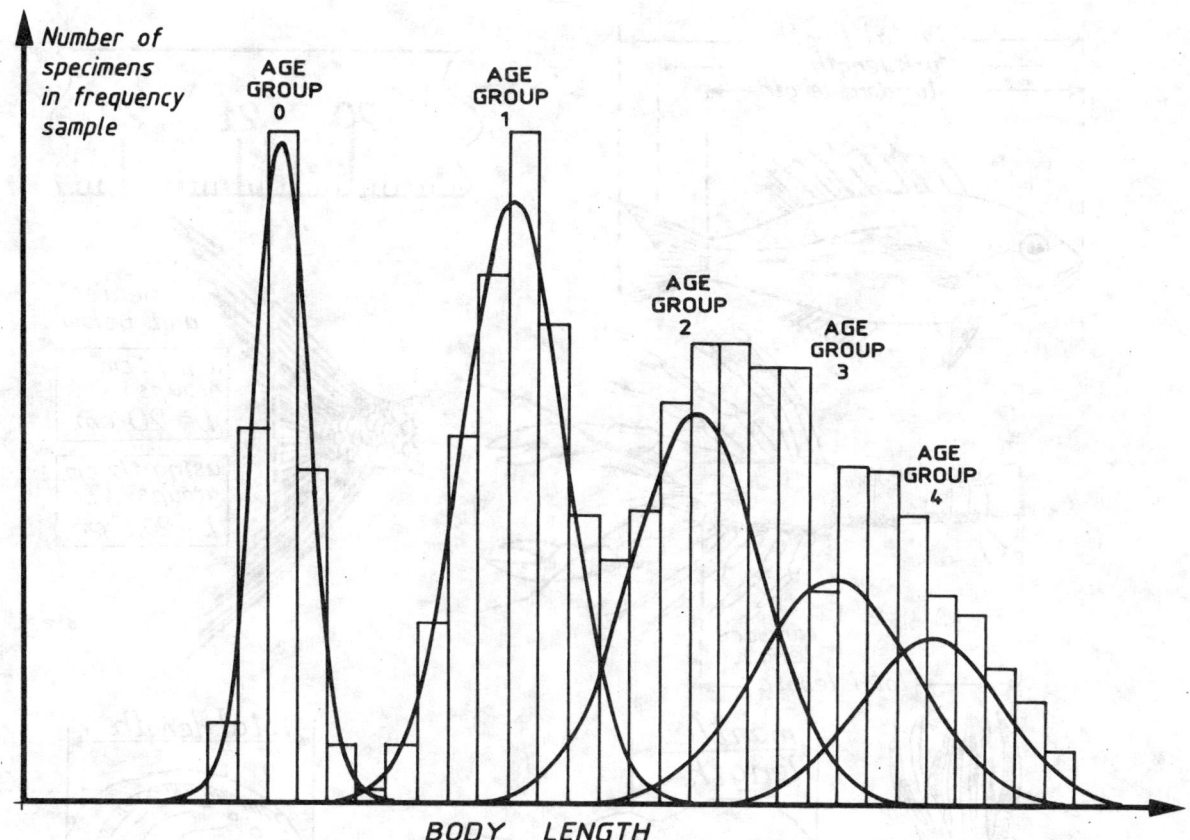

Fig. 1.4.1 Length-frequency sample resolved into age groups

In this manual, when explaining the theory behind the various methods we often start with the age-based version, because it is easier to explain and consequently also easier to understand. The next step is then to convert the age-based method into a length-based method by using the relationship between age and length.

1.5 DEFINITIONS OF BODY LENGTH

In the present context, "body length" means the average body length of a cohort. Individual fish are not considered in the models. When talking about "the length of an animal" in connection with a model it is always tacitly assumed that it is the "average length of the animals of a cohort". The estimate of average length, however, is derived from averaging the length measurements of individual specimens. The actual measure used for body length is not important as far as the theory behind the growth model is concerned. It is common practice to use the "total length" measured to the "nearest unit below" unless anatomical details make it not practicable (see Fig. 1.5.1). "Fork length" may be used for fish with stiff caudal fins (tunas) or special fin shapes (Nemipteridae). The "standard length" is not recommended for length-frequency sampling.

The most accurate measure for shrimps and lobsters is the "carapace length" (see Fig. 1.5.1). However, in many cases either total length or tail length will have been used. In such cases it is necessary to establish the relationship between the various measurements.

A really important thing is to specify exactly what kind of length measurement has been used, as one may otherwise run into difficulties when comparing results with those of other investigations.

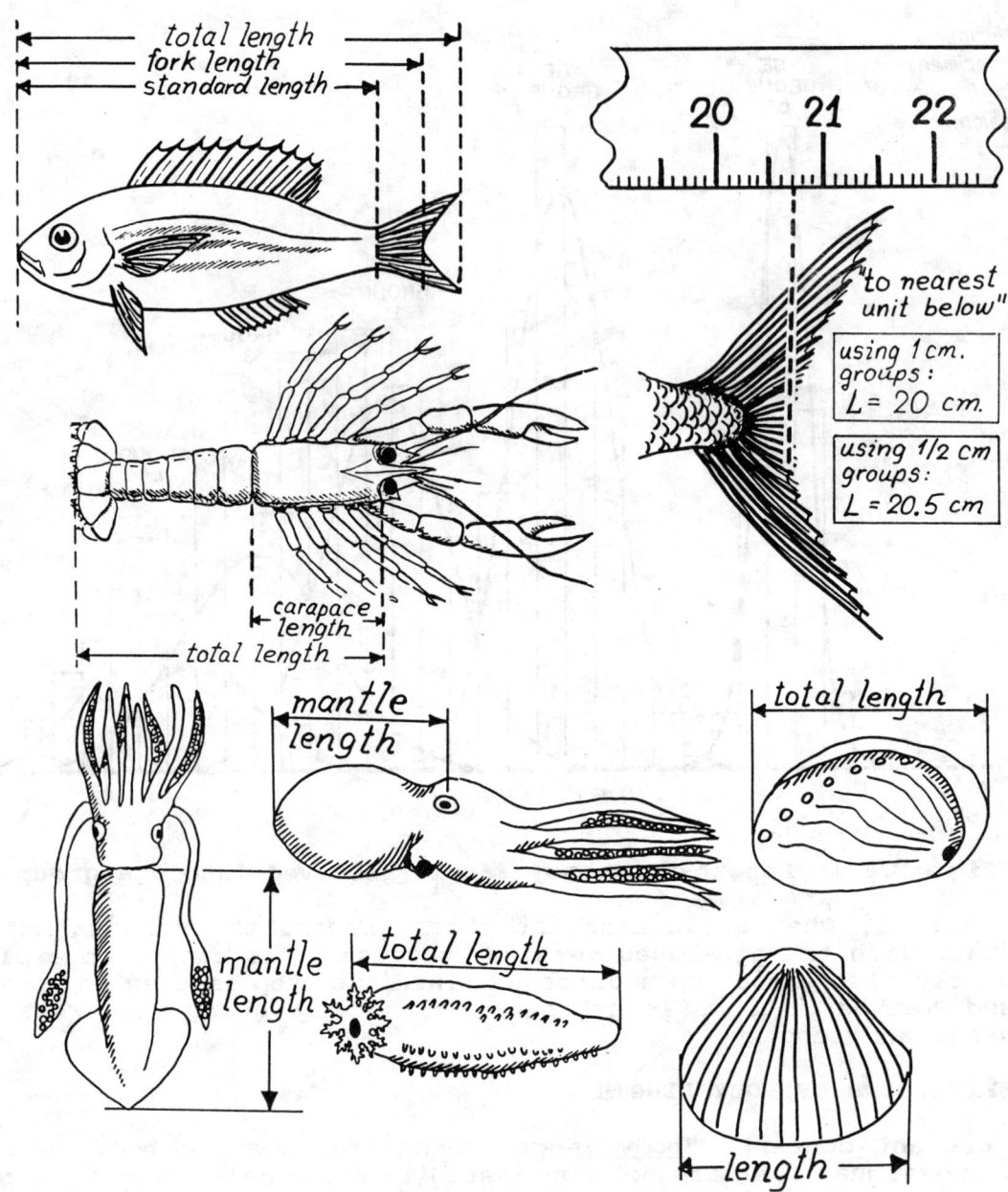

Fig. 1.5.1 Definitions of body length

Other examples given in Fig. 1.5.1 are squid, octopus, abalone, scallop and sea cucumber. For animals with a hard shell or skeleton it is not a problem to define a suitable length measure (fish, crustaceans and molluscs with shell). Also molluscs with a relatively constant body form (e.g. squid) create no major problem, but animals with a plastic body (e.g. octopus, sea cucumber or jellyfish) are problematic. It may in certain cases be preferable to work with body weight rather than length, as the former is obviously measureable with greater accuracy.

It is easy to transform one type of length measurement into another type for a single individual. In cases where a sample is grouped into length classes it is more cumbersome to change from one measurement to another as far as the computational aspects are concerned. One simple way of doing it by microcomputer is given in Sparre (1987).

1.6 AGE AND RECRUITMENT

When working with analytical models we need to define the concept of "<u>age</u>". As was said above in connection with body length, we do not operate at the individual specimen's level, so "age" means the average age of a cohort. To define age we must start with a definition of "<u>birthday</u>". The obvious biological definition of the day of birth is the day the larva hatches from the egg. We say that a newly hatched fish has age zero.

In the first part of their life the larvae (or juveniles) are usually little influenced by the fishery. We say that the fish is then in the unexploited phase of life. Because we are interested in the exploited phase of its life the unexploited phase is not important in the present context.

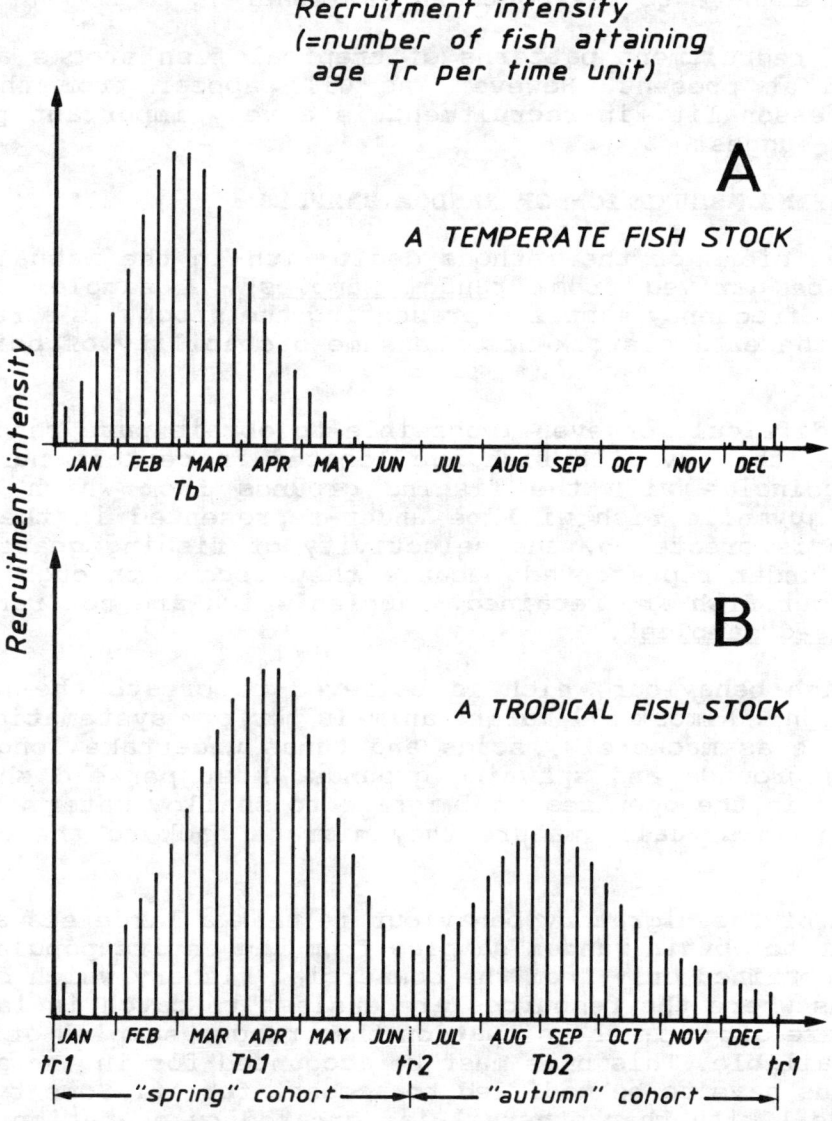

Fig. 1.6.1 Recruitment intensity during the year of typical temperate and tropical stocks

Let Tr be the youngest age at which the fish may be vulnerable to fishing gears. A fish of age Tr is called a "recruit". By "recruitment" we mean the number of recruits, i.e. the number of fish that have attained age Tr during a "recruitment season". The "recruitment intensity" is the number of recruits per time unit. The "recruitment pattern" of a temperate species could be as shown in Fig. 1.6.1A, where each line represents the recruitment intensity in one week. In most tropical fish stocks recruitment continues (more or less) all year round, but with seasonal oscillations, for example where monsoons occur (Pauly and Navaluna, 1983) (see Fig. 1.6.1B).

Let us tentatively define the recruitment season of a tropical fish stock by the dates (fractions of the year) tr1 and tr2 which correspond to the dates of minimum recruitment (see Fig. 1.6.1B). With $0 <= tr1 < tr2 <= 1.0$ we define the "spring cohort" as the fish recruited from time tr1 to tr2 and the "autumn cohort" as the fish recruited from time tr2 to tr1. ("Spring" and "autumn" refer here to the northern hemisphere).

In general, the recruitment patterns of tropical fish stocks are not very well understood at present. However, as will appear from the following chapters, the seasonality in recruitment is a very important prerequisite for the methods suggested.

1.7 THE UNDERLYING ASSUMPTION OF RANDOM SAMPLES

All the basic versions of the methods dealt with in the manual assume the input data to be derived from "random samples". A sample of fish, for example a length-frequency sample representing the stock, is a random sample if any fish in the entire stock has the same probability of being drawn as any other.

Usually, it is difficult or even impossible to obtain pure random samples. If, for example, the juvenile fish are located in certain nursery areas, which do not coincide with the fishing grounds from which our samples originate, the juvenile fish will be under-represented in the samples. A similar problem is created by the selectivity of fishing gears. Often the small fish are under-represented because they escape through the meshes, whereas the larger fish are retained. Samples which are not random samples are called "biased samples".

A feature of fish behaviour which is believed to create the most serious bias is "migration". Almost all marine animals perform systematic movements. Pelagic fish such as mackerels, scads and tunas undertake long migrations between feeding grounds and spawning grounds. Most penaeid shrimps start their life cycle in the open sea and migrate to shallow waters (lagoons and mangroves) and when sexually mature they migrate back to the open sea for reproduction.

The implication of the migratory behaviour is that a large sea area must be covered in order to obtain random samples from the entire population. Often samples can be obtained only from the commercial fishery which concentrates on those grounds where the resources are easiest to catch in large quantities. Thus, we are often in the situation that random samples of the population are not available. This bias must be accounted for in the analysis and the basic methods have to be modified to account for it. Some types of bias are easier to deal with than others. Bias created by migration can only be handled properly when the migration routes are known. When they are not we have to make certain assumptions about them in order to get on with the analysis. There are many serious problems in con-nection with bias. A few suggestions on how to get around them are presented, but the manual also leaves a number of relevant questions open, either because the author does not know the answer or the method is so complicated that it falls outside the scope of this manual. Unfortunately, one often comes across cases in practice, which are so heavily influenced by bias that they cannot be handled by the methods described here (see Chapter 11).

1.8 THE ORGANIZATION OF THE MANUAL

The complexity of fish stock assessment is reflected in the contents table of this manual. The various elements cannot be dealt with simultaneously and it has often been necessary to refer to earlier or later sections and chapters. In order to assist the reader (and teacher) a flow-chart for fish stock assessment as presented in this manual is given in Fig. 1.8.1. The flow-chart does not present the methods in the same sequence as they appear in the manual, but rather in the natural chronological order of a fish stock assessment. The numbers of the relevant chapters are given in brackets.

Before starting as a fish stock assessment worker there are a few general basic statistical techniques (for example linear regression analysis) one must master. These are dealt with in Chapter 2. They have been placed outside the proper flow-chart, because the methods are general and applied in many other scientific fields. Chapter 2 contains only a small selection of statistical methods and only those which are needed to follow the text in the subsequent chapters.

The flow-chart is divided into two parts. Part A deals with analytical methods and Part B deals with holistic methods. As the sizes of the two parts indicate, the main emphasis has been placed on the analytical methods. Both approaches follow the same main lines, namely the set-up given in Fig. 1.3.1.

Part A, the analytical methods

The first row shows the input to the estimation of growth parameters (the parameters by which we can predict the length of an animal for a given age). Although collection of data comes first chronologically, it is not dealt with in the beginning of the manual, because one cannot deal with data collection in a meaningful way before the objectives of the sampling scheme have been defined. To define the objectives we need the models used for the analysis of historical data, therefore, the main text on data collection is deferred to Chapter 7. The assumptions indicated as input are not dealt with in a particular chapter.

The theory behind the model for body growth and the estimation of growth parameters is dealt with in Chapter 3. Handling of bias problems is as mentioned above extremely complicated and only partly covered by the present manual. It has therefore been placed in Chapter 11 after the chapters dealing with the analytical methods in their basic versions. Although we should start the analysis with an evaluation of the bias, it has not been considered appropriate to start the manual with one of the most complicated subjects. Also the estimation or rather the qualified guessing on natural mortality is a tricky subject. It has been placed in Section 4.7.

The following chapters on analytical methods contain the theory for both age-based methods and length-based methods. The estimates of growth parameters are in fact only used for the length-based versions of the models, but in order to reduce the complexity of the flow-chart, no distinction has been made between length-based and age-based methods.

After the growth part (Chapter 3) the flow-chart branches. The two branches represent a grouping of methods according to their data requirements. Some analytical methods are based exclusively on <u>samples</u> from the commercial fishery, while the total catch is not known. In theory these methods could be used also on a single sample consisting of one bucket of fish sampled in the local fish market (although, of course, extensive sampling schemes are recommended). Such methods are called "<u>catch curve methods</u>".

Other methods are based on estimates of the <u>total catch</u>, i.e. on estimates of the total number landed in each length group from the entire stock. Such

PART A: ANALYTICAL METHODS *)

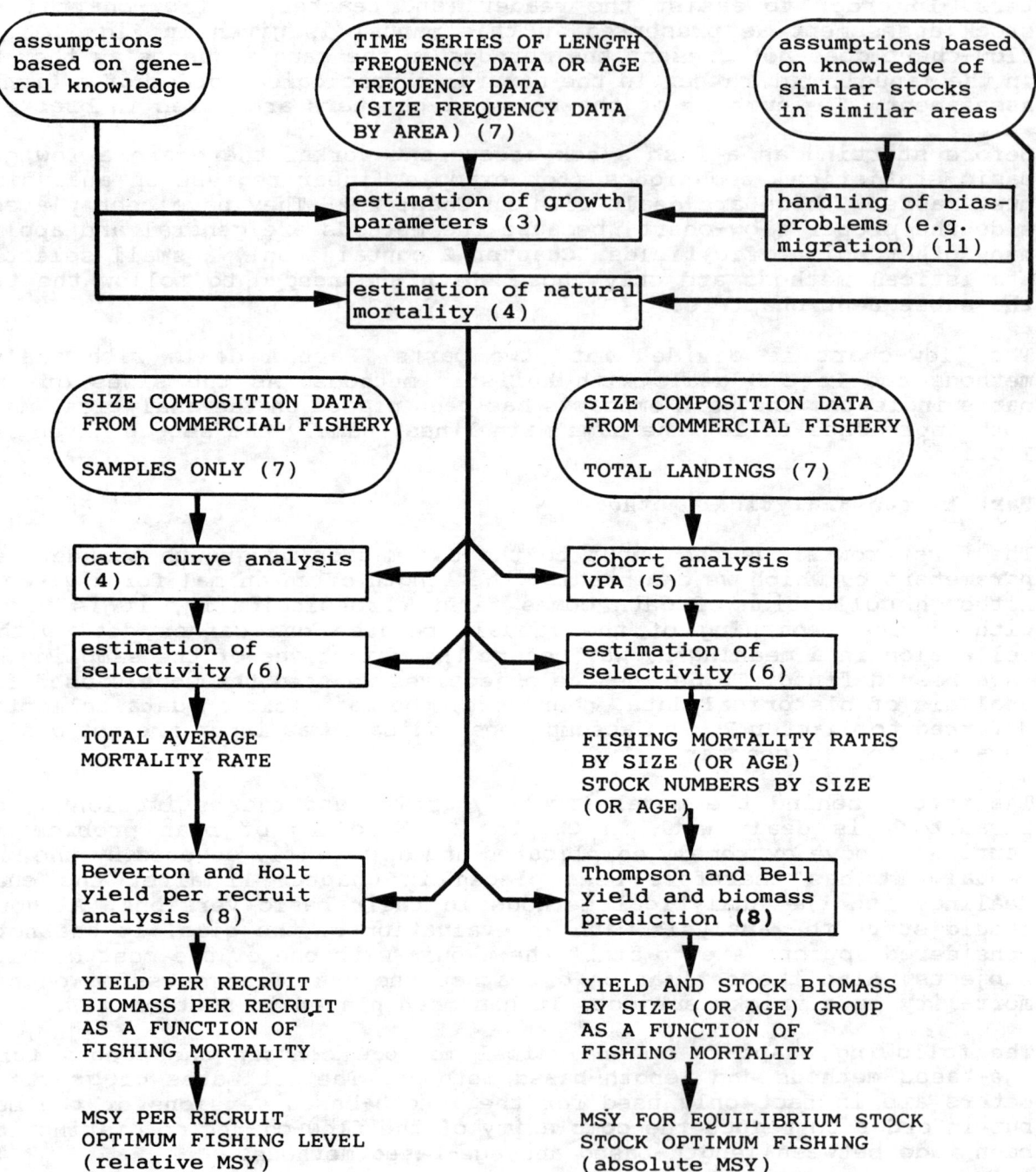

Fig. 1.8.1A The organization of the manual

PART B: HOLISTIC METHODS *)

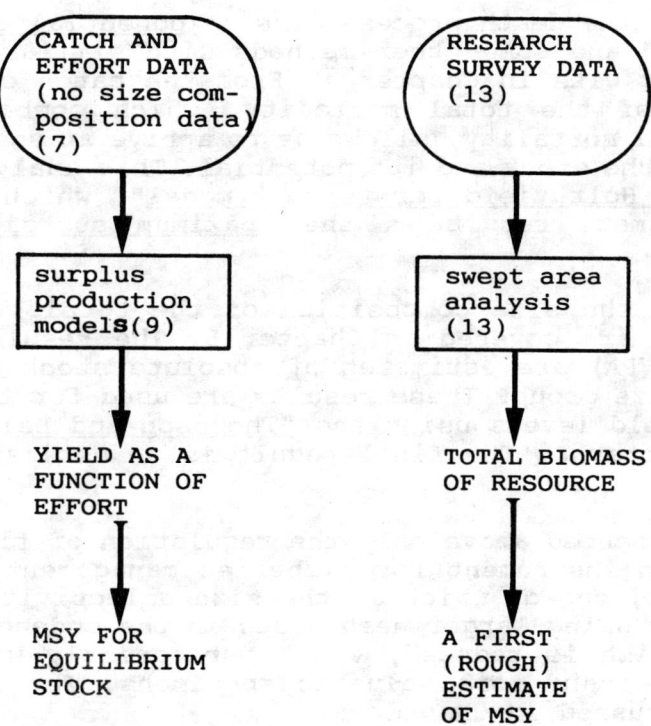

Fig. 1.8.1B The organization of the manual

*) The following symbols have been used:

    ⬭   :   Input data or assumptions

    ▭   :   Processing of data using a model

text not framed :   Output from a data processing stage; an output from
CAPITAL LETTERS   one processing stage acts as input for the subsequent processing stage

(n)   :   Figures in brackets refer to the Chapter with the relevant theory

⟶   :   Flow of data

numbers are derived from length-frequency samples by raising them to account for the entire catch using data on total landings. These methods are called "cohort analysis" or "Virtual Population Analysis" (VPA). Compared to the catch curve methods, they give more dependable estimates of the parameters and more reliable predictions of the future fishery. The sampling procedures to obtain the input data are discussed in Chapter 7 as mentioned above in connection with growth.

The general theory of the death process, the "exponential decay model", the "catch curve methods" and some other methods with similar or limited data requirements is dealt with in Chapter 4. From the catch curve analysis we obtain an estimate of the total mortality, which combined with growth parameters and natural mortality, allows us to arrive at some conclusions on the current state of the stock and its potential. This analysis is performed by the "Beverton and Holt yield per recruit model", which is presented in Chapter 8. The ultimate result is the "maximum sustainable yield per recruit" (MSY/R).

The methods based on the size composition of the total catch, the cohort analysis and the VPA are covered in Chapter 5. The results obtained from cohort analysis (or VPA) are estimates of absolute stock size and fishing mortality for each size group. These results are used for the prediction of stock biomass and yield levels using the "Thompson and Bell methods" which are described in Chapter 8. The final result is an estimate of the (absolute) MSY.

In the discussion presented above only the regulation of fishing effort has been considered as an instrument for fisheries management. However, there are other instruments, one of which is the size selectivity of the fishing gear. For example, by using larger mesh sizes in the codend of a trawl, the mortality on young fish is reduced, which subsequently will increase the catch of older, larger and more valuable specimens. The effects of "gear selectivity" are discussed in Chapter 6.

In all the chapters mentioned so far, the theory has been presented in its most simple form, i.e. only one stock of fish and only one type of fishing boat are considered. In reality there hardly exists any fishery where only one species is caught by only one type of boat. In most cases we have to deal with multispecies catches made by a variety of different boats. In theory it does not create any major problems to extend the analytical models to deal with the multispecies/multifleet case, and this has been demonstrated for the Thompson and Bell model in Chapter 10. Some other aspects of multispecies assessment are also briefly discussed in this chapter. With this we have reached the end of Part A of the flow-chart.

Part B, the holistic methods

This part of the flow-chart is less complicated, because it describes simpler methods. The "surplus production models" are presented in Chapter 9 and the "swept area method" in Chapter 13.

Chapters 12, 14 and 15 have not been included in the flow-chart. Chapter 12 deals with the stock/recruitment relationship - the question of a possible relationship between recruitment and the size of the parent stock. The problem is discussed in essay form and Chapter 12 does not suggest any models for practical application. Chapter 14 provides an overview of fish stock assessment, based among other things on the same flow-chart (Fig. 1.8.1). Chapter 15 describes briefly some microcomputer program packages, the LFSA (Length-based Fish Stock Assessment) package (Sparre, 1987), which matches the analytical models of this manual (Part A of the flow-chart), the COMPLEAT ELEFAN package (Gayanilo, Soriano and Pauly, 1988) and the FiSAT (FAO/ICLARM Stock Assessment Tools) package (in press) which also covers all the models. Some other programs developed by or in close cooperation with FAO have also been briefly described.

1.9 FURTHER READING

Since the scope of this manual is mainly limited to methods and their applications, it is advisable to supplement the knowledge derived from it, by reading additional textbooks and manuals on stock assessment and on the biology of the most important resources. Chapter 16 contains references to several recent publications, which will be of great use in providing a better insight to stock assessment problems, for example

1) The relationship between stock assessment and management (Pauly, 1979 and Gulland, 1988)

2) The biology and assessment of shrimps (Garcia and Le Reste, 1981; Gulland and Rothschild, 1984; Penn, 1984; Garcia, 1985; Rothlisberg, Hill and Staples, 1985, the Australian Journal of Marine and Freshwater Research, 1987 and Dall et al., 1990)

3) The biology and assessment of cephalopods and other invertebrates (Caddy, 1983, 1983a and 1989)

4) Resource mapping (Caddy and Garcia, 1986 and Butler et al., 1986)

5) Tunas (Sharp and Dizon, 1978; Kleiber, Argue and Kearney, 1983; I-ATTC, 1984 and Hunter et al., 1986)

6) Migration (Harden Jones, 1968, 1984 and Oxenford and Hunte, 1986)

7) Marine population regulation and speciation (Sinclair, 1988)

The various methods presented in this manual were used either directly or through the LFSA and COMPLEAT ELEFAN packages at FAO/DANIDA follow-up training courses, where participants processed their own data. The results of the analyses and the input data were published (Venema, Christensen and Pauly, 1988), with the aim of providing additional examples of the application of stock assessment methods on tropical resources. In some cases the papers demonstrate clearly the limitations of the data sets and of the methods used, but there are also several examples of successful applications of methods hitherto seldom used in tropical areas.

New text books on fish stock assessment are rare, and we are therefore pleased to draw your attention to recent books by Hilborn and Walters (1992) and Brêthes and O'Boyle (eds.)(1990). The latter is partly based on an earlier version of this manual.

A comprehensive overview of the several approaches on the use of length-frequency data for stock assessment purposes was prepared by J.A. Gulland and A.A. Rosenberg (1992). This document, John Gulland's last contribution to fisheries science, should be used in conjunction with or as a follow-up to this manual.

2 BIOSTATISTICS

This chapter contains a brief description of some statistical methods in common use in tropical fishery biology and introduces the statistical notation adopted in the manual. It serves as a refresher and reference point, but is not intended as a textbook in its own right.

The amount of literature on statistical methods is staggering, so there is no problem if you want to do further studies in biostatistics. Only two references are given here. The book "Biometry" by Sokal and Rohlf (1981) deals with the theory in a rather accessible way, while "Sampling techniques" by Cochran (1977) is perhaps a bit more complicated, but still recommended as an introduction. However, there are many other textbooks which may be equally useful.

2.1 MEAN VALUE AND VARIANCE

Let us consider a sample of n fish all of one species caught in one trawl haul and let x(i) be the length of fish no. i, i = 1,2,...,n. The "<u>mean length</u>" (in general the "<u>mean value</u>") of the sample is defined:

$$\bar{x} = [x(1) + x(2) + ... + x(n)]/n = \frac{1}{n} * \sum_{i=1}^{n} x(i) \qquad (2.1.1)$$

The two first columns of Table 2.1.1 show an example for n = 27.

The variance, which is a measure of the variability about the mean value is defined as follows:

$$s^2 = \frac{1}{n-1} * [(x(1)-\bar{x})^2 + (x(2)-\bar{x})^2 + ... + (x(n)-\bar{x})^2] =$$

$$\frac{1}{n-1} * \sum_{i=1}^{n} [x(i)-\bar{x}]^2 \qquad (2.1.2)$$

Thus, the variance, s^2 is the sum of the squares of the deviations from the mean divided by the number, n, minus one. The third and fourth column of Table 2.1.1 illustrate the calculation of the variance. Note that if all fish in the sample had the same length this would equal the mean length and the variance would be zero. The sum of the deviations (not squared) is always zero. The larger the deviations from the mean value, the larger the variance will be. The two largest values of the square of the deviations from the mean in Table 2.1.1 occurred for the smallest and the largest observations.

The square root of the variance, s, is called the "<u>standard deviation</u>". Often one is interested in the variance relative to the size of the mean length, and for that purpose s is the relevant quantity as it has the same unit as the mean. This leads to the relative standard deviation, $s/\bar{x}$, also called the "<u>coefficient of variation</u>".

When doing the calculations by hand it is easier to work with a rearranged form of Eq. 2.1.2, which is equivalent to

$$s^2 = \frac{1}{n-1} * \left[\sum_{i=1}^{n} x(i)^2 - \frac{1}{n} * [\sum_{i=1}^{n} x(i)]^2 \right] \qquad (2.1.3)$$

However, as most scientific pocket calculators contain an option for automatic calculation of mean and variance the calculations here are illustrated by Eq. 2.1.2, which is conceptually easier to understand.

For many purposes, e.g. for graphical representation, it is convenient to arrange the sample in the form of a "frequency table" by dividing the length range into a number of length intervals. The length range for the sample in Table 2.1.1 goes from 11.2 to 19.0 cm. With length groups of 1 cm we need nine length groups to cover the range. Using 10.5 as the lower limit of the first length interval, the intervals and the frequencies of lengths become those shown in the first four columns of Table 2.1.2, which is a so-called length-frequency table.

Table 2.1.1 Mean value, variance and standard deviation of a length-frequency sample

fish no.	length (cm)	deviation from mean	square of deviation from mean
i	x(i)	x(i)-$\bar{x}$	(x(i)-$\bar{x}$)^2
1	14.2	-0.87	0.75
2	16.3	1.23	1.52
3	14.8	-0.27	0.07
4	13.2	-1.87	3.48
5	16.9	1.83	3.36
6	12.4	-2.67	7.11
7	14.3	-0.77	0.59
8	15.7	0.63	0.40
9	15.3	0.23	0.05
10	11.2 (min.)	-3.87	14.95
11	12.9	-2.17	4.69
12	13.5	-1.57	2.45
13	18.2	3.13	9.82
14	11.6	-3.47	12.02
15	18.5	3.43	11.79
16	16.3	1.23	1.52
17	15.5	0.43	0.19
18	15.8	0.73	0.54
19	13.2	-1.87	3.48
20	19.0 (max.)	3.93	15.47
21	12.0	-3.07	9.40
22	17.1	2.03	4.13
23	15.4	0.33	0.11
24	14.6	-0.47	0.22
25	14.0	-1.07	1.14
26	18.1	3.03	9.20
27 = n	16.8	1.73	3.00
Total	406.8	0.00	121.48
	= Σ x(i)	= Σ(x(i)-$\bar{x}$)	= Σ(x(i)-$\bar{x}$)^2

mean length, $\bar{x}$: 406.8/27 = 15.07
variance, s^2 : 121.48/(27-1) = 4.67
standard deviation, s : $\sqrt{4.67}$ = 2.16
relative standard deviation, s/$\bar{x}$: 2.16/15.07 = 0.14
standard error, s/$\sqrt{n}$: 2.16/$\sqrt{27}$ = 0.41

(The concept of standard error is introduced in Section 2.3)

Table 2.1.2 Mean and variance from a length-frequency sample. (The sample is derived from Table 2.1.1 with a class interval, dL of 1 cm)

index	interval (cm)	mid-point (cm)	fre-quency			
j	L(j)-L(j)+dL	$\overline{L}(j)$	F(j)	$F(j)*\overline{L}(j)$	$(\overline{L}(j)-\overline{x})$	$F(j)*(\overline{L}(j)-\overline{x})^2$
1	10.5-11.5	11	1	11	-4.074	16.60
2	11.5-12.5	12	3	36	-3.074	28.35
3	12.5-13.5	13	3	39	-2.074	12.91
4	13.5-14.5	14	4	56	-1.074	4.61
5	14.5-15.5	15	4	60	-0.074	0.02
6	15.5-16.5	16	5	80	0.926	4.29
7	16.5-17.5	17	3	51	1.926	11.13
8	17.5-18.5	18	2	36	2.926	17.12
9	18.5-19.5	19	2	38	3.926	30.83
	total		27	407		125.86

mean length, $\overline{x}$: 407/27 = 15.074, say 15.07
variance, s^2 : 125.86/26 = 4.84
standard deviation, s : $\sqrt{4.84}$ = 2.20
relative standard deviation $s/\overline{x}$: 2.20/15.07 = 0.15

Let j be the index of a length group, and let the lower and upper class limit of length group no. j be denoted by respectively:

$L(j) = L(1) + (j-1)*dL$ and $L(j+1) = L(1) + j*dL$,

or $L(j+1) = L(j) + dL$

where dL is the "<u>interval size</u>". A fish of length x(j) then belongs to length group j when

$L(j) <= x(j) < L(j) + dL$

Let F(j) be the frequency of length group j, that is the number of fish observed in length group j. Let $\overline{L}(j) = L(j) + dL/2$ be the midpoint of length group no. j. The calculation of mean value and variance from a frequency table is then performed in the usual way using midpoints to represent the intervals:

$n = \sum_{j=1}^{m} F(j)$ is the total number of observations, where m is the number of length groups,

$\overline{x} = \frac{1}{n} * \sum_{j=1}^{m} F(j)*\overline{L}(j)$ is the mean value and

$s^2 = \frac{1}{n-1} * \sum_{j=1}^{m} F(j)*[\overline{L}(j)-\overline{x}]^2$ is the variance.

The calculation procedure is shown in Table 2.1.2. The class midpoint $\overline{L}(j)$, and the square of the deviations from the mean are weighted by the number of fish in each class, i.e. the frequency, F(j). The results of Table 2.1.2 deviate slightly from those of Table 2.1.1 because a representation in cm groups produces less precise results than a representation in mm groups.

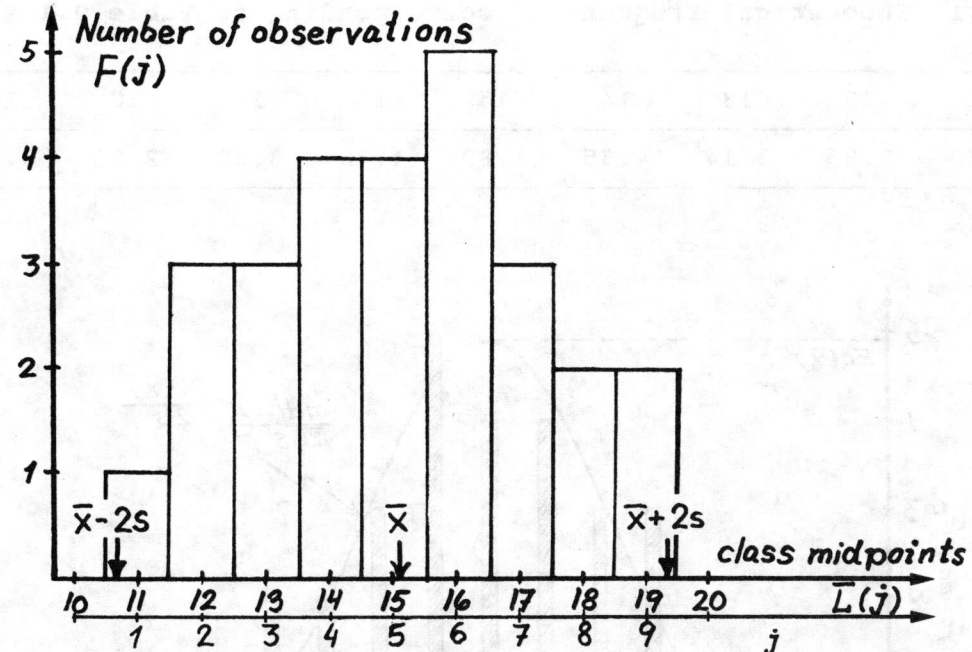

Fig. 2.1.1 Length-frequency diagram. Graphical representation of the length-frequency sample from Table 2.1.2

Fig. 2.1.1 shows a graphical representation of the frequency sample. Note that all observations lie in the interval from

$\bar{x} - 2*s$ to $\bar{x} + 2*s$

For the so-called normal distribution (discussed in the next section) we expect about 95% of the observations to be contained in that interval.

(See **Exercise(s)** in Part 2).

2.2 THE NORMAL DISTRIBUTION

Table 2.1.2 and Fig. 2.1.1 show an example of a small set of length-frequency data that approximately follows the so-called "<u>normal distribution</u>". The mathematical expression for a normal distribution is:

$$Fc(x) = \frac{n*dL}{s*\sqrt{2\pi}} * \exp\left[-\frac{(x-\bar{x})^2}{2s^2}\right] \qquad (2.2.1)$$

where Fc = "<u>calculated frequency</u>" or "<u>theoretical frequency</u>", n = number of observations, dL = interval size, s = standard deviation, $\bar{x}$ = mean length and π = 3.14159.

Using the values n = 27, dL = 1 cm, s = 2.20, $\bar{x}$ = 15.07 cm from Table 2.1.2 we get:

$$Fc(x) = \frac{27*1}{2.20*\sqrt{2*3.14159}} * \exp[-(x-15.07)^2/(2*4.84)]$$

$$= 4.896*\exp[-(x-15.07)^2/9.68]$$

The values of Fc for a number of different x-values are given in Table 2.2.1. Note that the notation is slightly modified as we now use the interval midpoint, x, as the argument in Fc instead of the interval index, j, as used for argument in F in Table 2.1.2.

Table 2.2.1 Theoretical frequencies corresponding to Table 2.1.2

x	11	12	13	14	15	16	17	18	19
Fc(x)	0.88	1.85	3.14	4.35	4.89	4.48	3.33	2.02	0.99

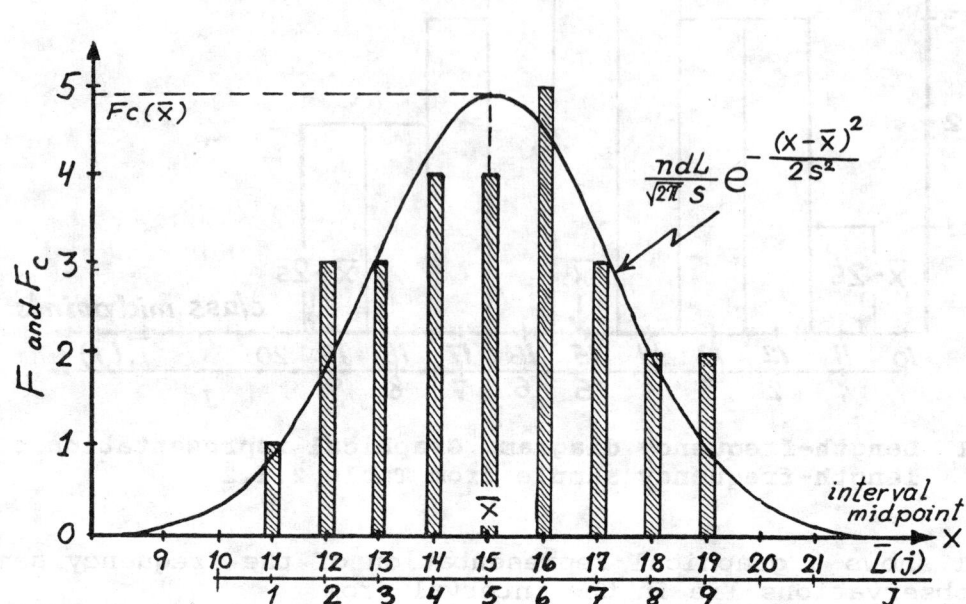

Fig. 2.2.1 The theoretical frequency, Fc, (the normal distribution curve) and the observed frequencies, F, (bars)

Fig. 2.2.1 shows the theoretical frequencies together with the bar diagram for F(j) from Fig. 2.1.1. As can be seen, Fc(x) gives a fair fit to the observed length-frequencies. This picture is often observed when recording length-frequencies of fish originating from one cohort, i.e. fish of the same age.

The normal distribution is observed in a great variety of different cases - hence the name. There are other types of probability distributions observed in fishery science. Examples are the "log-normal distribution", the "negative binomial distribution" and the "delta distribution". A conspicuous difference between these and the normal distribution is that they are skewed, whereas the normal distribution is symmetric. The delta distribution for example, is used to describe the probability distribution for the catch per hour by a trawl. It is composed of a log-normal distribution, which describes the distribution of the non-zero trawl catches and a special probability for zero catch (see Section 13.7, Fig. 13.7.2).

Perhaps the most important feature about normal distributions has to do with mean values. If you take, say, 50 random samples out of a certain population each of, say, 25 single observations, the 50 mean values will be (approximately) normally distributed. Thus, a mean value has a probability distribution. A mean value of any set of observations, is (approximately) normally distributed. This result is also valid for the mean values of log-normal distributions, delta distributions or any other type of distribution. This means that the mean values of all distributions observed in fishery biology are approximately normally distributed.

If we divide both sides of Eq. 2.2.1 by n (= sample size) we get:

$$Fc(x)/n = \frac{dL}{\sqrt{2\pi}} * \exp\left[-\frac{(x-\bar{x})^2}{2s^2}\right], \quad x = 1,2,3,\ldots \qquad (2.2.2)$$

the new found values, Fc(x)/n, will add up to nearly 1.0. Each value indicates the probability that a randomly drawn fish will belong to the corresponding length interval. That is, they can be interpreted as the probability of a randomly drawn fish to belong to the length interval from x-dL/2 to x+dL/2.

For the nine length intervals of Table 2.2.1 we find:

j	interval	probability
1	10.5-11.5	0.033
2	11.5-12.5	0.069
3	12.5-13.5	0.116
4	13.5-14.5	0.161
5	14.5-15.5	0.181
6	15.5-16.5	0.166
7	16.5-17.5	0.123
8	17.5-18.5	0.075
9	18.5-19.5	0.037
Total:		0.961

Thus for example, the odds are 181 to 1000 that a randomly drawn fish will be of a length between 14.5 and 15.5 cm. If we had included all length intervals and not only the nine for which we had observations, the probabilities would have added up to 1.000.

The normal distribution will be used in length-frequency analyses in the following chapters, because the length distribution of a single cohort of fish can be described by a normal distribution. As an introduction we shall study some of its aspects.

The procedures to calculate the mean and the standard deviation (Table 2.1.2) can be performed on any length-frequency data set. However, if for some reason, the observed frequency diagram does not represent the entire distribution, then the obtained values (from Eqs. 2.1.1 and 2.1.2) for sample mean and variance will be biased, i.e. the sample mean and variance may have no relation to the population mean and variance. The concept of "bias" will be further discussed in Section 7.1. If, for example, only the frequencies in the length interval from 10 to 15 cm are available (i.e. only the data for the left hand side) we are in a situation where Eq. 2.1.1 (mean value) and Eq. 2.1.2 (variance) do not represent the population. As will appear in Chapter 3 this is often the case when analyzing length-frequencies. However, there are a number of methods to overcome the problem.

(See **Exercises(s)** in Part 2)

2.3 CONFIDENCE LIMITS

In this section we shall also use the example of a length composition sample of fish from one cohort. We have estimated the mean length of the cohort, $\bar{x}$, from the sample. Such an estimate is usually different from the true population mean, the mean we would have obtained if all fish of that cohort in the sea had been measured. Usually the true mean length is unknown. If we were dealing with a population of cultured fish in a pond we might be able to measure the true mean length of that population, but for a wild fish stock it is impossible to measure the true value of any parameter. In practice this also applies to the population of fish caught in a fishery, since we will not be in a position to measure all fish caught. We shall deal with the precision of the estimate of the mean length, in other words how great the deviation between the estimate and the true mean is likely to be. This uncertainty about the true mean is expressed by the "confidence limits". In the case of a normal distribution, the lower and upper confidence limits are given by respectively:

$$\bar{x} - t_{n-1}*s/\sqrt{n} \text{ and } \bar{x} + t_{n-1}*s/\sqrt{n} \quad (2.3.1)$$

where n is the sample size, s the standard deviation and t_{n-1} the so-called fractiles in the "t-distribution" or "Student's distribution" (Table 2.3.1). The argument "f" in the t-distribution (Table 2.3.1) is called the "number of degrees of freedom". In general the number of degrees of freedom is the number of observations minus the number of parameters. In this case $\bar{x}$ is the only parameter, so f = n-1 and $t_f = t_{n-1}$ (see Table 2.3.1).

The confidence limits can be calculated at different levels of precision, usually 90%, 95% and 99%, as indicated in Table 2.3.1. The higher the level (percentage), the higher the fractiles and therefore the wider the interval between the lower and upper limits.

Returning to the example given in Section 2.1 (Table 2.1.2) we want to calculate, for example, the 95% confidence limits for the mean length of fish in the population from which the sample was drawn. We use the 95% fractile of the t-distribution (Table 2.3.1) with n-1 = 26 degrees of freedom and insert into Eq. 2.3.1:

$$t_{n-1}*s/\sqrt{n} = 2.06*2.20/\sqrt{27} = 0.87, \text{ while } \bar{x} = 15.07$$

the 95% confidence limits are:

lower limit: $\bar{x}$-0.87 = 15.07-0.87 = 14.20
upper limit: $\bar{x}$+0.87 = 15.07+0.87 = 15.94

Thus, we are "95% confident" that the true mean length lies somewhere between 14.20 and 15.94, or in other words, if sampling was repeated 100 times under the same conditions we would expect the means to lie 95 times between 14.20 and 15.94. The interval between the lower limit and the upper limit is called the "confidence interval".

For the example used above the confidence intervals at the 90% and 99% levels are respectively [14.35,15.79] and [13.89,16.25], of which the first is narrower and the second wider than the 95% interval.

The quantity $s/\sqrt{n}$ is the standard deviation of the estimate of the mean length (also called the "standard error") so that $\bar{x}$ has the variance (see Table 2.1.1):

$$VAR(\bar{x}) = s^2/n \quad (2.3.2)$$

Thus, the larger the sample, the more precise is the estimate of $\bar{x}$ (this subject will be discussed further in Section 7.2).

Eq. 2.3.2 follows from two general rules for random variables which are applied repeatedly in this manual. They are:

$$VAR(Cx) = C^2 * VAR(x) \tag{2.3.3}$$

$$VAR\left(\sum_{i=1}^{n} x\right) = n * VAR(x) \tag{2.3.4}$$

where C is a constant. For instance, when the variance of x is s^2 then the variance of 3x is $9s^2$; or, when the original observations are summed three by three, then the variance of $x_1+x_2+x_3$ is $3*s^2$.

The above statements about confidence limits apply only to "unbiased" estimates of the mean value. In cases when samples are biased, no matter how many fish we sample and measure we shall always get estimates of the mean value which are different from the true mean value.

Suppose we want to estimate the mean length of a certain fish species actually caught in a commercial fishery (note: fish caught are the fish landed plus the fish discarded at sea). Thus, if we sample only from the landings, and not the fish, usually below a certain size, which are discarded at sea, we get a biased estimate of the mean length of the fish caught. The mean length of the catch will be over-estimated, no matter how many fish we sample at the landing site. We can only get an unbiased estimate of the mean length of the fish that has been landed.

(See **Exercise(s)** in Part 2).

Table 2.3.1 Fractiles of the t-distribution (Student's distribution)*

degrees of freedom f	fractiles 90% t(f)	95% t(f)	99% t(f)	degrees of freedom f	fractiles 90% t(f)	95% t(f)	99% t(f)
1	6.31	12.71	63.66	15	1.75	2.13	2.95
2	2.92	4.30	9.93	16	1.75	2.12	2.92
3	2.35	3.18	5.84	17	1.74	2.11	2.90
4	2.13	2.78	4.60	18	1.73	2.10	2.88
5	2.02	2.57	4.03	19	1.73	2.09	2.86
6	1.94	2.45	3.71	20	1.73	2.09	2.85
7	1.90	2.37	3.50	25	1.71	2.06	2.79
8	1.86	2.31	3.36	30	1.70	2.04	2.75
9	1.83	2.26	3.25	40	1.68	2.02	2.70
10	1.81	2.23	3.17	50	1.67	2.01	2.68
11	1.80	2.20	3.11	60	1.67	2.00	2.66
12	1.78	2.18	3.06	80	1.67	1.99	2.64
13	1.77	2.16	3.01	100	1.66	1.98	2.63
14	1.76	2.15	2.98	∞	1.65	1.96	2.58

*) The use of the letter t in this context is universal. In this manual t is also used to represent the age of a fish. This table has been repeated on the last page of this volume for easy reference

2.4 ORDINARY LINEAR REGRESSION ANALYSIS

This method is used when we want to describe the variation of one quantity, e.g., the body depth of a fish, as a linear function of another quantity, e.g., the total length. The theory requires that the quantity on the horizontal axis (the independent variable) is measured with absolute precision. The method is often applied, however, when this requirement is violated. The effect of the inaccuracy of the values of the independent variable is that the slope of the line becomes flatter (closer to zero).

Suppose we have measured both the total length and the body depth of a sample of 7 fish.

Table 2.4.1 shows the total lengths, x(i), and the corresponding body depths, y(i), i = 1,2,...,7.

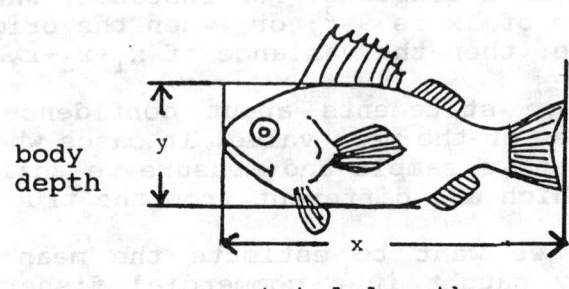

Table 2.4.1 Sample of total lengths, x, and corresponding body depths, y

i	1	2	3	4	5	6	7
x(i)	11.2	12.4	13.5	15.7	17.1	18.5	19.0
y(i)	3.0	3.2	4.0	4.8	4.8	4.9	5.6

As can be expected, the body depth tends to increase when the total length increases. If the body proportions of a fish would remain constant for all sizes, its body depth would be directly proportional to its length, and this could be described by the model:

$$y(i) = b*x(i) \qquad (2.4.1)$$

where b is a constant, also called a "parameter". The plot of this model always passes through the origin, the point where the x-axis and y-axis meet. We may allow for a deviation from proportionality between x and y by introducing a second parameter, a, and use instead of Eq. 2.4.1 the model:

$$y(i) = a + b*x(i) \qquad (2.4.2)$$

where a indicates the intercept with the y-axis of the line that fits to the points. Fig. 2.4.1 shows the "plot" (or the "scatter diagram") of y(i) against x(i).

An implication of Eq. 2.4.2 is that a fish of zero length has depth a, which makes no sense except when a is zero. However, if only lengths in a certain range are considered (e.g. only lengths above 5 cm), the two-parameter model may give a better fit to the observations than the one-parameter model, because the assumption of proportionality between length and depth is not strictly fulfilled.

The mathematical model of Eq. 2.4.2 is called a "linear model" because pairs (x,y) which conform to the model, lie on a straight line. With a = -0.32 and b = 0.30 we get the straight line shown in Fig. 2.4.1. With these values of a and b the line in Fig. 2.4.1 fits well to the observed pairs of (x,y).

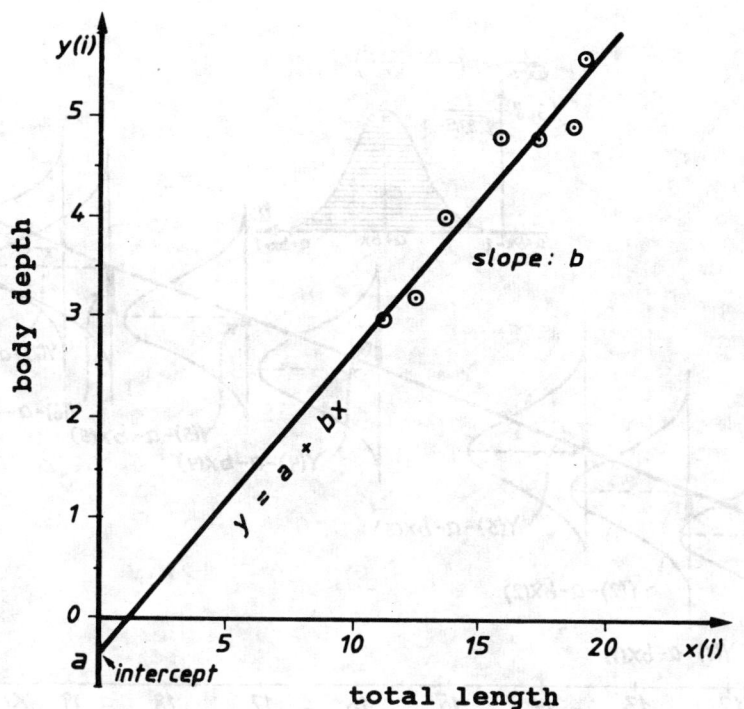

Fig. 2.4.1 Scatter diagram of body depth (y) against total length (x), also called the "plot of y on x"

We shall now look into the problem of determining the line, i.e. how to estimate the parameters a and b. Just as we did for the mean value (cf. Section 2.3) we shall also show how the confidence limits of the estimates of a and b are calculated. This procedure is called "<u>ordinary linear regression analysis</u>". This method is probably the most commonly used statistical technique in fishery biology. There are special names for the parameters: a is called the "<u>intercept</u>" and b is called the "<u>slope</u>". The inter-cept is the distance from the point (0,0) in the (x,y) diagram to the point where the "<u>regression line</u>"

$$y = a + b*x$$

intersects with the y-axis (see Fig. 2.4.1).

The slope, b, indicates how steep the line is. If b = 0 the line is parallel to the x-axis. If b is positive the slope is ascending. If b is negative the slope is descending.

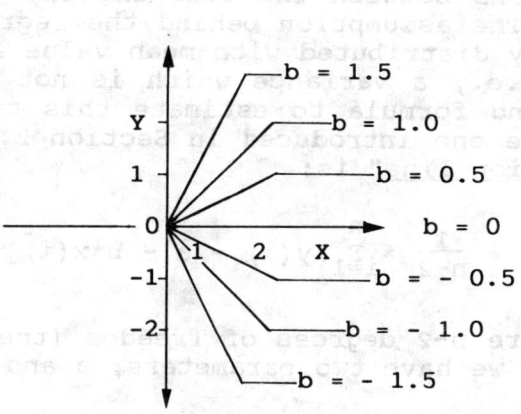

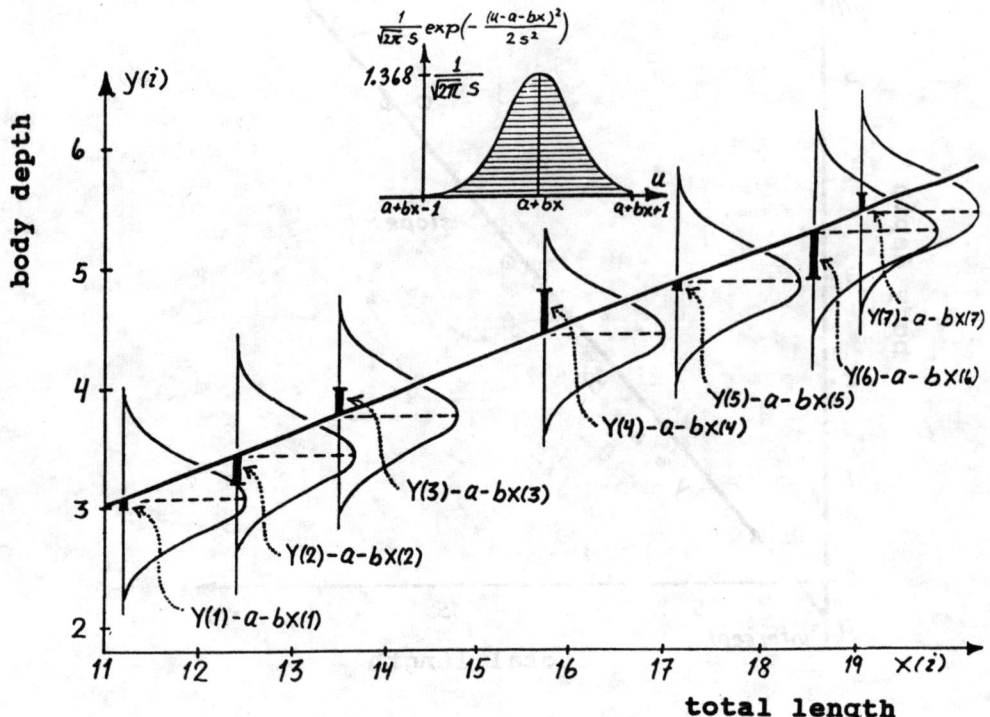

Fig. 2.4.2 **Illustration of the assumptions behind ordinary linear regression analysis. Each y(i) for a given x(i) is normally distributed with a common variance**

The variable on the horizontal axis, x, is called the "<u>independent variable</u>" and the variable on the vertical axis, y, is called the "<u>dependent variable</u>". The regression line is determined as the line which minimizes the sum of squares of deviations between the line y = a + b*x and the pairs of observations, (x(i),y(i)). We say that a and b are estimated by the "<u>least squares method</u>", i.e. we search for the values of a and b which minimize:

$$\sum_{i=1}^{n}[y(i) - a - b*x(i)]^2 \qquad (2.4.3)$$

where n is the number of pairs of observations (n = 7 in the example). The deviations between the line and the observations are illustrated in Fig. 2.4.2. The assumption behind the regression analysis is that each y(i) is normally distributed with mean value a + b*x(i), and with a constant variance, i.e., a variance which is not dependent on the value of x(i). The following formula to estimate this common variance differs only slightly from the one introduced in Section 2.1. The so-called "<u>variance about the regression line</u>" is:

$$s^2 = \frac{1}{n-2} * \sum_{i=1}^{n}[y(i) - a - b*x(i)]^2 \qquad (2.4.4)$$

There are n-2 degrees of freedom (the number by which the sum is divided) because we have two parameters, a and b.

Table 2.4.2 The calculation procedure for ordinary linear regression analysis. Results marked by #) are not used in the calculation of a and b, but are derived here for subsequent use

i	total length x(i)	$x(i)^2$	body depth y(i)	$y(i)^2$	x(i)*y(i)
1	11.2	125.44	3.0	9.00	33.60
2	12.4	153.76	3.2	10.24	39.68
3	13.5	182.25	4.0	16.00	54.00
4	15.7	246.49	4.8	23.04	75.36
5	17.1	292.41	4.8	23.04	82.08
6	18.5	342.25	4.9	24.01	90.65
7=n	19.0	361.00	5.6	31.36	106.40
Σ	107.4	1703.60	30.3	136.69	481.77
	Σx(i)	$Σx(i)^2$	Σy(i)	$Σy(i)^2$	Σx(i)*y(i)

$\bar{x} = 15.343$ $\qquad\qquad\qquad\qquad$ $\bar{y} = 4.329$

$\frac{1}{n}*(Σx(i))^2 = 1647.82$ $\qquad\qquad$ $\frac{1}{n}*(Σy(i))^2 = 131.16$ $\qquad$ #)

$Σx(i)^2 - \frac{1}{n}*(Σx(i))^2 = 55.78$ $\qquad$ $Σy(i)^2 - \frac{1}{n}*(Σy(i))^2 = 5.534$ #)

$sx^2 = 9.296$ #) $\qquad\qquad\qquad$ $sy^2 = 0.922$ #)
$sx = 3.049$ #) $\qquad\qquad\qquad$ $sy = 0.960$ #)

$\frac{1}{n}*Σx(i)*Σy(i) = 464.89$

$Σx(i)*y(i) - \frac{1}{n}*Σx(i)*Σy(i) = 16.88$ $\qquad$ $sxy = 2.814$ #)

$b = \frac{Σx(i)*y(i) - \frac{1}{n}*Σx(i)*Σy(i)}{Σx(i)^2 - \frac{1}{n}*(Σx(i))^2} = \frac{16.88}{55.78} = 0.303$

$a = \bar{y} - \bar{x}*b = 4.329 - 15.343*0.303 = -0.315$

Table 2.4.3 Calculation of variance about the line from Eq. 2.4.4

i	x(i)	y(i)	a+b*x(i)	$[y(i)-a-b*x(i)]^2$
1	11.2	3.0	3.079	0.0062
2	12.4	3.2	3.442	0.0587
3	13.5	4.0	3.776	0.0504
4	15.7	4.8	4.442	0.1281
5	17.1	4.8	4.866	0.0044
6	18.5	4.9	5.291	0.1525
7	19.0	5.6	5.442	0.0250

$s^2 = 0.4252/(7-2) = 0.085$ $\qquad\qquad$ sum: 0.4252

Estimates of the parameters a (intercept) and b (slope) are obtained by:

$$b = \frac{\sum_{i=1}^{n} x(i)*y(i) - \frac{1}{n}*\sum_{i=1}^{n} x(i) * \sum_{i=1}^{n} y(i)}{\sum_{i=1}^{n} x(i)^2 - \frac{1}{n}*\left[\sum_{i=1}^{n} x(i)\right]^2} \qquad (2.4.5)$$

$$a = \bar{y} - \bar{x}*b \qquad (2.4.6)$$

where $\bar{y}$ and $\bar{x}$ are the mean values of y and x as defined by Eq. 2.1.1.

In Table 2.4.2 the calculation procedures to estimate a and b are demonstrated using the data from Table 2.4.1. Thus, the estimated regression line becomes:

$$y = -0.315 + 0.303*x \qquad (2.4.7)$$

To calculate the confidence limits of a and b we need the sum of squares of deviations of x and y. The variances of x and y are defined by Eq. 2.1.3 as follows:

$$sx^2 = \frac{1}{n-1}*[\Sigma x(i)^2 - \frac{1}{n}*\{\Sigma x(i)\}^2] \qquad (2.4.8)$$

and a similar expression for sy^2. For use in the next section we introduce the "<u>covariance</u>":

$$sxy = \frac{1}{n-1}*[\Sigma x(i)*y(i) - \frac{1}{n}*\Sigma x(i)*\Sigma y(i)] \qquad (2.4.9)$$

The procedure for the calculation of variance about the regression line leading to Eq. 2.4.4 is demonstrated in Table 2.4.3. However, the variance about the line can be obtained more easily from sy and sx:

$$s^2 = \frac{n-1}{n-2}*[sy^2 - b^2*sx^2] \qquad (2.4.10)$$

Given the results from Table 2.4.2, Eq. 2.4.10 becomes:

$$s^2 = \frac{6}{5}*(0.922 - 0.303^2*9.297) = 0.085$$

The variances of the estimates of b and a are:

$$sb^2 = \frac{1}{n-2}*[(sy/sx)^2 - b^2] \qquad (2.4.11)$$

and

$$sa^2 = sb^2*[\frac{n-1}{n}*sx^2 + \bar{x}^2] \qquad (2.4.12)$$

Given the results from Table 2.4.2 we get:

$$sb^2 = \frac{1}{7-2}*[\frac{0.922}{9.297} - 0.303^2] = 0.00147, \qquad sb = 0.038$$

$$sa^2 = 0.00147*(\frac{7-1}{7}*9.297 + 15.343^2) = 0.3578, \quad sa = 0.598$$

The confidence limits for the intercept a and the slope b are respectively:

a: $[a - sa*t_{n-2}, a + sa*t_{n-2}]$ (2.4.13)

b: $[b - sb*t_{n-2}, b + sb*t_{n-2}]$ (2.4.14)

The 95% confidence limits of a and b for the example with n = 7 fish and t_{7-2} = 2.57 (Table 2.3.1) become:

a: [-0.315 - 0.598*2.57, -0.315 + 0.598*2.57] = [-1.85, 1.22]

b: [0.303 - 0.038*2.57, 0.303 + 0.038*2.57] = [0.21, 0.40]

Note that the confidence interval for the intercept a contains zero. This means that the hypothesis that body depth is directly proportional to length, (thus that "a = 0") cannot be rejected by the 95% confidence limits. We say that a is not significantly different from 0 at the 95% level.

If we have a good reason to assume that a = 0 then the estimated value should be replaced by 0 if the estimate is not significantly different from 0. Then, however, b must be recalculated as follows:

$$b = \frac{\Sigma x(i)*y(i)}{\Sigma x(i)^2}$$ (2.4.15)

Our present estimate is based on only seven fish. If we had measured 200 fish the estimate of the standard deviation, sa, would be smaller (cf. Eqs. 2.4.11 and 2.4.12). Let us assume for example, that $\bar{x}$, $\bar{y}$, sx, sy, a and b were the same for a sample size of n = 200 as those estimated for a sample size of n = 7 (which might well happen). Although the estimates of a and b turn out to have the same value, their standard deviations, sa and sb, will be different.

With n = 200 Eq. 2.4.11 gives sb = 0.006098, while Eq. 2.4.12 gives sa = 0.0091 and t_{198} = 1.97 (Table 2.3.1). Thus, sa and sb become smaller, and consequently the confidence interval of a becomes smaller:

a: [-0.315 - 0.0091*1.97, -0.315 + 0.0091*1.97] = [-0.33, -0.30]

The estimate of a would now be significantly different from 0. In that case we can conclude that the odds are less than 5% that the true value of a is larger than -0.30 or smaller than -0.33.

(See **Exercise(s)** in Part 2).

2.5 THE CORRELATION COEFFICIENT AND FUNCTIONAL REGRESSION

The "correlation coefficient", r, is a measure of the linear association between two quantities, both of which are subject to random variation. The total length and body depth sample from Section 2.4 is an example of two such quantities. In this case seven fish were drawn at random. By accident we could have drawn seven fish all of (nearly) the same length. In that case the sample would not be suitable for estimation of the length/depth relationship because the confidence limits of a and b would become very wide.

The correlation coefficient can be used only when both measurements are allowed to vary randomly. If we had selected seven fish with predetermined lengths rather than random lengths (e.g. had selected the lengths 4, 6, 8, 10, 12, 14 and 16 cm for the length/depth sample) the calculation of a correlation coefficient for this sample would be incorrect.

The correlation coefficient is defined as:

$$r = \frac{sxy}{sx \cdot sy} \qquad (2.5.1)$$

where sxy is defined by Eq. 2.4.9 and sx and sy are defined by Eq. 2.4.8.

Inserting the slope ($b = sxy/sx^2$), Eq. 2.5.1 becomes:

$$r = b \cdot sx/sy \qquad (2.5.2)$$

The range of r is: $-1.0 \leq r \leq 1.0$. r is negative if y tends to decrease when x increases and r is positive if y tends to increase when x increases. This statement also holds for the slope b and it follows from Eq. 2.5.2: As sx/sy is always positive (cf. the definition Eq. 2.4.8) r has the same sign as the slope b. The extreme cases, $r = 1$ or $r = -1$ occur when all pairs (x,y) lie exactly on a straight line. The closer r approaches zero the less pronounced is the linear association between y and x. When $r = 0$, x and y are independent of each other.

Fig. 2.5.1 shows four examples of scatter diagrams with different values of r. For the example of Table 2.4.2 we get:

$$r = \frac{2.814}{3.049 \cdot 0.960} = 0.961$$

Let us call r1 (lower) and r2 (upper) the 95% confidence limits for r. They can be calculated from the expressions:

$$r1 = \tanh[0.5 \cdot \ln(\tfrac{1+r}{1-r}) - 1.96/\sqrt{n-3}\,]$$

$$r2 = \tanh[0.5 \cdot \ln(\tfrac{1+r}{1-r}) + 1.96/\sqrt{n-3}\,] \qquad (2.5.3)$$

where "tanh" is the "hyperbolic tangent" which is standard on many scientific pocket calculators.

With r from the example ($r = 0.961$, $n = 7$) the 95% confidence limits become: [r1,r2] = [0.75,0.99]. The 99% confidence limits can be obtained by replacing the number 1.96 by 2.58 in Eq. 2.5.3.

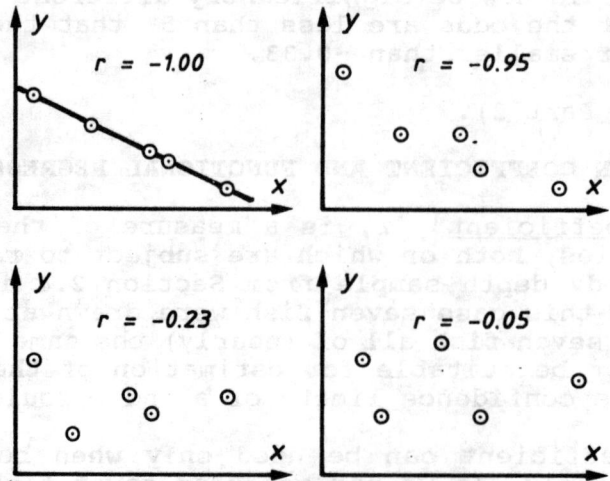

Fig. 2.5.1 Examples of correlation coefficients

Often we are interested in knowing whether zero lies in the confidence interval, viz. what the odds are that the linear association is due to chance. In this example the odds are less than 5% that the linear association is due to chance, because zero is not in the confidence interval.

In the example of regressing body depth on total length, the length was chosen as the independent variable and the body depth as the dependent variable. However, there is no special reason for this choice. Our sample consists of seven randomly chosen fish. We did not control what their lengths and body depths would be, thus, we could as well have made the opposite choice for dependent and independent variables.

One of the assumptions behind linear regression analysis is that the independent variable cannot be a random variable. The independent variable must be something of which we can determine the values beforehand. For example, if the independent variable is the time the sample is taken, it can be determined beforehand. We could decide to collect a sample on the first day of each month. If we measure time in units of years and start with time zero on the first of January, the independent variable would take the values: 0, 1/12, 2/12, 3/12 ... etc. These values are clearly not random variables.

In the case of the seven fish chosen at random in the example above the situation is that because they were chosen at random out of a normal distribution of fish lengths a correlation analysis can be performed on them. On the other hand, we are able to decide on the lengths beforehand. We could choose the four smallest fish and the three biggest. We could also decide, as we did, to take them as they come. Only in the latter case is it permissible to do both kinds of analysis. In the former case only regression analysis will do. On the other hand, it would probably be a more effective way of doing the regression analysis because of the great distance between the observations on the horizontal axis. This would cause a small variance of the slope. Choosing the fish at random, most of them are likely to be medium-sized and contribute little to the determination of the slope which might show a large variance.

Another question is whether we would have obtained a different result using the body depth as the independent variable, thus plotting fish length as a function of body depth. First it must be considered whether depth is as precisely measured as length. If it is not, the slope would be biased (flattened) as already mentioned. However, there are problems even if the two variables are measured with the same accuracy.

Taking now the body depth as the independent variable we get what is called an "inverse regression". Only in the exceptional case that all observations lie on the regression line (i.e. if $r = 1$ or $r = -1$) the same result would be obtained for the inverse regression as for the ordinary regression. The equation $y = a + b*x$ (Eq. 2.4.2) is mathematically equivalent to:

$$x = -a/b + y/b$$

or $\quad x = A + B*y \quad$ where $A = -a/b$ and $B = 1/b$ $\hfill(2.5.4)$

Carrying out the inverse regression (Eq. 2.5.4) we find that

$$A = 2.139 \quad \text{and} \quad B = 3.05$$

The equation: $x = 2.139 + 3.05*y \quad$ can be converted into:

$$y = -0.701 + 0.328*x$$

which can be compared with the result found for the original regression (Eq. 2.4.7: $y = -0.315 + 0.303*x$). Thus, the inverse regression gives a result that differs from that of the original regression analysis.

One way to circumvent the problem of choosing the independent variable when both variables are random variables is to use the so-called "<u>functional regression analysis</u>" (see Ricker, 1973). This method estimates a slope (which we call b' to distinguish it from slope b of the ordinary regression analysis) by the expressions:

$$b' = s_y/s_x \quad \text{if } r > 0$$
$$b' = -s_x/s_y \quad \text{if } r < 0 \tag{2.5.5}$$

and the intercept:

$$a' = \bar{y} - b'*\bar{x} \tag{2.5.6}$$

This type of analysis gives a result that may be considered a compromise between the original ordinary regression and its inverse counterpart.

With the results from Table 2.4.2 we get:

b' = 0.960/3.049 = 0.315 and a' = 4.329 - 0.315*15.343 = -0.504

and y = -0.504 + 0.315*x

Functional regression analysis is mentioned here for the sake of completeness. There are some rather intricate limitations to its applicability which we cannot go into here.

The following three regression lines have now been estimated:

1. Original ordinary regression analysis: y = -0.315 + 0.303*x
2. Functional regression analysis   : y = -0.504 + 0.315*x
3. Inverse ordinary regression analysis : y = -0.701 + 0.328*x

Fig. 2.5.2 shows the three regression lines. Note that all three lines pass through the point $(\bar{x},\bar{y})$ and that an increase in slope is partly balanced by a decrease of the intercept.

(See **Exercise(s)** in Part 2).

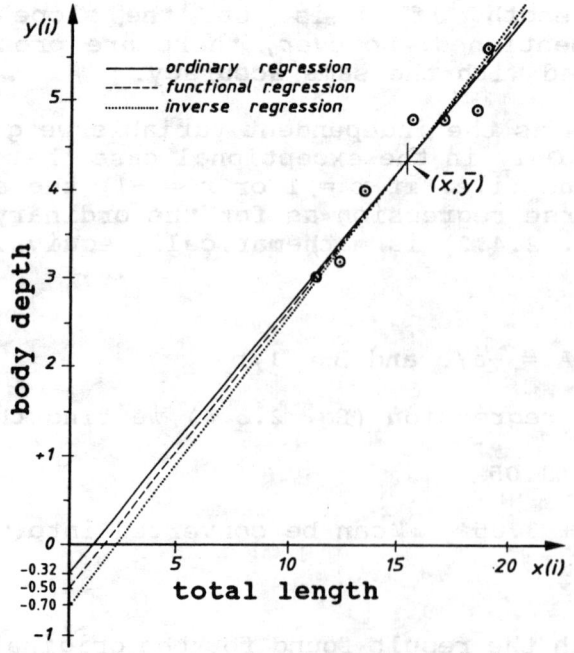

Fig. 2.5.2 Functional and inverse regression lines compared to the original regression line

2.6 LINEAR TRANSFORMATIONS

Linear functions are mathematically easy and also have the advantage that they can be graphically interpreted without any problem. However, many functional relationships observed in fishery biology are not linear. Fortunately, such non-linear functions can often be transformed into linear functions, which means that after transformation they can be dealt with in the way described in the foregoing sections. Several examples are given below of the application of transformations from non-linear functions to linear functions in fishery biology.

Example 1: Length-weight relationship

Here we consider a famous example, namely the functional relationship between total length and body weight of fish. Fig. 2.6.1 shows a plot of weight on length of the threadfin bream, <u>Nemipterus marginatus</u>. Clearly, this is not a linear relationship. The curve in Fig. 2.6.1 is of the function:

$$W(i) = q*L(i)^b \qquad (2.6.1)$$

where $W(i)$ is the body weight of fish no. i, $L(i)$ is the total length and q and b are parameters. Eq. 2.6.1 is usually called the "<u>length-weight relationship</u>". It can be transformed into a linear equation by taking logarithms on both sides:

$$\ln W(i) = \ln q + b*\ln L(i) \qquad (2.6.2)$$

or

$$y(i) = a + b*x(i) \qquad (2.6.2a)$$

where $y(i) = \ln W(i)$, $x(i) = \ln L(i)$ and $a = \ln q$.

With Eq. 2.6.2a we are now in a position to carry out the estimation of a and b by linear regression analysis. Input data are shown in Table 2.6.1 and the corresponding scatter diagram in Fig. 2.6.2. The results are:

$a = -4.538$, $b = 3.057$, $sx = 0.3311$, $sy = 1.0161$, $n = 16$,

$\bar{x} = 2.727$ and $\bar{y} = 3.799$

Since $a = \ln q$ we can obtain q of the original length-weight relationship (Eq. 2.6.1) by taking the antilog of a:

$q = \exp a = \exp(-4.538) = 0.0107$

Thus, the estimated relationship between W (in g) and L (in cm) becomes:

$$W = 0.0107*L^{3.057}$$

(The back-transformation from logarithms introduces a bias which we will not go into here.)

We can also calculate the 95% confidence limits of b, using the values of sx, sy, n and t_{14} (see Table 2.3.1) in Eq. 2.4.11:

$$sb^2 = \frac{1}{16-2} * \left[\left(\frac{1.0161}{0.3311}\right)^2 - 3.057^2\right] = 0.0052$$

$sb = 0.072$ and $sb*t_{n-2} = 0.072*2.15 = 0.155$

Table 2.6.1 Data for estimation of a length-weight relationship for the threadfin bream (<u>Nemipterus marginatus</u>) from the South China Sea (from Pauly, 1983)

i	L(i)	W(i)	ln L(i) x(i)	ln W(i) y(i)
1	8.1	6.3	2.092	1.841
2	9.1	9.6	2.208	2.262
3	10.2	11.6	2.322	2.451
4	11.9	18.5	2.477	2.918
5	12.2	26.2	2.501	3.266
6	13.8	36.1	2.625	3.586
7	14.8	40.1	2.695	3.691
8	15.7	47.3	2.754	3.857
9	16.6	65.6	2.809	4.184
10	17.7	69.4	2.874	4.240
11	18.7	76.4	2.929	4.336
12	19.0	82.5	2.944	4.413
13	20.6	106.6	3.025	4.669
14	21.9	119.8	3.086	4.786
15	22.9	169.2	3.131	5.131
16	23.5	173.3	3.157	5.155
		sum	43.629	60.786
		mean	2.7268	3.7991
		sx and sy	0.3311	1.0161

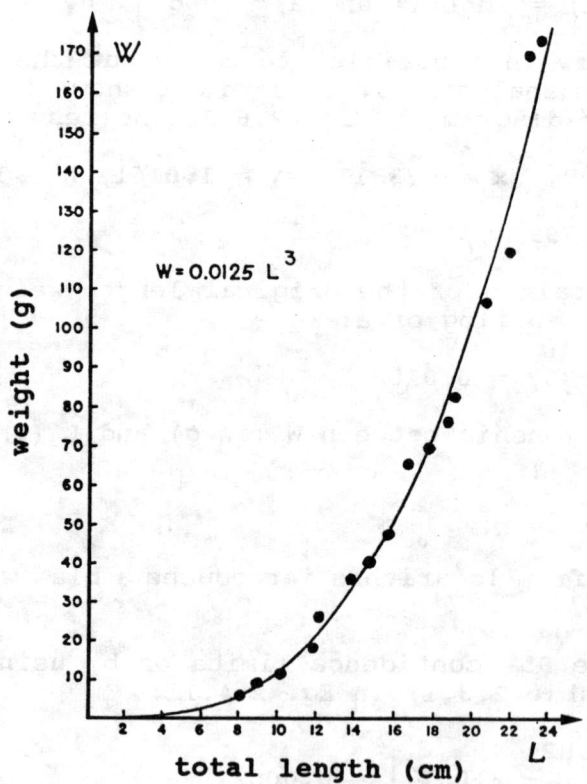

Fig. 2.6.1 Length-weight relationship of <u>Nemipterus marginatus</u> in the South China Sea. (Based on data from Table 2.6.1)

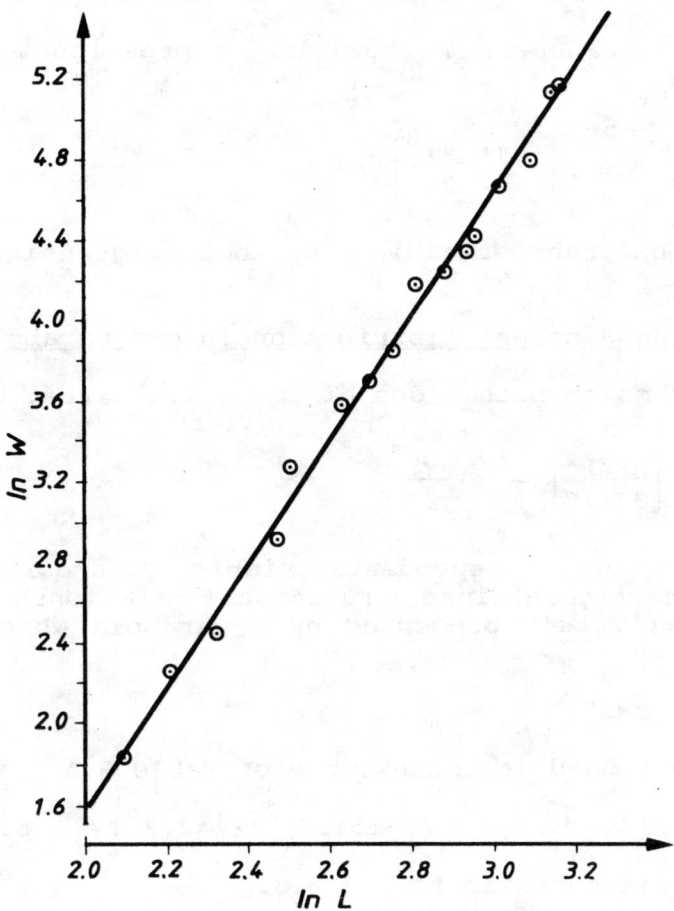

Fig. 2.6.2 The data from Fig. 2.6.1 converted to natural logarithms

The 95% confidence interval for b is [(3.057-0.155),(3.057+0.155)] or [2.90, 3.21]. These confidence limits tell us that only the first decimal in the estimate of b is significant (compare Section 2.3), thus the true value of b could just as well be 3.0.

Since the weight of a fish (in grammes) is approximately equal to its volume (in cubic cm), and since its volume is often proportional to the cube of its length, L^3, we would expect that the value of b in Eqs. 2.6.1 and 2.6.2 is close to 3.0.

Since the confidence interval calculated above supports this hypothesis we can simplify the length-weight relationship by replacing the estimate b = 3.057 by b = 3.0. This implies that a new estimate of the intercept a has to be obtained. Since the new straight line with b = 3.0 also passes through the point $(\bar{x},\bar{y})$ we can calculate the new intercept a using Eq. 2.6.2a:

$$a = \bar{y} - b*\bar{x} = 3.799 - 3.0*2.727 = -4.382$$

From a we obtain the corresponding new value for q

$$q = \exp(-4.382) = 0.0125$$

Thus, the new relationship becomes:

$$W = 0.0125*L^3$$

Example 2: Linearization of a normal distribution

In Section 2.2 (Eq. 2.2.1) the mathematical expression for a normal distribution is given as:

$$Fc(x) = \frac{n*dL}{s*\sqrt{2\pi}} * \exp\left[-\frac{(x-\bar{x})^2}{2s^2}\right]$$

This equation can be transformed into a linear regression in the following two stages:

Stage 1: <u>Converting a normal distribution into a parabola</u>

Taking the logarithms on both sides of Eq. 2.2.1 gives:

$$\ln Fc(x) = \ln\left[\frac{n*dL}{s*\sqrt{2\pi}}\right] - \frac{(x-\bar{x})^2}{2s^2} \qquad (2.6.3)$$

Considering ln Fc(x) as the dependent variable, y, and x as the independent variable, we have hereby obtained a functional relationship between y and x, which can graphically be represented by a parabola which has the general formula:

$$y = a + b*x + c*x^2$$

Inserting the values used in the example of Table 2.1.2 we obtain:

$$y = \ln[(27*1)/(2.2*\sqrt{2\pi})] - (x-15.07)^2/(2*2.2^2) = 1.59 - (x-15.07)^2/9.68$$

the graph of which is shown in Fig. 2.6.3.

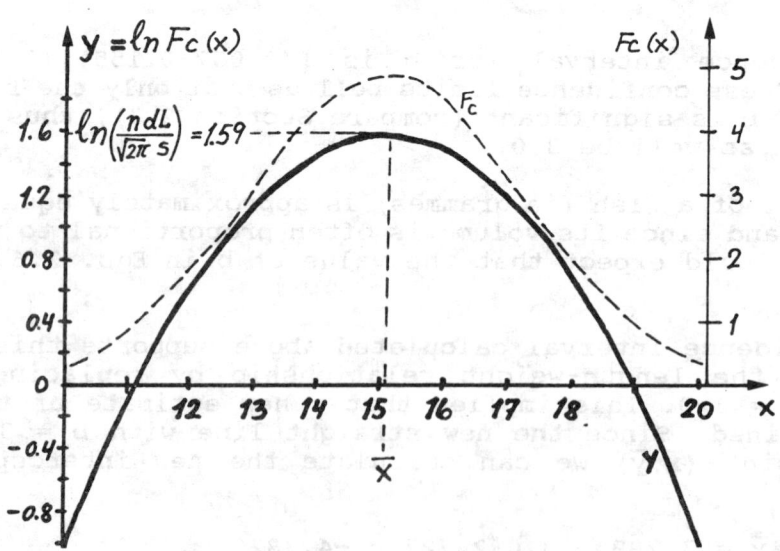

Fig. 2.6.3 The ln-transformed normal distribution (y) together with the original distribution (Fc)

Table 2.6.2 Estimation of mean value and variance from the Bhattacharya plot, illustrated by the theoretical frequencies, Fc(x), of Table 2.1.2, presented in Table 2.2.1

index j	$\bar{L}(j)$ (x)	interval x-dL/2, x+dL/2	Fc(x) (y)	ln Fc(x)	Δln Fc(z) (y')	x+dL/2 (z)
1	11	10.5-11.5	0.88	-0.128		
					0.743	11.5
2	12	11.5-12.5	1.85	0.615		
					0.529	12.5
3	13	12.5-13.5	3.14	1.144		
					0.326	13.5
4	14	13.5-14.5	4.35	1.470		
					0.117	14.5
5	15	14.5-15.5	4.89	1.587		
					-0.088	15.5
6	16	15.5-16.5	4.48	1.500		
					-0.297	16.5
7	17	16.5-17.5	3.33	1.203		
					-0.500	17.5
8	18	17.5-18.5	2.02	0.703		
					-0.713	18.5
9	19	18.5-19.5	0.99	0.010		
					a = 3.1237	(dL = 1)
					b = -0.2073	
$\bar{x} = -a/b = 15.07$		$s^2 = -dL/b = 4.82$		s = 2.20		

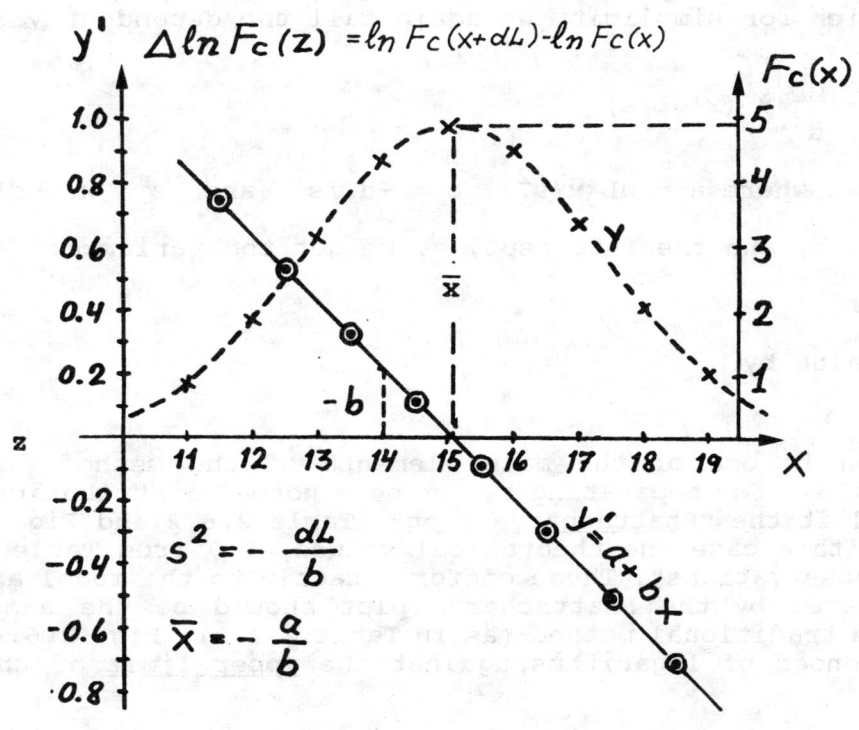

Fig. 2.6.4 Bhattacharya plot together with the original normal distribution. (Based on Table 2.6.2)

Stage 2: Converting a parabola into a straight line

In order to transform the parabola into a straight line we have to introduce a new dependent variable, y', which is the difference between the logarithm of the number in a certain length class and the logarithm of the number in the preceding class.

$$y' = \ln Fc(x+dL) - \ln Fc(x) \qquad (2.6.4)$$

This can also be expressed as

$$y' = \Delta \ln Fc(x+dL/2)$$

where Δ (delta) designates a "small" difference between two function values. y' is to be plotted against a new independent variable, z, which is equivalent to x plus half the length interval:

$$z = x + dL/2$$

We now have to insert Eq. 2.6.3 into Eq. 2.6.4 as follows:

$$y' = \Delta \ln Fc(x+dL/2) = \Delta \ln Fc(z) =$$

$$\left\{ \ln\left[\frac{n*dL}{s*\sqrt{2\pi}}\right] - \frac{(x+dL-\bar{x})^2}{2s^2} \right\} - \left\{ \ln\left[\frac{n*dL}{s*\sqrt{2\pi}}\right] - \frac{(x-\bar{x})^2}{2s^2} \right\} =$$

$$\left[\frac{-(x+dL-\bar{x})^2+(x-\bar{x})^2}{2s^2}\right]$$

After squaring and summing this can be converted into the relatively simple equation in which for simplicity we again call the dependent variable y:

$$y = \frac{dL*\bar{x}}{s^2} - \frac{dL}{s^2} * (x+dL/2) \qquad (2.6.5)$$

or $y = a + b*z$ where $a = dL*\bar{x}/s^2$, $b = -dL/s^2$ and $z = x + dL/2$.

From the slope, b, and the intercept, a, we get the variance:

$$s^2 = -dL/b \qquad (2.6.6)$$

and the mean value by

$$\bar{x} = -a/b \qquad (2.6.7)$$

This regression is one of the main elements of the method described by Bhattacharya (1967) for separating two or more normal distributions (Section 3.4.1). We call it the "Bhattacharya plot". Table 2.6.2 and Fig. 2.6.4 show an example. In this case the theoretical values, Fc, from Table 2.2.1 have been used as "observations". These conform exactly to the model and mean and variance estimated by the Bhattacharya plot should be the same as those obtained by the traditional method (as in Table 2.1.2). Fig. 2.6.4 shows the plot of differences of logarithms against the upper limit of the smallest length group.

Table 2.6.3 Bhattacharya plot corresponding to the length-frequency sample of Table 2.1.2

index	x (x)	x-dL/2, x+dL/2	F(x)	ln F(x) (y)	Δln F(z) (y')	x+dL/2 (z)
1-2	11.5	10.5-12.5	4	1.386		
					0.560	12.5
3-4	13.5	12.5-14.5	7	1.946		
					0.251	14.5
5-6	15.5	14.5-16.5	9	2.197		
					-0.588	16.5
7-8	17.5	16.5-18.5	5	1.609		
					-0.916	18.5
9	19.5	18.5-20.5	2	0.693		

$a = 3.909$ (dL = 2)
$b = -0.263$

$\bar{x} = -a/b = 14.8$ $s^2 = -dL/b = 7.605$ $s = 2.76$

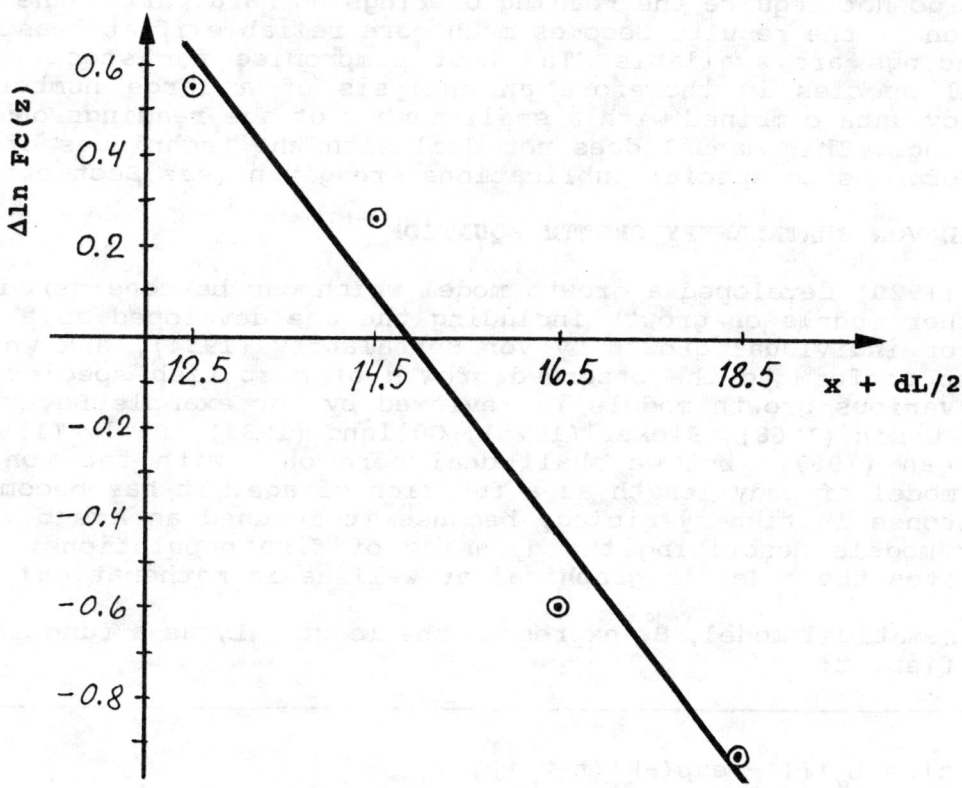

Fig. 2.6.5   Bhattacharya plot corresponding to Table 2.6.3

Table 2.6.3 shows the estimation of mean and variance by the Bhattacharya plot, but now with the actual observations given in Table 2.1.2. Because of the small sample size the observations have been grouped into 2 cm intervals. Fig. 2.6.5 shows the corresponding plot. The estimates of mean and variance obtained in Table 2.6.3 deviate from those calculated by the traditional method (Table 2.1.2) because of 1) the small sample size, 2) the bias introduced by large length intervals and 3) the use of a different statistical method (linear regression analysis).

(See **Exercise(s)** in Part 2).

3 ESTIMATION OF GROWTH PARAMETERS

The study of growth means basically the determination of the body size as a function of age. Therefore all stock assessment methods work essentially with age composition data. In temperate waters such data can usually be obtained through the counting of year rings on hard parts such as scales and otoliths. These rings are formed due to strong fluctuations in environmental conditions from summer to winter and vice versa. In tropical areas such drastic changes do not occur and it is therefore very difficult, if not impossible to use this kind of seasonal rings for age determination.

Only recently methods have been developed to use much finer structures, so-called daily rings, to count the age of the fish in number of days. These methods, however, require special expensive equipment and a lot of manpower, and it is therefore not likely that they will be applied on a routine basis in many places.

Fortunately several numerical methods have been developed which allow the conversion of length-frequency data into age composition. Although these methods do not require the reading of rings on hard parts, the final interpretation of the results becomes much more reliable if at least some direct age readings are available. The best compromise for stock assessment of tropical species is therefore an analysis of a large number of length-frequency data combined with a small number of age readings on the basis of daily rings. This manual does not deal with the techniques of age reading but references to special publications are given (see Section 3.2.1).

3.1 THE VON BERTALANFFY GROWTH EQUATION

Pütter (1920) developed a growth model which can be considered the base of most other models on growth including the one developed as a mathematical model for individual growth by von Bertalanffy (1934), and which has been shown to conform to the observed growth of most fish species. The theory behind various growth models is reviewed by for example Beverton and Holt (1957), Ursin (1968), Ricker (1975), Gulland (1983), Pauly (1984) and Pauly and Morgan (1987), but we shall deal here only with the von Bertalanffy growth model of body length as a function of age. It has become one of the cornerstones in fishery biology because it is used as a sub-model in more complex models describing the dynamics of fish populations. Fig. 3.1.0.1 illustrates the model in graphical as well as in mathematical form.

The mathematical model, B, expresses the length, L, as a function of the age of the fish, t:

$$L(t) = L_\infty * [1 - \exp(-K*(t-t_o))] \quad (3.1.0.1)$$

The right hand side of the equation contains the age, t, and some parameters. They are: "L_∞" (read "L-infinity"), "K" and "t_o" (read "t-zero"). Different growth curves will be created for each different set of parameters, therefore it is possible to use the same basic model to describe the growth of different species simply by using a special set of parameters for each species.

To illustrate the use of the model, assume that the three parameters have been estimated for some particular fish stock and that the values are:

L_∞ = 50 cm, K = 0.5 per year and t_o = -0.2 year

- 45 -

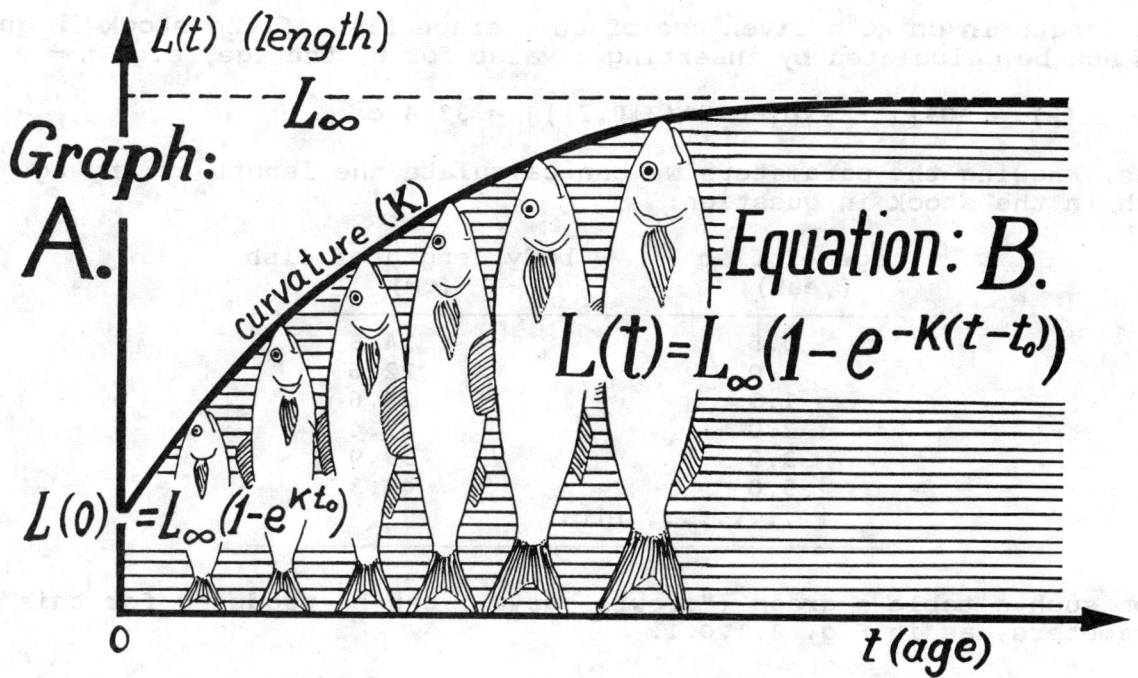

Fig. 3.1.0.1   The von Bertalanffy growth equation

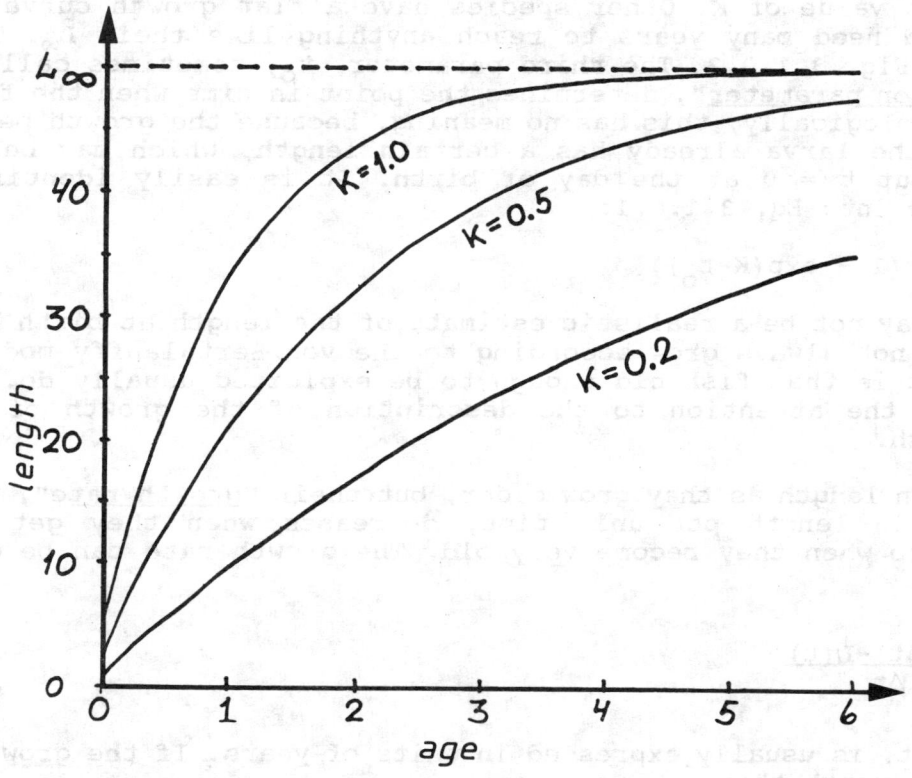

Fig. 3.1.0.2   A family of growth curves with different curvature parameters, different K values

Then we insert these parameter values into the von Bertalanffy growth equation (Eq. 3.1.0.1):

$$L(t) = 50*[1 - \exp(-0.5*(t+0.2))]$$

The length in cm at a given age of an average fish of the stock in question can now be calculated by inserting a value for t, the age, e.g. t = 2 years:

$$L(2) = 50*[1 - \exp(-0.5*(2+0.2))] = 33.4 \text{ cm}$$

Thus, knowing the parameters we can calculate the length at any age of the fish in the stock in question:

age of fish (year)	body length of fish (cm)
0.5	14.8
1.0	22.6
1.5	28.6
2.0	33.4
3.0	39.9
5.0	46.3
.... etc.	

From such a table a graph ("<u>growth curve</u>") can be produced for this set of parameters, as in Fig. 3.1.0.1.

The parameters can to some extent be interpreted biologically. L_∞ is interpreted as "<u>the mean length of very old (strictly: infinitely old) fish</u>", it is also called the "<u>asymptotic length</u>" (see Fig. 3.1.0.1). K is a "<u>curvature parameter</u>" which determines how fast the fish approaches its L_∞. Some species, most of them short-lived, almost reach their L_∞ in a year or two and have a high value of K. Other species have a flat growth curve with a low K-value and need many years to reach anything like their L_∞. This is illustrated in Fig. 3.1.0.2. The third parameter, t_o, sometimes called "<u>the initial condition parameter</u>", determines the point in time when the fish has zero length. Biologically, this has no meaning, because the growth begins at hatching when the larva already has a certain length, which may be called L(0) when we put t = 0 at the day of birth. It is easily identified by inserting t = 0 into Eq. 3.1.0.1:

$$L(0) = L_\infty*(1 - \exp(K*t_o))$$

However, L(0) may not be a realistic estimate of the length at birth because fish larvae do not always grow according to the von Bertalanffy model. The important point is that fish old enough to be exploited usually do. Let us therefore turn the attention to the description of the growth of larger (exploited) fish.

Fish increase in length as they grow older, but their "<u>growth rate</u>", that is the increment in length per unit time, decreases when they get older, approaching zero when they become very old. The growth rate can be defined by:

$$\frac{\Delta L}{\Delta t} = \frac{L(t+\Delta t) - L(t)}{\Delta t}$$

Time (or age), t, is usually expressed in units of years. If the growth rate is measured per month then

$$\Delta t = 1/12 \text{ years} = 0.0833 \text{ years}$$

Table 3.1.0.1 Growth rate as a function of age corresponding to the growth curve in Fig. 3.1.0.1. See also Fig. 3.1.0.3

A	B	C	D
age	length	growth rate	mean length
t	L(t)	$\dfrac{L(t+\Delta t)-L(t)}{1} = \dfrac{\Delta L}{\Delta t}$	$\dfrac{L(t+\Delta t)+L(t)}{2} = \bar{L}(t)$
years	cm	cm/year (y)	cm (x)
1	25.7		
		10.3	30.9
2	36.0		
		6.9	39.5
3	42.9		
		4.6	45.2
4	47.5		
		3.2	49.1
5	50.7		
		2.1	51.8
6	52.8		
		1.4	53.5
7	54.2		

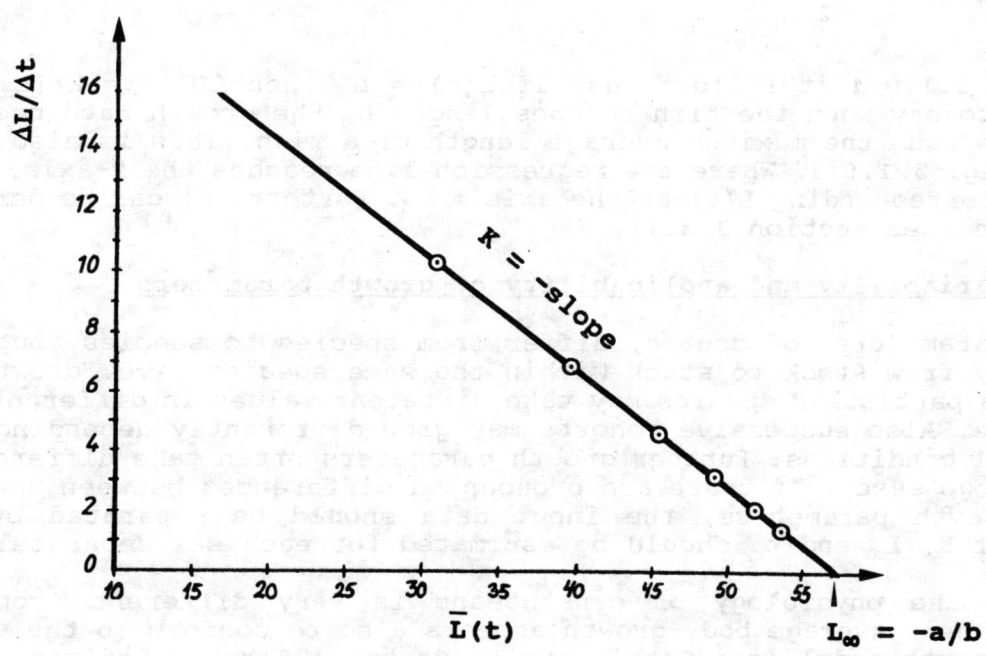

Fig. 3.1.0.3 Plot of growth rate against mean length. From columns C and D of Table 3.1.0.1

or per day then,

$\Delta t = 1/365$ years $= 0.00274$ years.

In Table 3.1.0.1 the ages (in years) and the lengths at the beginning of each year (in cm) corresponding to the example given in Fig. 3.1.0.1 are given in columns A and B, respectively. The growth rate is given in column C. It is evident that the growth rate decreases as the fish get older. The mathematical relationship between the length of a fish and the growth rate at a given time is a linear function:

$$\frac{\Delta L}{\Delta t} = a + b*L(t) \qquad (3.1.0.3)$$

This linear relationship can be derived from the von Bertalanffy growth equation, as follows:

$$\frac{\Delta L}{\Delta t} = K*(L_\infty - L(t)) \quad \text{cm/year} \qquad (3.1.0.4)$$

where $K = -b$ and $L_\infty = -a/b$

We shall not concern ourselves here with the mathematical proof. This linear relationship will be used in subsequent sections to determine the growth parameters K and L_∞. An example is already given in Fig. 3.1.0.3 where the growth rate $\Delta L/\Delta t$, as dependent variable, is plotted against the mean length, $\overline{L}(t)$, over the corresponding year, as the independent variable (see column D of Table 3.1.0.1):

$$\overline{L}(t) = \frac{L(t+\Delta t) + L(t)}{2}$$

From Eq. 3.1.0.4 it follows that if $\overline{L}(t) = L_\infty$ then $\Delta L/\Delta t = K*(L_\infty - L_\infty) = 0$, that is to say when the fish reaches length L_∞ the growth rate becomes zero and L_∞ is thus the maximum average length of a fish. This is also illustrated in Fig. 3.1.0.3. Where the regression line reaches the x-axis, $\Delta L/\Delta t = 0$ and the corresponding $L(t)$ at the axis $= L_\infty$. Further, K, can be derived from the slope (see Section 3.3.1).

3.1.1 Variability and applicability of growth parameters

Growth parameters, of course, differ from species to species, but they may also vary from stock to stock within the same species, i.e. growth parameters of a particular species may take different values in different parts of its range. Also successive cohorts may grow differently depending on environmental conditions. Further growth parameters often take different values for the two sexes. If there are pronounced differences between the sexes in their growth parameters, the input data should be separated by sex and values of K, L_∞ and t_o should be estimated for each sex separately.

Although the physiology of crustaceans is very different from that of fishes, their average body growth appears also to conform to the von Bertalanffy growth model (see Garcia and Le Reste, 1981). An individual crustacean (a shrimp or lobster) does not conform to the von Bertalanffy model, but to some "stepwise curve", with each step accounting for a moult (as illustrated in Fig. 3.1.1.1). However, members of a cohort moult at different times, and therefore the average growth curve of a cohort of crustaceans becomes a smooth curve (dotted line). For further discussion on the modelling of population dynamics of crustaceans see, for example, Jamieson and Bourne (1986) and Caddy (1987).

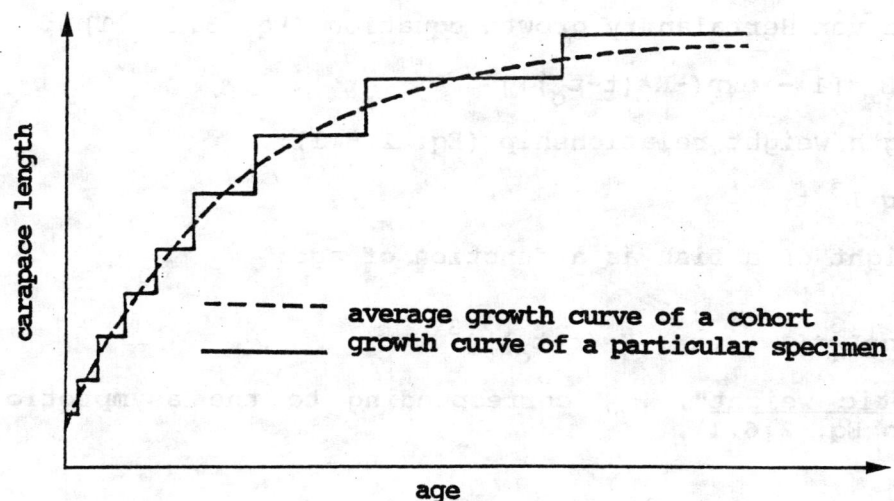

Fig. 3.1.1.1 Individual growth curve and average cohort growth curve of crustaceans

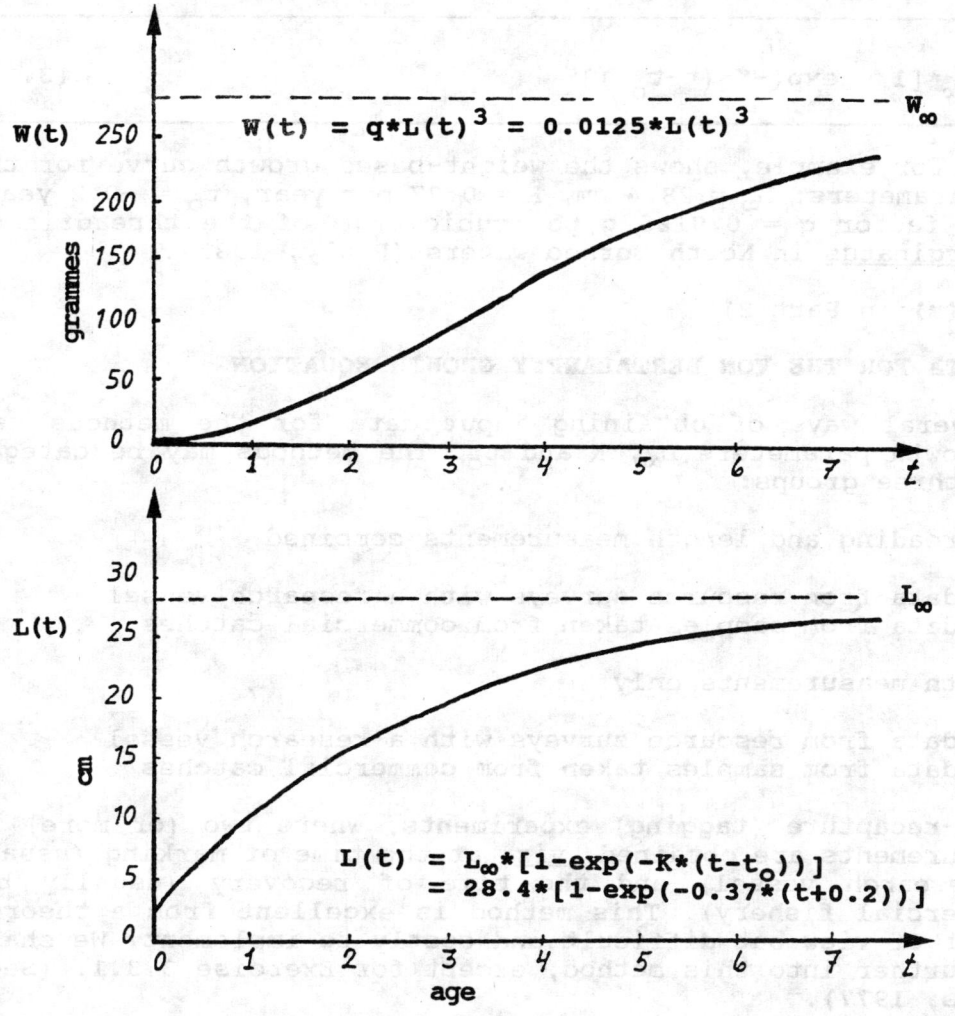

Fig. 3.1.2.1 A length-based growth curve and corresponding weight-based growth curve

3.1.2 The weight-based von Bertalanffy growth equation

Combining the von Bertalanffy growth equation (Eq. 3.1.0.1)

$$L(t) = L_\infty * [1 - \exp(-K*(t-t_o))]$$

with the length/weight relationship (Eq. 2.6.1):

$$W(t) = q*L^3(t)$$

gives the weight of a fish as a function of age:

$$W(t) = q*L_\infty^3 * [1 - \exp(-K*(t-t_o))]^3$$

The "<u>asymptotic weight</u>", W_∞, corresponding to the asymptotic length is (according to Eq. 2.6.1):

$$W_\infty = q*L_\infty^3$$

The parameter, q, is called the "<u>condition factor</u>". (Note that the letter q is also used in this manual to designate the catchability coefficient, Section 4.3.) Thus, "<u>the weight-based von Bertalanffy equation</u>" can be written:

$$W(t) = W_\infty * [1 - \exp(-K*(t-t_o))]^3 \qquad (3.1.2.1)$$

Fig. 3.1.2.1, for example, shows the weight-based growth curve for the von Bertalanffy parameters: L_∞ = 28.4 cm, K = 0.37 per year, t_o = -0.2 years and the condition factor q = 0.0125 g per cubic cm, of the threadfin bream, <u>Nemipterus marginatus</u> in North Borneo waters (Pauly, 1983).

(See **Exercise(s)** in Part 2)

3.2 INPUT DATA FOR THE VON BERTALANFFY GROWTH EQUATION

There are several ways of obtaining input data for the methods used to derive the growth parameters L_∞, K and t_o. The methods may be categorized roughly into three groups:

1) Age reading and length measurements combined

   a) data from resource surveys with a research vessel
   b) data from samples taken from commercial catches

2) Length measurements only

   a) data from resource surveys with a research vessel
   b) data from samples taken from commercial catches

3) Mark-recapture (tagging) experiments, where two (or more) length measurements are obtained, <u>viz</u>. at the time of marking (usually on a research vessel) and the time of recovery (usually by the commercial fishery). This method is excellent from a theoretical point of view but difficult and costly to implement. We shall not go further into this method, except for Exercise 3.3.1. (See also Jones, 1977).

Below we shall first consider 1a) in Section 3.2.1 and then 2b) in Section 3.2.2.

3.2.1 Data from age readings and length measurements

As has been stated in the introduction to this chapter, age reading is a relatively simple technique in the case of species from temperate waters, because their otoliths or scales show seasonal rings, one for the summer and one for the winter, which together form an annual ring. Sometimes such rings can be observed with the naked eye. In other cases simple techniques, such as burning can make them visible. The annual rings give sufficient information for most stock assessment purposes.

Unfortunately, tropical fish species seldom show clear annual rings in their otoliths or scales, because the strong seasonality which characterizes temperate zones is lacking. Recent discoveries, however, have created opportunities to also read ages of tropical fish, albeit within limited ranges and at a high cost in terms of manpower and initial investment. By going deeper into the formation of the rings in otoliths and scales it has been discovered that daily increments (or even increments caused by a certain food intake) can be detected by means of a strong microscope. The latest findings indicate that sometimes the daily rings are so thin that they defeat the ordinary microscope whose detection power is limited by the wavelength of the light. Such rings can be read only by a scanning electron microscope (Morales-Nin, 1991).

A large amount of literature has been produced on this subject in recent years, for example: Panella (1971), Bagenal (1974), Brothers (1980), Beamish and Mc Farlane (1983), Gjøsæter et al. (1983), Dayaratne and Gjøsæter (1986) and Williams (1986).

In a manual on tropical fish stock assessment it is necessary to concentrate on length measurements and consequently place less emphasis on age data. However, we are dealing with age here for two reasons. In the first place it may sometimes be possible to carry out a small number of age readings, which can be used to calibrate the findings obtained from length measurements alone. Secondly, it is easier to explain the concepts and theory on the basis of age and length data than on the basis of length data only. Also we use data from research vessels to avoid further complications at this stage. The first example deals with data from a single survey, while the second example deals with data from a time series of surveys.

Example 3: Age/length composition data from a single survey

Suppose we have a random sample of fish from the stock of species A. This sample was taken on a survey with a duration of, say, a fortnight, in which trawl hauls were made over the entire distribution area of the stock in such a manner that the pooled data from all hauls made up a random sample (see Section 7.1). Suppose that the survey took place in October 1983 and that pooled length-frequency data were obtained as presented in the last column of Table 3.2.1.1 (and also in Fig. 3.2.2.1). Suppose also that we have observed two annual peak recruitment seasons and therefore have decided to define two cohorts per year:

Spring cohort: Fish recruited from January to June
Autumn cohort: Fish recruited from July to December

A cohort was defined earlier as "a group of fish all of the same age belonging to the same stock" (see Section 1.3.1).

Table 3.2.1.1 Age/length composition (hypothetical example). Basic data for Table 3.2.1.2. The graph for the total length-frequency is shown in Fig. 3.2.2.1 ("-" stands for a zero observation)

recruitment season length interval cm	cohort spring 1983	autumn 1982	spring 1982	autumn 1981	spring 1981	autumn 1980	survey October 1983 total of all hauls
12-13	1	-	-	-	-	-	1
13-14	4	-	-	-	-	-	4
14-15	11	-	-	-	-	-	11
15-16	24	-	-	-	-	-	24
16-17	38	-	-	-	-	-	38
17-18	42	-	-	-	-	-	42
18-19	33	-	-	-	-	-	33
19-20	20	-	-	-	-	-	20
20-21	7	-	-	-	-	-	7
21-22	2	1	-	-	-	-	3
22-23	-	3	-	-	-	-	3
23-24	-	5	-	-	-	-	5
24-25	-	8	-	-	-	-	8
25-26	-	11	-	-	-	-	11
26-27	-	14	-	-	-	-	14
27-28	-	16	1	-	-	-	17
28-29	-	15	1	-	-	-	16
29-30	-	13	2	-	-	-	15
30-31	-	11	3	-	-	-	14
31-32	-	7	4	-	-	-	11
32-33	-	4	6	1	-	-	11
33-34	-	2	7	1	-	-	10
34-35	-	1	7	1	-	-	9
35-36	-	-	8	2	-	-	10
36-37	-	-	7	3	1	-	11
37-38	-	-	6	3	1	-	10
38-39	-	-	5	4	1	-	10
39-40	-	-	4	4	2	1	11
40-41	-	-	3	5	2	1	11
41-42	-	-	2	4	2	1	9
42-43	-	-	1	3	2	1	7
43-44	-	-	-	3	3	1	7
44-45	-	-	-	2	2	1	5
45-46	-	-	-	2	2	2	6
46-47	-	-	-	1	2	2	5
47-48	-	-	-	1	1	1	3
48-49	-	-	-	-	1	1	2
49-50	-	-	-	-	1	1	2
50-51	-	-	-	-	1	1	2
51-52	-	-	-	-	-	1	1
total	182	111	67	40	24	15	439
mean length	17.3	27.9	35.3	40.2	43.3	45.5	
std. dev.	1.7	2.7	3.4	3.6	3.8	3.6	
mean age (y)	0.64	1.16	1.65	2.10	2.64	3.21	

Now, also suppose that we are able to read the age of each fish, so that we can determine the day on which it was born. After having read the ages of all 439 fish of species A caught in the October 1983 survey, we can assign each fish to a specific cohort. It is then possible to make a length-frequency distribution for each cohort. Theoretically, these length distributions are normal distributions for which we can determine the mean length and the standard deviation.

The complex length-frequency table obtained after the survey has thus been split into six length-frequency tables for different cohorts, of which we also know the average age. The type of information contained in the first seven columns of Table 3.2.1.1 forms a so-called "age/length key" (this concept will be further discussed in Example 7). The main data for each cohort have been summarized in Table 3.2.1.2.

If we further assume that all the six cohorts have the same growth parameters, we can use the data contained in Table 3.2.1.2 to estimate the common growth parameters. In other words, we can determine the growth parameters which produce the growth curve that gives the best fit to the pairs of data of mean length and corresponding mean age. Exactly how this can be done is explained in subsequent sections.

Please note that the data presented in Table 3.2.1.1 are "hypothetical" or "faked" data. They were actually computed from a set of growth parameters (which determine the mean lengths for each cohort) and a set of standard deviations for the length distribution of each cohort. The mean age of the youngest cohort is 0.64 year or 234 days, this means that the birth date of this cohort was 234 days before 15 October 1983, or 23 February 1983 (northern spring). The other two spring cohorts were born one and two years earlier respectively, while the three autumn cohorts were born 6 months after each spring cohort. Due to random variations the birth dates vary slightly from year to year. The advantage of using such hypothetical data in the context of this manual is that the true parameters are known, which is not the case with data taken from a real stock. That puts us in a position to compare the results of various methods of parameter estimation with the real values. The data presented in Tables 3.2.1.1 and 3.2.1.2 will also be used as examples in Sections 3.4.1 and 3.4.2.

Table 3.2.1.2 Hypothetical example of age and length composition data of species A from one research survey in October 1983 (derived from the "raw data" in Table 3.2.1.1)

COHORT recruitment year	season	number observed	mean age (year)	mean length (cm)
1983	spring	182	0.64	17.3
1982	autumn	111	1.16	27.9
1982	spring	67	1.65	35.3
1981	autumn	40	2.10	40.2
1981	spring	24	2.64	43.3
1980	autumn	15	3.21	45.5
	total	439		

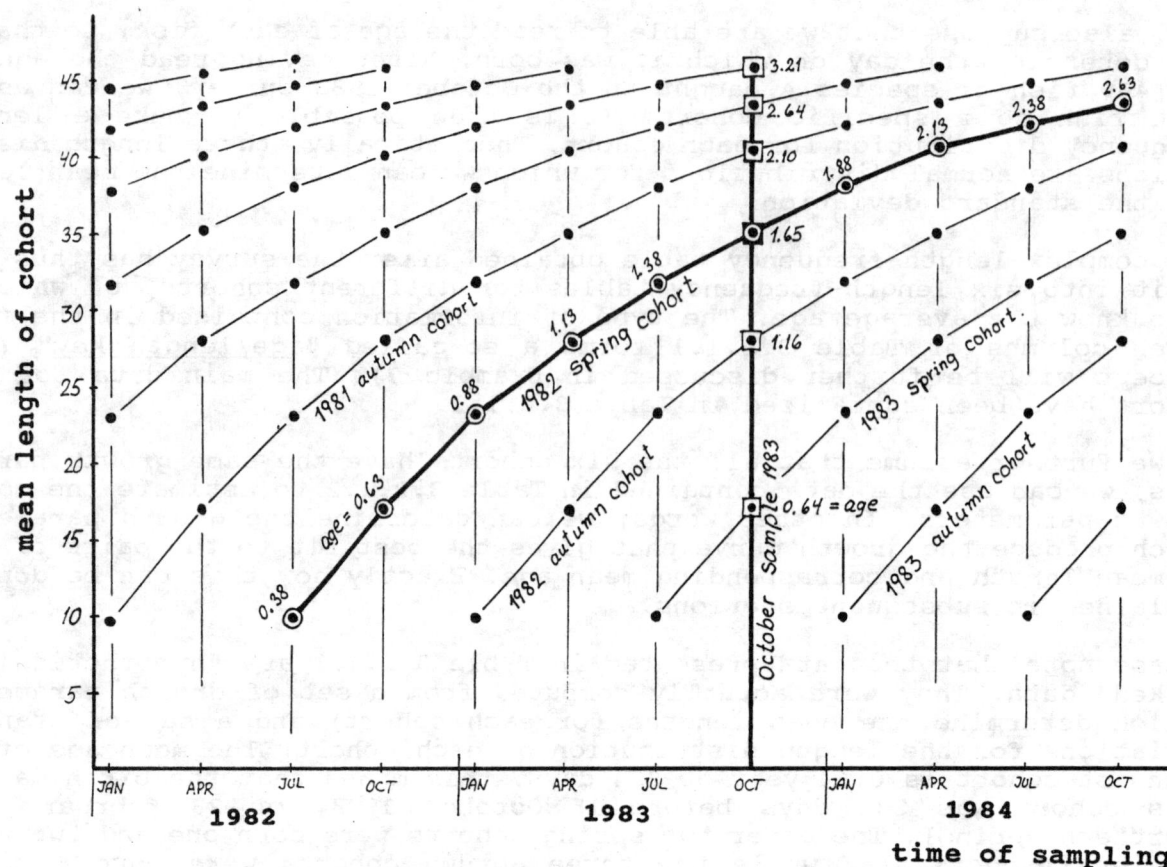

Fig. 3.2.1.1 Illustration of the data on age/length collected during a time series of surveys

Example 4: Age/length composition data from multiple surveys

If we now assume that the survey of Example 3 was only one of a series of 12 surveys carried out during the years 1982-1984, in the months of January, April, July and October each year, then such a survey programme would yield 12 tables like Table 3.2.1.1. By sampling the various cohorts regularly over a length of time, in this case three years, changes in the mean lengths can be determined by plotting them against the time of sampling as shown in Fig. 3.2.1.1. With this data set we are able to estimate the growth parameters for some of the cohorts individually. For the spring cohort in the recruitment year of 1982 for example, there are 10 pairs of age and length data which can be used to estimate the parameters for that particular cohort.

The difference between following one particular cohort in time as shown here, and the determination of mean lengths of different cohorts at a certain moment, as presented in Example 3, is illustrated in Fig. 3.2.1.1, where the two different data types are indicated by heavy lines. The curve starting July 1982 and running to October 1984 shows the "real" growth of a cohort. The vertical line "October sample 1983" shows a "cross section" of the stock at that date.

In case of a short-lived species (with a life span of, say, one to two years) we would have to follow a cohort in time as described in Example 4. The method based on a single sample would not be applicable because it would contain only one or two cohorts. Although there may be differences in the growth of different cohorts this difference is usually so small that it can be ignored. Data like those presented in Fig. 3.2.1.1 therefore could all be pooled into one data set and used in a way similar to the data of the October sample in 1983 (Table 3.2.1.1).

It is likely that bias will be less if sampling is done all the year round. Thus, although we can sometimes manage with a single sample, it is safer to use a time series of samples.

(See **Exercise(s)** in Part 2).

Example 5: The use of age/length keys

An age/length key is a table showing, for each length class of fish of a particular stock, the percentage or fractional age-frequency distribution, see Table 3.2.1.4. Once such a key is available, samples of fish which were only measured for length can be distributed over age groups according to the key.

The age/length key of Table 3.2.1.4 could be based on 182 randomly drawn fish with the following length distribution:

Length group (cm)	5-10	10-15	15-20	20-25	Total
Frequency	110	40	22	10	182

The next step is to age the fish in each length group. Let us assume that the results are as shown in Table 3.2.1.3. Table 3.2.1.4 is then derived from Table 3.2.1.3 simply by dividing each row entry by the row total for each length group.

Table 3.2.1.4 can then be used to assign ages to a much larger length-frequency sample of the same stock (for which the age composition is unknown), for example the length-frequency sample of 21041 fish given below:

Length group (cm)	5-10	10-15	15-20	20-25	Total
Frequency	12088	7035	1788	130	21041

By distributing the numbers in each length group over the age groups according to the proportions given in Table 3.2.1.4, we get the results presented in Table 3.2.1.5. Length group 10-15 cm, for example, is estimated to consist of 7035*0.25 = 1759 0-group fish and 7035*0.75 = 5276 1-group fish. By summing the column entries we finally arrive at the age composition given in the bottom row of Table 3.2.1.5.

Thus in order to estimate the age composition of the catch from a particular stock we only need to determine an age/length key based on a small sample of age readings and then we can restrict further sampling to the collection of length-frequency data. These lengths are converted to ages by means of the key. The same key may be used in consecutive years as long as there is no suspicion of major changes in the age composition of the stock. In a period of for instance, markedly increased effort the old fish may disappear from the catches and then a new age/length key will have to be prepared.

Table 3.2.1.3 Input data for estimation of an age/length key (hypothetical example)

length group cm	age group 0	age group 1	age group 2	total
5-10	110	0	0	110
10-15	10	30	0	40
15-20	0	11	11	22
20-25	0	1	9	10
total	120	42	20	182

Table 3.2.1.4 Hypothetical age/length key

length group cm	age group 0	age group 1	age group 2
5-10	1.0	0	0
10-15	0.25	0.75	0
15-20	0	0.5	0.5
20-25	0	0.1	0.9

Table 3.2.1.5 Age composition of a large length-frequency sample estimated by use of the age/length key in Table 3.2.1.4

length group cm	age group 0	age group 1	age group 2	total
5-10	12088	0	0	12088
10-15	1759	5276	0	7035
15-20	0	894	894	1788
20-25	0	13	117	130
total	13847	6183	1011	21041

Table 3.2.1.6 shows an age/length key for a long-lived tropical fish, the Spanish mackerel (Scomberomorus brasiliensis). To illustrate the limitations of an age/length key, consider the percentage age distribution of fish of 61-64 cm long, which are 4-7 years old. Now, if the fishing mortality (effort) increases markedly, most fish older than 5 years may be exterminated and the few fish of 61-64 cm still caught will be fast-growing 4-year-olds and a few remaining 5-year-olds, while the 6 and 7 years old fish have disappeared. Using the old key on the new length-frequency distributions would give the impression that the 61-64 cm fish are still 4-7 years old, with the 5-group dominating, when in fact they are only 4 years old.

When collecting samples for an age/length key it is important to include in the sample some very small, and some very large specimens. Otherwise, when large numbers of length measurements are to be distributed over age groups it will be found that some size classes represented in such length samples are not in the key at all. When small and large fish are deliberately over-represented in the key it is important to remember that the key data alone cannot be used for estimation of growth or mortality parameters.

Table 3.2.1.6 Age/length key for <u>Scomberomorus brasiliensis</u> based on otolith readings, in percentages per 3-cm length groups. From Sturm, 1974

length group cm	\multicolumn{10}{c|}{age group (year)}									
	0	1	2	3	4	5	6	7	8	9
13-16	100	-	-	-	-	-	-	-	-	-
16-19	-	100	-	-	-	-	-	-	-	-
19-22	-	100	-	-	-	-	-	-	-	-
22-25	-	100	-	-	-	-	-	-	-	-
25-28	-	96	4	-	-	-	-	-	-	-
28-31	-	55	45	-	-	-	-	-	-	-
31-34	-	5	95	-	-	-	-	-	-	-
34-37	-	-	91	9	-	-	-	-	-	-
37-40	-	-	73	27	-	-	-	-	-	-
40-43	-	-	33	63	2	-	-	-	-	-
43-46	-	-	15	77	8	-	-	-	-	-
46-49	-	-	5	65	29	-	-	-	-	-
49-52	-	-	1	47	50	2	-	-	-	-
52-55	-	-	-	38	51	11	-	-	-	-
55-58	-	-	-	10	62	21	7	-	-	-
58-61	-	-	-	3	50	25	22	-	-	-
61-64	-	-	-	-	19	44	31	6	-	-
64-67	-	-	-	-	-	66	17	17	-	-
67-70	-	-	-	-	-	-	75	25	-	-
70-73	-	-	-	-	-	-	-	33	33	33
> 73	-	-	-	-	-	-	-	-	50	50

The methodology of fish stock assessment can in fact be entirely based on age/length compositions alone. The application of mathematical growth models is not necessary. To a certain degree this is the case with the assessments made by the International Council for Exploration of the Seas (ICES) in the North Atlantic. However, since reliable age/length keys are not likely to become available for most tropical species in the near future, as well as for a number of other reasons which will be outlined in the following chapters, the highest priority has been given in this manual to the mathematical growth models.

3.2.2 Length composition data (without age compositions)

Assume that we have a data set consisting of length-frequencies of a certain species, but without age readings. The basic data set for a given sampling date would then look like the "total" (last) column of Table 3.2.1.1 or as drawn in Fig. 3.2.2.1. Is it possible to obtain a separation of the various cohorts which have contributed to this sample without using age reading techniques? The answer is that under certain conditions it is possible, except for parts where the ranges of length-frequencies of different cohorts overlap each other too much.

The hypothetical data set presented in Table 3.2.1.1 was created from a number of normally distributed components, representing cohorts, as shown in Fig. 3.2.2.2.

In Fig. 3.2.2.1 the youngest cohort, the spring cohort of 1983, can easily be distinguished from the rest of the sample. The next cohort, further to the right, is somewhat more difficult to see, while the remaining four cohorts may only be distinguished by using more sophisticated methods than visual inspection, or it may not be possible to separate them at all.

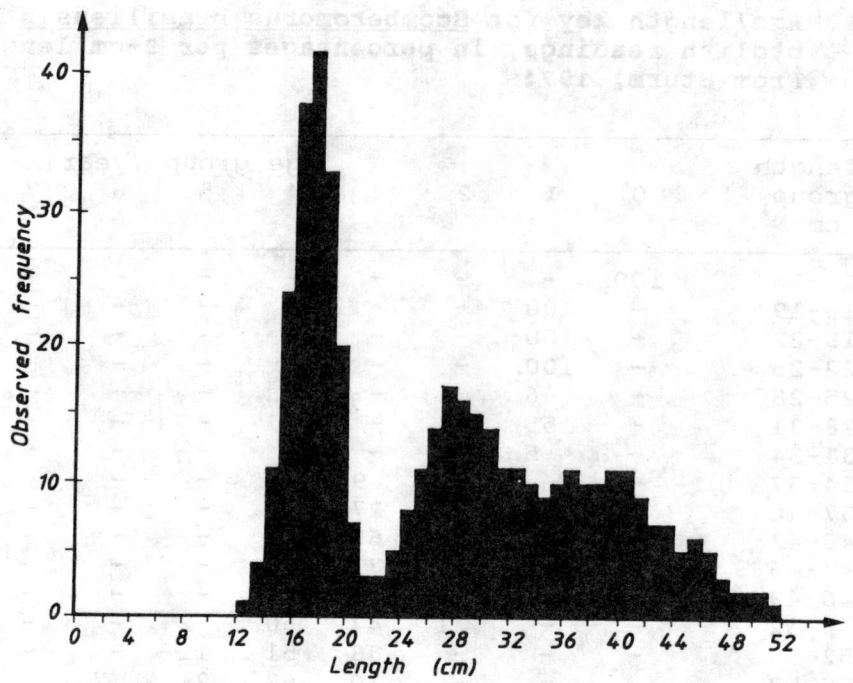

Fig. 3.2.2.1 The length-frequency sample, the only basic data in cases where age reading from hard parts is not possible. (Frequencies from the "total column" of Table 3.2.1.1)

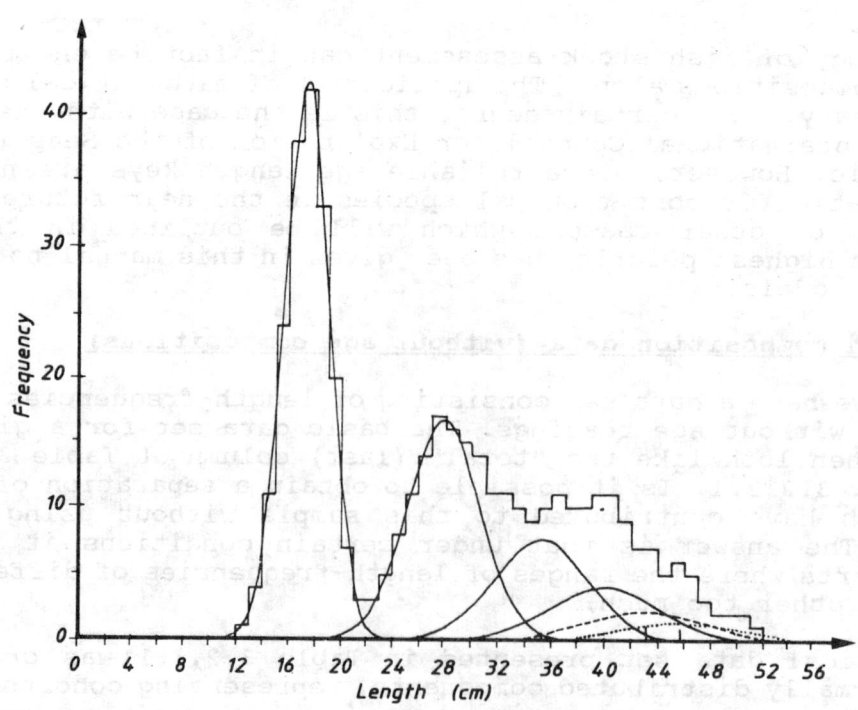

Fig. 3.2.2.2 The length-frequency sample of Fig. 3.2.2.1, separated into normally distributed components. (Frequencies from the "total" column of Table 3.2.1.1). This example is also used to illustrate the "Bhattacharya method" described in Section 3.4.1 and the "maximum likelihood method" discussed in Section 3.5.3

In Section 3.4 methods will be introduced which can be used to split length-frequency samples into normally distributed components, which are assumed to represent cohorts. It will be demonstrated, on the basis of the same data set, that it is not feasible in practice to separate more than three or four cohorts from the total data set. The overlap in the length composition of the older cohorts, the largest fish, clearly limits the analysis. Therefore, the conclusions that may be drawn from such a data set, compared to cases where the age of the fish can be determined, are also limited.

3.2.3 Data from commercial catches

Data for estimation of growth parameters may also be obtained from sampling the commercial catches. The basic principles in analysing samples from commercial landings are the same as for research survey data. The major difference lies in the bias problems. Commercial boats never attempt to collect a random sample of the stock, because they always go for the marketable sizes and try to find the areas with the highest concentrations of fish. However, keeping in mind sources of bias, and trying to stratify the sampling to minimize the bias, data from commercial fisheries can also be used for the estimation of growth parameters.

The major advantage of sampling commercial catches is that such samples are much cheaper to collect and thus sampling can be much more frequent than is possible with a single research vessel. In Chapter 7 problems relating to sampling of commercial catches are further elaborated.

3.3 METHODS FOR ESTIMATION OF GROWTH PARAMETERS FROM LENGTH-AT-AGE DATA

In this section we assume that pairs of observations of age and length are available. They may either be derived from readings of ring structures in hard parts or from length-frequency analysis (Sections 3.4 and 3.5). Input data are either in the detailed form of an age/length composition as in Table 3.2.1.1 or in the processed form shown in Table 3.2.1.2. They may or may not be derived from a time series of samples (cf. Fig. 3.2.1.1). In the following we use for simplicity the input data format illustrated by Table 3.2.1.2.

Growth parameters can be derived from such data by graphical methods or plots, which are always based on a conversion to a linear equation, as discussed in Chapter 2. These plots are named after the authors of the papers wherein they were first described, viz. Gulland and Holt (1959), Chapman (1961), Ford-Walford (1933 and 1946 respectively) and von Bertalanffy (1934). An other method that will be discussed is the "least squares method".

3.3.1 The Gulland and Holt plot

The Gulland and Holt (1959) plot was introduced in Section 3.1 by Eq. 3.1.0.4, which can also be written as:

$$\Delta L/\Delta t = K*L_\infty - K*\overline{L}(t) \qquad (3.3.1.1)$$

The length "L(t)" in Eq. 3.1.0.4 represents the length range from L(t) at age t to L(t+Δt) at age t+Δt. Thus, the natural quantity to enter into Eq. 3.3.1.1 is the mean length (cf. the example in Table 3.1.0.1):

$$\overline{L} = \frac{L(t+\Delta t)+L(t)}{2}$$

Only if Δt is small $\bar{L}(t)$ may be a reasonable approximation to the mean length. However, Δt does not need to be a constant, which is an important advantage over other methods.

Using $\bar{L}(t)$ as the independent variable and ΔL/Δt as the dependent variable Eq. 3.3.1.1 becomes a linear regression:

$$\Delta L/\Delta t = a + b*\bar{L}(t)$$

The growth parameters K and L_∞ are obtained from:

$$K = -b \quad \text{and} \quad L_\infty = -a/b$$

Table 3.1.0.1. contains an example of the input data (columns C and D) and Fig. 3.1.0.3 shows the corresponding plot. The length increment per year or growth rate is plotted against the mean length during the corresponding year. The regression analysis gives:

$$a = 22.40 \quad \text{and } b = -0.3923 \quad \text{from which we get}$$

$$K = -b = 0.39 \text{ say } 0.4 \text{ per year}, \quad \text{and} \quad L_\infty = -a/b = 57.1 \text{ cm}$$

Example 6: Estimating K and L_∞ with the Gulland and Holt plot

Another example of the Gulland and Holt plot can be derived from Table 3.2.1.2 as shown in Table 3.3.1.1. From the estimates of intercept and slope we get:

$$K = -b = 0.77 \text{ say } 0.8 \text{ per year},$$

$$L_\infty = -a/b = -38.52/-0.7670 = 50.2 \text{ cm}$$

The 95% confidence limits for K are the same as those for b but with the sign changed, i.e. [0.56 , 0.98] (cf. Table 3.3.1.1). The confidence limits of L_∞ are more complicated to obtain. However, we may say that for a given value of K, the confidence interval of L_∞ can be calculated from the confidence interval of a. If K = 0.8 and the confidence limits of a are [31.0/0.8, 46.0/0.8] = [38.7, 57.5]. These are not the confidence limits of L_∞, but only the limits conditioned on K = 0.8. (Actually, the confidence limits of a ratio are not defined, but the theory dealing with this is considered outside the scope of the present manual.) Fig. 3.3.1.1 shows the Gulland and Holt plot corresponding to Table 3.3.1.1.

In Section 3.1 it was stated that it can be proved mathematically that Eq. 3.1.0.4: $\Delta L/\Delta t = K*(L_\infty - L(t))$ is equivalent to the von Bertalanffy growth equation (Eq. 3.1.0.1):

$$L(t) = L_\infty*[1 - \exp(-K*(t-t_o))]$$

This, however, is correct only if the time interval, Δt, is infinitesimal. Thus, the Gulland and Holt plot, which is based on Eq. 3.1.0.4, is an approximation which is reasonable only for small values of Δt.

(See **Exercise(s)** in Part 2)

Table 3.3.1.1 Input data for the Gulland and Holt plot and regression analysis (data derived from Table 3.2.1.1)

t	Δt	L(t)	ΔL(t)	$\frac{\Delta L(t)}{\Delta t}$ (y)	$\frac{L(t+\Delta t)+L(t)}{2} = \bar{L}(t)$ (x)
0.64		17.3			
	0.52		10.6	20.4	22.6
1.16		27.9			
	0.49		7.4	15.1	31.6
1.65		35.3			
	0.45		4.9	10.9	37.7
2.10		40.2			
	0.54		3.1	5.7	41.8
2.64		43.3			
	0.57		2.2	3.9	44.4
3.21		45.5			

b (slope) = -0.7670,   a (intercept) = 38.52,   n = 5, $\bar{x}$ = 35.62

$sb^2 = \frac{1}{n-2} * [(sy/sx)^2 - b^2] = \frac{1}{3} * [(6.7727/8.7362)^2 - 0.7670^2] = 0.004216$

sb = 0.06493, t_{n-2} = 3.18, $sb*t_{n-2}$ = 0.2065

95 % confidence limits for b: [-0.974 , -0.561] (cf. Section 2.4)

K = -b = 0.77 ± 0.21

$sa^2 = sb*(\frac{n-1}{n}*sx^2 + \bar{x}^2) = 0.004216*(\frac{4}{5}*8.7362^2 + 35.62^2) = 5.607$

sa = 2.368 $sa*t_{n-2}$ = 7.53

95 % confidence limits for a: [31.0 , 46.0]

L_∞ = -a/b = -38.52/-0.7670 = 50.2 cm

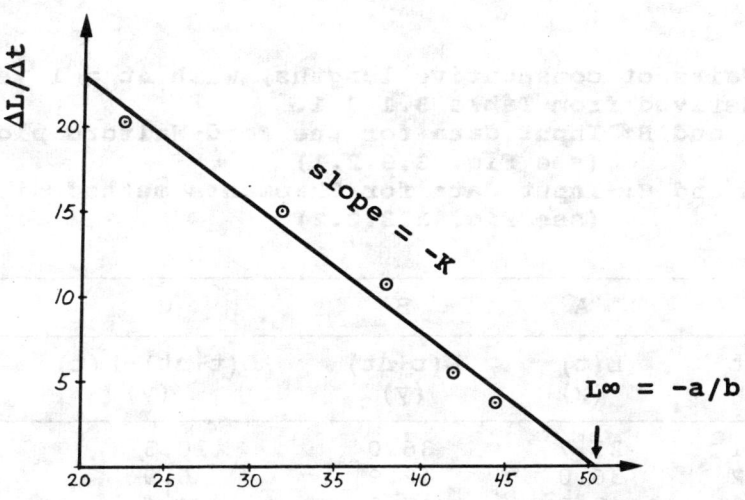

$\bar{L}(t) = [L(t+\Delta t)+L(t)]/2$

Fig. 3.3.1.1 Gulland and Holt plot corresponding to Table 3.3.1.1 (hypothetical example). The intersection point between the regression line and the x-axis gives L_∞

3.3.2 The Ford-Walford plot and Chapman's method

The method introduced by Ford (1933) and Walford (1946) has gained wide application because the plot could be used to obtain a quick estimate of L_∞, without calculations. Nowadays it is not used very much and it has been incorporated here mainly because it will often be found in older papers.

From the von Bertalanffy growth equation (Eq. 3.1.0.1) it follows from a series of algebraic manipulations that:

$$L(t+\Delta t) = a + b*L(t) \qquad (3.3.2.1)$$

where $a = L_\infty*(1-b)$ and $b = \exp(-K*\Delta t)$

Since K and L_∞ are constants, a and also b become constants <u>if Δt is a constant</u>. The growth parameters K and L_∞ are derived from:

$$K = -\frac{1}{\Delta t}*\ln b \quad \text{and} \quad L_\infty = \frac{a}{1-b}$$

To illustrate the use of Eq. 3.3.2.1 consider Table 3.3.2.1, where the figures in column A represent lengths, L(t), for a series of ages with a constant time interval of one year, while column B contains the lengths, $L(t+\Delta t)$, the length one year later.

Carrying out the regression analysis we get

$a = 18.70$ and $b = 0.6725$

from which we derive

$K = -(1/1)*\ln 0.6725 = 0.3968$, say 0.4 per year and

$L_\infty = 18.70/(1-0.6725) = 57.1$ cm

The actual Ford-Walford plot corresponding to these data is shown in Fig. 3.3.2.1. L_∞ can be estimated graphically from the intersection point of the 45° diagonal, where $L(t) = L(t+\Delta t)$ and the regression line, because for very old fish, which have stopped growing $L_\infty = L(t) = L(t+\Delta t)$.

Table 3.3.2.1   Pairs of consecutive lengths, with $\Delta t = 1$ year, derived from Table 3.1.0.1.
A and B: Input data for the Ford-Walford plot (see Fig. 3.3.2.1)
A and C: Input data for Chapman's method (see Fig. 3.3.2.2)

	A	B	C
t	L(t) (x)	$L(t+\Delta t)$ (y)	$L(t+\Delta t)-L(t)$ (y)
1	25.7	36.0	10.3
2	36.0	42.9	6.9
3	42.9	47.5	4.6
4	47.5	50.7	3.2
5	50.7	52.8	2.1
6	52.8	54.2	1.4

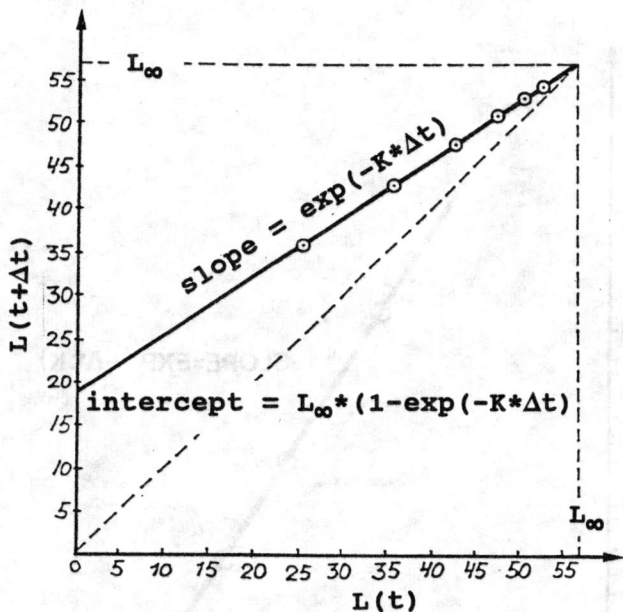

Fig. 3.3.2.1 Ford-Walford plot. Data from columns A and B of Table 3.3.2.1

Also the method described by Chapman (1961) and later by Gulland (1969) is based on a constant time interval Δt, that is to say the method is applicable if we have pairs of observations:

$(t, L(t))$, $(t+\Delta t, L(t+\Delta t))$, $(t+2\Delta, L(t+2\Delta t))$, etc.

It can be shown that the von Bertalanffy growth equation (Eq. 3.1.0.1) implies that:

$$L(t+\Delta t) - L(t) = c*L_\infty - c*L(t) \qquad (3.3.2.2)$$

where $c = 1 - \exp(-K*\Delta t)$

Thus, since K and L_∞ are constants, and if Δt remains constant, c will remain constant and consequently Eq. 3.3.2.2 becomes a linear regression

$$y = a + bx$$

where

$y = L(t+\Delta t)-L(t)$, $a = c*L_\infty$, $b = -c$ and $x = L(t)$

Note that the slope is negative and also that on the abscissa (x-axis) the smaller of the two lengths is used, instead of the mean value (cf. Section 3.3.1).

The growth parameters are derived from

$K = -(1/\Delta t)*\ln(1+b)$ and $L_\infty = -a/b$ or a/c

To illustrate the use of Eq. 3.3.2.2 consider once again Table 3.3.2.1, where $L(t) = x$ in column A and $L(t+1)-L(t) = y$, in column C. A regression analysis gives

a = 18.70 and b = -0.3275 and hence c = 0.3275
K = -(1/1)*ln(1-0.3275) = 0.3968, say 0.4 per year
L_∞ = 18.70/0.3275 = 57.1 cm

The plot is given in Fig. 3.3.2.2.

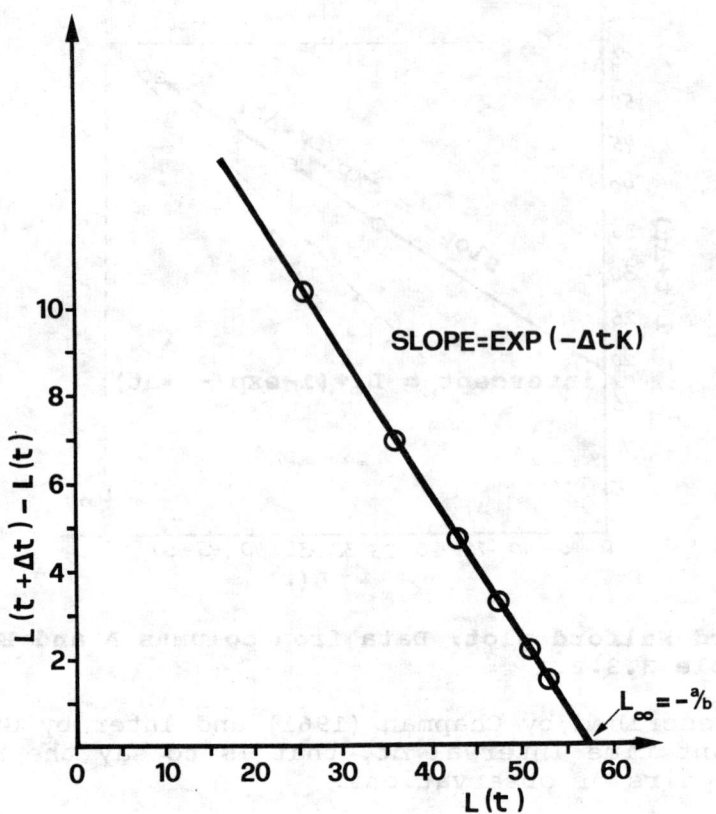

Fig. 3.3.2.2 Chapman's plot. Data from columns A and C of Table 3.3.2.1

The three methods described in Sections 3.3.1 and 3.3.2 give nearly the same results when applied to the data in Table 3.1.0.1. This is caused by the fact that the data conform exactly to the von Bertalanffy equation because they were obtained from the equation through back calculation. With real data one should expect to find some differences in the results.

The three methods described above can be used to estimate K and L_∞. A fourth method, the von Bertalanffy plot can be used to obtain an estimate of K and t_o. However, this method requires an estimate of L_∞ as input (see Section 3.3.3).

L_∞ can also be estimated by the "Powell-Wetherall method". Because this method can also be used to obtain an estimate of the total mortality coefficient, Z, it is presented in the next chapter in Section 4.5.4.

(See **Exercise(s)** in Part 2)

3.3.3 The von Bertalanffy plot

The first method for estimating the von Bertalanffy growth parameters was suggested by von Bertalanffy (1934). It can be used to estimate K and t_o from age/length data, while it requires an estimate of L_∞ as input.

The von Bertalanffy growth equation (Eq. 3.1.0.1) can be rewritten:

$$-\ln(1 - L(t)/L_\infty) = -K*t_o + K*t \quad (3.3.3.1)$$

With the age, t, as the independent variable (x) and the left-hand side as the dependent variable (y) the equation defines a linear regression, where the slope b = K and the intercept a = $-K*t_o$.

Example 7: Estimating K and t_o with the von Bertalanffy plot

Table 3.3.3.1 shows how to calculate the input data for the von Bertalanffy plot, based on data from Table 3.3.1.1 with L_∞ = 50 cm. The plot is shown in Fig. 3.3.3.1. Compare the K value (0.78 per year) with the estimate (K = 0.77 ± 0.21) obtained by the Gulland and Holt plot with the same data.

The von Bertalanffy plot is a more robust method than the Gulland and Holt plot (and the Ford-Walford plot) in the sense that it nearly always gives a reasonable estimate of K, given that a reasonable estimate of L_∞ is used in the computations (as illustrated in Exercise 3.3.3). One must ascertain, however, that the plot (Fig. 3.3.3.1) "looks" linear. On the other hand, one can say that the Gulland and Holt plot is stronger in the sense that it is better in bringing out cases where the observations are in conflict with the von Bertalanffy model.

Recalling the interpretation of L_∞ as the average length of a very old fish, there are various short cut methods for estimating L_∞ for use in the von Bertalanffy plot:

Table 3.3.3.1 Input data and regression for the von Bertalanffy plot (data derived from Table 3.3.1.1, L_∞ = 50 cm)

t (x)	L(t)	$-\ln(1-L(t)/L_\infty)$ (y)
0.64	17.3	0.425
1.16	27.9	0.816
1.65	35.3	1.224
2.10	40.2	1.630
2.64	43.3	2.010
3.21	45.5	2.408

a = -0.0680   b = 0.7825
K = b = 0.78 per year
t_o = -a/b = 0.087 year

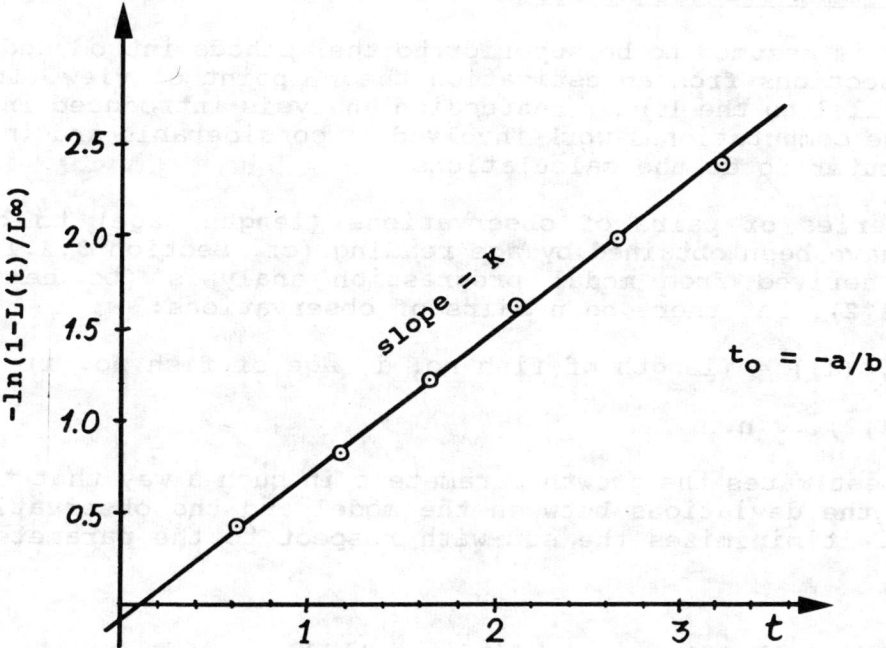

Fig. 3.3.3.1 Von Bertalanffy plot for the data in Table 3.3.3.1

1) In small samples you may simply use the largest fish.

2) In a very large sample you may take the average of the lengths of, say, the ten largest fish.

3) Perhaps the best way of estimating L_∞ is the Powell-Wetherall method described in Section 4.5.4.

It may not matter as much as one might think which estimate of L_∞ is used. If you over-estimate L_∞ the K will be under-estimated, and together they will balance out, so that the resulting growth curve remains nearly the same for the range of ages represented in the data set. (This aspect will be further discussed in Section 3.4.)

There is, however, a problem in using the von Bertalanffy plot in connection with the definition of L_∞. The argument of the logarithm in Eq. 3.3.3.1, i.e. $(1 - L(t)/L_\infty)$ must be positive as the logarithm would otherwise not be defined. Thus the von Bertalanffy plot cannot accept a length greater than L_∞, whereas with the definition of L_∞ as given in Section 3.1.4 it may well happen that for the very old fish, $L(t) > L_\infty$ because the observations $(t, L(t))$ fluctuate at random about the line. The von Bertalanffy plot actually uses the "<u>inverse von Bertalanffy growth equation</u>":

$$t(L) = t_o - \frac{1}{K} * \ln(1 - L/L_\infty) \qquad (3.3.3.2)$$

which is Eq. 3.1.0.1 solved for t. It may be necessary to omit the oldest fish to obtain that $1 - L/L_\infty > 0$.

The L_∞ concept as applied in the von Bertalanffy plot is different from that applied in the Gulland and Holt plot, for the same reasons as the parameters in the "inverse linear regression" differ from those of the "original linear regression".

(See **Exercise(s)** in Part 2).

3.3.4 The least squares method

This method is assumed to be superior to the methods introduced in the foregoing sub-sections from an estimation theory point of view. It is the non-linear parallel to the linear regression analysis introduced in Section 2.4. However, the computational work involved is considerable and in practice you need a computer to do the calculations.

Assume a series of pairs of observations (length, age) to be available. These may have been obtained by age reading (cf. Section 3.2.1) or they may have been derived from modal progression analysis (to be discussed in Section 3.4.2). Let there be n pairs of observations:

$(L(i), t(i))$ = (length of fish no. i, age of fish no. i)

where $i = 1, 2, ..., n$.

The method estimates the growth parameters in such a way that the sum of the squares of the deviations between the model and the observations is minimized, i.e. it minimizes the sum with respect to the parameters L_∞, K and t_o:

$$\sum_{i=1}^{n} [L(i) - L_\infty * [1 - \exp(-K*(t(i)-t_o))]]^2 \qquad (3.3.4.1)$$

Computer programs

The LFSA package of microcomputer programs for fish stock assessment (Sparre, 1987) contains the program "VONBER" which can do the least square estimation of growth parameters. The method used by this program is rather complicated and a full explanation falls outside the scope of this manual. However, conceptually, non-linear regression analysis is not more complex than linear regression, just as the square root of 3 is conceptually not more complex than the square root of 4, but the latter is much easier to calculate. FiSAT too contains such a program. Many other similar computer programs are available (see Chapter 15).

3.4 ESTIMATION OF AGE COMPOSITION FROM LENGTH-FREQUENCIES

In Section 3.3 we have dealt with methods for the estimation of the growth parameters of the von Bertalanffy growth equation. All these methods require input data on length and age. As has been stated earlier, it is difficult to determine the ages of tropical fish so, in most cases, only length-frequency data will be available. This section deals with the analysis of length-frequency data. The aim of the methods described below is to assign ages to certain length groups. In other words, the aim is to separate a complex length-frequency distribution into cohorts and to assign an arbitrary age to each of those cohorts. Since the mean length of each cohort can also be determined, we have then obtained the combination of length and age data which is necessary to determine the growth parameters using the methods described in Section 3.3. Before going into specific methods, the difficulties involved in this kind of analysis will be illustrated on the basis of an example from the tropics, after a short introduction of the first known application of these methods in Denmark.

Example 8: Estimating the age of species from temperate waters

The basic idea behind the techniques to be described in this section dates back to one of the first works on fishery biology, a paper on the eel-pout (Zoarces viviparus) by Petersen (1892). The length measurements of 156 fish are represented by dots in Fig. 3.4.0.1. Petersen divided the 156 fish into juveniles, males and females and he further divided the adult fish into two size groups:

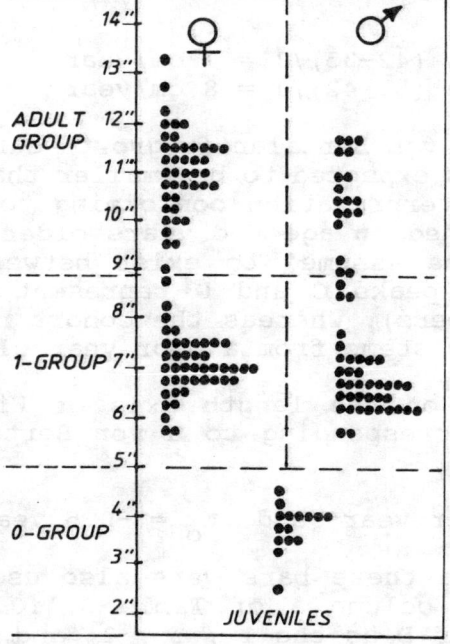

Fig. 3.4.0.1 Length-frequency sample of 156 eel-pout (Zoarces viviparus) in units of Danish inches. Collected in Holbaek Fjord (Denmark) 10-11 July 1890. (Redrawn from Petersen, 1892)

Medium sized: from 5 to 8 inches
Large sized :   from 9 inches and upwards

From earlier observations Petersen knew that in the winter juveniles of about 1.5" could be caught, while in summer the juveniles would all be from 3" to 5" long. Because certain length groups, depending on the season, appeared to be absent, he concluded that the three length groups in the July sample should be interpreted as follows:

less than 5"   :   0-group, born in winter 1889/90
from 5" to 8"   :   1-group, born in winter 1888/89
from 9" and upwards: 2+-group, born in winter 1887/88 or earlier

(The symbol "2+" stands for the "2-group plus older groups". We call it the "2 plus group".)

Petersen's findings indicated that <u>Zoarces</u> <u>viviparus</u> give birth once per year during a restricted period. For most temperate species propagation takes place during 2-4 months in winter or spring.

Such a breeding pattern makes it relatively easy to define a cohort. In temperate waters a cohort is simply a year-class of fish. Because all fish grow at approximately the same rate, a cohort can be followed during the first part of its life by tracing the peaks in the length-frequency samples. But when they approach their maximum size this is no longer possible, because by then fish of different ages have reached almost the same length.

Example 9: Estimating the age of coral trout, a tropical species

We shall now discuss a similar analysis with a species from a tropical area. Fig. 3.4.0.2a shows a length-frequency sample of the coral trout (<u>Plectropomus</u> <u>leopardus</u>) obtained by Goeden (1978), from Heron Island, Australia. This example appears easy to handle. There are four distinct peaks (A,B,C and D) and it is tempting to interpret these as age groups 1,2,3 and 4, as was done by Goeden. However, a closer examination shows that this interpretation does not conform to the von Bertalanffy model. The mean lengths of the peaks B,C and D are approximately 35 cm, 42 cm and 50 cm respectively. When we interpret these peaks as belonging to successive yearly age groups the growth rates become:

between peaks B and C: (42-35)/1 = 7 cm/year
between peaks C and D: (50-42)/1 = 8 cm/year

This does not conform to the von Bertalanffy growth curve, since the growth rate between peaks C and D is expected to be smaller than that between peaks B and C. Thus to give an interpretation conforming to the von Bertalanffy model, peak D must be assigned an age two years older than peak C, and an additional age group must be assumed to exist between peaks C and D. A likely explanation is that peaks C and D represent strong year classes (large number of cohort members), whereas the cohort represented by length groups between peaks C and D stems from a poor year class.

The small solid bars shown on the length axis of Fig. 3.4.0.2a are the lengths at age 1,2,...,7 corresponding to a von Bertalanffy growth curve with the parameters:

L_∞ = 57 cm,   K = 0.4 per year   and t_o = -0.5 years

The lengths corresponding to these bars were also used in Table 3.1.0.1, they have been repeated in column a of Table 3.4.0.1. These parameters interpret the peaks A,B,C and D as the 1-, 2-, 3- and 5-group respectively and place the 4-group between peaks C and D. This particular choice of growth parameters is not based on any fitting techniques or any other rational method. They were derived by a short series of trials with different parameters until a curve was obtained which placed the mean lengths of the various cohorts close to where the peaks are, except for age group 4.

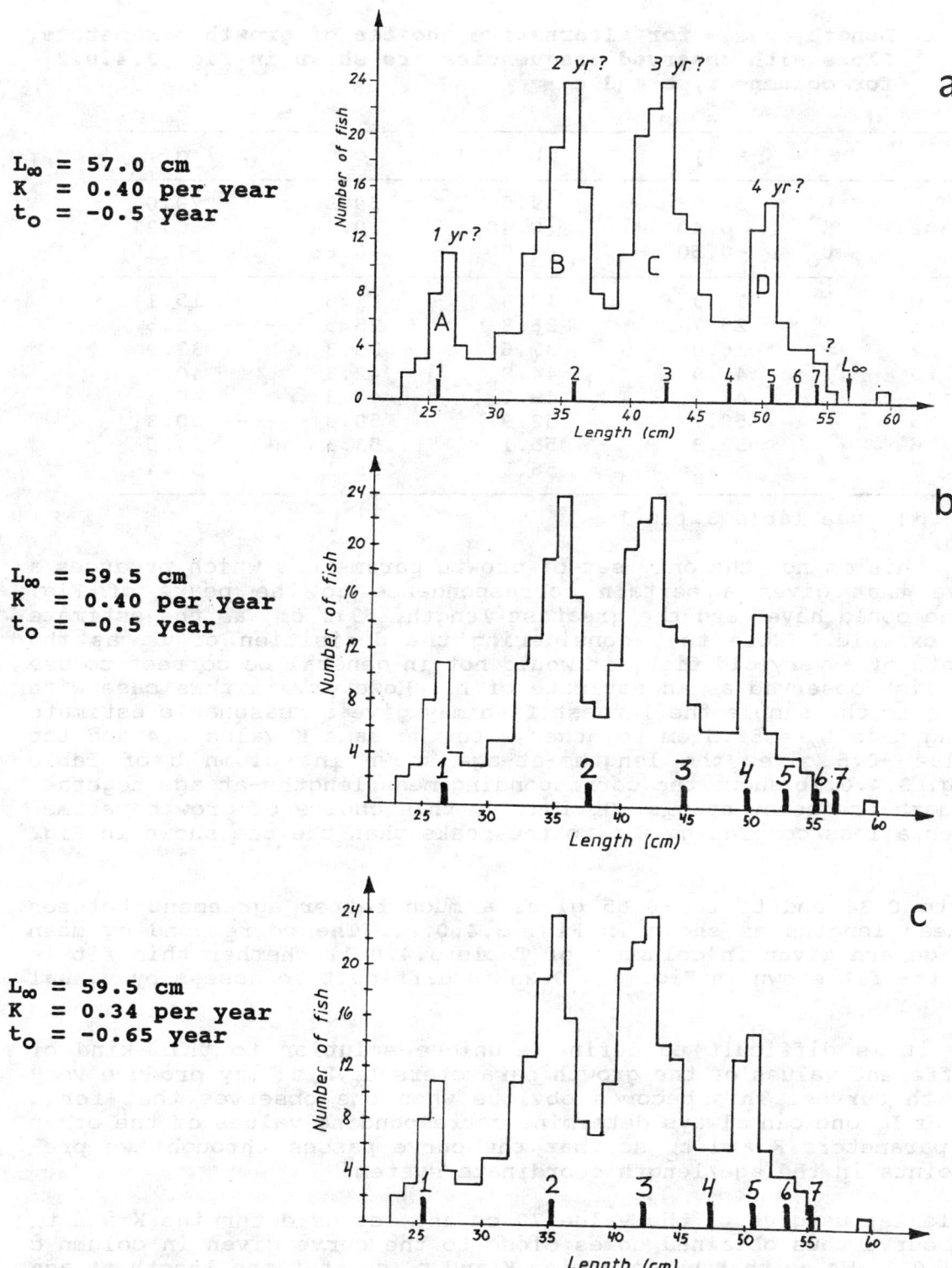

Fig. 3.4.0.2 Length-frequency sample of the coral trout (<u>Plectropomus leopardus</u>) from Goeden (1978). The small bars on the x-axis indicate the lengths-at-age corresponding to the growth parameters given in columns a, b and c of Table 3.4.0.1

Note: Mr. H. Weng, Brisbane, Australia, has drawn our attention to the fact that the coral trout (<u>Plectropomus leopardus</u>) changes sex, from male to female when reaching a length of 30 to 35 cm. This fact was mentioned by Goeden (1978), but overlooked by us. Although the results obtained in this example may not be the "real ones", the example still serves as an illustration of the method.

Table 3.4.0.1 Length-at-age for alternative choices of growth parameters. Plots with observed frequencies are shown in Fig. 3.4.0.2 for columns a, b and c

		a *)	b	c	d
	L_∞	57.0	59.5	59.5	70.0
age	K	0.40	0.40	0.34	0.21
	t_0	-0.50	-0.50	-0.65	-1.15
0		10.3	10.8	11.8	15.1
1		25.7	26.8	25.5	25.5
2		36.0	37.6	35.3	33.9
3		42.9	44.8	42.3	40.8
4		47.5	49.7	47.3	46.3
5		50.7	52.9	50.8	50.8
6		52.8	55.1	53.3	54.5
7		54.2	56.5	55.1	57.4

*) see Table 3.1.0.1

Most likely, this is not the only set of growth parameters which produces a growth curve that gives a certain correspondence to the peaks of Fig. 3.4.0.2a. One could have used the greatest length, 59.5 cm, as the estimate of L_∞, for example. (Note that considering the definition of L_∞ as the average length of a very old fish, it would not in general be correct to use the largest fish observed as an estimate of L_∞. However, in this case with only 312 fish in the sample the largest fish may give a reasonable estimate of L_∞.) Using this L_∞ = 59.5 cm together with the same K value 0.4 and the same t_0 value -0.5 gives the lengths-at-age shown in column b of Table 3.4.0.1. Fig. 3.4.0.2b shows the corresponding mean lengths-at-age together with the length-frequency sample. Obviously, this choice of growth parameters produces a less convincing fit to the peaks than the one shown in Fig. 3.4.0.2a.

Reducing K to 0.34 and t_0 to -0.65 gives a much better agreement between peaks and mean lengths as shown in Fig. 3.4.0.2c. The corresponding mean lengths-at-age are given in column c of Table 3.4.0.1. Whether this fit is better than the fit shown in Fig. 3.4.0.2a is difficult to assess by visual inspection only.

In general, it is difficult to define a unique solution to this kind of problem. Different values of the growth parameters L_∞, K, t_0 may produce very similar growth curves. This becomes obvious when one observes that for a given value of L_∞ one can always determine corresponding values of the other two growth parameters K and t_0 so that the curve passes through two pre-specified points in the age/length coordinate system.

As an example let us give L_∞ the value 70 cm and let us determine K and t_0 so that the curve thus obtained comes close to the curve given in column c of Table 3.4.0.1. We do that by selecting K and t_0 so that the length at age t = 1, L(1) = 25.5 cm and L(5) = 50.8 cm.

Formulas for K and t_0 can be derived from Eq. 3.3.3.1 as follows:

$$-\ln(1 - \frac{L(t1)}{L_\infty}) = -K*t_0 + K*t_1 \quad \text{(a)}$$

$$-\ln(1 - \frac{L(t2)}{L_\infty}) = -K*t_0 + K*t_2 \quad \text{(b)}$$

Subtract (b) from (a).

Since $\ln a - \ln b = \ln \frac{a}{b}$ and after some rearranging, we get

$$\ln \frac{L\infty - L(t1)}{L\infty - L(t2)} = K*(t_2 - t_1) \quad \text{or}$$

$$K = \frac{1}{t2-t1} * \ln \frac{L\infty - L(t1)}{L\infty - L(t2)} \tag{3.4.0.1}$$

The formula for t_o is simply obtained by rearranging Eq. 3.3.3.1. In the case of $t = t1$ it becomes

$$t_o = t1 + \frac{1}{K} * \ln(1 - \frac{L(t1)}{L_\infty}) \tag{3.4.0.2}$$

Thus, for t1 = 1 and t2 = 5 corresponding to L(1) and L(5) respectively we get:

$$K = \frac{1}{5-1} * \ln \frac{70.0 - 25.5}{70.0 - 50.8} = 0.21 \text{ per year} \quad \text{and}$$

$$t_o = 1 + \frac{1}{0.21} * \ln(1 - \frac{25.5}{70.0}) = -1.15 \text{ year}$$

These growth parameters produced the lengths-at-age shown in column d of Table 3.4.0.1. The two growth curves corresponding to columns c and d of Table 3.4.0.1 are shown in Fig. 3.4.0.3. It appears, that it is difficult to decide which of the two growth curves gives the best fit to the length-frequency sample in Fig. 3.4.0.2.

It is often extremely difficult to obtain an unambiguous interpretation of a data set of length-frequencies of tropical fish, in particular when there is only one complex length-frequency sample available and not a time series (see Venema, Christensen and Pauly, 1988). Additional information on the biology of the species in question may help a lot in correctly interpreting the data.

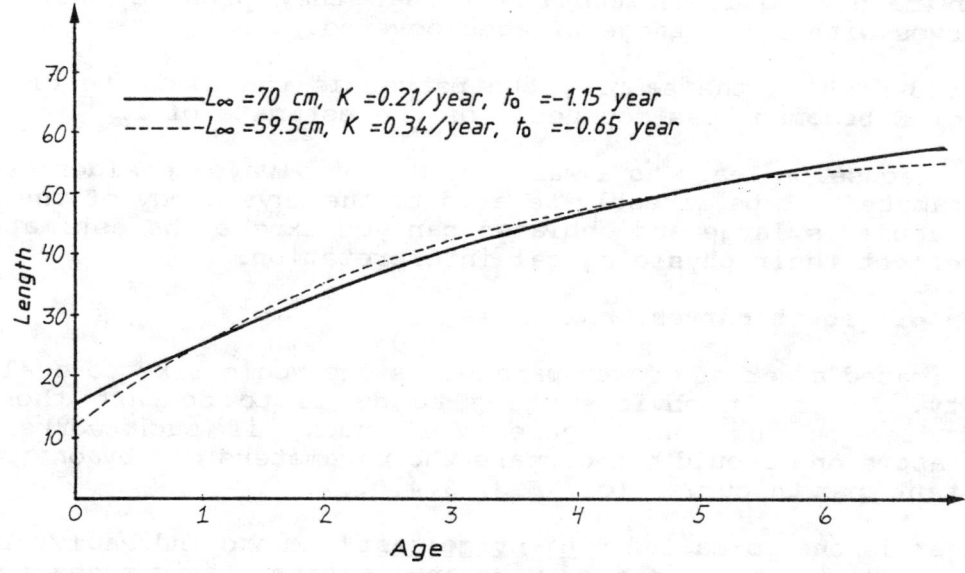

Fig. 3.4.0.3 Example of two growth curves which are approximately equal, but have quite different growth parameters. Derived from columns c and d of Table 3.4.0.1

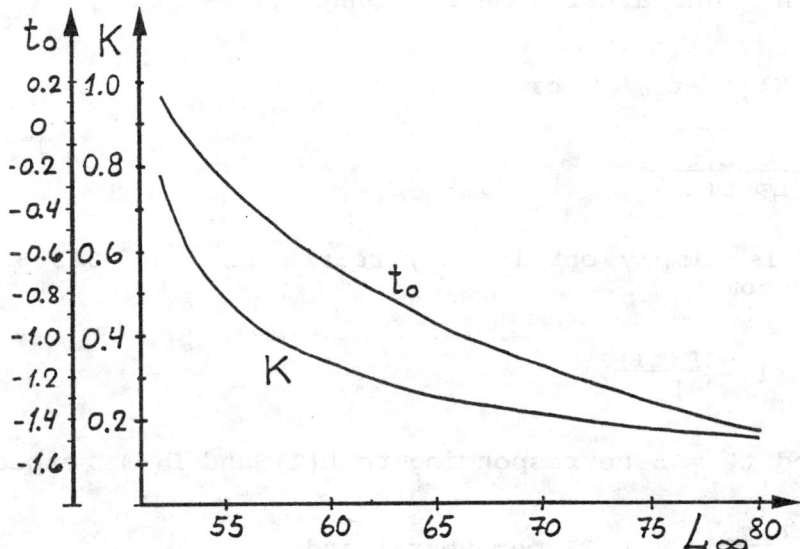

Fig. 3.4.0.4 The growth parameters K and t_o as a function of L_∞ for growth curves fulfilling the condition L(1) = 25.5 cm and L(5) = 50.8 cm

The interrelationship between the growth parameters, L_∞, K and t_o, is further demonstrated in Fig. 3.4.0.4 which shows K and t_o as a function of L_∞ for growth curves fulfilling the condition L(1) = 25.5 cm and L(5) = 50.8 cm. Note that K and t_o decrease as L_∞ increases. Thus, when comparing different estimates of K, L_∞ and t_o the comparison should not be made on the basis of one individual parameter but on the basis of the resulting growth curves. In the example of columns c and d in Table 3.4.0.1 we would say that the two parameter sets:

$(L_\infty, K, t_o) = (59.5, 0.34, -0.65)$ and

$(L_\infty, K, t_o) = (70.0, 0.21, -1.15)$

are approximately equal in the sense that they produce nearly the same growth curves within the range of ages covered.

The more old fish in the sample, the better is the estimate of L_∞ and the estimate of K becomes less dependent of the estimate of L_∞.

The above discussion leads to a warning: do not always consider estimates of growth parameters to be directly related to the physiology of the fish. Only when the sample is large and unbiased can you expect the estimated parameters to reflect their physiological interpretation.

Comparison of growth curves, phi prime

Having estimated a set of growth parameters one would like to evaluate their reliability. The first obvious thing to do is to compare them to other growth studies on the same species (or stock) if such works exist. As discussed above one should not compare the parameters one by one but compare the resultant growth curves (cf. Fig. 3.4.0.3).

Another test is the so-called "phi prime test" (Munro and Pauly, 1983, Pauly and Munro, 1984) which has gained wide application. It is suggested that the "overall growth performance" is reflected by:

ϕ' (phi prime) = ln K + 2*ln L_∞ (3.4.0.3)

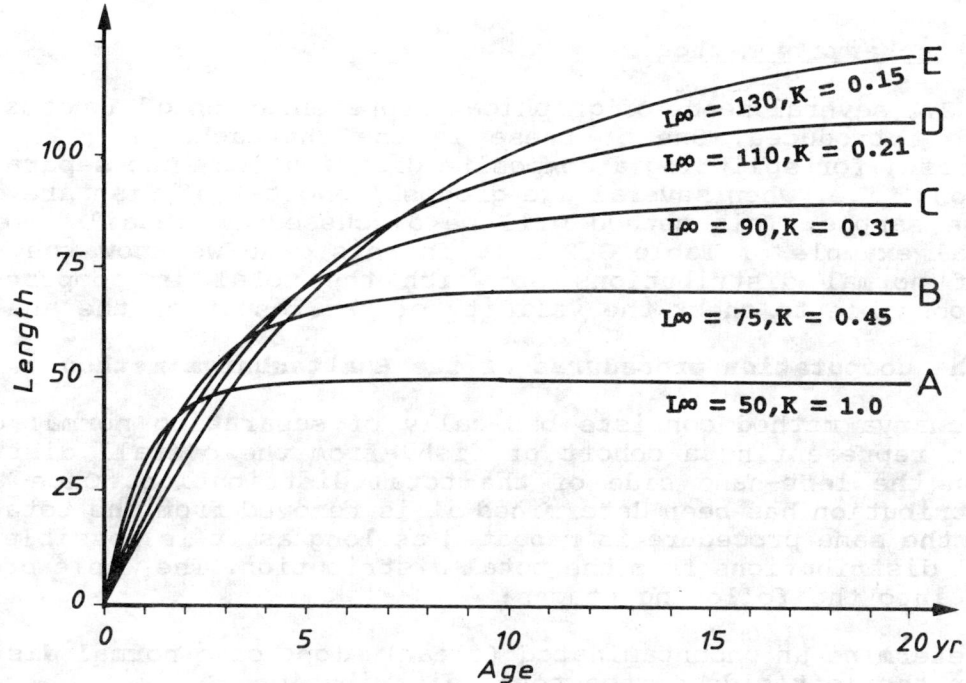

Fig. 3.4.0.5 **Five different growth curves with the same ϕ' (phi prime)**

(Note, the original paper gives ϕ' in base 10 logarithms. Using natural logarithms, as has been done throughout this manual, the ϕ' values will be ln 10 = 2.3026 times higher than when base 10 logarithms are used.)

This test is based on the discovery by Pauly (1979a) that phi prime values are very similar within related taxa and have narrow normal distributions. For example, Moreau, Bambino and Pauly (1986) analysed 100 different Tilapia populations and calculated four alternative indices for overall growth performance. They found that ϕ' was the best index of overall growth performance, in the sense that it had minimum variance. If growth parameters are available for a number of stocks of the same species (or species within the same genus), it is possible to evaluate the reliability of an estimated growth curve by means of ϕ'.

The alternative indices of overall growth performance were

$K*L_\infty$, $\ln(K*L_\infty)$ and $\phi = \ln K + 0.67*\ln W_\infty$

They concluded that the first index, $K*L_\infty$, is poor (has large variance within the family) and should not be used.

Another use of Eq. 3.4.0.3 is to get a first rough estimate of K for a species which has not previously been studied, but which belongs to a family, some members of which have been analysed for growth parameters. If an estimate of L_∞ is available and the average ϕ' for the family members with known growth curves is available, a first estimate of K is given by:

$$K = \exp(\phi')/L_\infty^2 \qquad (3.4.0.4)$$

which follows from solving Eq. 3.4.0.3 with respect to K.

Growth curves estimated for the same stock can have the same ϕ' and yet be very different as illustrated in Fig. 3.4.0.5 which shows five very different growth curves, all with the same ϕ'. However, if two phi primes representing alternative estimates of growth parameters for the same stock differ greatly it indicates that one (or both) of the estimates may be biased.

3.4.1 Bhattacharya's method

In Section 2.2 several means of graphical representation of a normal distribution were introduced. One of these is the Bhattacharya (1967) method, which is useful for splitting a composite distribution into separate normal distributions, i.e. when several age groups (cohorts) of fish are contained in the same sample. This method will be discussed in detail based on the hypothetical example of Table 3.2.1.1. In this case we know the solution: the set of normal distributions of which the total is composed. It is therefore possible to check the validity of the result of the analysis.

Basis of the computation procedures of the Bhattacharya method

The Bhattacharya method consists basically of separating normal distributions, each representing a cohort of fish, from the overall distribution, starting on the left-hand side of the total distribution. Once the first normal distribution has been determined it is removed from the total distribution and the same procedure is repeated as long as it is possible to separate normal distributions from the total distribution. The whole process can be divided into the following stages:

Stage 1: Determine an uncontaminated (clean) slope of a normal distribution on the left side of the total distribution.

Stage 2: Determine the normal distribution of the first cohort by means of a transformation into a straight line.

Stage 3: Determine the numbers of fish per length group belonging to that first cohort and then subtract them from the total distribution.

Stage 4: Repeat the process for the next normal distribution from the left, until no more clean normal distributions can be found.

Stage 5: Relate the mean lengths of the cohorts determined in stages 1 and 4 to the age difference between the cohorts.

As has already been shown in Section 2.6, a normal distribution is transformed into a straight line when: 1) numbers are replaced by their logarithms and 2) differences are calculated between consecutive logarithmic values. Let N designate the number in a length-frequency sample belonging to the length group:

$$[x - dL/2, x + dL/2]$$

where dL is the interval size, x is the interval midpoint, $x - dL/2$ is the lower and $x + dL/2$ is the upper limit of the interval.

If a certain length range in the sample contains only one cohort, this part of the frequency sample should conform to a normal distribution (e.g. from 10 cm to 22 cm in the sample shown in Fig. 3.2.2.2). In that case the linear relationship (cf. Eq. 2.6.5):

$$\Delta \ln N = a + b*(x + dL/2)$$

would hold between the difference of the logarithms of the number in a certain length class and the logarithm of the number in the preceding class or

$$\Delta \ln N = \ln N(x + dL/2, x + 3dL/2) - \ln N(x - dL/2, x + dL/2)$$

as the dependent variable, y,

and the upper limit of the smallest length group:

$x + dL/2$

as the independent variable x (compare Figs. 2.6.4 and 2.6.5).

Recall that the standard deviation of the normal distribution and the mean are obtained by:

$s = \sqrt{-dL/b}$ and $\bar{x} = -a/b$ (compare Eqs. 2.6.6 and 2.6.7)

Example 10: A Bhattacharya analysis of a constructed data set

The computation procedures related to steps 1 to 5 in the previous section will be illustrated below on the hand of the constructed data set presented earlier in Table 3.2.1.1 and the related graphical representation in Fig. 3.2.2.1. This data set was created from 6 normally distributed components, as shown in Fig. 3.2.2.2.

We will now try to use the Bhattacharya method to analyse the "total" column of Table 3.2.1.1, trying to break it up into the six normal contributions from which is has been composed. The advantage of using a constructed data set is that it is then possible to compare the results of the Bhattacharya analyses with the exact input data. The possibilities and limitations of the method can thus be illustrated. The computation procedure will be followed step by step in general terms, with examples drawn from the data set of Table 3.2.1.1. Unless indicated otherwise, the examples refer to Table 3.4.1.1, which is the first of a number of work sheets.

Step 1: Create a work sheet like Table 3.4.1.1 and complete column A, the length groups and column B, the corresponding frequencies from the available data set.

Example: Columns A and B in Table 3.4.1.1, taken from Table 3.2.1.1. Column B is labelled "N1+", because it contains the distribution of the first cohort (N1) plus all the other cohorts. In general, the symbol "Na+" stands for the number in the a'th plus older cohorts.

Step 2: Create column C by taking logarithms of the frequencies of N1+ (column B).

Examples: ln 1 = 0
ln 4 = 1.386

Step 3: Column D contains the differences between the logarithms of two adjacent frequencies

Δln N1+ = ln N1+ of the current line minus
ln N1+ of the previous line

Complete column D. Start on the <u>second</u> line, subtract the ln value of the first line of column C from that of the second line of column C and place it on the second line of column D. The first place remains open, since a difference between the first point and a foregoing point does not exist. Take care to use at least three decimal points. Continue by determining the differences (Δln) between the third and the second line, etc.

Table 3.4.1.1 Bhattacharya method: Estimation of first cohort, N1 (the 1983 spring cohort).
in columns B, C, G and H indicates where to start the calculations of N1 (cf. Fig. 3.4.1.2)
Column I contains the remainder of the sample
(N1+ - N1 = N2+)

A	B	C	D	E	F	G	H	I
L1-L2	N1+	ln N1+	Δln N1+ (y)	L (x)	Δln N1	ln N1	N1	N2+
12-13	1	0.000	-	-	-	-	1	0
13-14	4	1.386	1.386	13*	1.375	-	4	0
14-15	11	2.398	1.012	14*	1.059	-	11	0
15-16	24	3.178	0.780	15*	0.743	-	24	0
16-17	38#	3.638#	0.460	16*	0.427	3.638#	38#	0
17-18	42	3.738	0.100	17*	0.111	3.749	(42.48)	(-0.48)
18-19	33	3.497	-0.241	18*	-0.205	3.545	(34.61)	(-1.61)
19-20	20	2.996	-0.501	19*	-0.521	3.023	(20.55)	(-0.55)
20-21	7	1.946	-1.050	20	-0.837	2.186	(8.90)	(-1.90)
21-22	3	1.099	-0.847	21	-1.153	1.033	2.81	0.19
22-23	3	1.099	0.000	22	-1.469	-0.436	0.65	2.35
23-24	5	1.609	0.511	23	-1.785	-2.211	0.11	4.89
24-25	8	2.079	0.470	24	-	-	-	8
25-26	11	2.398	0.318	25	-	-	-	11
26-27	14	2.639	0.241	26	-	-	-	14
27-28	17	2.833	0.194	27	-	-	-	17
28-29	16	2.773	-0.060	28	-	-	-	16
29-30	15	2.708	-0.065	29	-	-	-	15
30-31	14	2.639	-0.069	30	-	-	-	14
31-32	11	2.398	-0.241	31	-	-	-	11
32-33	11	2.398	0.000	32	-	-	-	11
33-34	10	2.303	-0.095	33	-	-	-	10
34-35	9	2.197	-0.106	34	-	-	-	9
35-36	10	2.303	0.106	35	-	-	-	10
36-37	11	2.398	0.095	36	-	-	-	11
37-38	10	2.303	-0.095	37	-	-	-	10
38-39	10	2.303	0.000	38	-	-	-	10
39-40	11	2.398	0.095	39	-	-	-	11
40-41	11	2.398	0.000	40	-	-	-	11
41-42	9	2.197	-0.201	41	-	-	-	9
42-43	7	1.946	-0.251	42	-	-	-	7
43-44	7	1.946	0.000	43	-	-	-	7
44-45	5	1.609	-0.337	44	-	-	-	5
45-46	6	1.792	0.183	45	-	-	-	6
46-47	5	1.609	-0.183	46	-	-	-	5
47-48	3	1.099	-0.510	47	-	-	-	3
48-49	2	0.693	-0.406	48	-	-	-	2
49-50	2	0.693	0.000	49	-	-	-	2
50-51	2	0.693	0.000	50	-	-	-	2
51-52	1	0.000	-0.693	51	-	-	-	1

Class interval, dL = 1.   Total number in cohort N1: 183.57
*) points used in the regression analysis, with results:

$a = 5.4834$, $b = -0.3160$, $\bar{L}(N1) = -a/b = 17.35$,
$s(N1) = \sqrt{(-dL/b)} = 1.78$

Example:

A L1-L2	B N1+	C ln N1+	D Δln N1+	E L
12-13	1	0	-	
13-14	4	1.386	1.386-0 = 1.386	13
14-15	11	2.398	2.398-1.386 = 1.012	14

Step 4: Complete column E. Recall from Section 2.6 that Δln N1+ should be plotted against the upper limit of the smallest length group of the two from which Δln N1+ is calculated. Insert the mid-point, the upper limit of the smallest of the two classes, or the lower limit of the largest of the two at the same level as the corresponding Δln N.

See example Step 3.

Step 5: Make a complete plot of the length (column E, at the x-axis) against Δln N1+ (column D, at the y-axis).

Example: Fig. 3.4.1.1

Step 6: Inspect the plot and determine which points lie on a straight line. Mark these points in column E. Do not include points that may be affected by the next distribution. The further the points are lying to the right, the higher the chance is that they are influenced by the next distribution.

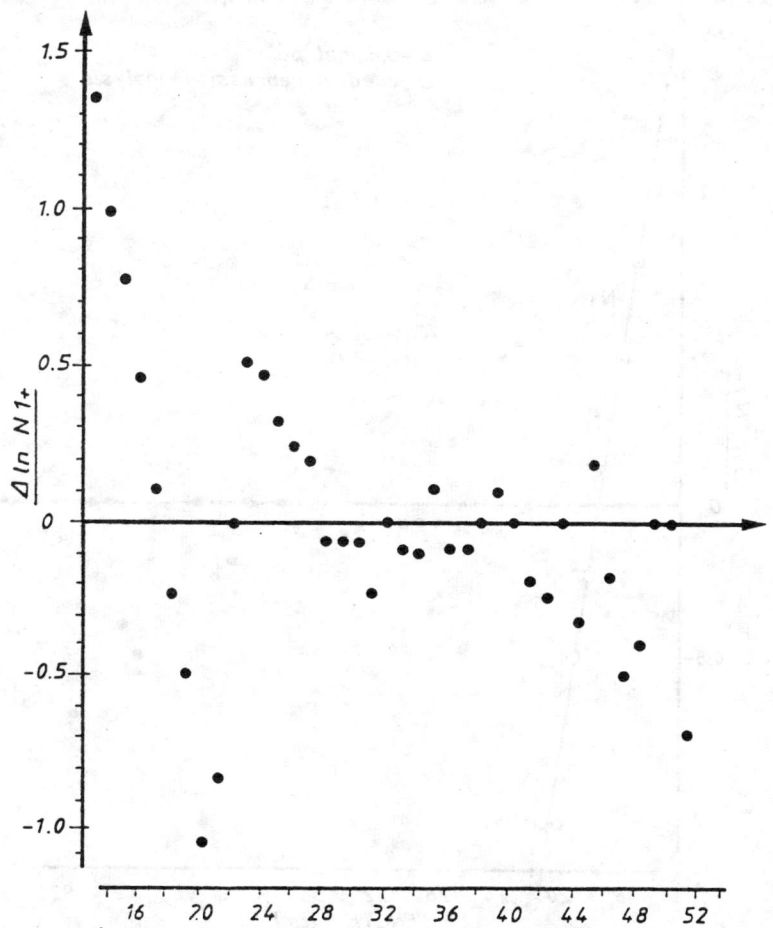

Fig. 3.4.1.1 Bhattacharya method: plot corresponding to columns D (y-axis) and E (x-axis) of Table 3.4.1.1

Example: Visual inspection of Fig. 3.4.1.1 shows that a straight line can be fitted to the first seven points (indicated by "*" in Table 3.4.1.1). Even the eighth point lies on the same line, but because it may be influenced by the next distribution it was not included in the subsequent calculations.

This straight line corresponds to the first normally distributed component, N1, which is interpreted as the 1983 spring cohort. That the straight line corresponding to N1 comes out so nicely is not surprising, since the first component has very little overlap with the next component, as can be observed in Figs. 3.2.2.1 and 3.2.2.2.

Step 7: Calculate the straight line that fits to the points by regressing column E against column D for the selected points (asterisks). Determine a (intercept) and b (slope) and calculate the mean length

$\overline{L}(N1) = -a/b$ and the standard deviation $s(N1) = \sqrt{-1/b}$

Example: $y = a + b*x$
where $y = \Delta \ln N1+$ (in column D) and $x = L$ (in column E)

a (intercept) = 5.4834, b (slope) = -0.3160

$\overline{L}(N1) = -a/b = 17.35$ cm and $s(N1) = \sqrt{-1/b} = 1.78$

The regression line is shown in Fig. 3.4.1.2.

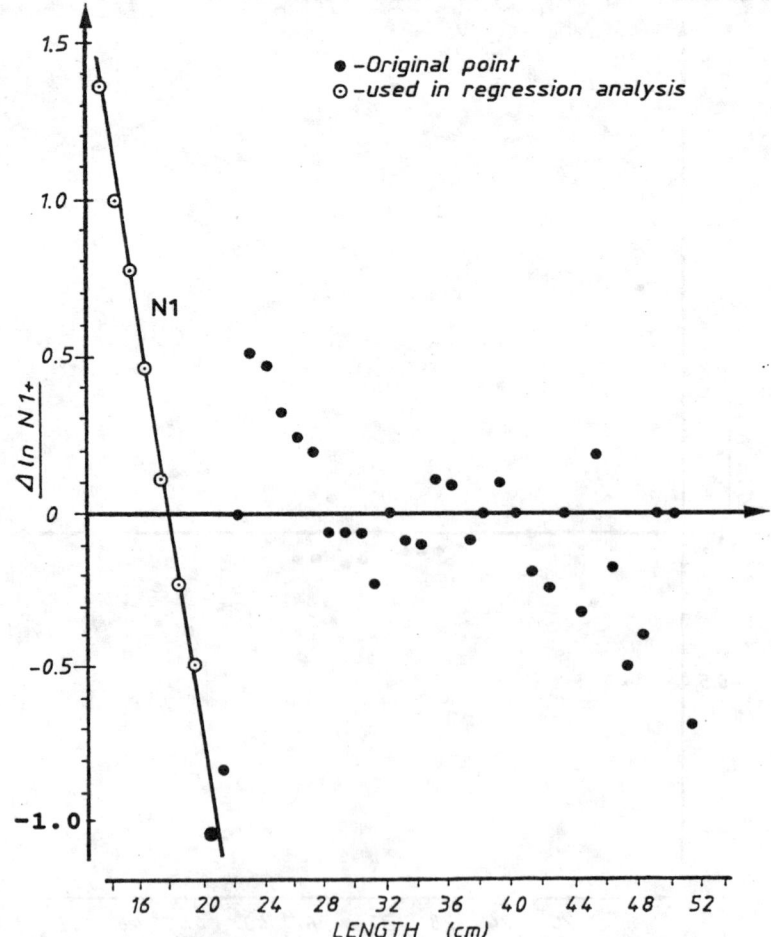

Fig. 3.4.1.2 Bhattacharya method: regression line estimated for the first cohort (compare to columns D and E in Table 3.4.1.1)

We have now determined the line that represents a normal distribution, which should to a large extent correspond with the left side of the actual distribution we have in our sample. The line should represent the first cohort, N1. In order to determine how far this is true we must first calculate the theoretical values of $\Delta \ln N1$, those corresponding to the line we have just determined and then reverse the process and convert the differences ($\Delta \ln N1$) into ln N1 and then into numbers (N1). This process is illustrated in columns F, G and H of Table 3.4.1.1.

The second part of the computation procedure consists of the following steps:

Step 8: The formula $\Delta \ln N = a + bL$ can now be used to calculate the theoretical value $\Delta \ln N1$. This is done for as many length groups as one can expect to find in the first cohort (normal distribution).

Example: $\Delta \ln N1$ for the length groups 12-13 and 13-14 with mid length 13 is determined from

$$a + b*13 = 5.4834 - 0.3160*13 = 1.375,$$

which is the first value in column F, the next value is

$$a + b*14 = 1.059, \text{ etc.}$$

Step 9: In order to be able to convert "a difference", $\Delta \ln N$, into its two components, ln N of a certain length group and ln N of the length group above it, we need a starting point. This starting point should be based on a frequency that is not contaminated by overlap with the following cohort (normal distribution). Therefore a frequency should be chosen on the left side of the first normal distribution. Preferably the frequency should also not be too low.

Example: The frequency 38 of the length group 16-17 cm was chosen as the clean starting point, as indicated by "#". It is placed in column H as the first entry for N1, the numbers in the length-frequency distribution of the first cohort. The real starting point is actually the logarithm of 38, viz., 3.638 (see column C). This value is inserted in column G. The choice of 38 as a "clean" frequency also implies that the frequencies lying to its left, those above it in the table, viz., 1, 4, 11 and 24, are also considered to be clean. In other words none of these frequencies are supposed to overlap with the next cohort, so they are all clean frequencies of the first normal distribution N1 (the 1983 spring cohort).

Step 10: We now have $\Delta \ln N1$ corresponding to two adjacent length classes in column F and the first ln N1 of the lower length class in column G, which permits us to calculate the ln N1 of the next length class up using the formula

> ln N1 (upper length class) = ln N1 (lower length class) + the corresponding $\Delta \ln N1$

Example:

> ln N1(17-18) = ln N1(16-17) + $\Delta \ln N1$(17-18 and 16-17)
> ln N1(17-18) = 3.638 + 0.111 = 3.749
> ln N1(18-19) = 3.749 + (-0.205) = 3.544

The new values are entered in column G.

Step 11: By taking antilogs the numbers corresponding to ln N1 in column G can be found and inserted in column H. This column is stopped when the number in column H approaches zero.

Examples:

for length group 17-18: N1(17-18) = exp(3.749) = 42.48
for length group 18-19: N1(18-19) = exp(3.544) = 34.61

The results (in column H) are not exactly the same as the observed frequencies given in column B, because observations always deviate somewhat from the theoretical values. In the present case of a hypothetical data set, the deviations are due to rounding errors. With "real" data there are also deviations caused by "random noise". Even if the sample is a perfect random sample the observations will fluctuate around the true length distribution (the length distribution of the population).

Step 12: The numbers of fish per length group belonging to the youngest (1983 spring) cohort or N1, in column G, can now be subtracted from the Total distribution, or N1+, in column B. The new distribution obtained is placed in column I and called N2+, the frequency distribution of fish in the second cohort plus all the subsequent cohorts.

N1+ minus N1 = N2+ or column B - column H = column I

In practice it may well happen that the figures in column I become negative because of random variation of the observations. However, this can be adjusted. Whenever the estimate of N2+ numbers becomes negative we assign the value zero to N2+ (in column I) while the N1 is given the value of column B.

Examples:

42-42.48 = -0.48, which is adjusted to 0 in column I, and to 42 in column H

33-34.61 = -1.61, which is adjusted to 0 in column I and to 33 in column H

The results of the whole analysis of the first normal distribution will be:

A L1-L2	B N1+	H N1	I N2+
12-13	1	1	0
13-14	4	4	0
14-15	11	11	0
15-16	24	24	0
16-17	38	38	0
17-18	42	42	0
18-19	33	33	0
19-20	20	20	0
20-21	7	7	0
21-22	3	2.81	0.19
22-23	3	0.65	2.35
23-24	5	0.11	4.89
24-25	8	0	8

Total number in cohort N1 =   183.57

$\overline{L}(N1) = 17.35$ and $s(N1) = 1.78$

Since this is based on a constructed data set we can compare these results with the real values, which are given in Table 3.2.1.1 (spring 1983):

Total number $(N1) = 182$, $\overline{L}(N1) = 17.3$ and $s(N1) = 1.7$

In this case the results obtained from the analysis are very close to the real values. What we have obtained are all the necessary elements to describe the first normal distribution, viz.,

$\overline{L}(N1)$, $s(N1)$ and $n(N1)$

The whole process is now repeated in order to obtain those values for the next normal distribution, the one pertaining to the cohort that was born in the autumn of 1982 (see Table 3.2.1.1).

We have come to the end of the use of Table 3.4.1.1. By eliminating all values pertaining to N1 we create the next work sheet (Table 3.4.1.2) with N2+ (column I of Table 3.4.1.1) as the new column B. The whole procedure can then be repeated.

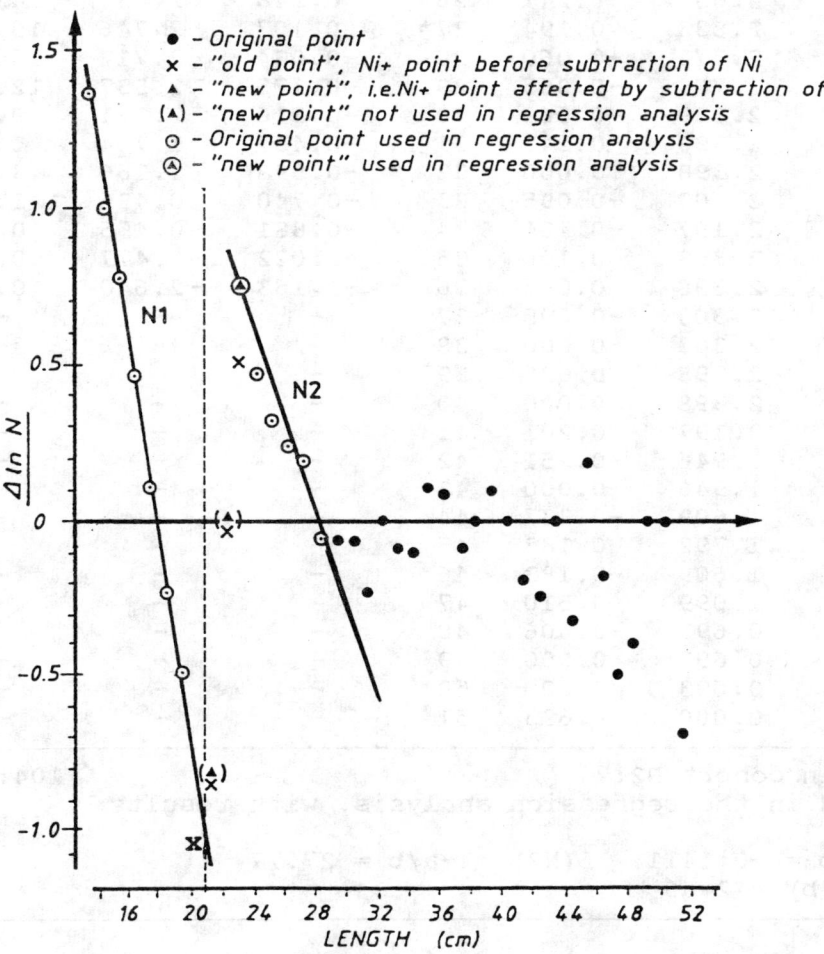

Fig. 3.4.1.3 Bhattacharya method: regression line estimated for the second cohort (compare columns D and E in Table 3.4.1.2)

Table 3.4.1.2 Bhattacharya method: Estimation of the second cohort, N2 (the 1982 autumn cohort).
\# in columns B, C, G and H indicates where to start the calculations of N2 (compare Fig. 3.4.1.3)

A	B	C	D	E	F	G	H	I
L1-L2	N2+	ln N2+	Δln N2+ (y)	L (x)	Δln N2	ln N2	N2	N3+
12-13	0	-	-	-	-	-	0	0
13-14	0	-	-	13	-	-	0	0
14-15	0	-	-	14	-	-	0	0
15-16	0	-	-	15	-	-	0	0
16-17	0	-	-	16	-	-	0	0
17-18	0	-	-	17	-	-	0	0
18-19	0	-	-	18	-	-	0	0
19-20	0	-	-	19	-	-	0	0
20-21	0	-	-	20	-	-	0	0
21-22	0.19	-1.661	-	21	-	-	0.19	0
22-23	2.35	0.854	2.515	22	-	-	2.35	0
23-24	4.89	1.587	0.733	23*	-	-	4.89	0
24-25	8	2.079	0.492	24*	-	-	8	0
25-26	11	2.398	0.319	25*	-	-	11	0
26-27	14#	2.639#	0.241	26*	0.248	2.639#	14#	0
27-28	17	2.833	0.194	27*	0.107	2.746	15.58	1.42
28-29	16	2.773	-0.060	28*	-0.034	2.712	15.06	0.94
29-30	15	2.708	-0.065	29	-0.175	2.537	12.64	2.36
30-31	14	2.639	-0.069	30	-0.316	2.221	9.22	4.78
31-32	11	2.398	-0.241	31	-0.457	1.764	5.84	5.16
32-33	11	2.398	0.000	32	-0.598	1.166	3.21	7.79
33-34	10	2.303	-0.095	33	-0.740	0.426	1.53	8.47
34-35	9	2.197	-0.106	34	-0.881	-0.455	0.63	8.37
35-36	10	2.303	0.106	35	-1.022	-1.477	0.23	9.77
36-37	11	2.398	0.095	36	-1.163	-2.640	0.07	10.93
37-38	10	2.303	-0.095	37	-	-	-	10
38-39	10	2.303	0.000	38	-	-	-	10
39-40	11	2.398	0.095	39	-	-	-	11
40-41	11	2.398	0.000	40	-	-	-	11
41-42	9	2.197	-0.201	41	-	-	-	9
42-43	7	1.946	-0.251	42	-	-	-	7
43-44	7	1.946	0.000	43	-	-	-	7
44-45	5	1.609	-0.337	44	-	-	-	5
45-46	6	1.792	0.183	45	-	-	-	6
46-47	5	1.609	-0.183	46	-	-	-	5
47-48	3	1.099	-0.510	47	-	-	-	3
48-49	2	0.693	-0.406	48	-	-	-	2
49-50	2	0.693	0.000	49	-	-	-	2
50-51	2	0.693	0.000	50	-	-	-	2
51-52	1	0.000	-0.693	51	-	-	-	1

Total number in cohort N2: 104.44
*) points used in the regression analysis, with results:

$a = 3.9168$, $b = -0.1411$, $\bar{L}(N2) = -a/b = 27.77$
$s(N2) = \sqrt{(-dL/b)} = 2.66$

Fig. 3.4.1.3 shows the Bhattacharya plot for N2+ together with the estimated line for N1. Only the points to the right of the broken line of Fig. 3.4.1.2 are used in the analysis now. The N1-line is shown for comparison only. Some points have been moved due to the subtraction of N1. The "old" points (i.e. those from the N1+ plot) are indicated by "x" and the "new" points by a triangle, in cases where the movement is visible. The two first points (corresponding to lengths 21 and 22 cm) are disregarded, because they refer to very small numbers of specimens.

The selection of points to fit a straight line is now a bit more difficult than in the case of the first cohort. In Fig. 3.4.1.3 the six points from lengths 23 to 28 cm were chosen. One can question why these points were chosen in preference to the points, for example, from lengths 24 to 29 cm or from lengths 24 to 28 cm. The choice is a subjective one. The results of the Bhattacharya method may sometimes be dependent on the person who actually performs the analysis. If, for example, only the points from lengths 24 to 27 cm were used the estimated mean length would be 28.7 cm and the standard deviation 3.2 cm. The actual choice made in Fig. 3.4.1.3 gives a mean value of $\bar{L}(N2) = 27.77$ cm and a standard deviation $s(N2)$ of 2.66 cm which are both very close to the true values (see Tables 3.2.1.1 and 3.4.1.2). However, this cannot be used as a justification for this choice, because in real life we would not know what the true values should be. Also the selection of the "clean" value of ln N2 from which N2 and N3+ are calculated is a subjective one. The more the observations deviate from the calculated frequencies the more pronounced the element of subjectivity.

In summary, the results obtained so far are:

cohort N1: mean length 17.35 cm, standard deviation 1.78 cm
(Table 3.4.1.1)

cohort N2: mean length 27.77 cm, standard deviation 2.66 cm
(Table 3.4.1.2)

Now that the first two mean lengths of cohorts have been estimated we are in a position to obtain a first rough estimate of the von Bertalanffy parameter K, provided we also have an estimate of the age difference between the two cohorts. We use Eqs. 3.4.0.1 and 3.4.0.2 with a time difference between the two cohorts equal to t2-t1 = 0.5 year. Further a rough estimate of L_∞ is obtained from the length-frequency sample, which tells us that the fish rarely get longer than 50 cm, so it is assumed that L_∞ = 50 cm. From Eq. 3.4.0.1 we get:

$$K = \frac{1}{t2-t1} * \ln \frac{L_\infty - \bar{L}(t1)}{L_\infty - \bar{L}(t2)} = \frac{1}{0.5} * \ln \frac{50-17.35}{50-27.77} = 0.77$$

and from Eq. 3.4.0.2:

$$t_0 = t1 + \frac{1}{K} * \ln(1 - \frac{\bar{L}(t1)}{L_\infty}) = 0.5 + \frac{1}{0.77} * \ln(1 - \frac{17.35}{50}) = -0.05$$

where the value t1 = 0.5 is an arbitrary age.

Thus, as a first rough estimate of the growth curve we have

$$L(t) = 50*[1 - \exp(-0.77*(t+0.05))]$$

The estimation procedure presented above is not generally recommended. It has been given here to demonstrate how little data are actually required to roughly estimate a growth curve. Such a first estimate, however, may be used to predict the next mean length, i.e. the mean length of cohort N3.

Table 3.4.1.3 Bhattacharya method: Estimation of the third cohort, N3 (the 1982 spring cohort).
in columns B, C, G and H indicates where to start the calculations of N3 (compare Fig. 3.4.1.4)

A	B	C	D	E	F	G	H	I
L1-L2	N3+	ln N3+	Δln N3+	L	Δln N3	ln N3	N3	N4+
25-26	0	-	-	25	-	-	0	0
26-27	0	-	-	26	-	-	0	0
27-28	1.42	0.351	-	27	-	-	1.42	0
28-29	0.94	-0.062	-0.413	28	-	-	0.94	0
29-30	2.36	0.859	0.921	29	-	-	2.36	0
30-31	4.78	1.564	0.705	30 *	-	-	4.78	0
31-32	5.16	1.641	0.077	31 *	-	-	5.16	0
32-33	7.79#	2.053#	0.412	32 *	-	2.053#	7.79#	0
33-34	8.47	2.137	0.084	33 *	0.111	2.164	8.71	0
34-35	8.37	2.125	-0.012	34 *	-0.032	2.132	8.43	0
35-36	9.77	2.279	0.154	35	-0.175	1.957	7.08	2.69
36-37	10.93	2.392	0.113	36	-0.318	1.639	5.15	5.78
37-38	10	2.303	-0.089	37	-0.460	1.179	3.25	6.75
38-39	10	2.303	0.000	38	-0.603	0.576	1.78	8.22
39-40	11	2.398	0.095	39	-0.746	-0.170	0.84	10.16
40-41	11	2.398	0.000	40	-0.888	-1.058	0.35	10.65
41-42	9	2.197	-0.201	41	-1.031	-2.089	0.12	8.88
42-43	7	1.946	-0.251	42	-1.174	-3.263	0.04	6.96
43-44	7	1.946	0.000	43	-	-	-	7
44-45	5	1.609	-0.337	44	-	-	-	5
45-46	6	1.792	0.183	45	-	-	-	6
46-47	5	1.609	-0.183	46	-	-	-	5
47-48	3	1.099	-0.510	47	-	-	-	3
48-49	2	0.693	-0.406	48	-	-	-	2
49-50	2	0.693	0.000	49	-	-	-	2
50-51	2	0.693	0.000	50	-	-	-	2
51-52	1	0.000	-0.693	51	-	-	-	1

Total number in cohort N3: 58.20
*) points used in the regression analysis, with results:

$a = 4.8196$, $b = -0.1427$, $\bar{L}(N3) = -a/b = 33.77$,
$s(N3) = \sqrt{(-dL/b)} = 2.65$

Table 3.4.1.4 Estimation of K and t_o by the von Bertalanffy plot using arbitrary input ages and the mean lengths estimated in Tables 3.4.1.1 to 3.4.1.3 (compare Table 3.3.3.1)

t (x)	$\bar{L}(t)$	$-\ln(1-\bar{L}(t)/50)$ (y)
0.5	17.4	0.428
1.0	27.8	0.812
1.5	33.8	1.127

a (intercept) = 0.09
b (slope) = 0.699, K = 0.7 per year
$t_o = -a/b = -0.13$ year

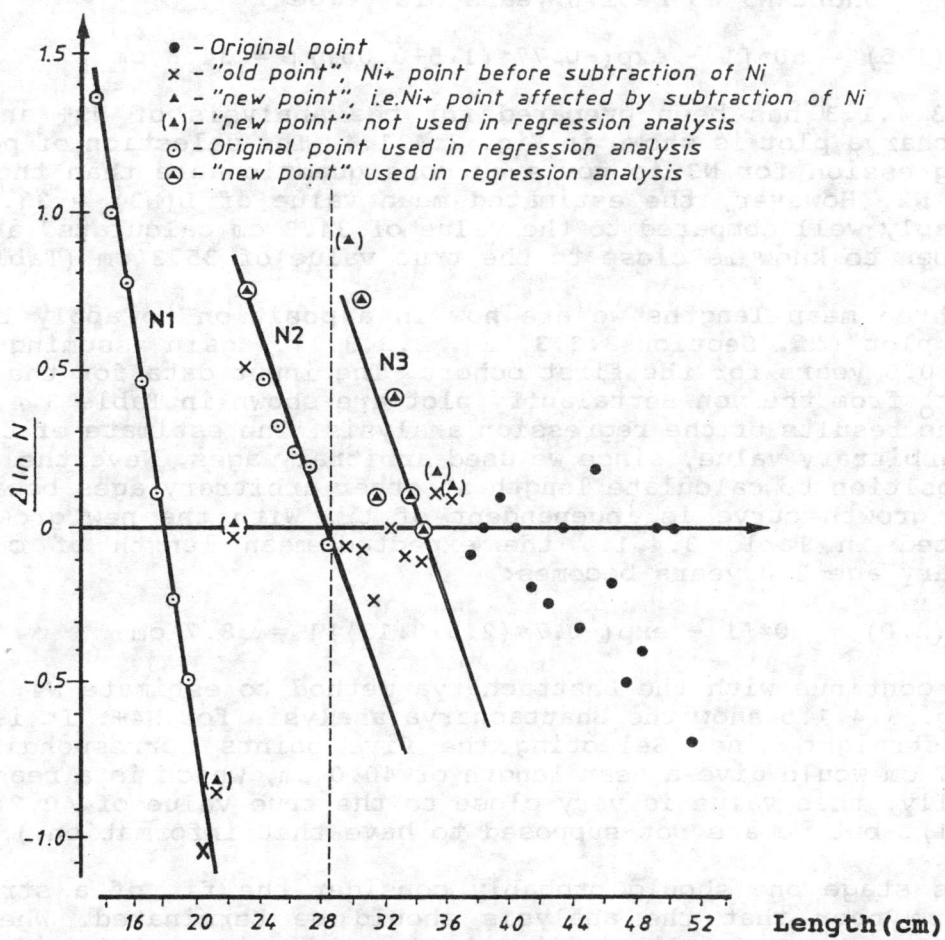

Fig. 3.4.1.4 Bhattacharya method: regression line estimated for the third cohort (compare columns D and E in Table 3.4.1.3)

Table 3.4.1.5 Bhattacharya method: Attempt to estimate the fourth cohort, N4 (the 1981 autumn cohort), compare Fig. 3.4.1.5

A	B	C	D	E	F	G	H	I
L1-L2	N4+	ln N4+	Δln N4+	L	Δln N4	ln N4	N4	N5+
34-35	0	–	–	34	–	–	0	0
35-36	2.69	0.990	0.990	35	?	?	?	?
36-37	5.78	1.754	0.764	36	?	?	?	?
37-38	6.75	1.910	0.156	37 (*)	?	?	?	?
38-39	8.22	2.107	0.197	38 (*)	?	?	?	?
39-40	10.16	2.318	0.211	39 (*)	?	?	?	?
40-41	10.65	2.366	0.048	40 (*)	?	?	?	?
41-42	8.88	2.184	-0.184	41 (*)	?	?	?	?
42-43	6.96	1.940	-0.244	42	?	?	?	?
43-44	7	1.946	0.006	43	?	?	?	?
44-45	5	1.609	-0.337	44	?	?	?	?
45-46	6	1.792	0.183	45	?	?	?	?
46-47	5	1.609	-0.183	46	?	?	?	?
47-48	3	1.099	-0.510	47	?	?	?	?
48-49	2	0.693	-0.406	48	?	?	?	?
49-50	2	0.693	0.000	49	?	?	?	?
50-51	2	0.693	0.000	50	?	?	?	?
51-52	1	0.000	-0.693	51	?	?	?	?

Assuming cohort N3 to be 1.5 years old we get:

$$L(1.5) = 50*[1 - \exp(-0.77*(1.5+0.05))] = 34.8 \text{ cm}$$

Table 3.4.1.3 has been prepared for the analysis of N3+ and the related Bhattacharya plot is shown in Fig. 3.4.1.4. The selection of points used for the regression for N3 is now even more questionable than the one made for cohort N2. However, the estimated mean value of $\bar{L}(N3) = 33.8$ cm came out reasonably well compared to the value of 34.8 cm calculated above and which we happen to know is close to the true value of 35.3 cm (Table 3.2.1.1).

With three mean lengths we are now in a position to apply the von Bertalanffy plot (cf. Section 3.3.3, Eq. 3.3.3.1), again assuming the arbitrary age of 0.5 years for the first cohort. The input data for the estimation of K and t_o from the von Bertalanffy plot are shown in Table 3.4.1.4, together with the results of the regression analysis. The estimate of $t_o = -0.13$ year is an arbitrary value, since we used arbitrary ages. Nevertheless it puts us in a position to calculate length at other arbitrary ages because the shape of the growth curve is independent of t_o. With the new growth parameters estimated in Table 3.4.1.4 the expected mean length of cohort N4 with arbitrary age 2.0 years becomes:

$$L(2.0) = 50*[1 - \exp(-0.7*(2.0+0.13))] = 38.7 \text{ cm}$$

We now continue with the Bhattacharya method to estimate N4. Table 3.4.1.5 and Fig. 3.4.1.5 show the Bhattacharya analysis for N4+. It is difficult to see a straight line. Selecting the five points corresponding to lengths 37 - 41 cm would give a mean length of 40.0 cm, which is a reasonable value. (Actually, this value is very close to the true value of 40.2 cm (cf. Table 3.2.1.1), but we are not supposed to have that information.)

At this stage one should probably consider the fit of a straight line as being so poor that the analysis should be terminated. When to stop is largely a matter of taste, although some objective criteria for limitations of the Bhattacharya method can be devised as will be discussed in Section 3.5.4. Anyway, we stop here to bring this example to an end.

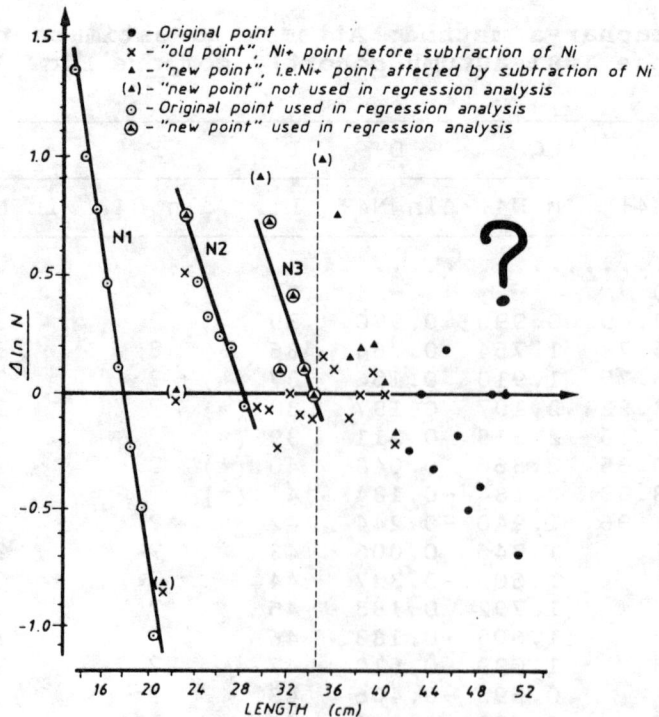

Fig. 3.4.1.5 **Bhattacharya method: plot for estimation of the fourth cohort (compare columns D and E in Table 3.4.1.5). In this case the fit was considered too dubious**

Bias

Input data for the Bhattacharya analysis are often biased due to gear selection and recruitment, i.e. the small fish are under-represented in the frequency samples, either because they escape through the meshes of the gear, or because they have not yet migrated from the nursery grounds to the fishing grounds (cf. Section 2.7). Aspects connected with bias caused by selection will be discussed in Chapter 6, where also a method to adjust length-frequency samples for selection will be presented. In many cases the Bhattacharya analysis should be preceded by an adjustment for selection.

Another source of bias is observed for migratory fish species. Sometimes components are lacking because the cohort was not present in the area where the samples were taken. This aspect will be discussed in Chapter 11.

Computer programs

As you may have noticed, the Bhattacharya exercise takes some time to do by "paper-and-pencil". With the aid of a computer (which may be a microcomputer), however, the method is not hard to work with in practice.

The computer program "BHATTAC" in the LFSA package of microcomputer programs (see Chapter 14) closely follows the set-up explained in the foregoing. With a little experience you can do the exercise of Section 3.4.1 with BHATTAC in a few minutes. The program has a number of additional features: Whenever you have estimated a component BHATTAC displays a graph like Fig. 3.2.2.2 on the screen, allowing you to evaluate the fit to the original data. BHATTAC also checks whether your results are reasonable or not by calculating the "separation index", described in Section 3.5.4. Perhaps the most important feature of BHATTAC, compared to the "paper-and-pencil-method", is that it allows you to do the analysis several times, each time with a different set of input data. You may for example want to try out a range of alternative ways to fit the straight lines in the Bhattacharya plot.

One of the weak points of the "paper-and-pencil" version of the Bhattacharya method is the estimation of the numbers of fish in each cohort, since it is based on the subjective selection of one "clean point", from which the values of $\ln Na+$ are calculated. A more rigorous statistical approach would be to apply all points used for the estimation of the regression line. In fact this more correct procedure is applied in BHATTAC.

When doing the Bhattacharya analysis on the computer you should always, as a matter of routine, try out different length class intervals (cf. Exercise 3.4.1), since it often happens that the structure of the points on the Bhattacharya plot emerges only for an optimal length class interval, which you may find simply by trying out various alternatives. Similar improvements may be obtained by pooling data over longer periods. In most cases you will be working with time series of length-frequencies (to be dealt with in Section 3.4.2), for example in the form of monthly length-frequency samples. You will then have the choice between, say, working with samples representing one month or to pool the data of three months to represent a quarter of the year. Such alternative aggregations of the basic data can easily be made by computer.

The "COMPLEAT ELEFAN" package contains a program "MPA", which also does the Bhattacharya exercise. FiSAT contains the same program.

Pauly and Caddy (1985), have developed a slightly different version of the Bhattacharya method for use with a programmable calculator. In their version the lines are determined by three successive points only, which are chosen so that they have the highest negative correlation coefficient. Their version is an attempt to turn the Bhattacharya method into an objective method, i.e. a method producing results independent of the person carrying out the analysis.

(See **Exercise(s)** in Part 2).

3.4.2 Modal progression analysis

Example 10 used in Section 3.4.1 was based on one length-frequency sample collected during one survey. It was demonstrated that a somewhat rough estimate of the growth equation could be obtained from such a data set. L_∞ and K could be estimated, whereas t_o could only be determined relative to the arbitrary ages chosen for the cohorts.

Example 11: Modal progression analysis, based on the data of Example 4

Now suppose we had the type of data described in Example 4 in Section 3.2.1, i.e. length-frequency samples from each month or quarter during one or several years. The example illustrated in Fig. 3.2.1.1 consists of 12 length-frequency samples collected during surveys carried out in the months January, April, July and October during three years (1982 to 1984). Such a time series puts us in a much better position to estimate growth parameters than in the case of a single sample (October) as used in Example 10 to illustrate the Bhattacharya analysis.

Suppose that each of the twelve samples of the time series is given the same treatment as the single October 1983 sample. The results of the twelve Bhattacharya analyses could then be those given in Table 3.4.2.1. In each of the first nine samples three components have been found (as was the case in Section 3.4.1), whereas the last three samples were more difficult to analyse so that only two components could be identified. Please note that the number of components (cohorts) that could be identified is much lower than the actual number present, which are represented by dots in Fig. 3.2.1.1.

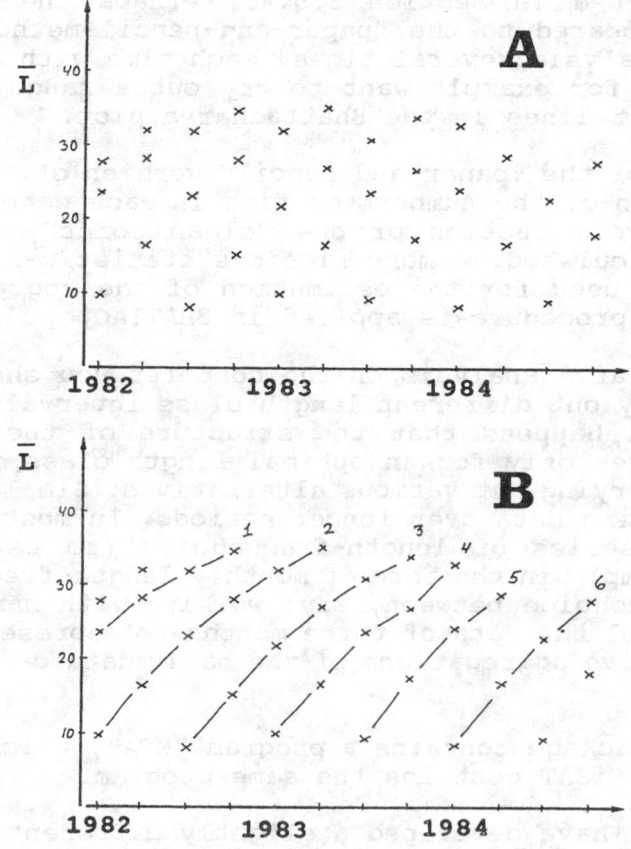

Fig. 3.4.2.1 Modal progression based on the results of the Bhattacharya analyses
  A: Mean lengths of components from Bhattacharya plots
  B: Mean lengths connected to represent growth curves of assumed cohorts

Table 3.4.2.1 Results of Bhattacharya analyses of the time series of length-frequency samples illustrated in Fig. 3.2.1.1

date of sample	third component	second component	first component
JAN 82	27.9	23.5	9.8
APR 82	32.0	28.1	16.5
JUL 82	31.8	23.1	8.0
OCT 82	34.6	28.0	15.3
JAN 83	32.0	21.8	10.0
APR 83	35.1	27.0	16.5
JUL 83	30.9	23.5	9.2
OCT 83 *)	33.8	27.8	17.4
JAN 84	32.9	24.0	8.3
APR 84	-	28.2	16.8
JUL 84	-	22.9	9.0
OCT 84	-	27.9	18.0

*) from Table 3.4.1.4.

Table 3.4.2.2 The mean lengths from Table 3.4.2.1 rearranged into cohorts (see Fig. 3.2.1.1)

	COHORTS, $\bar{L}(t)$ in cm					
	1	2	3	4	5	6
date of sample	spring 1981	autumn 1981	spring 1982	autumn 1982	spring 1983	autumn 1983
JAN 82	23.5	9.8	-	-	-	-
APR 82	28.1	16.5	-	-	-	-
JUL 82	31.8	23.1	8.0	-	-	-
OCT 82	34.6	28.0	15.3	-	-	-
JAN 83	-	32.0	21.8	10.0	-	-
APR 83	-	35.1	27.0	16.5	-	-
JUL 83	-	-	30.9	23.5	9.2	-
OCT 83	-	-	33.8	27.8	17.4	-
JAN 84	-	-	-	32.9	24.0	8.3
APR 84	-	-	-	-	28.2	16.8
JUL 84	-	-	-	-	-	22.9
OCT 84	-	-	-	-	-	27.9

We may assume that the various cohorts remain in the sea for some time and thus that they are sampled at different stages of growth from the time of recruitment to the fishing (or sampling) area to their extinction. We may also assume that a mean length of a cohort, as determined for example by the Bhattacharya method, will correspond to a somewhat larger mean length in a sample taken a few months later and so forth. By plotting those mean lengths from a series of samples against a time axis and connecting them a growth curve can be obtained.

In Fig. 3.4.2.1A the mean lengths of the components have been plotted against the sample date. In Fig. 3.4.2.1B those mean lengths which we believe to correspond to the same cohorts, have been connected. Excluding the two first and the two last points we have thus identified six cohorts. The connection of points to produce cohorts is a subjective process although in the present case the choice appeared quite easy to make. In practice it may not always be so simple.

It appears from Fig. 3.4.2.1B that there are two cohorts per year, for instance cohorts No. 3 and No. 4 which recruited in 1982. Assuming seasons of the northern hemisphere, No. 3 will be called the 1982 spring cohort and No. 4 the autumn cohort. The various growth curves drawn for each cohort enable us to interpret and rearrange the results of the twelve Bhattacharya analyses (Table 3.4.2.1) by cohorts as shown in Table 3.4.2.2.

The estimation of K and L_∞

The data in Table 3.4.2.2 are of the type that make it possible to apply the Gulland and Holt plot (cf. Section 3.3.1) by calculating:

$$\frac{\Delta L}{\Delta t} = \frac{L(t+\Delta t)-L(t)}{(t+\Delta t)-t} \quad \text{and}$$

$$\bar{L}(t,t+\Delta t) = \frac{L(t)+L(t+\Delta t)}{2}$$

The time difference $\Delta t = 0.25$ years, remains constant in this case, so it would also be possible to apply Chapman's method (Eq. 3.3.2.2).

The values of $\Delta L/\Delta t$ and $\bar{L}(t,t+\Delta t)$ are shown in Table 3.4.2.3. To illustrate the calculations we consider cohort No. 1, recruited in the spring of 1981 (see Fig. 3.4.2.1). For the two first samples we get:

$$\frac{\Delta L}{\Delta t} = \frac{L(\text{Apr } 82)-L(\text{Jan } 82)}{t(\text{Apr } 82)-t(\text{Jan } 82)} = \frac{28.1-23.5}{0.25} = 18.4 \quad \text{and}$$

$$\bar{L}(\text{Jan } 82, \text{Apr } 82) = \frac{L(\text{Jan } 82)+L(\text{Apr } 82)}{2} = (23.5+28.1)/2 = 25.8$$

It would be possible to make separate Gulland and Holt plots for each of the six cohorts, each with two to four points only. However, under the assumption that the growth parameters remain constant over the entire sampling period, all the 23 data pairs given in Table 3.4.2.3 may be combined into one single Gulland and Holt plot.

The regression of all 23 $\Delta L/\Delta t$ values on $\bar{L}(t,t+\Delta t)$ values gives the following results:

a (intercept) = 41.84 and b (slope) = -0.8740

from which we get

L_∞ = -a/b = 47.9 say 48 cm and

K = -b = 0.87 per year with a 95% confidence interval [0.72, 1.02] (see Table 3.4.2.3)

The Gulland and Holt plot is shown in Fig. 3.4.2.2. Estimates of L_∞ and K have thus been obtained based on the entire time series.

Table 3.4.2.3 Input data and regression analysis for Gulland and Holt plot derived from Table 3.4.2.2. Note $\Delta t = 0.25$ year

cohort sample date	1 spring 1981		2 autumn 1981		3 spring 1982		4 autumn 1982		5 spring 1983		6 autumn 1983	
	$\bar{L}$	$\frac{\Delta L}{\Delta t}$	$\bar{L}$	$\frac{\Delta L}{\Delta t}$	$\bar{L}$	$\frac{\Delta L}{\Delta t}$	$\bar{L}$	$\frac{\Delta L}{\Delta t}$	$\bar{L}$	$\frac{\Delta L}{\Delta t}$	$\bar{L}$	$\frac{\Delta L}{\Delta t}$
JAN 82	25.8	18.4	13.2	26.8	-	-	-	-	-	-	-	-
APR 82	30.0	14.8	19.8	26.4	-	-	-	-	-	-	-	-
JUL 82	33.2	11.2	25.6	19.6	11.7	29.2	-	-	-	-	-	-
OCT 82	-	-	30.0	16.0	18.6	26.0	-	-	-	-	-	-
JAN 83	-	-	33.6	12.4	24.4	20.8	13.3	26.0	-	-	-	-
APR 83	-	-	-	-	29.0	15.6	20.0	28.0	-	-	-	-
JUL 83	-	-	-	-	32.4	11.6	25.7	17.2	13.3	32.8	-	-
OCT 83	-	-	-	-	-	-	30.4	20.4	20.7	26.4	-	-
JAN 84	-	-	-	-	-	-	-	-	26.1	16.8	12.6	34.0
APR 84	-	-	-	-	-	-	-	-	-	-	20.0	24.4
JUL 84	-	-	-	-	-	-	-	-	-	-	25.4	20.0
OCT 84												

$a = 41.84$ $\quad\quad b = -0.8740$

$sb^2 = \frac{1}{n-2}*((sy/sx)^2 - b^2) =$

$\frac{1}{23-2}*((6.62333/7.08467)^2 - (-0.8740)^2) = 0.005242$

$K = -b = 0.87$ per year, $\quad L_\infty = -a/b = 47.9$ cm

$sb = 0.072$, $\quad t_{21} = 2.09$ (see Table 2.3.1)

95% confidence interval of b (= -K): [-1.02 , -0.72] and of K: [0.72, 1.02]

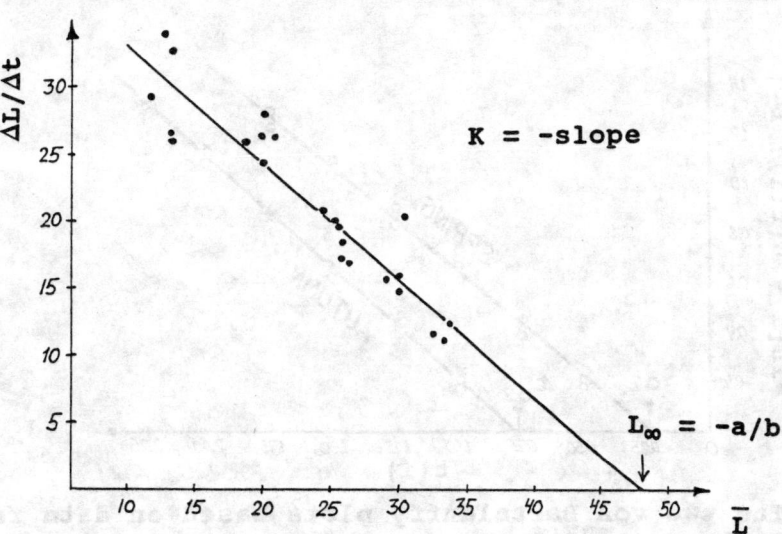

Fig. 3.4.2.2 Gulland and Holt plot based on data in Table 3.4.2.3

The estimation of t_o

The next step is to estimate the arbitrary initial condition parameters t_{o1} for the spring cohorts and t_{o2} for the autumn cohorts using the von Bertalanffy plot. We allot an arbitrary age of one year to the spring cohort of 1981 in January 1982, 1.25 years in April 1982, etc. The spring cohort of 1982 No. 3 is similarly allotted an age of one year in January 1983, etc. the procedure is the same for the autumn cohorts.

Table 3.4.2.4 contains the arbitrary ages, $t(i)$ of each cohort together with the dependent variable of the von Bertalanffy plot:

$$y = -\ln(1 - \overline{L}(t)/L_\infty)$$

Values of $\overline{L}(t)$ are taken from Table 3.4.2.2. There are two regression analyses to be carried out:

Spring cohorts: $y = -K*t_{o1} + K*t(i)$, $i = 1,3,5$
Autumn cohorts: $y = -K*t_{o2} + K*t(i)$, $i = 2,4,6$

where $t(i)$, the independent variable, is the arbitrary age of cohort no. i, as defined in Table 3.4.2.4. In this case six cohorts are considered simultaneously, and we believe that there are three spring cohorts and three autumn cohorts. As shown in Table 3.4.2.4 the two regression analyses gave the results:

	a (intercept)	b (slope)	t_{o1} = -a/b year	K (per year)
spring cohorts:	-0.2055	0.8433	0.24	0.84
autumn cohorts:	-0.7305	0.9169	0.80	0.92

As expected, the difference between t_{o1} and t_{o2} became close to half a year, as explained in Section 3.2.1 (see Table 3.2.1.2) for this example. The mean of the two K-values is 0.88 (close to the value of 0.87 estimated from the Gulland and Holt plot). A statistical test would show that the two estimates are not significantly different and we would therefore use the common value K = 0.88 per year. Thus the two equations:

Spring cohorts: $L(t) = 48*[1 - \exp(-0.88*(t-0.24))]$
Autumn cohorts: $L(t) = 48*[1 - \exp(-0.88*(t-0.80))]$

can be used to calculate the length of spring cohorts and autumn cohorts for different arbitrary ages. We may stop the analysis at this stage, or we may continue trying to estimate the birthday of the cohorts.

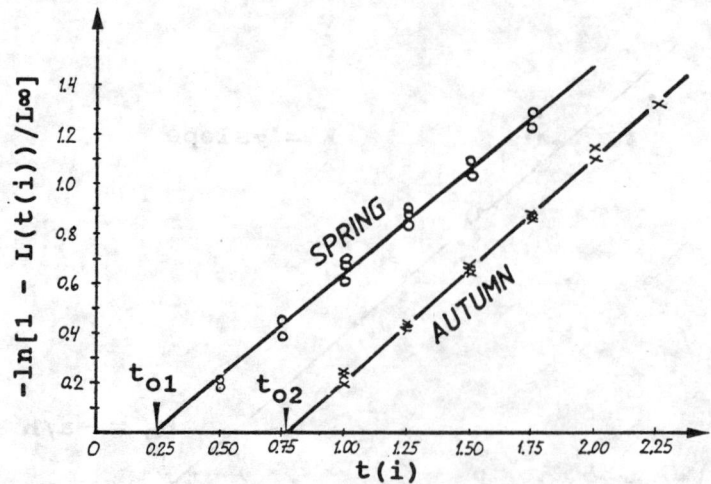

Fig. 3.4.2.3   The two von Bertalanffy plots based on data from Table 3.4.2.4

Table 3.4.2.4 Input data and regression analysis for von Bertalanffy plot. Mean lengths of the components, L(t) derived from Table 3.4.2.2, $L_\infty = 48$ cm

A: spring cohorts

date of sample	no. 1 spring 1981 t(1)	y*)	no. 3 spring 1982 t(3)	y*)	no. 5 spring 1983 t(5)	y*)	time of sampling T = (x)
JAN 82	1.00	0.673	–		–		1982.00
APR 82	1.25	0.880	–		–		1982.25
JUL 82	1.50	1.086	0.50	0.182	–		1982.50
OCT 82	1.75	1.276	0.75	0.384	–		1982.75
JAN 83	–		1.00	0.605	–		1983.00
APR 83	–		1.25	0.827	–		1983.25
JUL 83	–		1.50	1.032	0.50	0.213	1983.50
OCT 83	–		1.75	1.218	0.75	0.450	1983.75
JAN 84	–		–		1.00	0.693	1984.00
APR 84	–		–		1.25	0.886	1984.25
JUL 84	–		–		–		1984.50
OCT 84	–		–		–		1984.75

spring cohorts: n = 14
a = -0.2055, b = 0.8433, so K = 0.84 per year
t_{o1} = -a/b = 0.24 year

$$sb^2 = \frac{1}{n-2} \ast [(sy/sx)^2 - b^2] = \frac{1}{14-2} \ast [(0.35016/0.41313)^2 - 0.84332^2] = 0.0006005$$

sb = 0.0245, t_{12} = 2.18 (see Table 2.3.1)
95% confidence interval of b (= K): [0.79 , 0.90]

B: autumn cohorts

date of sample	no. 2 autumn 1981 t(2)	y*)	no. 4 autumn 1982 t(4)	y*)	no. 6 autumn 1983 t(6)	y*)	time of sampling T = (x)
JAN 82	1.00	0.228	–		–		1982.00
APR 82	1.25	0.241	–		–		1982.25
JUL 82	1.50	0.656	–		–		1982.50
OCT 82	1.75	0.875	–		–		1982.75
JAN 83	2.00	1.099	1.00	0.234	–		1983.00
APR 83	2.25	1.314	1.25	0.421	–		1983.25
JUL 83	–		1.50	0.673	–		1983.50
OCT 83	–		1.75	0.866	–		1983.75
JAN 84	–		2.00	1.157	1.00	0.190	1984.00
APR 84	–		–		1.25	0.431	1984.25
JUL 84	–		–		1.50	0.648	1984.50
OCT 84	–		–		1.75	0.870	1984.75

autumn cohorts: n = 15
a = -0.7305, b = 0.9169, so K = 0.92 per year
t_{o2} = -a/b = 0.80 year

$$sb^2 = \frac{1}{n-2} \ast [(sy/sx)^2 - b^2] = \frac{1}{15-2} \ast [(0.36593/0.39491)^2 - 0.91692^2] = 0.0001375$$

sb = 0.037, t_{13} = 2.16 (see Table 2.3.1)
95% confidence interval of b (= K): [0.84 , 1.00]

*) $y = -\ln(1 - \overline{L}(t)/L_\infty)$

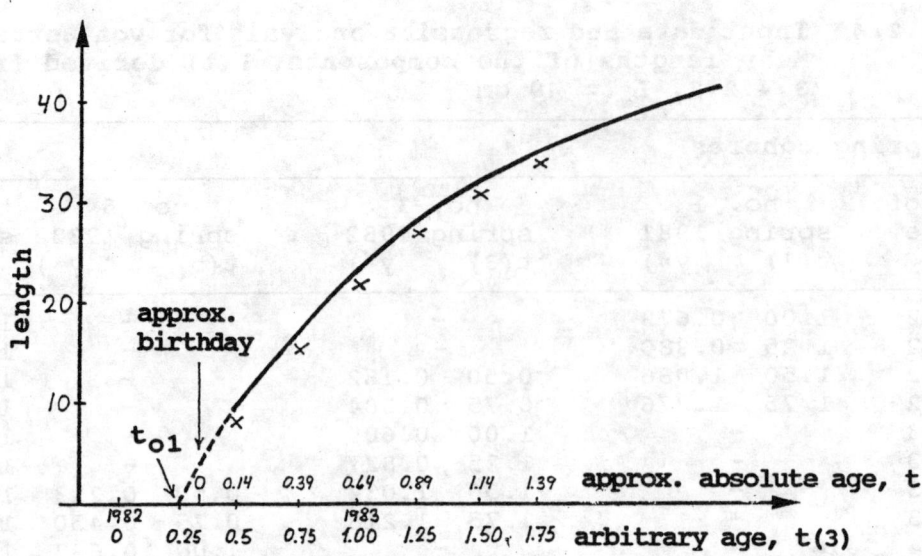

Fig. 3.4.2.4   Illustration of how the approximate birthday is estimated

Estimation of the birthday

To estimate the birthday, the idea is to extrapolate the growth curve beyond the first data point and see where it intersects with the time axis as illustrated in Fig. 3.4.2.4. This figure shows cohort no. 3 as an example. The curve cuts the time axis at the point 1982.24. On the arbitrary age axis the intersection point is t_{o1} = 0.24. The point 1982.24 (29th of March) must be somewhere in the neighbourhood of the birthday. Because the von Bertalanffy growth curve does not conform to the early life stages of fish (cf. Section 3.1) this is an approximation. An alternative way of finding the approximate birthday is to use gonadal maturity stage data.

The use of data on gonadal maturity

Another method of estimating the birth day is to estimate the spawning season from maturity stages of the adults. Fig. 3.4.2.5 shows an example of maturity stage data (from Wyatt, 1983). In this case the percentages of the three main stages of gonadal maturity are presented.

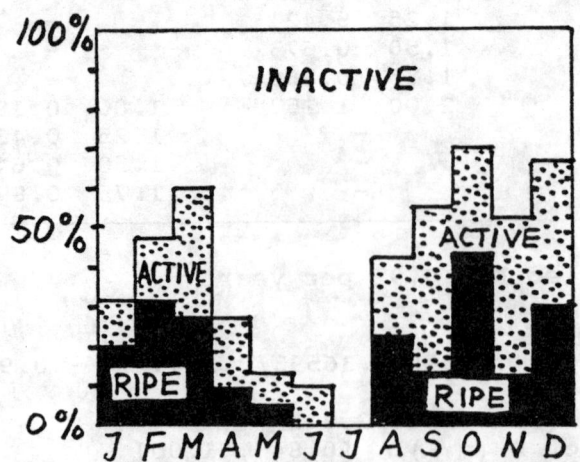

Fig. 3.4.2.5   Maturation stages observed for the squirrel fish
(<u>Holocentrus</u> <u>rufus</u>) from Wyatt (1983).
Based on samples of 1331 fish

From maturation stage data, e.g., a graph of the percentage of ripe fish, we can define one (or two) mean spawning day(s), in the same way as the mean recruitment day was defined in Chapter 1, if the graph is unimodal (or bimodal). The histogram for the percentage of ripe fish in Fig. 3.4.2.6 could be interpreted as two spawning seasons with peaks in February and October. The mean spawning day may then be used as an estimate of the birth day (perhaps corrected for a time lag). However, the results of such analyses should be treated with a certain reservation as fluctuations in spawning are not the only factor which determine the fluctuations of recruitment. The success of a larva to feed and grow into a recruit and at the same time to avoid being eaten by predators is a complex process affected by a variety of environmental (biotic and abiotic) factors. The survival rate could, for instance, be almost nil for one spawning season and high for another. For a discussion of these matters see, for example, Bakun et al. (1982).

The application of modal progression analysis

The estimates obtained by following the progression of the modes (= cohorts) in the length-frequencies is considered superior to the method based on one single sample (Section 3.4.1). Further, there are cases where the single sample approach is not applicable at all. This is the case for short-lived species where there is only one (or only two) cohorts in a length-frequency sample. Such an example is shown in Fig. 3.4.2.6. It deals with commercial catches of the shrimp Penaeus semisulcatus in Kuwait waters (from Mohamed et al., 1979). This species has a life span of one to two years and there are two cohorts per year. Most of the samples contain only one mode, so that the single sample approach is not applicable. However, to follow the progression of the modes appears a simple thing in this case. Modal progression analysis is especially useful for such short-lived species.

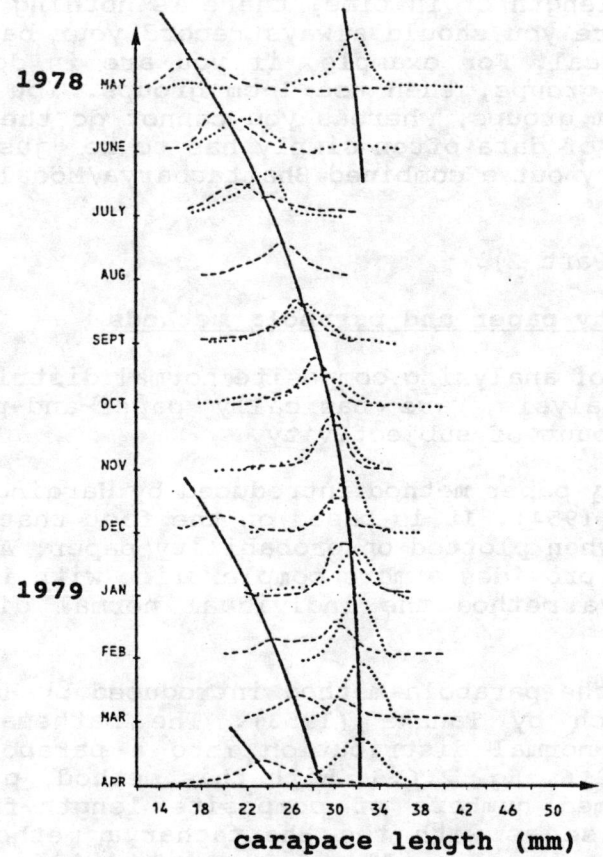

Fig. 3.4.2.6 Example of modal progression analysis. Size distributions of catches of Penaeus semisulcatus in the artisanal (----) and industrial (....) catches in Kuwait waters. (From Mohamed et al., 1979)

Computer programs

The program "MODALPR" in the LFSA package can execute the modal progression analysis as described above. The LFSA package also allows you to continue from the Bhattacharya analysis (program "BHATTAC") with a least squares estimation of the growth parameters (program "VONBER", cf. Section 3.3.4) instead of the Gulland and Holt plot. The "COMPLEAT ELEFAN" package contains a program "MPA" to do the modal progression analysis. A similar program has been incorporated in FiSAT. There are several other computer programs available which attempt to solve the problem dealt with in this section, some of which will be discussed in Section 3.5.

Data massage

Running the Bhattacharya analysis and the modal progression analysis on a computer, one should always as a routine try out different aggregations of data, i.e. the so-called "data-massage" or "data-squeezing". Table 3.4.2.5 illustrates the process of data-massage. Part A contains the original data, i.e. a time series of fourteen monthly length-frequency samples grouped into sixteen 1-cm groups. From part A to part B the data have been squeezed into eight 2-cm length groups. From part B to part C data have been further squeezed into five 3-monthly groups. Sometimes a data-massage makes the structure of the data more apparent. (With "structure" is meant the straight lines in the Bhattacharya plots and the modal progression.)

If the data are grouped in such small classes that the "random noise" within each cell of the table hides the structure of the data one should massage the data. We may also observe the opposite problem, namely that the data are grouped in class intervals which are too large so that the structure becomes concealed behind the grouping. If your basic data are grouped in such large class intervals (in length or in time) there is nothing you can do to solve the problem. Therefore you should always record your basic data in as fine a grouping as practical. For example, if you are in doubt whether to use 1-cm groups or 2-cm groups, then use 1-cm groups. You can easily convert 1-cm groups into 2-cm groups, whereas you cannot do the opposite transformation. The grouping of data often simply has to be "just right" before you can successfully carry out a combined Bhattacharya/Modal Progression Analysis.

(See **Exercise(s)** in Part 2).

3.4.3 The probability paper and parabola methods

There are other ways of analysing composite normal distributions which, like the Bhattacharya analysis, are basically paper-and-pencil methods and contain a certain amount of subjectivity.

One is the probability paper method introduced by Harding (1949) and further developed by Cassie (1954). It is based on the fact that a normal distribution becomes linear when plotted on probability paper. A mixture of several normal distributions provides a more complex line with inflexion points. As with the Bhattacharya method the individual normal distributions can be removed one by one.

Another approach is the parabola method introduced by Hald (1952) and used in fisheries research by Tanaka (1953). The mathematical base is the transformation of a normal distribution into a parabola by taking logarithms, see Section 2.6, Eq. 2.6.3. With this method, parabolas are fitted to the log-transformed numbers of composite length-frequency data. The procedure is otherwise as with the Bhattacharya method which is a more sophisticated version based on the fact that differences between equidistanced points on a parabola form a straight line.

The Bhattacharya method seems to leave less to subjective decisions on the researcher's part than the other methods do. However, persons skilled in the application of either the probability paper method or the parabola method also seem to reach plausible results.

Table 3.4.2.5 Illustration of the process of "data-massage".
             For further explanation, see text

A: BASIC DATA: 1-cm length groups by month

Length class	MAR	APR	MAY	JUN	JUL	AUG	SEP	OCT	NOV	DEC	JAN	FEB	MAR	APR
4- 5														
5- 6														
6- 7														
7- 8														
8- 9				18	24	12								
9-10				21	51	16								
10-11														
11-12														
12-13														
13-14														
14-15														
15-16														
16-17														
17-18														
18-19														
19-20														

B: MASSAGED DATA: 2-cm length groups by month

Length class	MAR	APR	MAY	JUN	JUL	AUG	SEP	OCT	NOV	DEC	JAN	FEB	MAR	APR
4- 6														
6- 8														
8-10				39	75	28								
10-12														
12-14														
14-16														
16-18														
18-20														

C: MASSAGED DATA: 2-cm length groups by 3 months

Length class	MAR-MAY 1981	JUN-AUG	SEP-NOV	DEC-FEB	MAR-APR 1982
4- 6					
6- 8					
8-10		142			
10-12					
12-14					
14-16					
16-18					
18-20					

3.5 FITTING GROWTH CURVES BY MEANS OF COMPUTER PROGRAMS

The methods presented in Section 3.4, the "paper-and-pencil" methods and their computer-based counterparts basically treat the data sample by sample. Often the tracing of the growth curves becomes easier when the entire time series is considered. Some samples may be easy to resolve into cohort components and to interpret in terms of growth in an unambiguous way. By using the findings from the "easy" samples we may also be able to give unambiguous interpretations of samples we would otherwise not be able to interpret.

Figs. 3.5.0.1 and 3.5.0.2 illustrate this feature. The January sample in Fig. 3.5.0.1 seems easy to resolve into two components as shown in Fig. 3.5.0.2, whereas the September sample shows no structure whatsoever. The May sample appears more problematic than the January sample, but it is still possible to interpret. However, together the January and the May samples show a clear picture from which a growth curve can be estimated. By extrapolating the growth curve to the September sample we are now also in a position to split that into cohorts.

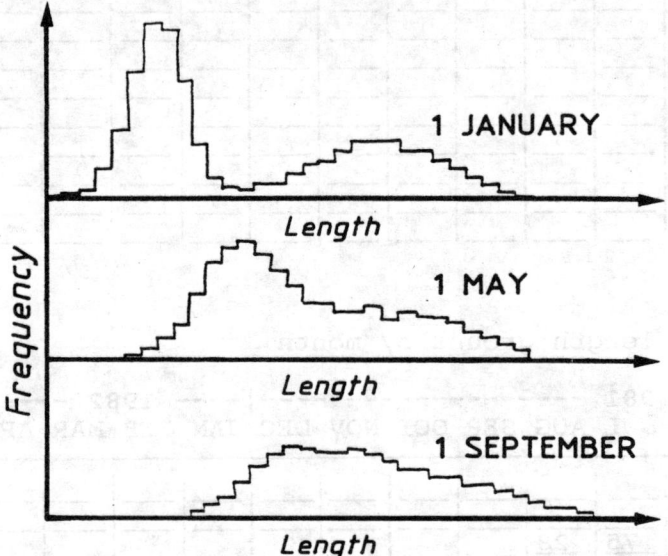

Fig. 3.5.0.1 Examples of an "easy" sample (January) and a "difficult" sample (September)

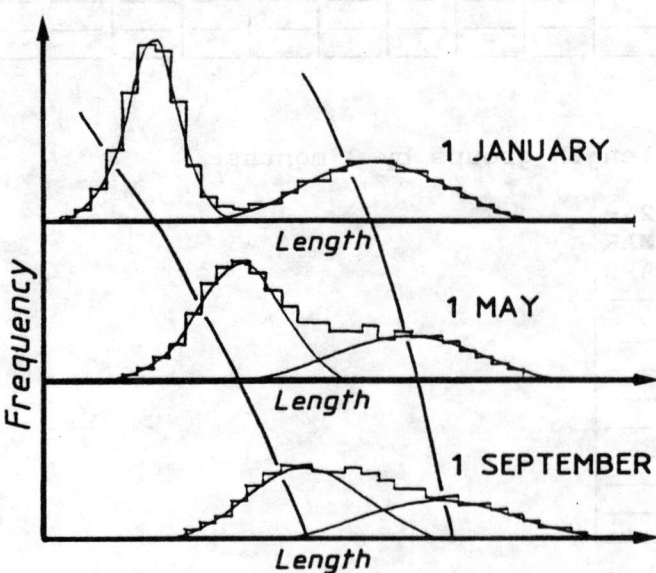

Fig. 3.5.0.2 Hypothetical example of how an "easy" sample (January) is used to treat a "difficult" sample (September)

This approach may be applied when using the "paper-and-pencil" method, especially when aided by a computer. It is however possible to leave more work to the computer and to let it do the analysis using a more sophisticated technique, such as a least squares estimation technique, (cf. Section 3.3.4).

The computer-based methods to be dealt with here require so many computations that it is almost impossible to do them by paper-and-pencil. We present two alternative approaches:

1. The "ELEFAN I" method (Electronic LEngth-Frequency ANalysis)
2. The "maximum-likelihood" method.

The first was introduced by Pauly and David (1981). The second may be considered a computerized version of the Bhattacharya method. It is based on the traditional theory on statistical analysis of frequency samples - a method which you may consider a generalized version of linear regression analysis. The basic philosophies behind the two methods are similar.

A detailed discussion of computer-based methods is considered outside the scope of the manual. The main purpose is to present some basic features of the methods which hopefully encourage the reader to go into further studies in this field.

3.5.1 ELEFAN I

The "ELEFAN I" program deals with estimation of growth parameters using length-frequency analysis (Pauly and David, 1981; and Pauly, 1987). The most recent description of the entire package will be found in Pauly (1987).

Example 12: The application of ELEFAN I to the coral trout data

To illustrate ELEFAN I we use the data on coral trout shown in Fig. 3.4.0.2. ELEFAN I consists of two major stages:

Stage 1: Restructuring of length-frequencies

Stage 2: Fitting of a growth curve

Stage 1, the restructuring process is illustrated in Fig. 3.5.1.1 where part "a" shows the original data as presented by Goeden (1978) in 0.5 cm length groups. To smooth out small irregularities the data have been rearranged in 2 cm length groups as shown in part "b". The curve in part "b" is the "moving average frequency" over 5 length groups. The method to obtain a moving average is illustrated for the length interval 26-28 cm:

interval	frequency	
18-20	0 *	
20-22	0 *	
22-24	2	
24-26	11	
26-28	15	> moving average = $\frac{2+11+15+6+10}{5}$ = 8.8
28-30	6	
30-32	10	
......		

The values, for the first length groups 22-24 and 24-26 cm are calculated by adding two zeroes and one zero respectively as indicated by "*". (A similar procedure is applied to the last length groups.) The curve that results from this procedure is used to emphasize peaks (shaded bars above moving average) and intervening troughs. In part "c" the original frequencies of part "b" have been divided by the moving average and 1 has been subtracted. Consider again as an example length group 26-28 cm. Here we get:

15/8.8 - 1 = 0.7 "points"

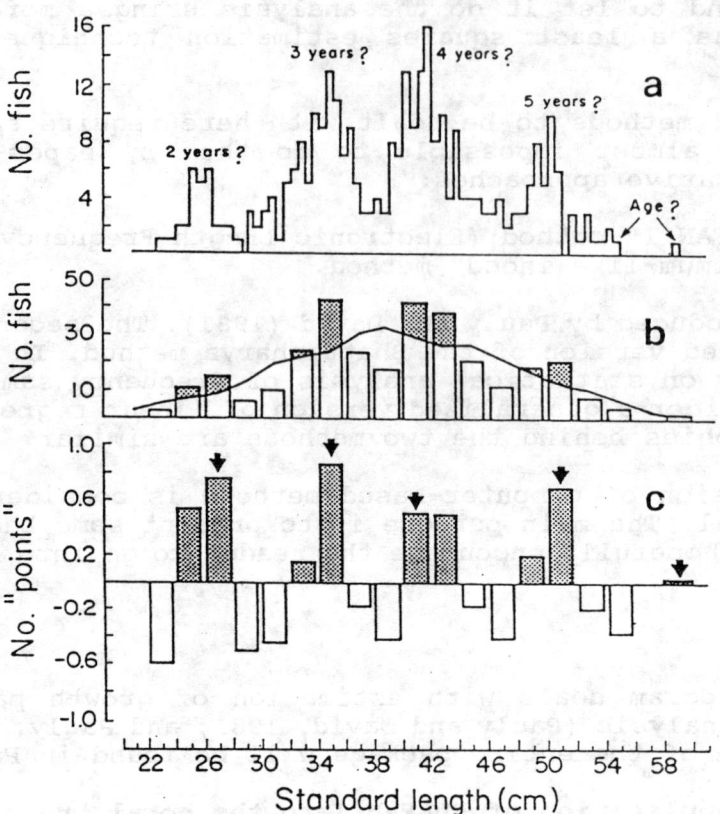

Fig. 3.5.1.1 Example of the ELEFAN I restructuring of a length-frequency sample (from Pauly & David, 1981). Data from Goeden, (1978), on the coral trout (Plectropomus leopardus)

Actually, some additional minor adjustments have also been made but we shall not go into that. Using the restructuring process the peaks and the troughs became well-structured and easy to identify by the "points" allotted. Note that clear peaks have been allotted a similar number of points irrespective of the number of fish they represent.

Stage 2, the fitting of a growth curve is illustrated in Figs. 3.5.1.2 and 3.5.1.3.

In the present example for coral trout only one sample was used. To do the ELEFAN I type of fitting growth curves we should preferably have a time series of samples. Basically, ELEFAN I is a modal progression analysis. However, if a time series is not available we can circumvent the problem by assuming one, simply by repeating the sample for a suitable range of years, the assumption being that all cohorts follow the same growth curve. Thus, ELEFAN I can be applied to both the single sample case and the time series case.

If the constructed time series over the ten years shown in Fig.3.5.1.2 had been a real time series we would have got slightly different frequencies each year. Fig. 3.5.1.3 shows eight repetitions of the restructured sample arranged similarly to Fig. 3.5.1.2. It is difficult to fit a curve to the original frequencies in Fig. 3.5.1.2 and it is not possible to give an objective criterion whether one curve fits better than another if one uses an eye fit only. The restructured samples in Fig. 3.5.1.3 however, are easier to fit because peaks and troughs have been exaggerated.

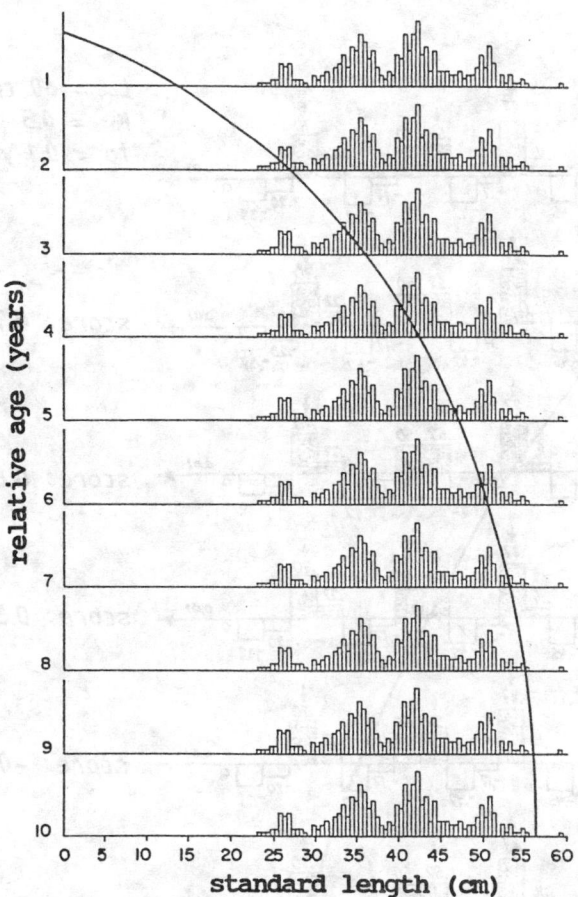

Fig. 3.5.1.2 The length-frequency sample of Fig. 3.5.1.1a repeated over 10 years for simulation of time series of samples (compare Fig. 3.5.1.3)

With the restructured data (the "points" shown in Fig. 3.5.1.1c) it has become possible to define an objective measure for goodness of fit, for which Pauly and David (1981) suggested the ratio "ESP/ASP", where "ESP" stands for "Explained Sum of Peaks" and "ASP" for "Available Sum of Peaks".

To understand the concept of "ESP" consider Fig. 3.5.1.3. The most convincing fit of a growth curve is one which hits all the peaks indicated by arrows. However, there may not exist such a von Bertalanffy growth curve, and therefore a "score" concept has been introduced to measure how close a curve can come to the best fit. Whenever a curve hits a bar at the axis, either positive or negative it scores "points" (cf. Fig. 3.5.1.1). The total score of a growth curve is the sum of the points scored from each sample as shown in Fig. 3.5.1.3.

"ASP" (available sum of peaks) is the maximum score a curve can reach, i.e. the sum of the positive peaks indicated with arrows. Such an arrow occurs whenever there is a sequence of positive bars. (In this connection a "sequence" may be a single bar.) The ratio ESP/ASP thus becomes a measure for how close a curve is to the best possible fit.

The computational procedure described so far may be carried out by paper and pencil for a single growth curve within a reasonable time. But after that it is no longer possible (in practice) to follow ELEFAN I by paper and pencil. One of the main features of ELEFAN I is that many (say, thousands) of different growth curves are tested in the way described in Fig. 3.5.1.3. Among the thousands of possible growth curves the one that produces the highest value of ESP/ASP is selected.

(See **Exercise(s)** in Part 2).

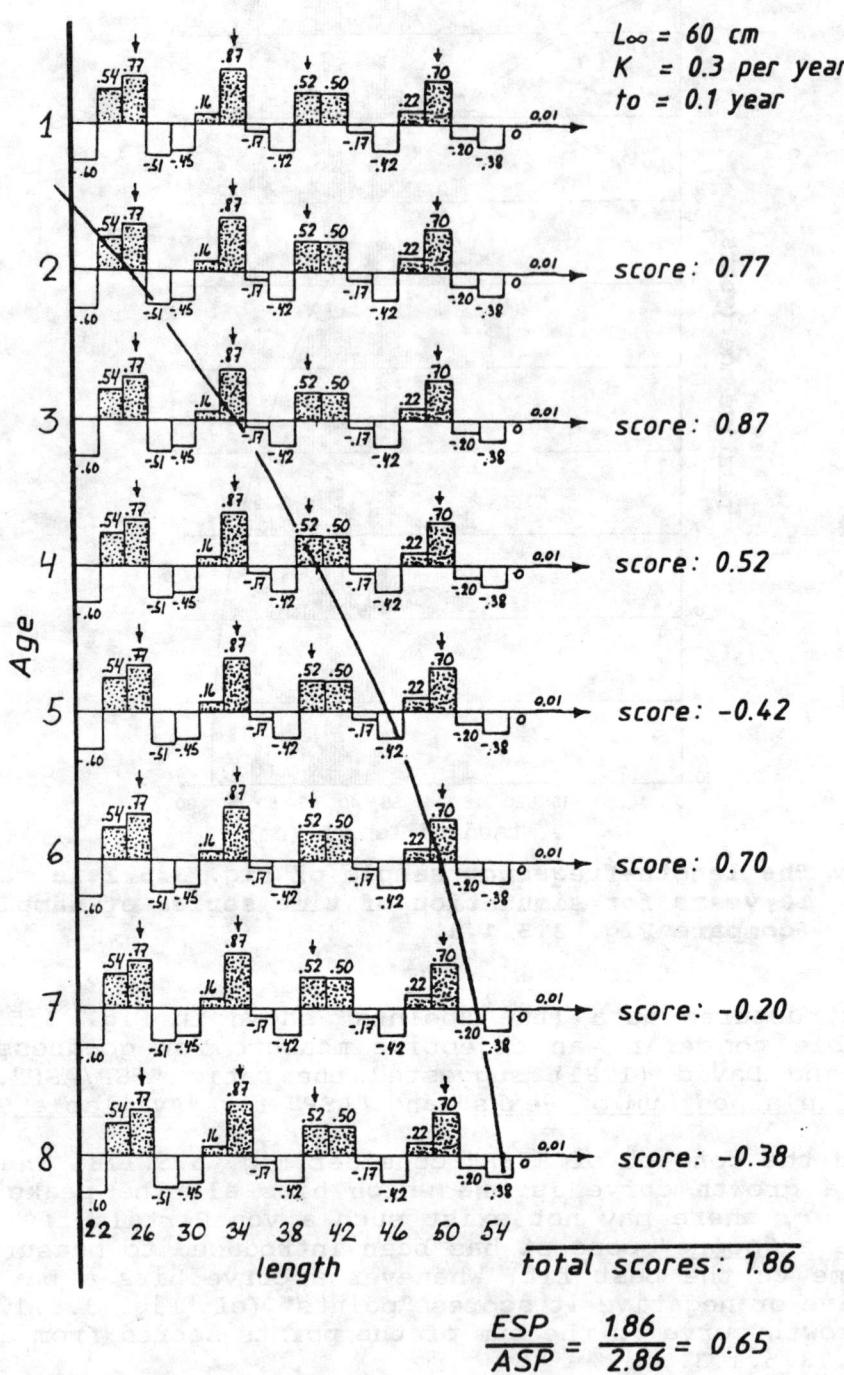

Fig. 3.5.1.3 The restructured length-frequency sample of Fig. 3.5.1.1c repeated over eight years to simulate a time series of samples (compare Fig. 3.5.1.2). A single growth curve determined by the parameters $L_\infty = 60$ cm, $K = 0.3$ per year is tested for goodness of fit (ESP/ASP)

3.5.2 The seasonalized von Bertalanffy growth equation

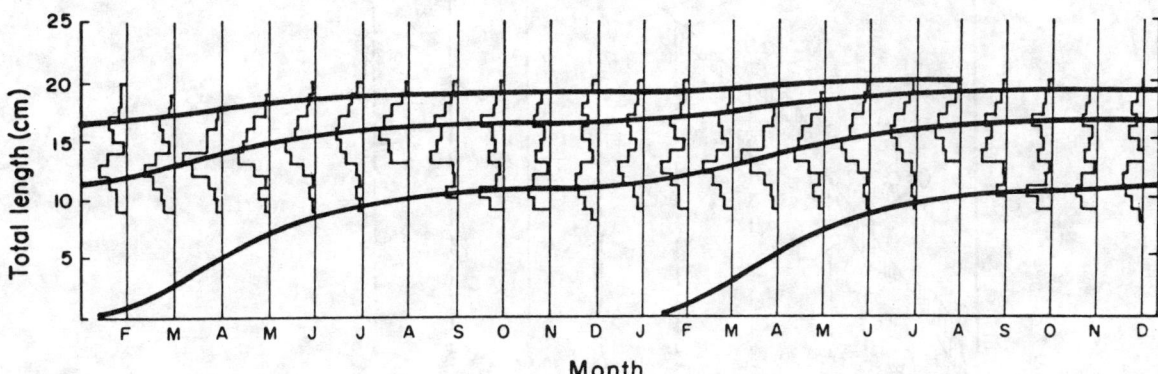

Fig. 3.5.2.1 Example of a seasonally oscillating growth curve estimated by ELEFAN I (from Pauly, 1981). Data from Rodriguez (1977) on female shrimp (<u>Penaeus kerathurus</u>) off Cadiz, Spain. Note that data were available for one year, and these have been repeated to simulate two years of sampling. Estimated parameters are: L_∞ = 21.0 cm (total length), K = 0.8 per year, C = 0.9, tw = 0.8 (winter point), ESP/ASP = 0.46

Fig. 3.5.2.1 shows an application of ELEFAN I to a penaeid shrimp. This growth curve estimated by ELEFAN I is clearly not a von Bertalanffy growth curve because $\Delta L/\Delta t$ does not decrease linearly with age (cf. Section 3.1). The explanation is that ELEFAN I works with the "seasonalized von Bertalanffy growth equation" (Pitcher and Macdonald, 1973; Cloern and Nichols, 1978 and Pauly and Gaschütz, 1979):

$$L(t) = L_\infty * [1 - \exp\{-K*(t-t_o) - (CK/2\pi)*\sin(2\pi*(t-ts))\}] \qquad (3.5.2.1)$$

This is the usual von Bertalanffy equation (Eq. 3.1.0.1) with an extra term:

$$(CK/2\pi)*\sin(2\pi*(t-ts)) \qquad \text{(where } \pi = 3.14159..\text{)}$$

This term produces seasonal oscillations of the growth rate, actually by changing t_o during the year. The parameter "ts" is called the "<u>summer point</u>" and takes values between 0 and 1. At the time of the year when the fraction ts of the year has elapsed the growth rate is highest. At time tw = ts+0.5, the "<u>winter point</u>", the growth rate is lowest. The parameter C, the "<u>amplitude</u>", also usually takes values between 0 and 1. If C = 0 Eq. 3.5.2.1 reduces to the ordinary von Bertalanffy equation, that is C = 0 implies that there is no seasonality in the growth rate. The higher the value of C the more pronounced are the seasonal oscillations. If C = 1 the growth rate becomes zero at the winter point. Fig. 3.5.2.2 shows a seasonalized growth curve with C = 1 together with an ordinary von Bertalanffy curve (C = 0). All other seasonalized curves with different C's (but with other parameters kept constant) will be in the shaded area.

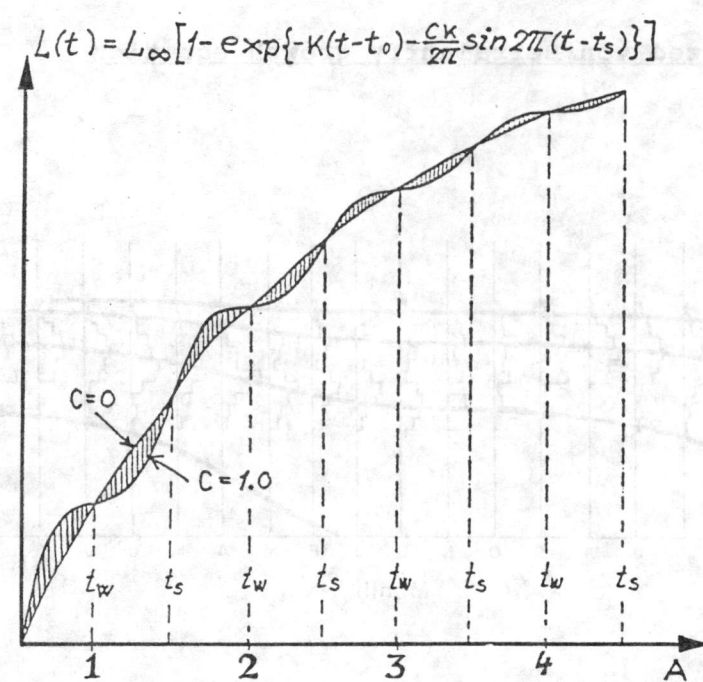

Fig. 3.5.2.2 **The seasonalized von Bertalanffy growth equation. Note that for C = 1 the growth rate is zero (curve is horizontal) at the winter points**

3.5.3 Maximum likelihood methods

The calculation of a mean value as described in Section 2.1 and the least squares method described in Section 3.3.4 are applications of the "maximum likelihood principle".

The method to be described in this section aims at solving the same problem as the ELEFAN I method and some other problems. The main difference lies in the definition of the goodness of fit. ELEFAN I uses the ratio ESP/ASP (cf. Section 3.5.1) whereas the "maximum likelihood method" uses the (weighted) sum of the squares of the deviations between model and observations (or measures with similar properties). In principle this measure of goodness of fit is the same as the one used in linear regression analysis (cf. Eq. 2.4.3 and Fig. 2.4.2).

The full statistical theory behind this method is complicated and so is the computer program. However, a fishery scientist running the program does not need to know all the technical details. If the basic principles behind the method are understood, few difficulties in using the program should be encountered.

The basic idea of ELEFAN I, to follow the progression of modes and test a large number of alternative combinations of growth parameters, is also the basic idea behind the maximum likelihood approach. The measure for goodness of fit used in the maximum likelihood method is closely related to the so-called "chi-squared criterion" which is conceptually simple and therefore used in the following explanation of the method.

In Fig. 3.5.3.1 a length-frequency sample is presented that we assume to be composed of two cohorts. When using the maximum likelihood computer program on that sample, we would obtain a result as illustrated in Fig. 3.5.3.2, where the dotted curves represent the two cohorts and the full line the sum of the calculated frequencies of the two cohorts. The dots indicate the original, observed frequencies, and the bars the differences between observed and calculated frequencies.

In addition to the growth parameters the maximum likelihood method also works with the following parameters (in the case of two cohorts):

    N1 = total number of observations in first cohort
    N2 = total number of observations in second cohort
    s1 = standard deviation of first cohort
    s2 = standard deviation of second cohort

The mean lengths, $\bar{L}1$ and $\bar{L}2$ follow from the growth parameters (cf. Fig. 3.5.3.2, where L1 and L2 corresponding to arbitrary ages t1 and t2 are shown as an example). From the parameters the calculated (theoretical) frequency of each cohort, $fc_1(L)$ and $fc_2(L)$ and the total frequency

$$fc_{total}(L) = fc_1(L) + fc_2(L)$$

of each length group can be calculated as explained in Section 2.2.

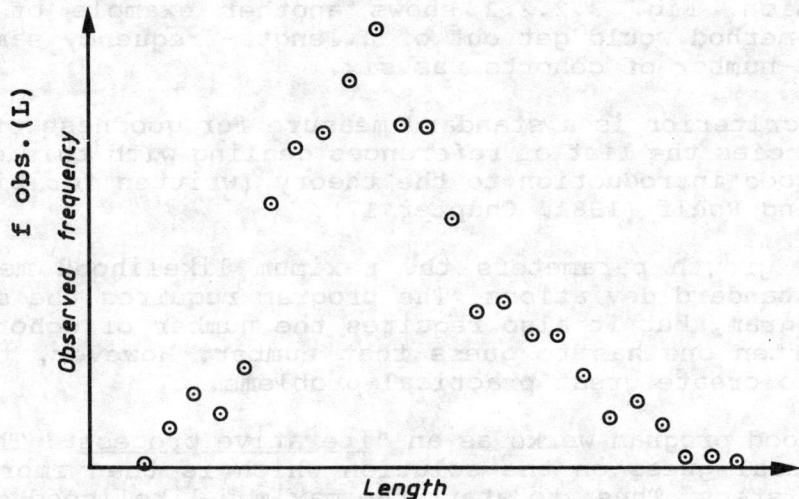

Fig. 3.5.3.1 The basic data from which the resolution into normally distributed components in Fig. 3.5.3.2 is derived

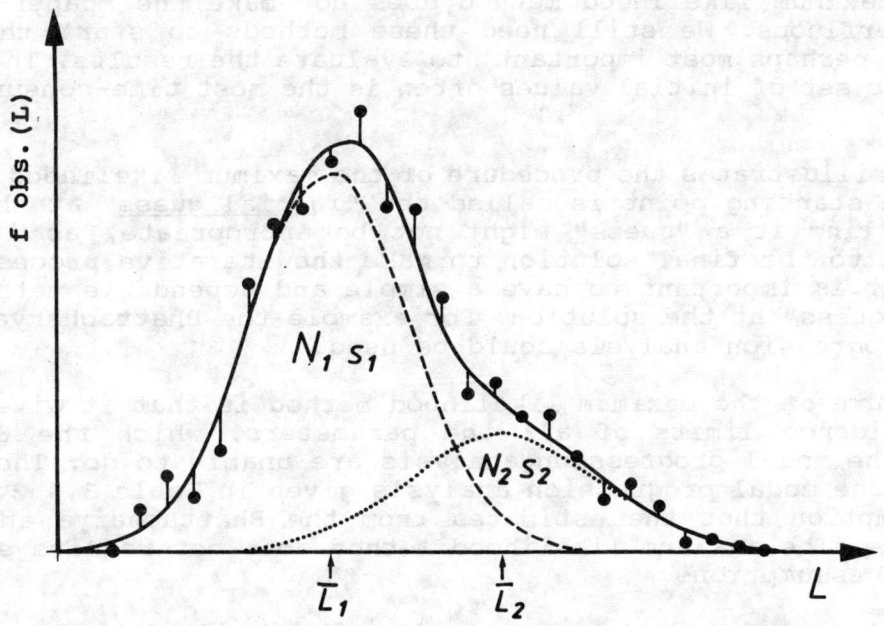

Fig. 3.5.3.2 Illustration of the chi-squared criterion. Input data are from Fig. 3.5.3.1. Also the number of cohorts must be given as input

The measure of goodness of fit, the "chi-squared criterion", is defined as:

$$\chi^2 = \sum \frac{[f_{obs}(L) - fc_{total}(L)]^2}{fc_{total}(L)} \qquad (3.5.3.1)$$

which is the sum over all $fc_{total}(L)$ values > 0

where $f_{obs}(L)$ stands for the observed frequency in length group L (= interval midpoint). It is used to minimize the differences between observed and calculated frequencies over the entire length range of the sample. The maximum likelihood program determines that set of parameters (L_∞, K, t_o, N1, N2, s1 and s2) which minimizes the chi-squared criterion. A comparison with Eq. 2.4.3 ("fc_{total}" and "f_{obs}" correspond to "a + b*x(i)" and "y(i)", respectively) illustrates the relationship between the chi-squared criterion and linear regression. Fig. 3.2.2.2 shows another example of what the maximum likelihood method would get out of a length-frequency sample if it were given that the number of cohorts was six.

As the chi-squared criterion is a standard measure for goodness of fit when dealing with frequencies the list of references dealing with this concept is nearly endless. A good introduction to the theory (written for biologists) is given in Sokal and Rohlf (1981, Chapter 17).

In addition to the growth parameters the maximum likelihood method also gives numbers and standard deviations. The program requires the same input as the ELEFAN I program, but it also requires the number of cohorts in the sample as input. Often one has to guess that number. However, this extra input appears not to create great practical problems.

The maximum likelihood program works as an "iterative process". That is, it must be fed an initial guess on the solution which is then improved in a number of iterative steps. Thus, to start the maximum likelihood estimation procedure we need an approximation to the solution of the exercise. Such an initial solution can be obtained from, for example, the Bhattacharya analysis and the modal progression analysis described in Sections 3.4.1 and 3.4.2. The maximum likelihood method does not make the "paper-and-pencil" methods superfluous. We still need these methods to start the iteration process and, perhaps most important, to evaluate the results. The search for an acceptable set of initial values often is the most time-consuming part of the task.

Fig. 3.5.3.3 illustrates the procedure of the maximum likelihood estimation. Usually, the starting point is called the "initial guess" at the solution. However, calling it a "guess" might not be appropriate, as it has to be rather close to the final solution to make the iterative process converge. Therefore, it is important to have a simple and dependable method to get a first "good guess" at the solution. For example the Bhattacharya method and the modal progression analysis could be used.

Another feature of the maximum likelihood method is that it gives estimates of the confidence limits of all the parameters, which the Bhattacharya method and the modal progression analysis are unable to do. The confidence limits from the modal progression analysis given in Table 3.4.2.3 are based on the assumption that the estimates from the Bhattacharya analysis have zero variance. The maximum likelihood method does not require such (highly unrealistic) assumptions.

We conclude this brief discussion of the maximum likelihood method with a few words on its historical development. The first work in the field is nearly as old as Petersen's pioneering work on length-frequencies of fish (cf. Section 3.4.) as Pearson in 1894 presented his work on separation of frequencies into normally distributed components.

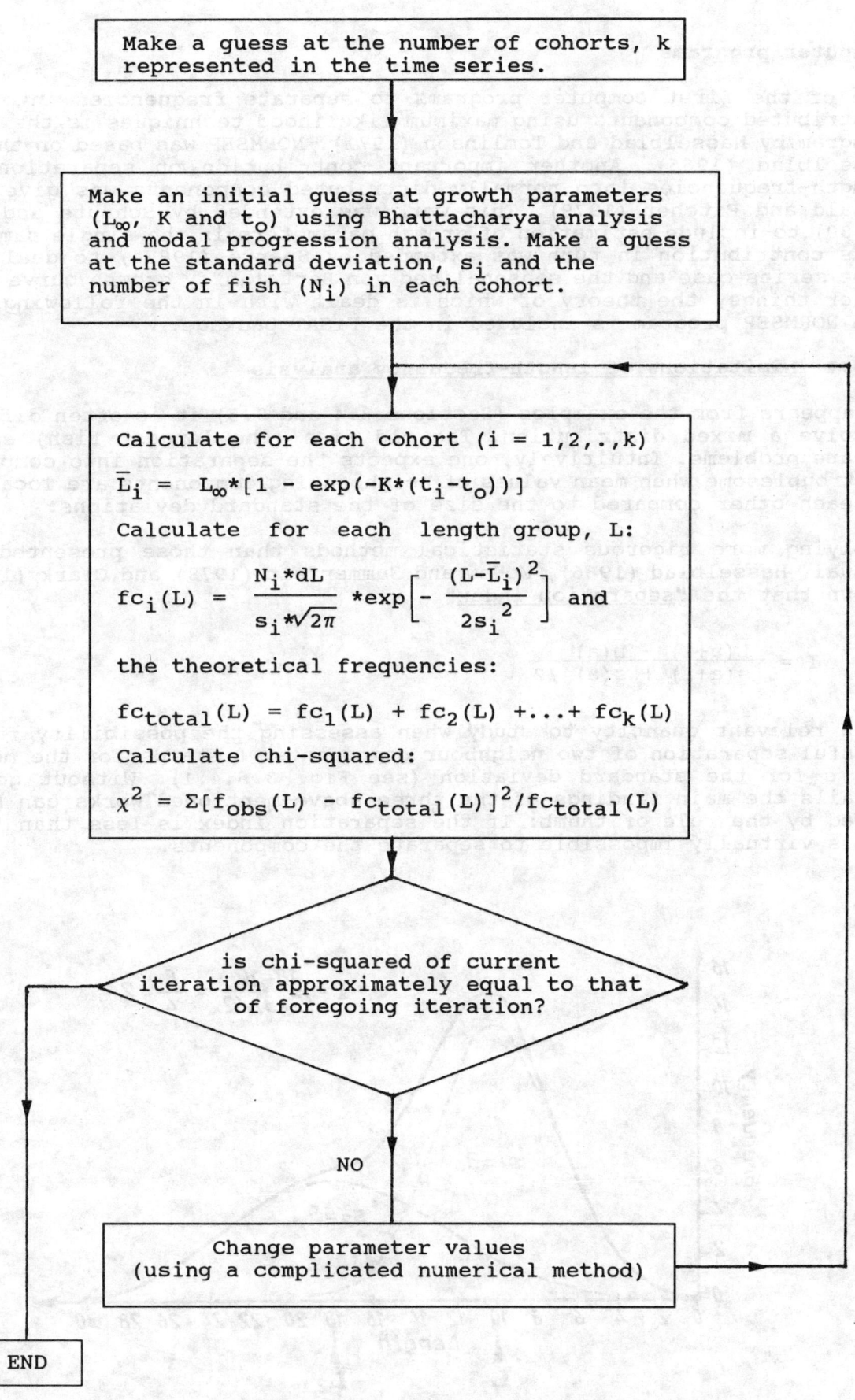

Fig. 3.5.3.3 The iterative process of the maximum likelihood estimation procedure (see also Fig. 3.5.3.2)

Computer programs

One of the first computer programs to separate frequencies into normally distributed components using maximum likelihood techniques is the "NORMSEP" program by Hasselblad and Tomlinson (1971). NORMSEP was based on the work by Hasselblad (1966). Another important contribution on separation of fish length-frequencies into normally distributed components was given by Macdonald and Pitcher (1979). This work was extended by Schnute and Fournier (1980) to include estimation of growth parameters in the single sample case. This contribution in turn was extended by Sparre (1987a) to deal with the time series case and the seasonalized von Bertalanffy growth curve and a few other things, the theory of which is dealt with in the following section. The NORMSEP program is included in the FiSAT package.

3.5.4 Limitations of length-frequency analysis

As appears from the examples (Sections 3.4 and 3.5) it is often difficult to resolve a mixed distribution. The old fish (the longest fish) especially create problems. Intuitively, one expects the separation into components to be troublesome when mean values of neighbouring components are located close to each other compared to the size of the standard deviations.

Applying more rigorous statistical methods than those presented in this manual, Hasselblad (1966), McNew and Summerfelt (1978) and Clark (1981) have shown that the "separation index"

$$I = \frac{\bar{L}(a+1) - \bar{L}(a)}{[s(a+1) + s(a)]/2} \qquad (3.5.4.1)$$

is a relevant quantity to study when assessing the possibility for a successful separation of two neighbour components. L stands for the mean value and s for the standard deviation (see Fig. 3.5.4.1). Without going into details the main findings of the three above-mentioned works can be summarized by the rule of thumb: If the separation index is less than two, I<2, it is virtually impossible to separate the components.

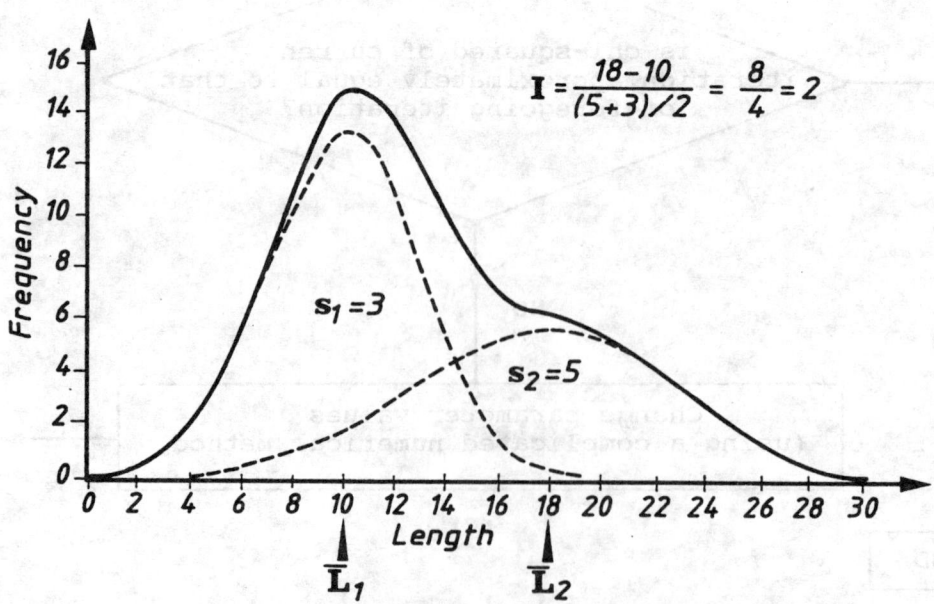

Fig. 3.5.4.1 Example of two normally distributed components with the critical separation index, I-value of 2

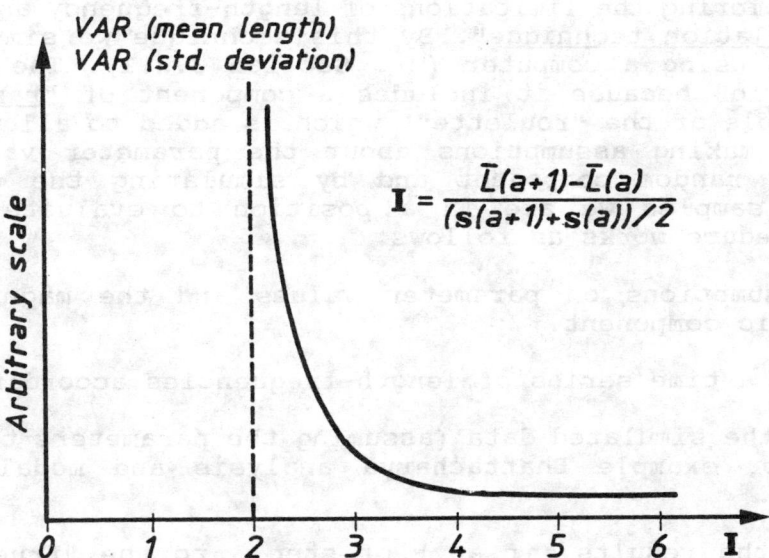

Fig. 3.5.4.2 General description of the functional relationship between separation index, I, and variances of estimates

Fig. 3.5.4.1 shows an example of two normally distributed components with I = 2. Fig. 3.5.4.2 shows the typical functional relationship between separation index and variance of the estimates. (For further details see, for example Hasselblad, 1966.)

As an example consider Table 3.2.1.1 (i.e. the hypothetical data used to illustrate the paper and pencil methods). In Table 3.5.4.1 the separation indices have been calculated for the six components. These are known because the data are hypothetical or constructed. Suppose the data had been real data for which we did not know the true parameters. In that case there would be hope for estimation of only three components with separation indices 4.82 and 2.43 respectively. This conclusion holds for <u>all methods</u>, including the most sophisticated computerized ones.

Table 3.5.4.1 Separation indices calculated for the example of Section 3.4.1. The parameters marked by "*" cannot be estimated from length-frequency data alone (cf. Table 3.2.1.1 and Fig. 3.2.2.2)

a	$\overline{L}(a)$	s(a)	$I = 2*\dfrac{\overline{L}(a+1) - \overline{L}(a)}{s(a+1) + s(a)}$
1	17.3	1.7	
			4.82
2	27.9	2.7	
			2.43
3	35.3	3.4	
			1.40
4	40.2 *	3.6 *	
			0.84
5	43.3 *	3.8 *	
			0.59
6	45.5 *	3.6 *	

Another way of exploring the limitations of length-frequency analysis is the "Monte Carlo simulation technique". By this technique we simulate length-frequency samples using a computer (cf. Section 3.2.1). The technique is called "Monte Carlo" because it includes a component of "random variability", the principle of the "roulette", which is added to all the simulated observations. By making assumptions about the parameter values and the magnitude of the random component and by simulating the corresponding length-frequency samples we are in a position to evaluate the various methods. The procedure works as follows:

Step 1: Make assumptions on parameter values and the magnitude of the stochastic component.

Step 2: Simulate a time series of length-frequencies according to step 1.

Step 3: Analyse the simulated data (assuming the parameters to be unknown) using for example Bhattacharya analysis and modal progression analysis.

Step 4: Compare the results (if any) of step 3 to the "true" parameters from step 1.

Using this procedure we will be able to give statements like: If a fish stock has length distributions with certain parameters then we are able or not able to estimate the growth parameters with a certain prespecified accuracy.

Also difficulties in obtaining unbiased samples should be mentioned in connection with the limitations of length-frequency analysis. Probably the most important source of bias stems from the migration of fish. Limitations of length-based methods applied to migratory fish stocks are discussed in Chapter 11.

4 ESTIMATION OF MORTALITY RATES

Chapter 3 dealt with growth, the positive aspects in the dynamics of a fish stock. This chapter deals with its negative counterpart, the death process or mortality (cf. Fig. 1.3.2). The growth was described with the aid of a model and a number of parameters, and so is the death process going to be. The key parameters used when describing death are called the "mortality rates".

This chapter deals with the definition of mortality, and it introduces some of the methods to estimate mortality rates which require input data from random samples representative of a certain part of the population, namely the exploited part. These methods do not require estimates of total catch from the population. On the other hand, they do not provide estimates of population size. Methods dealing with those aspects will be introduced in Chapter 5.

The easiest way to describe the change in numbers in a fish stock is often to follow the fate of fish spawned at approximately the same time, a cohort. We shall consider the mortality of a cohort as composed of the mortality caused by fishing and that due to all other causes lumped together as the "natural mortality". The latter covers events such as predation, disease and deaths due to old age.

4.1 THE CONCEPT OF A COHORT AND SOME BASIC NOTATION

A "cohort" is a batch of fish all of approximately the same age and belonging to the same stock (cf. Section 1.3.1). The concepts "day of recruitment of a cohort" and "recruitment" were introduced in Section 1.6. In all the following derivations we assume (with Beverton and Holt, 1957) that a cohort consists of "average fish" only. This means that all fish of a cohort are assumed to have the same age at a given time so that they all attain the "recruitment age", Tr, at the same time (cf. Section 1.6). In Chapter 3 we similarly used the average length of a cohort to describe growth. In the context of mortality rates we are interested in the number of survivors from a cohort as a function of time (cf. Fig. 1.3.2A).

The symbol "N(t)" is used to designate the "number of survivors from a cohort attaining age t". The age is usually measured in units of years. Thus, N(Tr) is the "number of recruits" to the fishery. Often the symbol "R" is used to designate the "recruitment", R = N(Tr). The actual choice of Tr is not critical, since all calculations are based on relative ages (age differences). In many applications we do not need to define the recruitment age. Tr is the minimum age at which the fish can enter the fishery, i.e. become liable to encounter with fishing gears (Beverton and Holt, 1957). The age at which they actually enter the fishery, Tc, is dependent on the mesh size. Tc is called the "age at first capture" and marks the beginning of the "exploited phase".

Fig. 4.1.1 illustrates the basic features of cohort dynamics. Due to mortality (fishing or natural causes) there is a continuous decrease in the number of survivors. At birth the cohort has age zero. From age 0 to Tr the cohort is in the "pre-recruitment phase". In the present context we are not concerned about what happens before age Tr. After age Tr the fish may be caught if a suitable small meshed gear is used. At age Tc the fish start to be caught with the mesh size actually in use. By definition we must have Tr ≤ Tc.

In some applications we consider several cohorts at a time as illustrated by Fig. 4.1.2, where a situation with two cohorts per year during a period of two years is depicted. Fig. 4.1.2 is a parallel to the multi cohort case for growth which was illustrated in Fig. 3.2.1.1. If more than one cohort is considered at a time, the symbol N(t) is not sufficient, and some index to indicate the cohort is then required. However, in the following only one cohort at a time is usually considered, so the short symbol N(t) will do.

- 112 -

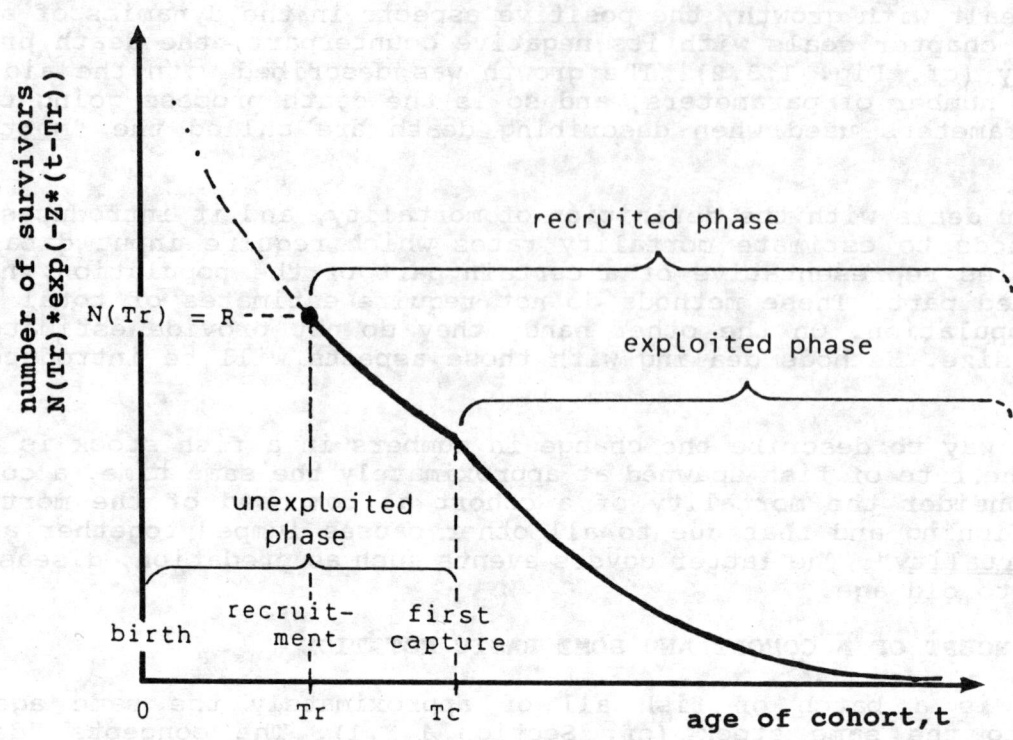

Fig. 4.1.1   Basic features of cohort dynamics

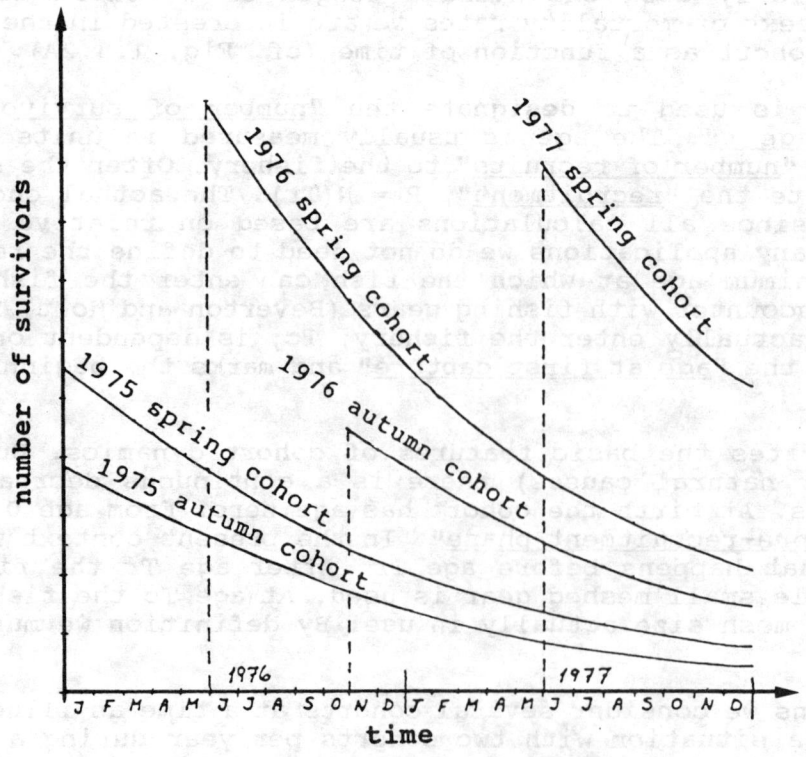

Fig. 4.1.2   Illustration of the situation with more than one cohort (compare Fig. 3.2.1.1)

4.2 THE DYNAMICS OF A COHORT, THE EXPONENTIAL DECAY MODEL

As an example, consider the number of survivors at age t = 0.5 year, N(0.5), and the number of survivors one day later, N(0.50274) (1 day = 1/365 year = 0.00274 year). The number of specimens lost during that day is:
N(0.5) - N(0.50274)

To designate the change in numbers during a relatively short time period (say, 1 day) we use the symbol ΔN:

$\Delta N(0.5) = N(0.50274) - N(0.5)$

Note that ΔN is negative because it represents a loss from the cohort. The rate of change in numbers is written:

$\Delta N(t)/\Delta t$

where Δt is the length of the time period (1 day in this case).

Suppose N(0.5) = 1000 and N(0.50274) = 997, then

$$\frac{\Delta N(0.5)}{\Delta t} = \frac{997-1000}{0.00274} = -1095 \text{ per year}$$

Obviously, the rate of change in numbers per year, $\Delta N/\Delta t$, depends on the number of survivors, N. The more survivors there are, the more will die. The high number of deaths, 1095 per year, more than N(0.5), is due to the fact that it is calculated as if we started every day with N(0.5) = 1000 fish. $\Delta N/\Delta t$ is not dependent on Δt, as a reduction of Δt will reduce ΔN accordingly. Thus it is natural to assume $\Delta N/\Delta t$ to be proportional to N:

$$\frac{\Delta N(t)}{\Delta t} = -Z*N(t) \qquad (4.2.1)$$

where Z is the coefficient of proportionality.

For the example above: $\frac{\Delta N(0.5)}{\Delta t} = -Z*N(0.5)$ or $\frac{-3}{0.00274} = -1.095*1000$,

Z, becomes 1.095 per year. Z is called the "<u>instantaneous rate of total mortality</u>", the "<u>total mortality coefficient</u>" or simply the "<u>total mortality rate</u>". The unit of Z is "per year", or in general "per time unit". If Z remains constant throughout the life of the cohort it can be shown that Eq. 4.2.1 is mathematically equivalent to:

$$N(t) = N(Tr)*\exp[-Z*(t-Tr)] \qquad (4.2.2)$$

Eq. 4.2.2 is called the "<u>exponential decay model</u>" and (together with the growth equation) it is a corner-stone of the theory of exploited fish stocks (cf. Baranov, 1918; Thompson and Bell, 1934; Fry, 1949 and Beverton and Holt, 1957). Fig. 4.2.1 shows a family of exponential decay curves for different Z-values. The higher the value of Z the faster the decrease in numbers and the lower the maximum age.

It is usually easier to understand the concept of mortality and survival if the number of survivors at a certain moment is expressed as a percentage of the original number. The following table gives the percentages of survivors after one and two years respectively of two populations subject to two different total mortality rates, Z = 0.5 and Z = 2.0.

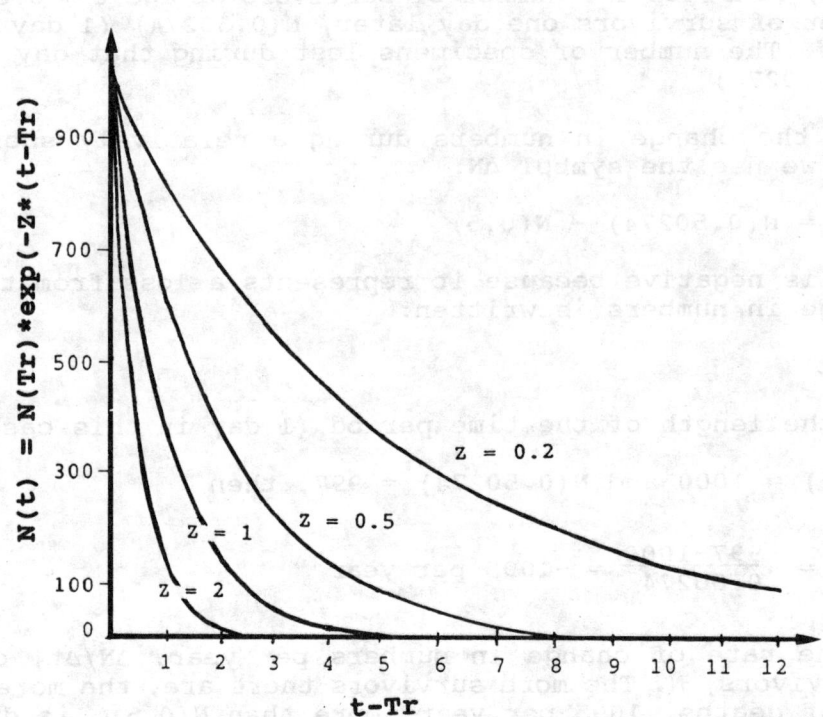

Fig. 4.2.1 Exponential decay curves, for Z = 0.2, 0.5, 1 and 2 per year, with recruitment, N(Tr) = 1000 fish

total mortality rate Z	percentage of survivors *)	
	after 1 year 100*N(Tr+1)/N(Tr)	after 2 years 100*N(Tr+2)/N(Tr)
0.5	61%	37%
2.0	14%	2%

*) Percentage of survivors = 100*exp(-Z*(t-Tr)), and t-Tr = 1 and 2 respectively

It is evident that Z = 2.0 represents a high mortality rate and that a cohort subjected to such a high rate of total mortality will have been practically exterminated in two years.

Table 4.2.1 and Fig. 4.2.2 show an example of the exponential decay of a cohort subject to a constant total motality Z = 1.5 per year. In order to simplify the example, it has been assumed that the cohort starts to be caught as soon as it is recruited to the fishery, so that Tr = Tc (cf. Fig. 4.1.1).

The number of survivors of the cohort of 100000 recruits, N(Tr), at a given time, t, can be calculated from Eq. 4.2.2 as shown in the second column of Table 4.2.1, for example:

N(Tr+0.4) = 100000*exp(-1.5*(Tr+0.4-Tr)) = 100000*exp(-0.6) = 54881

Table 4.2.1 Exponential decay of a cohort with recruitment N(Tr) = 100,000 and total (constant) mortality Z = 1.5 per year. It is demonstrated that the equation $\Delta N/\Delta t = -Z*N$ is fulfilled for various ages of the cohort (t). The corresponding graph is shown in Fig. 4.2.2. Δt is one day (= 1/365 year)

age of cohort t years	number of survivors N(Tr)*exp(-Z*(t-Tr))	change in numbers during one day, ΔN	$-\frac{\Delta N}{\Delta t} * \frac{1}{N} = Z$
Tr	100000		
Tr+0.0+Δt	99590	-410	1.5
Tr+0.2	74081		
Tr+0.4	54881		
Tr+0.6	40657		
Tr+0.6+Δt	40490	-167	1.5
Tr+0.8	30119		
Tr+1.0	22313		
Tr+1.5	10540		
Tr+1.5+Δt	10497	-43	1.5
Tr+2.0	4978		
Tr+2.5	2351		
Tr+3.0	1111		
Tr+3.5	524.75		
Tr+3.5+Δt	522.60	-2.15	1.5
Tr+4.0	247.88		
Tr+5.0	55.31		
Tr+8.0	0.61		

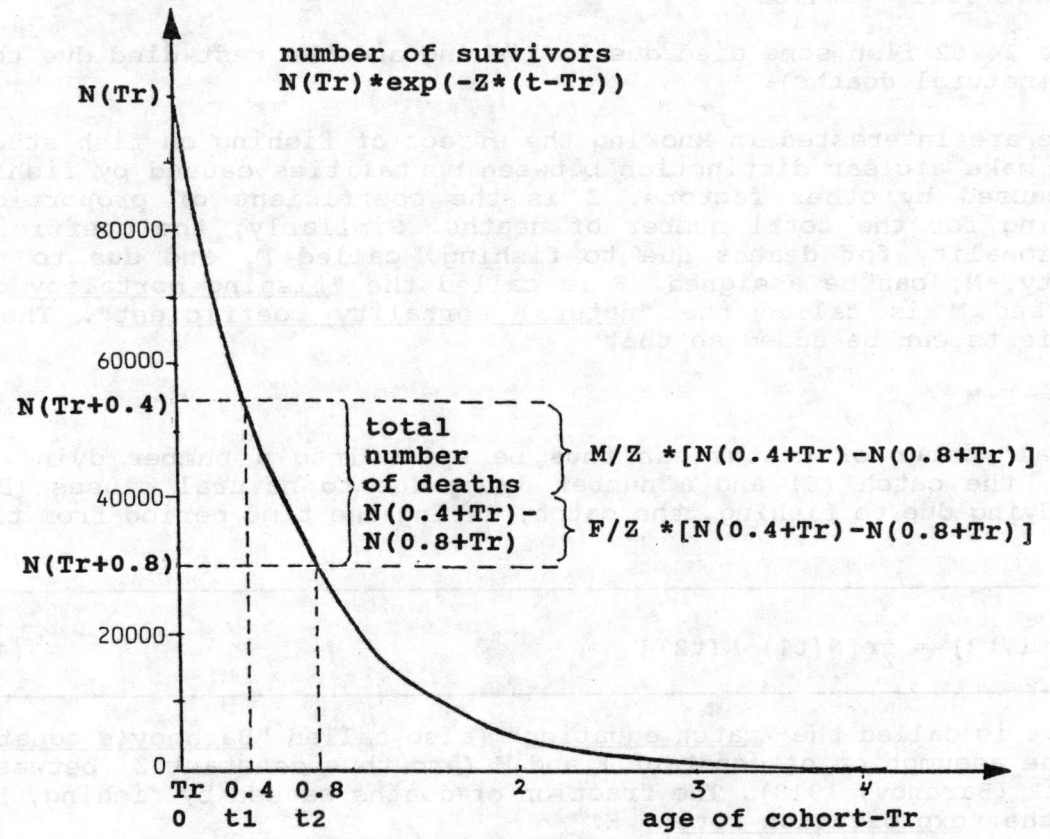

Fig. 4.2.2 Exponential decay curve with N(Tr) = 100,000 recruits and Z = 1.5 (data from Table 4.2.1). Z remains constant for all ages, t

As shown in the third column of Table 4.2.1, the decrease in numbers during one day changes during the life span of a cohort, because the total number of surviving fish becomes smaller every day. However, it can be demonstrated that Eq. 4.2.1

$$\frac{\Delta N(t)}{\Delta t} = -Z*N(t)$$

is fulfilled for different ages of the cohort, throughout its life span.

Ricker (1975) defines the "<u>survival rate</u>", S, as "the number of fish alive after a specified time interval, divided by the initial number, usually on a yearly basis".

The survival rate after one year is:

$$S = \frac{N(Tr+1)}{N(Tr)} = \frac{N(Tr)*\exp(-Z*(Tr+1-Tr))}{N(Tr)} = \exp(-Z)$$

Let t1 and t2 be two ages with t1 < t2. The total number of deaths during the time interval from t1 to t2 is

N(t1)-N(t2)

In Table 4.2.1 and Fig. 4.2.2, when t1 = Tr+0.4 and t2 = Tr+0.8 years, the number of deaths over this period is the difference between the number of survivors at t1, N(Tr+0.4) and t2, N(Tr+0.8):

54881-30119 = 24762

Of these 24762 fish some died due to fishing and the rest died due to other causes (natural deaths).

Since we are interested in knowing the effect of fishing on fish stocks, we have to make a clear distinction between mortalities caused by fishing and those caused by other factors. Z is the coefficient of proportionality accounting for the total number of deaths. Similarly, the coefficient of proportionality for deaths due to fishing, called F, and due to natural mortality, M, can be assigned. F is called the "<u>fishing mortality coefficient</u>" and M is called the "<u>natural mortality coefficient</u>". These two coefficients can be added so that

Z = F+M (4.2.3)

The total number of deaths can thus be split into a number dying due to fishing, the catch (C) and a number dying due to natural causes (D). The number dying due to fishing, the catch, during the time period from t1 to t2 is:

$$C(t1,t2) = \frac{F}{Z}*[N(t1)-N(t2)] \quad (4.2.4)$$

Eq. 4.2.4 is called the "<u>catch equation</u>" (also called "<u>Baranov's equation</u>"), under the assumption of constant F and M (and thus constant Z) between ages t1 and t2 (Baranov, 1918). The fraction of deaths caused by fishing, F/Z, is called the "<u>exploitation rate</u>", E.

Correspondingly the number dying due to natural causes is:

$$D(t1,t2) = \frac{M}{Z}*[N(t1)-N(t2)] \quad (4.2.5)$$

Note that $N(t1)-N(t2) = C(t1,t2)+D(t1,t2)$ because $F/Z+M/Z = 1$.

In the example given above, based on Table 4.2.1 and Fig. 4.2.2, the total number of deaths was 24762, at a constant total mortality rate of $Z = 1.5$.

Now, suppose that between the ages $t1 = Tr+0.4$ and $t2 = Tr+0.8$ the fishing mortality coefficient was $F = 0.6$, and the natural mortality coefficient $M = 0.9$. The numbers of deaths due to fishing and due to natural causes can then be calculated as follows:

$$C(Tr+0.4,Tr+0.8) = (0.6/1.5)*(54881-30119) = 9905$$
$$D(Tr+0.4,Tr+0.8) = (0.9/1.5)*(54881-30119) = 14857$$
$$\overline{}$$
$$N(Tr+0.4)-N(Tr+0.8) = 54881-30119 = 24762$$

In real life the mortalities usually vary with the age of the cohort. Small (young) fish are exposed to a greater natural mortality because more predators can eat them. On the other hand small fish may suffer less fishing mortality than large (old) fish because they either have not yet migrated to the fishing grounds or they escape through the meshes of the gear. However, if the time span from t1 to t2 is not too large it is considered a fair approximation to assume F and M to remain constant within this period.

The "catch equation" (Eq. 4.2.4) is one of the most important mathematical expressions in fisheries biology. For many applications, however, it is convenient to rearrange its terms, where the catch is related to the number present at the beginning of the time span, N(t1). To do so Eq. 4.2.2 is applied to N(t2):

$N(t2) = N(Tr)*\exp[-Z*(t2-Tr)]$ which is equivalent to

$N(t2) = N(Tr)*\exp[-Z*(t1-Tr)]*\exp[-Z*(t2-t1)]$ and to

$$N(t2) = N(t1)*\exp[-Z*(t2-t1)] \qquad (4.2.6)$$

Inserting Eq. 4.2.6 into Eq. 4.2.4 gives:

$$\boxed{C(t1,t2) = N(t1)*\frac{F}{Z}*[1 - \exp(-Z*(t2-t1))]} \qquad (4.2.7)$$

Eq. 4.2.7 is the most commonly used version of the catch equation. Another is convenient for special applications:

$$C(t1,t2) = (t2-t1)*F*\overline{N}(t1,t2) \qquad (4.2.8)$$

where $\overline{N}(t1,t2)$ is the "average number of survivors during the time period from t1 to t2". To get consistency between Eq. 4.2.7 and Eq. 4.2.8 we must have:

$$\boxed{\overline{N}(t1,t2) = N(t1) * \frac{1 - \exp(-Z*(t2-t1))}{Z*(t2-t1)}} \qquad (4.2.9)$$

The mathematical proof of Eq. 4.2.9 is outside the scope of this manual. However, in Fig. 4.2.3 it is demonstrated that Eq. 4.2.9 conforms to the intuitive concept of "average number".

For the example used above (Fig. 4.2.2 and Table 4.2.1) we find:

$$\bar{N}(Tr+0.4, Tr+0.8) = 54881*\frac{1}{1.5*(0.8-0.4)}*[1 - \exp(-1.5*(0.8-0.4))] = 41269$$

$$C(Tr+0.4, Tr+0.8) = (0.8-0.4)*0.6*41269 = 9905,$$

which is also the answer we got from applying Eq. 4.2.4.

Eq. 4.2.8 has the advantage of being easy to interpret. It says that the number caught during the time period from t_1 to t_2 depends on:

1) The length of the time period
2) The fishing mortality
3) The average number of fish in the sea

and each of these statements is easy to accept.

If $\Delta t = t_2 - t_1$ is very small, it can be shown that

$$C(t, t+\Delta t) = \Delta t * F * N(t) \tag{4.2.10}$$

is approximately correct. If Δt is small the number of survivors will change very little during the period and N and $\bar{N}$ will be approximately equal. Eq. 4.2.10 then follows from Eq. 4.2.8. Eq. 4.2.10 is another version of the catch equation which is convenient for reasoning about the fishing mortality created by a single fishing operation, one trawl haul, for example, as will be demonstrated in Section 4.3.

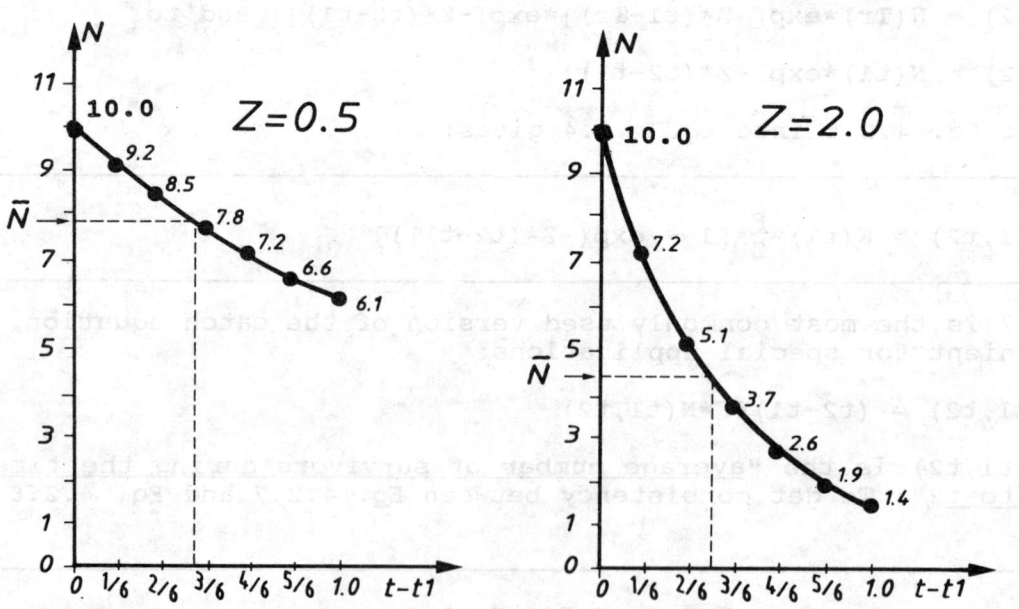

$Z = 0.5$	Approximation:	$\bar{N} = (10.0+9.2+8.5+7.8+7.2+6.6+6.1)/7 = 7.9$
	Exact expression:	$\bar{N} = 10.0*[1 - \exp(-0.5)]/0.5 = 7.9$
$Z = 2.0$	Approximation:	$\bar{N} = (10.0+7.2+5.1+3.7+2.6+1.9+1.4)/7 = 4.5$
	Exact expression:	$\bar{N} = 10.0*[1 - \exp(-2.0)]/2.0 = 4.3$

Fig. 4.2.3   Illustration of Eq. 4.2.9:
$\bar{N}(t_1, t_2) = N(t_1)*[1 - \exp(-Z*(t_2-t_1))]/Z$ if $t_2-t_1 = 1$ year and $Z = 0.5$ or 2.0. Note that the larger the Z is the larger is the deviation between the approximation and the exact value

Table 4.2.2 Example of the dynamics of a cohort with different mortality rates during its life span

start of period t1	end of period t2	natural mort. M	fishing mort. F	total mort. Z	comments
Tr	Tr+0.5	2.0	0	2.0	The cohort has been recruited to the fishing grounds but is not yet exploited. Exposed to great predation.
Tr+0.5	Tr+1.0	1.0	0.5	1.5	The cohort has migrated to fishing grounds, but 50% escape through the meshes. Predation mortality reduced.
Tr+1.0	Tr+7.0	0.5	1.0	1.5	The cohort under full exploitation (all fish are retained by the meshes). Predation mortality further reduced.

To apply the exponential decay model (Eq. 4.2.2) and the catch equations, it is not necessary to assume that M, F and Z remain constant during the entire life span of the cohort. The life span may be divided into a number of shorter time periods within which mortalities are assumed to remain constant whereas they may vary from period to period. As an example, consider a cohort with a life span of seven years. The seven years may be divided into three periods with different mortality rates as shown in Table 4.2.2. Suppose that N(Tr) = 100000 then, using Eq. 4.2.6:

$$N(t2) = N(t1)*\exp[-Z*(t2-t1)]$$
$$N(Tr+0.5) = 100000*\exp(-2.0*0.5) = 36788$$
$$N(Tr+1.0) = 36788*\exp(-1.5*0.5) = 17377$$
$$N(Tr+6.0) = 17377*\exp(-1.5*5.0) = 9.61$$
$$N(Tr+7.0) = 9.61*\exp(-1.5*1.0) = 2.14$$
$$N(Tr+8.0) = 2.14*\exp(-1.5*1.0) = 0.48$$

Thus, after seven years the cohort has died out. The number caught in the first period (from Tr to Tr+0.5) is zero because F = 0. The number caught in the second period (from Tr+0.5 to Tr+1.0) is (using Eq. 4.2.7):

$$C(Tr+0.5,Tr+1) = N(Tr+0.5)*\frac{F}{Z}*[1 - \exp(-Z*0.5)] =$$

$$36788*\frac{0.5}{1.5}*[1 - \exp(-1.5*0.5)] = 6470$$

Alternatively, Eq. 4.2.4 could have been used to calculate the number caught, C:

$$C(Tr+0.5,Tr+1.0) = \frac{F}{Z}*[N(Tr+0.5)-N(Tr+1.0)] =$$

$$\frac{0.5}{1.5}*(36788-17377) = 6470$$

The third period is treated in a similar way. The results may be summarized as follows:

t1	t2	t2-t1	Z	N(t1)	F	C(t1,t2)
Tr	Tr+0.5	0.5	2.0	100000	0	0
Tr+0.5	Tr+1.0	0.5	1.5	36788	0.5	6470
Tr+1.0	Tr+7.0	6.0	1.5	17377	1.0	11583

In the subsequent sections various methods for the estimation of Z, F and M will be discussed.

(See **Exercise(s)** in Part 2).

4.3 ESTIMATION OF Z FROM CATCH PER UNIT OF EFFORT DATA AND THE CONCEPT OF THE CATCHABILITY COEFFICIENT

The total mortality coefficient Z can be estimated when estimates of the number of fish in the cohort are available for two different moments during its exploited phase, t1 and t2. To calculate Z from such data Eq. 4.2.6 may be rewritten:

$$Z = \frac{1}{t2-t1} * \ln \frac{N(t1)}{N(t2)} \qquad (4.3.0.1)$$

For the estimation of Z with this formula, it is not necessary to know the absolute values of N(t1) and N(t2), only their ratio is required. Thus, if only an estimate of some quantity proportional to N(t) is available Eq. 4.3.0.1 can still be applied.

The catch per unit of effort (CPUE), for example, the numbers caught per trawl hour, is a quantity which can be assumed proportional to the number of fish in the sea, N. Intuitively, it is not hard to accept that if there are twice as many fish in the sea, twice as many will be caught per fishing operation. Let CPUE(t) be the number caught of a cohort per unit of effort at time t, then mathematically, the assumption is expressed:

$$CPUE(t) = q*N(t) \qquad (4.3.0.2)$$

where q is a parameter called the "<u>catchability coefficient</u>". The more efficient the gear is, the higher the value of q because q is a measure of the ability of the gear to catch fish.

CPUE data from research surveys

Assume that we have obtained CPUE data for a certain species with a research vessel during a trawl survey in different periods, t1 and t2, using the same gear without any modifications and thus with a constant catchability coefficient, q. In that case it follows from Eq. 4.3.0.2 that:

$$\frac{N(t1)}{N(t2)} = \frac{q*N(t1)}{q*N(t2)} = \frac{CPUE(t1)}{CPUE(t2)}$$

Inserting the above into Eq. 4.3.0.1 gives:

$$\boxed{Z = \frac{1}{t2-t1} * \ln \frac{CPUE(t1)}{CPUE(t2)} \qquad (4.3.0.3)}$$

Eq. 4.3.0.3 can be used to obtain an estimate of Z from time series of CPUE data from research surveys if age compositions have been obtained, either from reading ring structures in hard parts or from length-frequency analysis. When age data are available, the numbers of fish caught per unit of effort of each age group (cohort) can be determined directly. So when such numbers are available for a time series the numbers of a certain cohort present at different moments can be used in Eq. 4.3.0.3. Recall that the results of a length-frequency analysis, e.g. the Bhattacharya method (Section 3.4.1), also include estimates of the numbers of fish belonging to each cohort. If such numbers are available for the same cohort at different times, then the progressive decline in numbers represents the total mortality and Eq. 4.3.0.3 can be applied to obtain an estimate of Z. An illustration of this method is provided in Exercise 4.3, based on the data presented in Table 3.2.1.2.

CPUE data from commercial fisheries

Sometimes CPUE is derived as the average value over a longer period. This is the case when data are collected from a commercial fishery, where data are most often given on a quarterly or yearly basis. The model to be used is mathematically equivalent to Eq. 4.3.0.3, but conceptually it is slightly different. The parallel to Eq. 4.3.0.2 corresponding to a longer time period is:

$$\overline{CPUE}(t1,t2) = q*\overline{N}(t1,t2) \qquad (4.3.0.4)$$

This is the model relevant to data on commercial fisheries. The mean CPUE $\overline{CPUE}(t1,t2)$ is usually calculated as the catch of a cohort during the period from t1 to t2 divided by the effort during that period. $\overline{N}(t1,t2)$ is the average number of survivors during the period from t1 to t2.

The theory behind Eqs. 4.3.0.2 and 4.3.0.4 will be used several times in the following chapters. We therefore will discuss the reasoning leading to these expressions.

Let f be the effort during one year (t2-t1 = 1). Eq. 4.3.0.4 may then be written:

$$C/f = q*\overline{N} \qquad (4.3.0.5)$$

When t2-t1 = 1 year we have a special version of Eq. 4.2.8:

$$C = F*\overline{N} \qquad (4.3.0.6)$$

Inserting Eq. 4.3.0.6 into Eq. 4.3.0.5 gives

$$F = q*f \qquad (4.3.0.7)$$

Eq. 4.3.0.7 can be shown to apply to any time period [t2,t1]. It is intuitively easy to accept. It says that the more effort (e.g. boat days) goes into the fishery, the higher will be the fishing mortality, or the fishing mortality is proportional to the effort. As Eq. 4.3.0.7 is so simple we could have taken that as the starting point for the derivation of Eq. 4.3.0.2 or Eq. 4.3.0.4.

Fig. 4.3.0.1 illustrates Eq. 4.3.0.7 in a simplified system, consisting of 2 million fish uniformly distributed over a coastal area. The figure shows only 200 fish (1/10000 of the system).

The rectangle labelled "A" indicates the area swept by the trawl in, say, one hour = Δt. The catch in this short time, $\Delta C1$, consists of six fish. For simplicity, we assume that no fish dies due to natural causes during the towing period and that all fish encountered by the trawl are retained. Thus, the fishing mortality, $\Delta t*F1$, created by haul A must satisfy Eq. 4.2.10 (in this case Δt is small):

$$\Delta C1 = \Delta t * F1 * N \quad \text{or} \qquad (4.3.0.8)$$

$$6 = \Delta t * F1 * 2000000 \quad \text{so that}$$

$$\Delta t * F1 = 0.000003$$

Similarly, for two and three tows during one hour we get:

two tows (A+B)   : $\Delta t * F2 = 0.000006$
three tows (A+B+C) : $\Delta t * F3 = 0.000009$

Let one trawl hour be one effort unit. Then from Eq. 4.3.0.7 we have:

one tow   : $\Delta t * F1 = \Delta t * q * 1$
two tows   : $\Delta t * F2 = \Delta t * q * 2$
three tows : $\Delta t * F3 = \Delta t * q * 3$

and q thus remains constant for different F-values:

$$q = \frac{\Delta t * F1}{1 * \Delta t} = \frac{\Delta t * F2}{2 * \Delta t} = \frac{\Delta t * F3}{3 * \Delta t}$$

The value of q on a "per year" basis is

$$\frac{0.000003}{1/(365*24)} = 0.02628$$

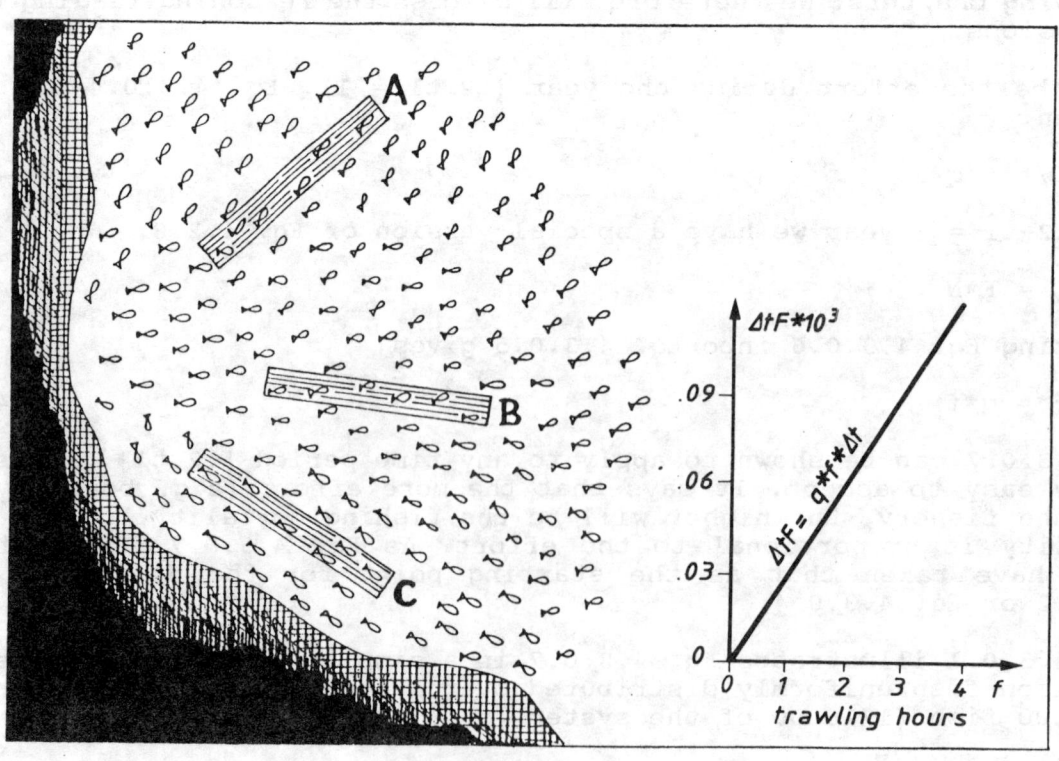

Fig. 4.3.0.1   Illustration of Eq. 4.3.0.7: $F = q*f$.
(For further explanation, see text)

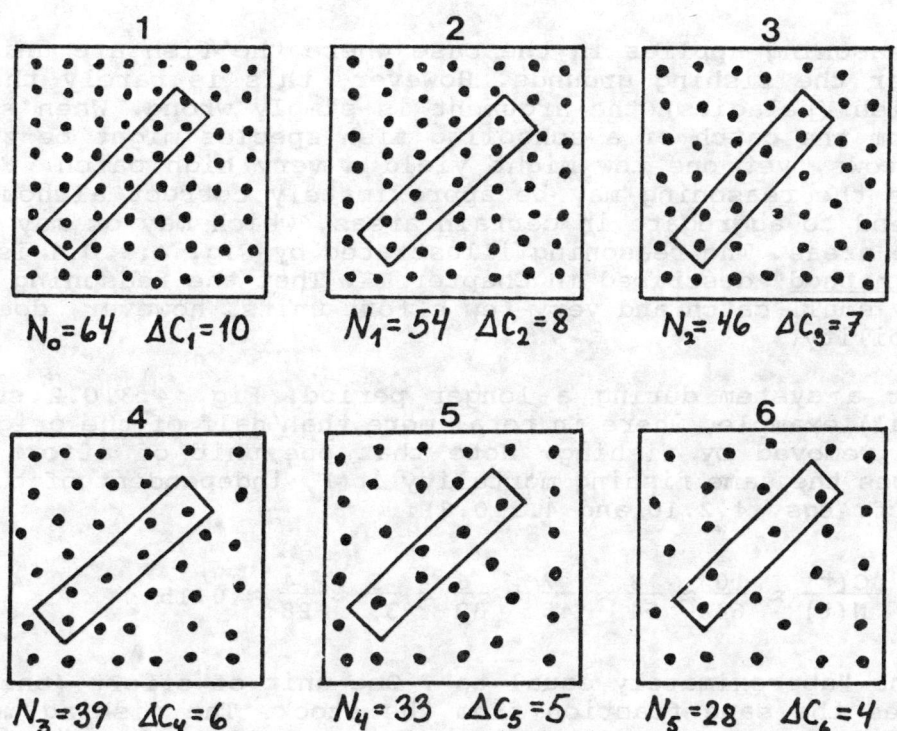

Fig. 4.3.0.2 Illustration of Eq. 4.3.0.7, F = q*f, applied several times to a system during a longer time period.
(For further explanation, see text)

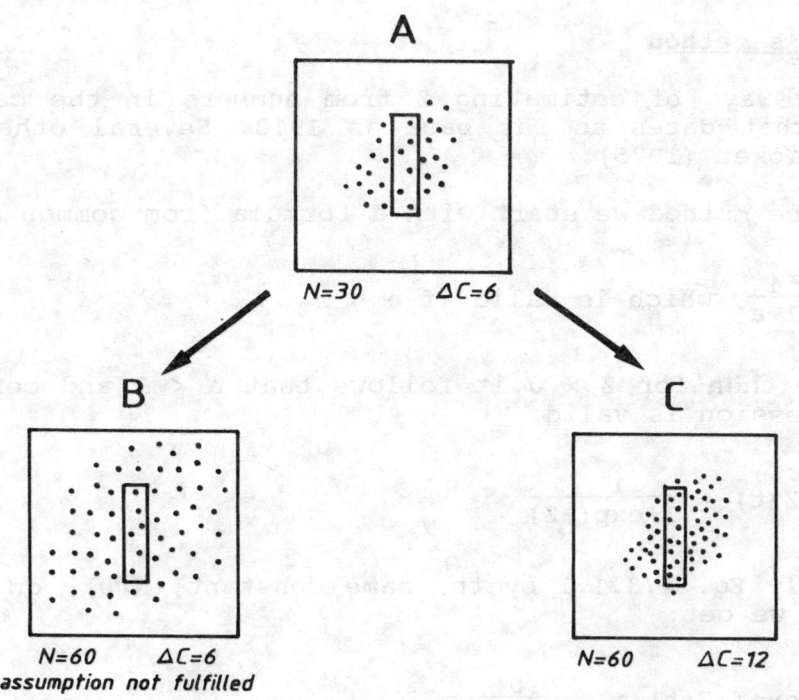

Fig. 4.3.0.3 Illustration of one of the assumptions behind Eq. 4.3.0.7: F = q*f. (For further explanation, see text)

The above reasoning applies in the case where the fish are uniformly distributed over the fishing grounds. However, this is rarely the case. For schooling fish (pelagics) the argument is simply wrong. When shooting the net at random the catch of a schooling fish species might be zero in nine out of ten tows, yet one tow might yield a very high catch. For demersal fish species the reasoning may be approximately correct although demersal fish also tend to aggregate in certain areas, which may or may not also be the fishable areas. The reasoning illustrated by Fig. 4.3.0.1 is behind the "<u>swept area method</u>" described in Chapter 13. That the reasoning is based on a relatively small catch and very few effort units, however, does not limit its applicability.

Now consider a system during a longer period. Fig. 4.3.0.2 shows such a (hypothetical) example, where in total more than half of the original number of fish are removed by fishing. Note that one unit of effort (one trawl haul) produces the same fishing mortality $\Delta t*F$, independent of the number of survivors (cf. Eqs. 4.2.10 and 4.3.0.8):

$$\Delta t*F \approx \frac{\Delta C(t)}{N(t)} \approx \frac{10}{64} \approx \frac{8}{54} \approx \frac{7}{46} \approx \frac{6}{39} \approx \frac{5}{33} \approx \frac{4}{28} \approx 0.15$$

where $\approx$ means "approximately equal to". One unit of effort (one trawl haul or tow) takes the same fraction from the stock. The fishing mortality is independent of the number of survivors, i.e., all hauls give equal $\Delta t*F$. Note also that $\Delta t*F$ will remain constant only if the fish redistribute uniformly over the area after each removal by fishing.

Figs. 4.3.0.3A and B show an example in which Eq. 4.3.0.7, $F = q*f$, is not applicable, <u>viz</u>. when an increase in numbers produces also an increase of the area occupied by the stock. Then one effort unit (one trawl haul) would remove a smaller fraction of the stock and thus produce a smaller fishing mortality. Figs. 4.3.0.3A and C illustrate the way a stock must behave to conform to Eq. 4.3.0.7, namely that the area occupied remains constant for different stock numbers.

4.3.1 Heincke's method

There are many ways of estimating Z from numbers in the catch. We shall describe one that dates as far back as 1913. Several other methods are described in Ricker (1975).

To introduce the method we start with a formula from common algebra:

$$\sum_{i=0}^{\infty} a^i = \frac{1}{1-a}, \text{ which is valid if } a < 1$$

If $a = \exp(-Z)$ then for $Z > 0$ it follows that $a < 1$ and consequently the following expression is valid:

$$\sum_{t=0}^{\infty} \exp(-Z*t) = \frac{1}{1-\exp(-Z)} \qquad (4.3.1.1)$$

When we multiply Eq. 4.3.1.1 by the same constant, $N(0)$, on both sides of the equal sign we get:

$$\sum_{t=0}^{\infty} N(0)*\exp(-Z*t) = \frac{N(0)}{1-\exp(-Z)} \qquad (4.3.1.2)$$

Assuming total mortality, Z, to remain constant we have (cf. Eq. 4.2.6):

$N(t) = N(0)*\exp(-Z*t)$ and Eq. 4.3.1.2 can be written as

$$\sum_{t=0}^{\infty} N(t) = \frac{N(0)}{1-\exp(-Z)} \qquad (4.3.1.3)$$

It can be shown that Eq. 4.3.1.3 is equivalent to:

$$Z = -\ln \frac{N(1) + N(2) + N(3) + N(4) + \ldots}{N(0) + N(1) + N(2) + N(3) + N(4) + \ldots}$$

or

$$Z = -\ln \left[\sum_{t=1}^{\infty} N(t) / \sum_{t=0}^{\infty} N(t) \right] \qquad (4.3.1.4)$$

Eq. 4.3.1.4 is "<u>Heincke's formula</u>" (Heincke, 1913, in Ricker, 1975).

Eq. 4.3.1.4 should still hold if CPUE can be assumed proportional to the stock number so, replacing the N's by CPUE(t), Eq. 4.3.1.4 can be used in the form:

$$Z = -\ln \frac{CPUE(1) + CPUE(2) + CPUE(age\ 3\ and\ older)}{CPUE(0) + CPUE(1) + CPUE(2) + CPUE(age\ 3\ and\ older)}$$

where the CPUE of the oldest age groups is lumped. The reason for lumping them is that it is often so that age groups can be relatively easily separated for the first two or three (or perhaps four) age groups, whereas it is difficult to separate the older age groups (Ricker, 1975). Also it is sometimes neces-sary to exclude the youngest age group when it is not fully exploited by the fishery. In that case CPUE(0) should not be included in the denominator and CPUE(1) not in the numerator.

4.3.2 Robson and Chapman's method

Another method was introduced by Robson and Chapman, 1961 (in Ricker, 1975). They showed that the best estimate of Z from age composition data i.e. the numbers caught per age group, is:

$$Z = -\ln \frac{A}{B+A-1} \qquad (4.3.2.1)$$

where $A = N(1)+2*N(2)+3*N(3)+ \ldots$
$B = N(0)+N(1)+N(2)+N(3)+ \ldots$

The variance of the survival rate, $S = \exp(-Z)$, is

$$VAR(S) = S*\{(S-(A-1))/(B+A-2)\}$$

4.4 ESTIMATION OF Z FROM A LINEARIZED CATCH CURVE

A "linearized catch curve" is a graphical representation of the logarithms of numbers caught plotted against age.

The methods to be described in this section are essentially based on age composition data and as such they have been applied to temperate fish stocks. Because direct age reading is problematic for tropical fish stocks (see Section 3.2.1) the methods have to be converted into versions which take length composition data as input. This is possible when growth parameters are available so that lengths can be converted into ages by using the inverse von Bertalanffy equation (Eq. 3.3.3.3). The theoretical derivations start by using the age structured model, which will subsequently be converted into a corresponding length structured model.

4.4.1 The constant parameter system

The linearized catch curve methods for estimating Z are based on the assumption of a "constant parameter system". This section is therefore used to explain this concept.

Consider, as an example, a fish stock during the period 1971-1975 with a life span of five years. Let the numbers of survivors be those given in Table 4.4.1.1A. For simplicity we assume only one cohort per year, recruiting on 1 January. The figures in Table 4.4.1.1A are the numbers of survivors per cohort and per age group on 1 January. Because we consider many cohorts simultaneously, an index to specify the cohort, y, is needed. $N(y,t)$ is the number of survivors attaining age t from the cohort recruited at the start of year y-t. In this particular case t only takes the values t = 0,1,2,3 or 4. Note that a specific cohort can be followed diagonally across the Table. Every year the survivors of that cohort advance to a new age group. Each column, on the other hand, contains the numbers of survivors of five _different_ cohorts on 1 January of a particular year.

Table 4.4.1.1A contains parts of nine different cohorts, of which only one is given for its entire life span (the 1971 cohort). It appears that recruitment, $N(y,0)$, y = 1971, 1972, 1973, 1974 and 1975, has varied from year to year. The 1971 cohort and the 1973 cohort are strong ones, whereas the 1972 cohort is a weak one ("strong" means that $N(y,0)$ is considerably greater than the average recruitment).

Now, suppose that the recruitment, $N(y,0)$, remains constant every year, as is the case in Table 4.4.1.1B. Assume further that also F and M remain constant, then the number of survivors and the numbers caught would be the same for all cohorts. An inspection of Table B shows that in this case the number of survivors per year during the life span of a cohort equals the number of survivors within a particular year for each age group. Thus, in the case of constant recruitment it does not matter whether we consider one cohort over its entire life span, or if we consider all the different cohorts (the entire stock) in one particular year. (The major part of the theory of Beverton and Holt (1957) is based on the assumption of constant recruitment.)

We know for sure that the assumption of a constant parameter system is never strictly fulfilled in real life. However, we are often in a situation which forces us to make assumptions, which are known to be crude approximations to reality. It often happens that only by making such assumptions are we able to carry out an analysis of available data, and it is better to do a crude analysis than none at all.

We speak of a "pseudo-cohort" when working with data from one year assuming these to resemble those of a cohort during its life span. Thus, the numbers 2105, 2575, 155, 102 and 6 in Table 4.4.1.1A form a pseudo-cohort, whereas the numbers 2105, 736, 281, 109 and 29 form a real cohort.

Table 4.4.1.1 Illustration of the "constant parameter system" and the "variable parameter system". Note that the column of numbers in boxes consists of five different cohorts in the year 1971, whereas the diagonal of numbers in boxes consists entirely of the 1971 cohort. (For further explanation, see text)

A: VARIABLE PARAMETER SYSTEM

age*\year t	1971 N(71,t)	1972 N(72,t)	1973 N(73,t)	1974 N(74,t)	1975 N(75,t)
0	2105	1111	9560	1869	1236
1	2575	736	405	3817	618
2	155	1097	281	142	1193
3	102	58	298	109	59
4	6	38	18	138	29

COHORTS, number of survivors

B: CONSTANT PARAMETER SYSTEM

age*\year t	1971 N(71,t)	1972 N(72,t)	1973 N(73,t)	1974 N(74,t)	1975 N(75,t)
0	2560	2560	2560	2560	2560
1	942	942	942	942	942
2	346	346	346	346	346
3	127	127	127	127	127
4	47	47	47	47	47

COHORTS, number of survivors

*) Age from recruitment data

4.4.2 The linearized catch curve equation

Suppose that an estimate of the age composition of the catch during a year is available, i.e. pairs of observations:

age interval	numbers caught
t1-t2	C(t1,t2)
t2-t3	C(t2,t3)
t3-t4	C(t3,t4)
.	.
.	.

To develop a method to estimate Z from such data, the starting point is the catch equation (Eq. 4.2.7):

$$C(t1,t2) = N(t1) * \frac{F}{Z} * [1 - \exp(-Z*(t2-t1))]$$

Although the time interval from t1 to t2 is the first interval it will also be used in the following as the symbol for the general time interval, i.e. t2-t3, t3-t4, etc. Eq. 4.2.7 is not linear in t1 (or t2), so some transformation is required to make the analysis a linear regression (cf. Section 2.6). The first step is to replace N(t1) using Eq. 4.2.2 which gives:

$$C(t1,t2) = N(Tr) * \exp[-Z*(t1-Tr)] * \frac{F}{Z} * [1 - \exp(-Z*(t2-t1))]$$

The second step is to take logarithms on both sides. After rearranging the terms we get:

$$\ln C(t1,t2) = \ln N(Tr) + \ln \frac{F}{Z} + Z*Tr - Z*t1 + \ln[1 - \exp(-Z*(t2-t1))]$$

At least t1 now appears in one linear term (namely: -Z*t1). The terms ln N(Tr) + ln (F/Z) + Z*Tr form a constant, as N(Tr), Tr, F and Z are assumed to remain constant. To simplify the notation, this constant is named "d", and the equation then reads:

$$\ln C(t1,t2) = d - Z*t1 + \ln[1 - \exp(-Z*(t2-t1))] \qquad (4.4.2.1)$$

A linear expression in t1 has been obtained, except for the last term. Various methods suggest different ways of dealing with this term.

The first uses of this "linearized catch curve equation" date back to Edser (1908), Heincke (1913) and Baranov (1918) as reviewed in Ricker (1975). In the remaining part of Section 4.4 we shall discuss various special applications of Eq. 4.4.2.1.

4.4.3 The linearized catch curve based on age composition data

One of the methods commonly applied in temperate waters to estimate total mortality is the "linearized catch curve method with constant time intervals", which has been reviewed in Beverton and Holt (1956), Chapman and Robson (1960), Robson and Chapman (1961) and Ricker (1975).

If t2-t1 in Eq. 4.4.2.1 remains constant (e.g. when age groups are considered, so that t2-t1 = t3-t2 = t4-t3 = ... = 1 year), the non-linear term becomes a constant, which can be included in the intercept term. Thus, if the constant g is defined:

$$g = d + \ln[1 - \exp(-Z*(t2-t1))]$$

then Eq. 4.4.2.1 reads:

$$\ln C(t1,t2) = g - Z*t \quad \text{or}$$

$$\ln C(t,t+\Delta t) = g - Z*t \qquad (4.4.3.1)$$

Eq. 4.4.3.1 is the "<u>linearized catch curve equation with constant time intervals</u>", where the slope is -Z.

Example 13: Catch curve with constant time intervals, North Sea whiting

Table 4.4.3.1 shows an example of a linearized catch curve analysis based on age composition data. Most often such examples come from temperate waters where direct age reading is possible and where spawning is confined to a short period, once per year. This is also the case in this example which deals with whiting (<u>Merlangius merlangus</u>) caught in the North Sea.

Table 4.4.3.1 gives the annual age composition of whiting catches during the period 1974-1980 (from ICES, 1981a). The figures in Table 4.4.3.1 are numbers caught per year per age group (cohort), where

$C(y,t,t+1)$ = number caught in year y of age between t and t+1 years (in millions). For instance $C(1976,3,4) = 159$

For each year the total mortality, Z, (bottom row of Table 4.4.3.1) has been calculated using Eq. 4.4.3.1. These are linear regressions where x = age, y = ln C(y,t,t+1) and b = -Z. Thus, recruitment and other parameters are assumed to have remained constant from 1974 to 1980.

However, we could also calculate an overall estimate of Z for the seven year period by using the average number caught per age group, as shown in the last column of Table 4.4.3.1. By doing so we (more or less) circumvent the problem of assuming parameters to remain constant, as e.g. variation in recruitment is levelled out in the averaging process. Note that in this particular case the assumption of constant recruitment is not crucial. Except for the year 1974 the annual estimates of Z do not deviate much from the overall estimate.

Table 4.4.3.1 Linearized catch curve analysis based on age composition data. Number caught per year by age group (in millions caught per year) of North Sea whiting.
(From ICES, 1981a, see Fig. 4.4.3.1)

year y age t	1974	1975	1976	1977 C(y,t,t+1)	1978	1979	1980	average 1974-80	
0	599	239	424	664	685	478	330	488	not
1	678	860	431	1004	418	607	288	612	used
2	1097	390	1071	532	335	464	323	601	
3	275	298	159	269	203	211	243	237	used
4	40	54	75	32	69	86	80	62	in
5	6	9	13	18	8	25	31	15.7	ana-
6	1	8	3	5	5	3	8	4.7	ly-
7+	(6)	(0)	(1)	(0)	(1)	(1)	(1)	1.4	sis
Z	1.88	1.26	1.36	1.25	1.33	1.40	1.12	1.28	

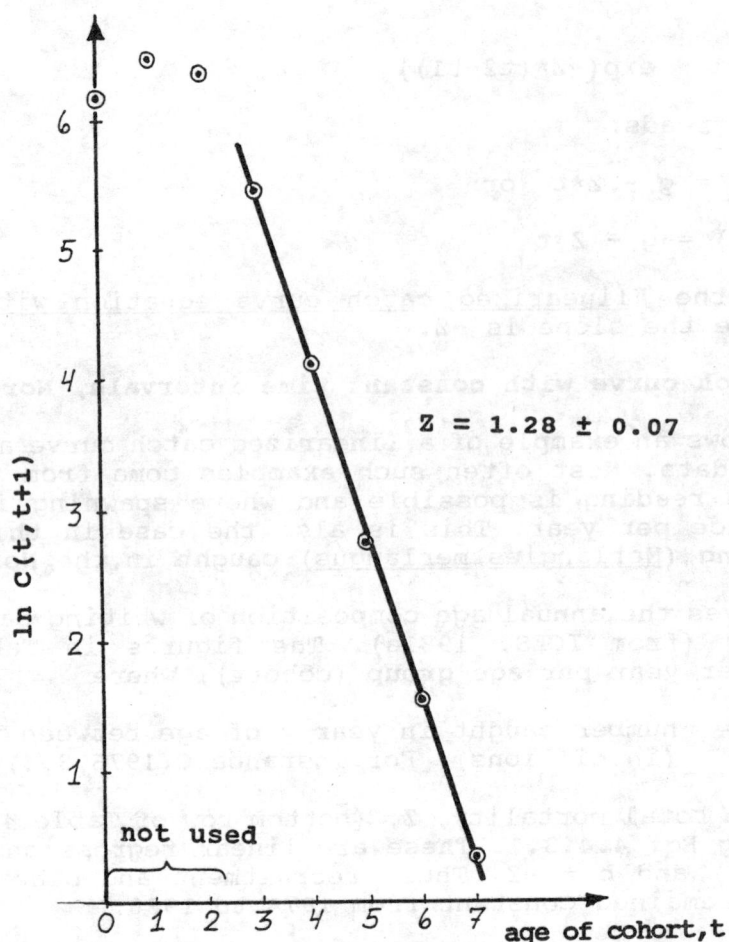

Fig. 4.4.3.1 Linearized catch curve based on average age composition of North Sea whiting catches 1974-80 (ICES, 1981a)

Fig. 4.4.3.1 shows the graph for the estimation of Z based on the average age composition. As appears from Table 4.4.3.1 data for the three first age groups have been excluded from all the regression analyses, since they do not conform to the straight line (see Fig. 4.4.3.1). The reason is that these age groups are not under full exploitation. The fishing mortality is lower than that of the larger (older) fish because the fish of age groups 0 to 2 are so small that they escape through the meshes of the trawl-net. (The major part of the North Sea whiting are caught by trawls with codend mesh sizes from 70 mm to 110 mm.) Another reason for expecting a reduced fishing mortality on the small fish is that all of these may not yet have recruited to the fishing grounds and are still on the nursery grounds.

Thus total mortality, $Z = M+F$, does not remain constant for all age groups although this was one of the assumptions behind the linearized catch curve analysis. One way to circumvent the problem is to exclude those age groups we suspect not to be under full exploitation, as was done in this example. Which age groups to exclude is a somewhat subjective choice. Usually the points systematically deviating from the straight line are excluded. However, it is hard to give a general rule to decide when this deviation is sufficiently large to justify the exclusion of the point. As a rule of thumb: whenever you are in doubt about a point, exclude it. In the case of Fig. 4.4.3.1 it is easy to make the choice since the five points used provide a very convincing fit to the model.

In the estimations of Z's for each year (bottom row of Table 4.4.3.1) age group 7+ has been excluded as well (indicated by brackets). The reason is that the number of specimens of age 7 or more (7+ groups) in each year has been estimated from small samples only. Because few fish survive to age 7 these estimates are subject to great uncertainty. However, for the overall estimate of Z the age group 7+ has been used, because in this case the samples cover seven years and the sample size is accordingly larger.

Using the procedure described in Section 2.4 the 95% confidence limits of Z can be calculated:

$$Z = -\text{slope} = -b = 1.284 \quad sy = 2.0316 \quad sx = 1.5811$$

variance of the slope:

$$sb^2 = \frac{1}{n-2}*((sy/sx)^2 - b^2) = \frac{1}{5-2}*((2.0316/1.5811)^2 - 1.284^2) = 0.0005$$

$$sb = \sqrt{0.0005} = 0.0227$$

$sb*t_{5-2} = 0.0227*3.18 = 0.07$, so that the confidence interval becomes

$$[1.21, 1.35] \quad \text{or:} \quad Z = 1.28 \pm 0.07$$

With the data of Table 4.4.3.1 the calculations could have been made on a cohort basis as well. Exercise 4.4.3 deals with this aspect.

(See **Exercise(s)** in Part 2).

4.4.4 The linearized catch curve based on age compositions with variable time intervals

Let us return to the general linearized catch curve model (Eq. 4.4.2.1)

$$\ln C(t1,t2) = d - Z*t1 + \ln[1 - \exp(-Z*(t2-t1))]$$

In case the time intervals t2-t1, t3-t2, t4-t3, ... are not constant the non-linear term, $\ln[1 - \exp(-Z*(t2-t1))]$ takes different values for different values of the time interval. As will appear from Section 4.4.5 this case is often relevant to tropical fish stocks. Two alternative ways of getting rid of the non-linear term have been suggested.

Cumulated catch curve equation

Jones and van Zalinge (1981) suggested to let t2 take a very high value (actually t2 = ∞ (infinity)). With t2 very large, the term $\exp(-Z*(t2-t1))$ would be close to zero, consequently $\ln(1 - \exp(-Z*(t2-t1))) = \ln 1 = 0$. Thus, if C(t,∞) designates all fish caught of age t and older, a linear relationship is achieved by

$$\ln C(t,\infty) = d - Z*t \qquad (4.4.4.1)$$

C(t,∞) is called the "<u>cumulated catch</u>", and Eq. 4.4.4.1 the "<u>cumulated catch curve equation</u>".

Linearized catch curve equation with variable time intervals

Van Sickle (1977) suggested a different approach which was further developed by Pauly (1983a). It is based on the observation that for small values of x, (x < 1.0):

$$\ln[1 - \exp(-x)] = \ln(x) - x/2 \quad \text{(approximately)}$$

If $Z*(t2-t1)$ is small it follows that approximately:

$$\ln[1 - \exp(-Z*(t2-t1))] = \ln[Z*(t2-t1)] - Z*(t2-t1)/2$$

Inserting this into Eq. 4.4.2.1 and rearranging terms gives:

$$\ln \frac{C(t1,t2)}{t2-t1} = d + \ln Z - Z*t1 - Z*(t2-t1)/2$$

Because t2-t1 must be small, it is called Δt. To simplify notation the constant $c = d + \ln Z$ is introduced. Then the equation reads:

$$\ln \frac{C(t,t+\Delta t)}{\Delta t} = c - Z*(t+\Delta t/2) \qquad (4.4.4.2)$$

which is linear in the interval midpoint, $(t+\Delta t/2)$. Eq. 4.4.4.2 is the so-called "<u>linearized catch curve equation with variable time intervals</u>". Note that only the slope has significance to the present analysis. The intercept is not used.

The name of the equation might lead to the misunderstanding that the method is applicable only when data for total catches are available. However, Eq. 4.4.4.2 can still be used even if only the percentage composition of the catches is known, because from these we get the absolute catches by multiplying all the percentages by a constant. A multiplication of all the values for C in Eq. 4.4.4.2 by a constant, K, will change the intercept, c, but not the slope, Z, since

$$\ln \frac{K*C(t,t+\Delta t)}{\Delta t} = c - Z*(t+\Delta t/2) \qquad \text{is equivalent to:}$$

$$\ln \frac{C(t,t+\Delta t)}{\Delta t} = c1 - Z*(t+\Delta t/2) \qquad \text{where}$$

$$c1 = c - \ln K$$

Thus, if you have a sample from a fish population (for instance from a research survey or a market), but the total catch is unknown, you can use Eq. 4.4.4.2. Therefore the linearized catch curve analysis is often used for survey data, whereas cohort analysis (introduced in Chapter 5) is used for the analysis of total catch data. It should be noted that the example used in the next section to illustrate a linearized catch curve analysis based on length composition data (see Table 4.4.5.1) is a sample of only 3816 fish, which was not raised to the total catch.

4.4.5 <u>The linearized catch curve based on length composition data</u>

The method to be described here does not, as Eq. 4.4.3.1, assume a direct age reading but uses the von Bertalanffy growth equation to convert length into age. This model is discussed in Pauly (1983a, 1984a and b). It is often referred to as the "<u>length-converted catch curve</u>" or the "<u>linearized length-converted catch curve</u>".

To make the catch curve usable for length data it is in principle necessary to replace in Eq. 4.4.4.2 the t's (ages) by L's (lengths):

$$\ln \frac{C(t,t+\Delta t)}{\Delta t} = c - Z*(t+\Delta t/2)$$

However, what is actually done is to convert length data into age data, using the inverse von Bertalanffy growth equation (Eq. 3.3.3.2)

$$t(L) = t_o - \frac{1}{K}*\ln(1 - \frac{L}{L_\infty})$$

The conversion of lengths into ages is fairly complex, because the amount of time needed for a fish to grow through a given length interval increases continuously as it gets older. Fig. 4.4.5.1 shows the transformation of length groups into age intervals. The growth curve used is the average growth curve for the entire cohort, from which large individual deviations can occur. These deviations are relatively small for small fish, but may be relatively large for large fish. In other words: the larger the fish, the longer time it will take to grow through a length group and therefore the relationship between length and age becomes more inaccurate for large fish. For example, it takes less than half a year for Upeneus vittatus to grow from 4 to 8 cm, but much more than a year to grow from 16 to 20 cm.

The biggest fish in the catch are often bigger because they grow faster, not because they are older as we assume in the inverse von Bertalanffy equation.

As an example consider the resolution of the length-frequency sample in Fig. 3.2.2.2 into normally distributed cohorts. The length group 15-16 cm, consists only of members of the 1983 spring cohort (as appears from Table 3.2.1.1), whereas the length group 45-46 cm contains three cohorts:

1. Fast growing members of the 1981 autumn cohort
2. Medium/fast growing members of the 1981 spring cohort
3. Medium growing members of the 1980 autumn cohort

The age corresponding to a certain length can be calculated, when t_o is ignored or assumed to be 0.

Let age t correspond to L1 and age t+Δt to L2:

L1 = L(t) and L2 = L(t+Δt)

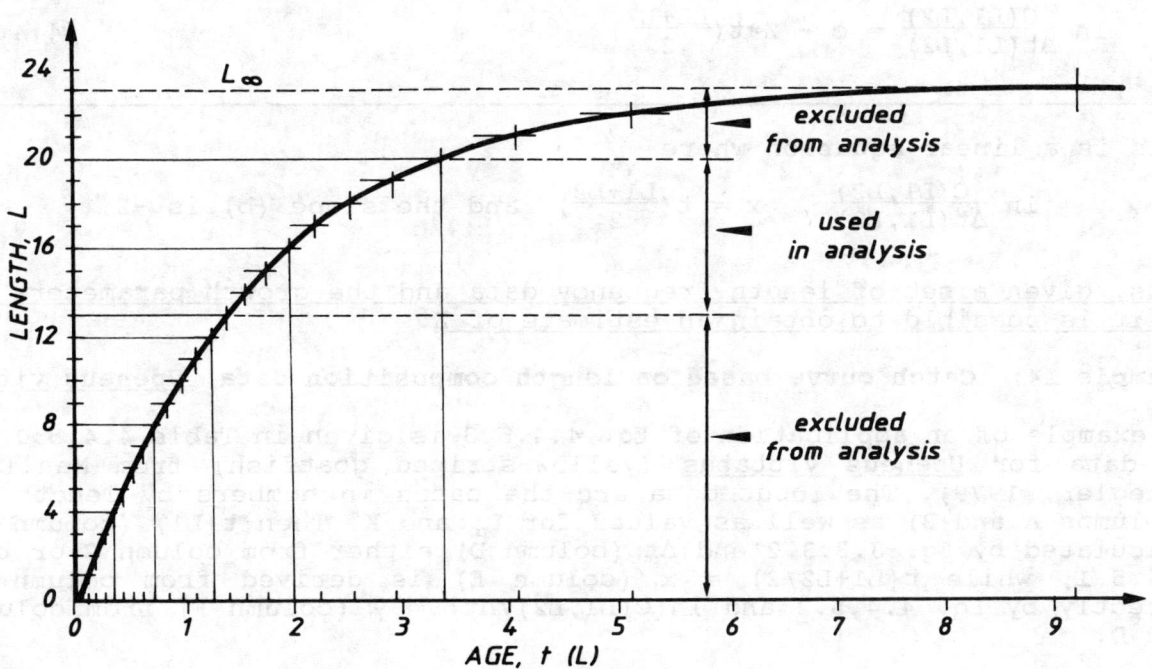

Fig. 4.4.5.1 Conversion of 1-cm length groups of Upeneus vittatus into age intervals, by the equation:

$$t(L) = t_o - \frac{1}{K} \ast \ln(1 - L/L_\infty) \quad (t_o = 0, K = 0.59 \text{ and } L_\infty = 23.1)$$

With this one-to-one correspondence between the age interval [t,t+Δt] and the length interval [L1,L2] we can change the notation for the number caught:

$$C(t,t+\Delta t) = C(L1,L2)$$

In the present context Δt is the time it takes for an average fish to grow from length L1 to length L2, so we obtain Δt by subtracting the two inverse von Bertalanffy equations (Eq. 3.3.3.2) corresponding to L2 and L1 respectively and obtain:

$$\Delta t = t(L2) - t(L1) = \frac{1}{K} * \ln \frac{L\infty - L1}{L\infty - L2} \qquad (4.4.5.1)$$

It is now possible to calculate Δt directly from the growth parameters K and L_∞ and the corresponding L1 and L2, as well as, of course by subtracting t(L1) from t(L2).

The term (t+Δt/2) of Eq. 4.4.4.2 can also be converted into an expression in length (L1 and L2), based on a suggestion by Pauly, viz., that the age interval midpoint (t+Δt/2) can be converted into a length-based midpoint by assuming that

t(L1) + Δt/2 is approximately equal to

$$t(\frac{L1+L2}{2}) = t_o - \frac{1}{K} * \ln(1 - \frac{L1+L2}{2L_\infty}) \qquad (4.4.5.2)$$

From these equations the so-called "<u>linearized length-converted catch curve</u>" can be derived:

$$\boxed{\ln \frac{C(L1,L2)}{\Delta t(L1,L2)} = c - Z * t(\frac{L1+L2}{2}) \qquad (4.4.5.3)}$$

This is a linear equation where

$$y = \ln \frac{C(L1,L2)}{\Delta t(L1,L2)}, \quad x = t(\frac{L1+L2}{2}) \quad \text{and the slope (b) is } -Z.$$

<u>Thus, given a set of length-frequency data and the growth parameters K and $L\infty$ it is possible to obtain an estimate of Z.</u>

Example 14: Catch curve based on length composition data, <u>Upeneus vittatus</u>

An example of an application of Eq. 4.4.5.3 is given in Table 4.4.5.1 based on data for <u>Upeneus vittatus</u> (yellow-striped goatfish) from Manila Bay (Ziegler, 1979). The input data are the catch in numbers by length group (columns A and B) as well as values for L_∞ and K. Then t(L1) (column C) is calculated by Eq. 3.3.3.2 and Δt (column D) either from column C or by Eq. 4.4.5.1, while t(L1+L2/2) = x (column E) is derived from column C or directly by Eq. 4.4.5.2 and ln(C(L1,L2)/Δt) = y (column F) from columns B and D.

The next step is to plot x against y (Fig. 4.4.5.2) and to decide which points should be used for a regression analysis, of which the slope b corresponds to -Z.

The confidence limits are calculated for various numbers of observations, so as to determine the best estimate of Z. (A similar procedure should be followed in Exercise 4.4.5.)

Table 4.4.5.1 Linearized catch curve based on length composition data for *Upeneus vittatus* from Manila Bay, Philippines (from Ziegler, 1979).
$L_\infty = 23.1$ cm, $K = 0.59$ per year (see Fig. 4.4.5.1)

A	B	C	D	E	F	G	H
L1-L2	C(L1,L2)	t(L1)	Δt	$t(\frac{L1+L2}{2})$	$\ln \frac{C(L1,L2)}{\Delta t}$	Z	remarks
		a)	b)	c)			
				(x)	(y)		
6- 7	3	0.510	0.102	0.56	3.38	-	not
7- 8	143	0.612	0.109	0.67	7.18	-	used;
8- 9	271	0.721	0.116	0.78	7.76	-	not
9-10	318	0.837	0.125	0.90	7.86	-	under
10-11	416	0.961	0.134	1.03	8.04	-	full
11-12	488	1.096	0.146	1.17	8.11	-	exploi-
12-13	614	1.242	0.160	1.32	8.25	-	tation
13-14	613	1.402	0.177	1.49	8.15	-	portion
14-15	493	1.579	0.197	1.67	7.83	-	used
15-16	278	1.776	0.223	1.88	7.13	2.64	in the
16-17	93	2.000	0.257	2.12	5.89	3.61	regres-
17-18	73	2.257	0.303	2.40	5.48	3.20	sion
18-19	7	2.560	0.370	2.74	2.94	4.03	analy-
19-20	2	2.930	0.473	3.15	1.44	4.19	sis
20-21	2	3.404	0.659	3.70	1.11	-	not used;
21-22	0	4.064	1.094	4.53	-	-	too close
22-23	1	5.160	4.047	6.19	-1.40	-	to L_∞
23-24	1	-	-	-	-	-	

Details of the five regression analyses:

L1-L2	slope b Z	number of obs. n	Student's distrib. 95% t_{n-2}	variance of slope sb^2	confidence limits $Z \pm t_{n-2}*sb$
13-14	-	1	-	-	-
14-15	-	2	-	-	-
15-16	2.64	3	12.70	0.198	2.64 ± 5.65
16-17	3.61	4	4.30	0.283	3.61 ± 2.28
17-18	3.20	5	3.18	0.121	3.20 ± 1.11
18-19	4.03	6	2.78	0.198	4.03 ± 1.24
19-20	4.19	7	2.57	0.087	4.19 ± 0.76

a) Eq. 3.3.3.2
b) Eq. 4.4.5.1
c) Eq. 4.4.5.2

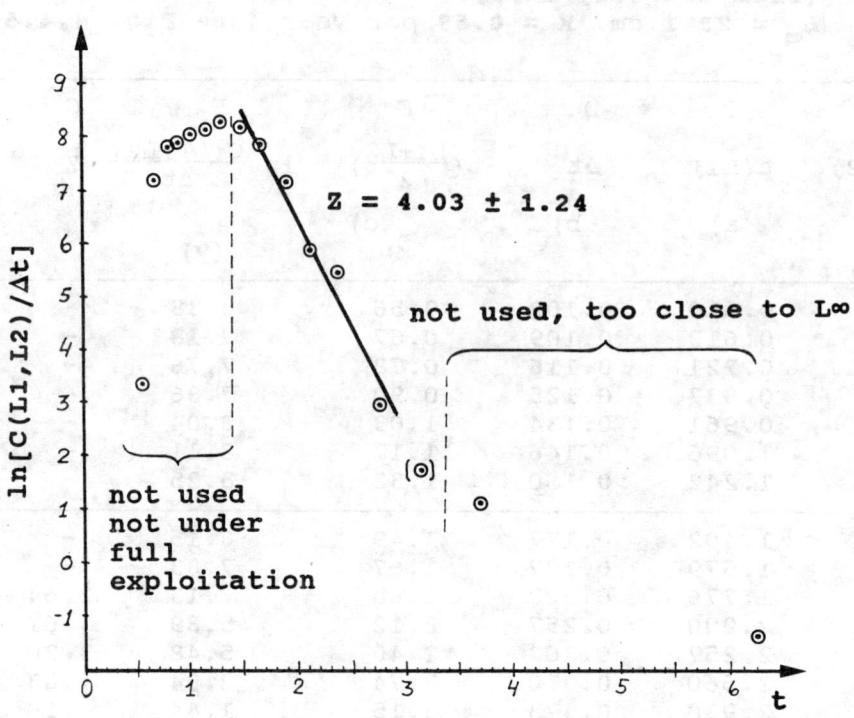

Fig. 4.4.5.2 **Linearized catch curve based on length composition data for Upeneus vittatus (from Ziegler, 1979) compare Table 4.4.5.1**

As was the case for the age-based linearized catch curve (compare Table 4.4.3.1), some of the observations have been excluded from the regression analysis in Table 4.4.5.1. The first seven length groups (6-13 cm) form the ascending part of the curve. These fish are considered not yet fully recruited to the fishery. The last four length groups are excluded as well, the reasons being:

1) Small numbers in the samples (as for the age-based linearized catch curve)

2) When approaching L_∞ the relationship between age, $t(L)$, and length, L, becomes uncertain (compare with the problems with the von Bertalanffy plot discussed in Section 3.3.3).

Point two above is an important one for length-based linearized catch curves. It is a good reason for never using the largest length groups.

Let us go back to the example of Table 4.4.5.1. Only the length groups 13-20 cm are used in the analysis of the data. To assess the sensitivity due to the choice of points to be excluded, the slope $b = Z$ has been calculated from 3 points, 4 points, ..., and 7 points as shown at the bottom of Table 4.4.5.1. The first Z is calculated for the lengths 13-16 cm, the next Z for lengths 13-17 cm, ..., etc. Confidence limits have been calculated using the method described in Section 2.4. In the present case the conclusion is that Z is somewhere between 4.03-1.24 = 2.8 and 4.03+1.24 = 5.2 per year, using the six intervals from 13 to 19 cm. For the length groups 16-20 cm, $Z*\Delta t > 1$ which is not small, therefore the approximation behind the linearized catch curve ($\ln[1 - \exp(-x)] = \ln x - x/2$, see Section 4.4.4) is a crude one in this case.

(See **Exercise(s)** in Part 2).

4.4.6 The cumulated catch curve based on length composition data. (The Jones and van Zalinge method)

To convert the "cumulated catch curve" (Eq. 4.4.4.1) into an equation that can be used for a length-based analysis, the inverse von Bertalanffy equation (Eq. 3.3.3.2)

$$t(L) = t_o - \frac{1}{K}*\ln(1 - \frac{L}{L_\infty})$$

is inserted into the cumulated catch curve equation (Eq. 4.4.4.1):

$$\ln C(t,\infty) = d - Z*t$$

which gives the following results:

$$\ln C(L,L_\infty) = d - Z*[t_o - \frac{1}{K}*\ln(1-L/L_\infty)]$$

which can be converted into

$$\ln C(L,L_\infty) = d - Z*t_o - \frac{Z}{K}*\ln L_\infty + \frac{Z}{K}*\ln(L_\infty-L)$$

The first three terms are constants which can conveniently be renamed:

$$a = d - Z*t_o - \frac{Z}{K}*\ln L_\infty$$

The "Jones and van Zalinge equation" has thus been derived:

$$\ln C(L,L_\infty) = a + \frac{Z}{K}*\ln(L_\infty-L) \qquad (4.4.6.1)$$

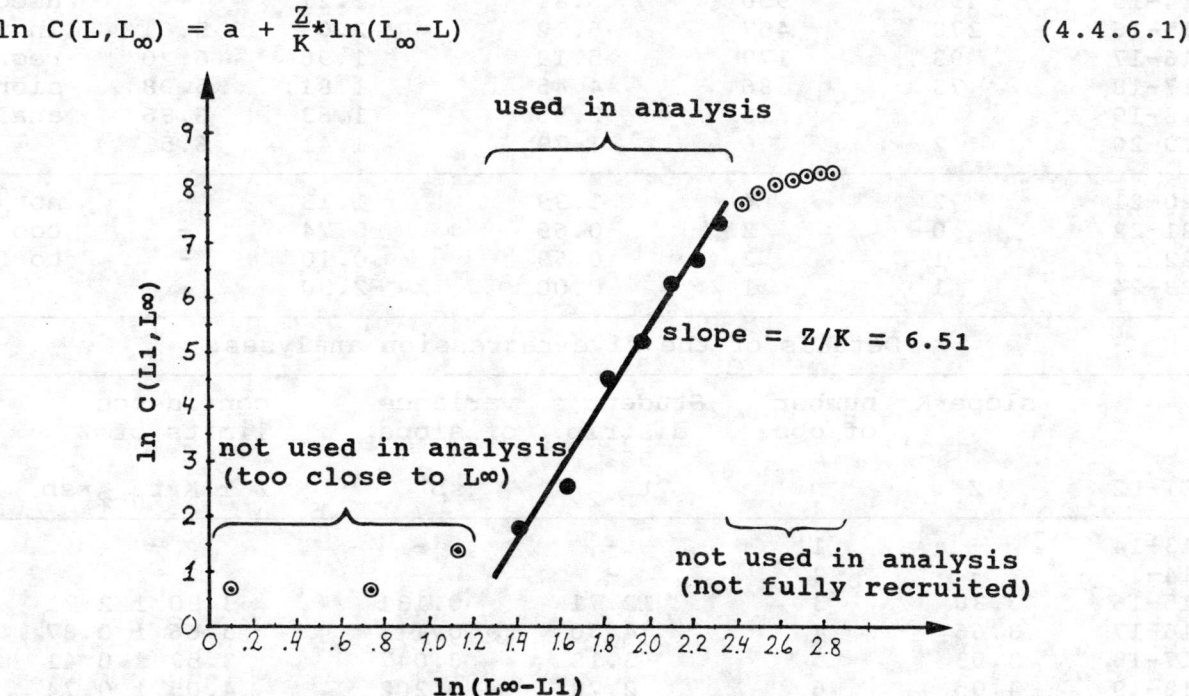

Fig. 4.4.6.1 Cumulated catch curve based on length composition data for *Upeneus vittatus*. The line drawn is estimated from seven points. Data from Table 4.4.6.1

where $C(L,L_\infty)$ stands for the cumulated catch of fish of length L and above. The slope estimated by the linear regression Eq. 4.4.6.1 is Z/K so that an estimate of Z is obtained from:

$$Z = K*slope$$

Example 15: The Jones and van Zalinge method, <u>Upeneus vittatus</u>

Table 4.4.6.1 and Fig. 4.4.6.1 show an application of Eq. 4.4.6.1 to <u>Upeneus vittatus</u> (the data used in the foregoing section). Note that the basic data also for this method is the sample - not the total catch (compare Sections 4.4.4-4.4.5). The results of the Jones and van Zalinge analysis using the length groups 13 and 19 cm are similar to the results from the linearized catch curve analysis (Table 4.4.5.1).

Table 4.4.6.1 Cumulated catch curve based on length composition data for <u>Upeneus vittatus</u> from Manila Bay, Philippines (from Ziegler, 1979).
$L_\infty = 23.1$ cm, $K = 0.59$ per year (see Fig. 4.4.6.1)

L1-L2	catch $C(L1,L2)$	cumulated catch $C(L1,L_\infty)$	ln cumulated catch ln $C(L1,L_\infty)$ (y)	ln$(L_\infty-L1)$ (x)	slope Z/K	remarks
6- 7	3	3816	8.25	2.84	-	not
7- 8	143	3813	8.25	2.78	-	used,
8- 9	271	3670	8.21	2.71	-	not
9-10	318	3399	8.13	2.65	-	under
10-11	416	3081	8.03	2.57	-	full
11-12	488	2665	7.89	2.49	-	exploi-
12-13	614	2177	7.69	2.41	-	tation
13-14	613	1563	7.35	2.31	-	portion
14-15	493	950	6.86	2.21	-	used
15-16	278	457	6.12	2.09	5.61	in the
16-17	93	179	5.19	1.96	6.20	regres-
17-18	73	86	4.45	1.81	5.98	sion
18-19	7	13	2.56	1.63	6.86	analysis
19-20	2	6	1.79	1.41	6.51	
20-21	2	4	1.39	1.13	-	not used
21-22	0	2	0.69	0.74	-	too close
22-23	1	2	0.69	0.10	-	to L_∞
23-24	1	1	0.00	-2.30	-	

Details of the five regression analyses:

L1-L2	slope*K Z	number of obs. n	Student's distrib. 95% t_{n-2}	variance of slope sb^2	confidence limits of Z $Z \pm K*t_{n-2}*sb$
13-14	-	1	-	-	-
14-15	-	2	-	-	-
15-16	3.30	3	12.71	0.131	3.30 ± 2.71
16-17	3.66	4	4.30	0.118	3.66 ± 0.87
17-18	3.53	5	3.18	0.047	3.53 ± 0.41
18-19	4.05	6	2.78	0.202	4.05 ± 0.74
19-20	3.84	7	2.57	0.110	3.84 ± 0.50

(See **Exercise(s)** in Part 2).

GENERAL LINEARIZED CATCH CURVE MODEL

$$\ln C(t1,t2) = d - Z*t1 + \ln[1 - \exp(-Z*(t2-t1))]$$

(Eq. 4.4.2.1)

Models based on age composition data	$\Delta t = t2-t1$ is constant **LINEARIZED CATCH CURVE EQUATION WITH CONSTANT TIME INTERVALS** $$\ln C(t, t+\Delta t) = g - Z*t$$ (Eq. 4.4.3.1)	
	$\Delta t = t2-t1$ is variable	
	Δt = small **LINEARIZED CATCH CURVE EQUATION WITH VARIABLE TIME INTERVALS** $$\ln \frac{C(t, t+\Delta t)}{\Delta t} = c - Z*(t+\Delta t/2)$$ (Eq. 4.4.4.2)	$\Delta t = \infty$ **CUMULATED CATCH CURVE EQUATION** $$\ln C(t, \infty) = d - Z*t$$ (Eq. 4.4.4.1)
Conversion of age into length	$$\Delta t(L1, L2) = \frac{1}{K} * \ln \frac{L_\infty - L1}{L_\infty - L2}$$ (Eq. 4.4.5.1) and $t + \Delta t/2 \approx$ $$t_o - \frac{1}{K} * \ln\left(1 - \frac{L1+L2}{2L_\infty}\right)$$ (Eq. 4.4.5.2)	$$t(L) = t_o - \frac{1}{K} * \ln\left(1 - \frac{L}{L_\infty}\right)$$ (Eq. 3.3.3.2)
Models based on length composition data	**LINEARIZED LENGTH CONVERTED CATCH CURVE** $$\ln \frac{C(L1, L2)}{\Delta t(L1, L2)} = c - Z*t\left(\frac{L1+L2}{2}\right)$$ (Eq. 4.4.5.3)	**JONES AND VAN ZALINGE METHOD** $$\ln C(L, L_\infty) = a + \frac{Z}{K} * \ln(L_\infty - L)$$ (Eq. 4.4.6.1)

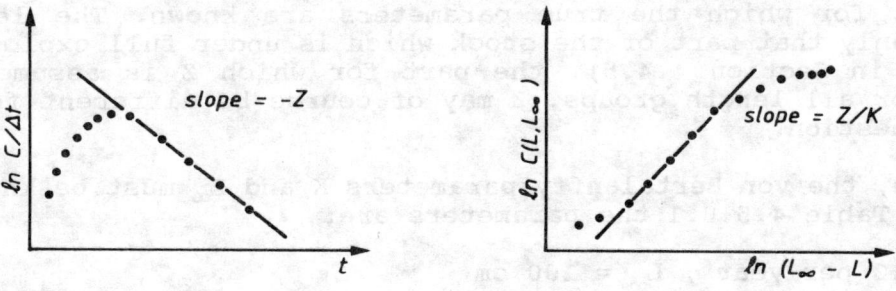

Fig. 4.4.7.1 Summary of models presented in Section 4.4

4.4.7 Summary of the linearized catch curve methods

Fig. 4.4.7.1 shows a summary of the versions of the linearized catch curve (Eq. 4.4.2.1) discussed in Section 4.4. These all originated from the catch equation (or Baranov's equation), Eq. 4.2.7. The following five models were introduced, three based on <u>age composition data</u>:

1) The linearized catch curve with constant time intervals. Eq. 4.4.3.1

2) The linearized catch curve with variable time intervals. Eq. 4.4.4.2 (generalizes Eq. 4.4.3.1)

3) The cumulated catch curve. Eq. 4.4.4.1

and two based on <u>length composition data</u>:

4) The linearized length-converted catch curve. Eq. 4.4.5.3, derived from Eq. 4.4.4.2

5) The Jones and van Zalinge method. Eq. 4.4.6.1, derived from Eq. 4.4.4.1

Sparre (1990) pointed out that there are problems with length-based stock assessment methods when growth is seasonal, and in particular the length-converted catch curve method. In the same issue of Fishbyte, Pauly (1990) describes a new method for the construction of length-converted catch curves, which takes the seasonality of growth into account. The bias in the estimation of Z for small short-lived species, which are exposed to strong seasonal fluctuations in growth due to changing environmental conditions, is to a large extent removed by this new method. It gives much lower values for Z than those obtained with the length-converted methods described in sub-sections 4.4.5 and 4.4.6.

4.5 BEVERTON AND HOLT'S Z-EQUATIONS

The first method dealt with in this section estimates Z from the mean length of the fish in the catch and the von Bertalanffy parameters K and L_∞. Thus, the data requirements are even less than those for the length-based linearized catch curve methods, for which the relative (not necessarily the absolute) size composition of the catch should be known (see Sections 4.4.4 to 4.4.6). The derivation of Beverton and Holt's Z equations, however, is somewhat more complicated from a mathematical point of view. In the following we attempt to explain and justify the methods by aid of a numerical example, skipping the mathematics. It is finally shown that even L_∞ can be estimated together with an estimate of Z/K.

Assume that samples of length compositions of catches have been collected from a fishery in 1960, 1970 and 1980, and that the results are those shown in Table 4.5.0.1. The example used here is a hypothetical data set, that is, a data set for which the true parameters are known. The length groups represent only that part of the stock which is under full exploitation (cf. discussion in Section 4.4.5), the part for which Z is assumed to remain constant for all length groups. Z may of course be different for the three years in question.

In addition, the von Bertalanffy parameters K and L_∞ must be known. For the example in Table 4.5.0.1 the parameters are:

$K = 0.3$ per year , $L_\infty = 100$ cm

- 141 -

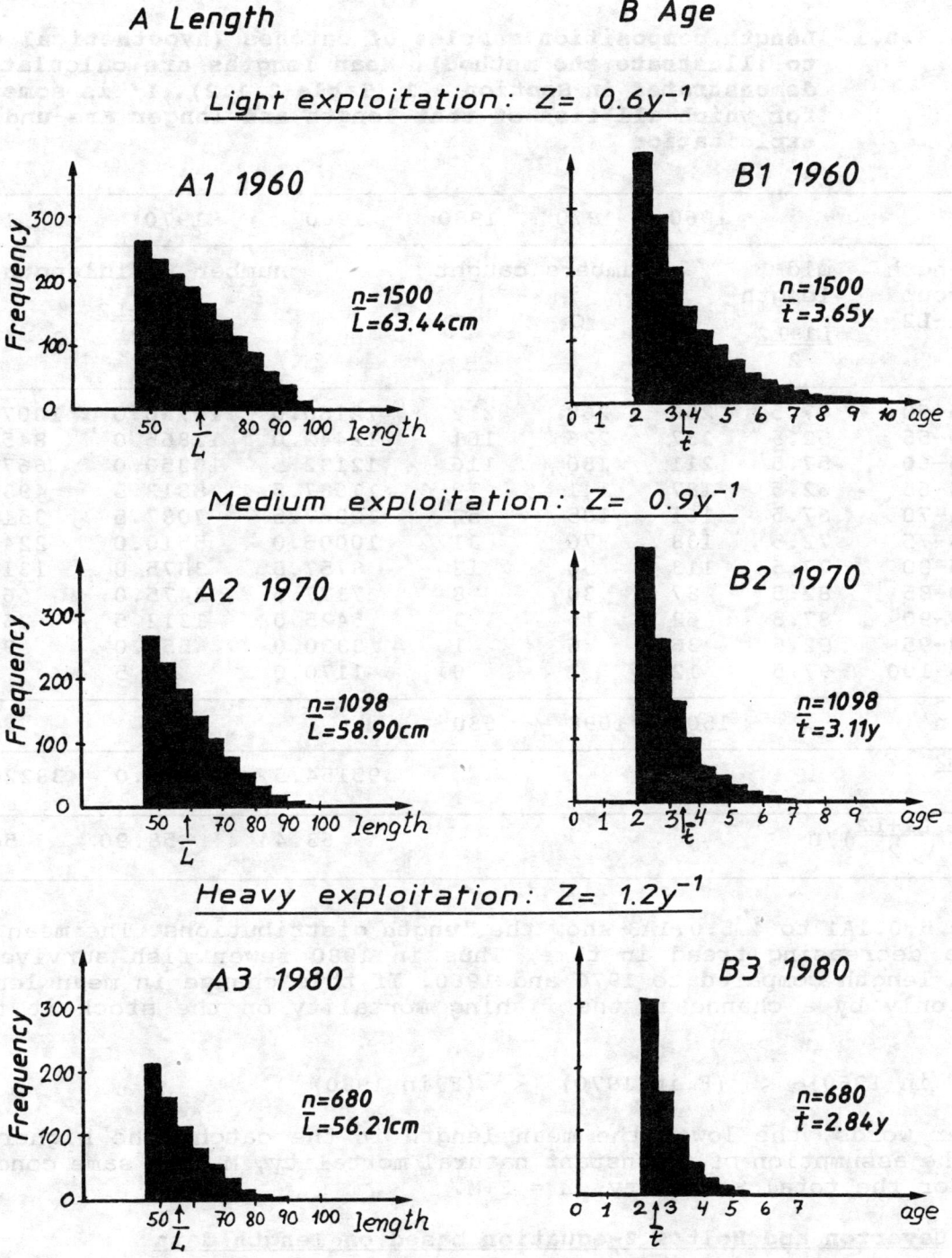

Fig. 4.5.0.1 Hypothetical example to illustrate Beverton and Holt's Z-equation (Eq. 4.5.0.1). K = 0.3, L' = 45 cm and L_∞ = 100 cm

Table 4.5.0.1 Length composition samples of catches (hypothetical example to illustrate the method). Mean lengths are calculated as demonstrated in Section 2.1 (Table 2.1.2). L' is some length for which all fish of that length and longer are under full exploitation

year			1960	1970	1980	1960	1970	1980
	length group L_1-L_2	mid-length $\frac{L_1+L_2}{2}$	numbers caught C			number * midlength $C*\frac{L_1+L_2}{2}$		
$L'=$ 45-50		47.5	256	268	212	12160.0	12730.0	10070.0
50-55		52.5	237	226	161	12442.0	11865.0	8452.5
55-60		57.5	211	180	116	12132.5	10350.0	6670.0
60-65		62.5	187	141	79	11687.5	8812.5	4937.5
65-70		67.5	161	105	52	10867.5	7087.5	3510.0
70-75		72.5	138	76	31	10005.0	5510.0	2247.5
75-80		77.5	113	50	17	8757.5	3875.0	1317.5
80-85		82.5	87	30	8	7177.5	2475.0	660.0
85-90		87.5	62	15	3	5425.0	1312.5	262.5
90-95		92.5	36	6	1	3330.0	555.0	92.5
95-100		97.5	12	1	0	1170.0	97.5	0.0
Total, n			1500	1098	680			
$\Sigma C * \frac{L_1+L_2}{2}$						95154.5	64670.0	38220.0
$\overline{L} = (\Sigma C * \frac{L_1+L_2}{2})/n$						63.44	58.90	56.21

Figs. 4.5.0.1A1 to 4.5.0.1A3 show the length distributions. The mean length $\overline{L}$ has a decreasing trend in time. Thus in 1980 fewer fish survived to a certain length compared to 1970 and 1960. If this change in mean length is caused only by a change in the fishing mortality on the stock it follows that:

(F in 1960) < (F in 1970) < (F in 1980)

In other words, the lower the mean length in the catch, the higher is F. Under the assumption of a constant natural mortality, M, the same conclusion holds for the total mortality, Z = F+M.

4.5.1 Beverton and Holt's Z-equation based on length data

Beverton and Holt (1956) showed that the functional relationship between Z and $\overline{L}$ is:

$$Z = K * \frac{L\infty - \overline{L}}{\overline{L} - L'} \qquad (4.5.1.1)$$

where $\overline{L}$ is the mean length of fish of length L' and longer, while L' is "some length for which all fish of that length and longer are under full exploitation". Note that L' is the lower limit of the corresponding length interval. For the choice of L', the same comments as those given in Section 4.4.2 and 4.4.6 can be applied. In the example the value of L' = 45 cm has been chosen.

The Z-values for the three years in the example become (compare Table 4.5.0.1 and Fig. 4.5.0.1A):

$$Z(1960) = 0.3 * \frac{100.0 - 63.44}{63.44 - 45.0} = 0.6 \text{ per year}$$

$$Z(1970) = 0.3 * \frac{100.0 - 58.90}{58.90 - 45.0} = 0.9 \text{ per year}$$

$$Z(1980) = 0.3 * \frac{100.0 - 56.21}{56.21 - 45.0} = 1.2 \text{ per year}$$

The method is refined in Section 4.5.4.

(See **Exercise(s)** in Part 2).

4.5.2 Beverton and Holt's Z-equation based on age data

As was the case for the linearized catch curve method the length-based Beverton and Holt formula has an age-based parallel. It is mentioned here mainly because it illustrates a basic feature about the Beverton and Holt theory (cf. Section 1.3).

The age compositions corresponding to the length compositions in Table 4.5.0.1 are shown in Table 4.5.2.1 and Fig. 4.5.0.1B. The same reasoning as was used for the mean length can be used for the mean age: the larger the fishing mortality the smaller is the mean age, t, in the catch, or, the higher the mean age the smaller is F. This may be considered an implication of the functional relationship between age and length. However, the exact relationship between mean age and mean length is somewhat complicated. It should be noticed that the von Bertalanffy equation does not transform the mean age into the mean length, because the growth equation is not linear.

Table 4.5.2.1 Age composition of the samples given in Table 4.5.0.1

age group	1960	1979	1980
2.0- 2.5	390	399	308
2.5- 3.0	289	255	169
3.0- 3.5	214	162	92
3.5- 4.0	150	103	52
4.0- 4.5	118	66	28
4.5- 5.0	97	42	15
5.0- 5.5	64	27	8
5.5- 6.0	48	17	6
6.0- 6.5	35	11	2
6.5- 7.0	26	7	-
7.0- 7.5	19	4	-
7.5- 8.0	14	3	-
8.0- 8.5	11	2	-
8.5- 9.0	8	-	-
9.0- 9.5	6	-	-
9.5-10.0	4	-	-
10.0-10.5	3	-	-
10.5-11.0	2	-	-
11.0-11.5	2	-	-
sample size	1500	1098	680
Σ number*age	5475	3415	1931
mean age, $\bar{t}$	3.65	3.11	2.84

The age-based parallel to Eq. 4.5.1.1 is somewhat simpler. Beverton and Holt showed that:

$$Z = \frac{1}{\bar{t}-t'} \tag{4.5.2.1}$$

where $\bar{t}$ is the mean age of all fish of age t' and older, and where t' is "<u>some age for which all fish of that age and older are under full exploitation</u>". For the example age, t', corresponding to $L' = 45$ cm is

$$t'_{45} = t_o - \frac{1}{K}*\ln(1-L/L_\infty) = 0 - \frac{1}{0.3}*\ln(1-45/100) = 1.99 \text{ years}$$

and

$$Z(1960) = \frac{1}{3.65-2.0} = 0.6 \text{ per year}$$

$$Z(1970) = \frac{1}{3.11-2.0} = 0.9 \text{ per year}$$

$$Z(1980) = \frac{1}{2.84-2.0} = 1.2 \text{ per year}$$

4.5.3 Beverton and Holt's Z-equation based on length-at-first-capture

A third version of the equation exists which is mathematically equivalent to Eq. 4.5.1.1, but conceptually different. Consider a "<u>gear selection curve</u>" for, say, a trawl-net as shown in Fig. 4.5.3.1. The vertical axis shows the percentage of the fish entering the gear (trawl) that is retained by the meshes. Lc, or L50%, is the length at which 50% are retained and 50% escape through the meshes. Note that Lc < L'. (In Chapter 6 we shall further discuss properties of selection curves.)

This version of the Beverton and Holt Z-equation reads:

$$Z = K*\frac{L_\infty - \overline{Lc}}{\overline{Lc}-Lc} \tag{4.5.3.1}$$

where Lc is the "<u>length at which 50% of the fish entering the gear are retained</u>" and $\overline{Lc}$ is the "<u>average length of the entire catch</u>".

Eq. 4.5.3.1 has proved useful in many cases where estimates of Lc and $\overline{Lc}$ are available, but where $\bar{L}$ and L' are not known.

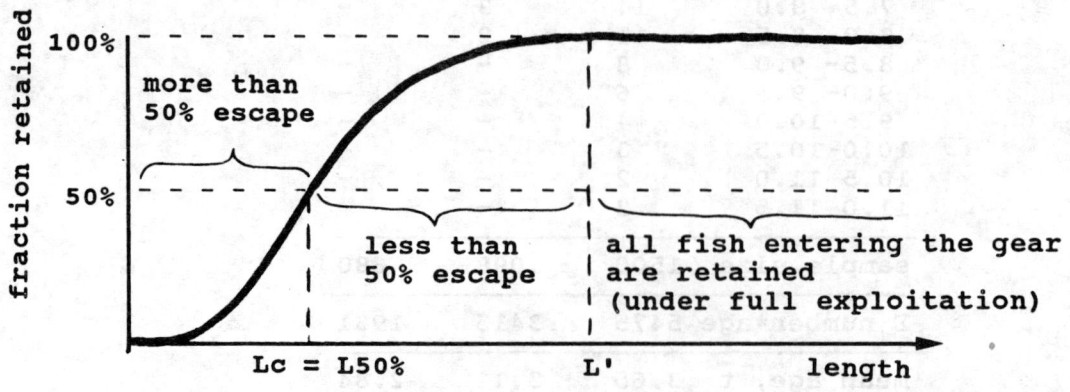

Fig. 4.5.3.1 Gear selection curve

4.5.4 The Powell-Wetherall method

Powell (1979), discussed in Wetherall et al. (1987), suggested a special application of Eq. 4.5.1.1 by which L_∞ and Z/K can be estimated. As L' can take any value equal to and above the smallest length under full exploitation, Eq. 4.5.1.1 can give a series of estimates of Z, namely one for each choice of L'. This makes it possible to turn Eq. 4.5.1.1 into a regression analysis with L' as the independent variable. A series of algebraic manipulations shows that Eq. 4.5.1.1 is equivalent to:

$$\overline{L} - L' = a + b*L' \qquad (4.5.4.1)$$

where

$$Z/K = -(1+b)/b \quad \text{and} \quad L_\infty = -a/b$$

or $\quad b = -K/(Z+K) \quad$ and $\quad a = -b*L_\infty$

Thus, plotting $\overline{L} - L'$ against L' gives a linear regression from which a and b can be estimated and hence L_∞ and Z/K (see Fig. 4.5.4.1). Powell (1979) actually gave a whole suite of different formulas for Z/K, of which Eq. 4.5.4.1 represents the simplest approach. This method is especially suitable for situations where little or nothing is known about the fish stock in question. The estimation of L_∞ is especially useful (compare Section 3.3.2).

It should be remembered that like the Beverton and Holt formula (Eq. 4.5.1.1) the method is based on the assumption of a constant parameter system which reduces its applicability.

Example 16: The Powell-Wetherall method

An example of the Powell-Wetherall method is given in Table 4.5.4.1 and Fig. 4.5.4.2. The calculation of $\overline{L}$ starts with the mid-length of the largest length class, in this case 46 cm.

The next value of $\overline{L}$ is calculated as follows: $(46*3+44*10)/13 = 44.462$. L' is the lower limit of each length-class. The values for $\overline{L} - L'$ are obtained by subtracting the corresponding value of L' from $\overline{L}$, for example $46.000-45 = 1.000$ and $44.462-43 = 1.462$. These values are then first plotted against L', in order to be able to judge which points lie on the straight line and should be included in the linear regression analysis. In this example the first two points were excluded, because they represent very few fish. (Note, the same data set has been used to illustrate the estimation of the selection ogive from a catch curve, see Section 6.5.)

Computer programs

The program "BHZWET" in the LFSA package of microcomputer programs (Sparre, 1987) can execute the Powell-Wetherall analysis as well as the estimation of Z from Beverton and Holt's Z-equation. The "COMPLEAT ELEFAN" package of microcomputer programs (Gayanilo, Soriano and Pauly, 1988) and FiSAT also contain such programs.

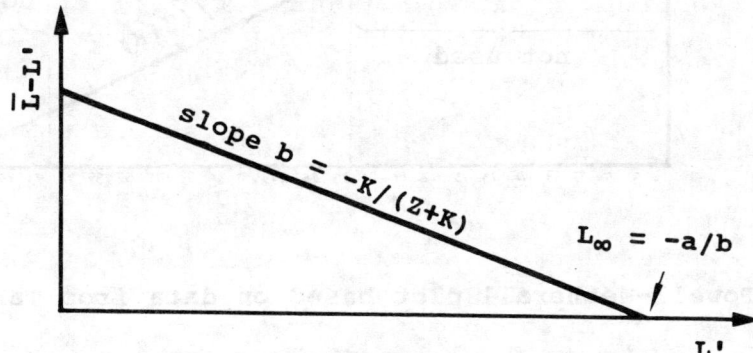

Fig. 4.5.4.1 Powell-Wetherall plot

Table 4.5.4.1 Powell-Wetherall method (see Fig. 4.5.4.2). The same data are used in Table 6.5.1

length interval L1-L2 (x) = L1	number caught C(L1,L2)	mean length of fish longer than L', $\bar{L}$	$\bar{L}-L'$ (y)	
3- 5	37	21.101	18.101	
5- 7	56	21.291	16.291	
7- 9	86	21.552	14.552	not
9-11	129	21.917	12.917	used
11-13	188	22.419	11.419	in
13-15	258	23.101	10.101	analysis
15-17	319	23.998	8.998	
17-19	352	25.108	8.108	
19-21	351	26.394	7.394	
21-23	324	27.801	6.801	
23-25	283	29.279	6.279	
25-27	239	30.792	5.792	
27-29	196	32.320	5.320	
29-31	158	33.852	4.852	used
31-33	123	35.392	4.392	in
33-35	93	36.926	3.926	analysis
35-37	69	38.447	3.447	
37-39	48	39.982	2.982	
39-41	31	41.516	2.516	
41-43	18	43.032	2.032	
43-45	10	44.462	1.462	not
45-47	3	46.000	1.000	used

intercept a: 11.671   confidence limits of a: [11.64 , 11.70]
slope b   : -0.2349 confidence limits of b: [-0.2359 , -0.2340]

$L_\infty = -a/b = 49.7$ $Z/K = -(1+b)/b = 3.26$

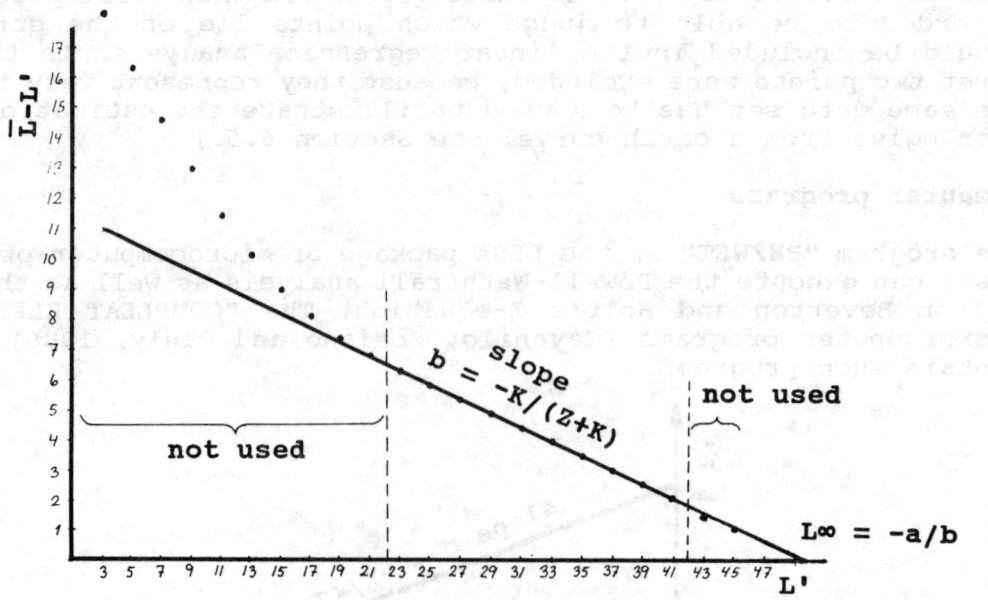

Fig. 4.5.4.2 Powell-Wetherall plot based on data from Table 4.5.4.1

4.6 A PLOT OF Z ON EFFORT FOR SEPARATE ESTIMATES OF F AND M

The estimate of Z derived by the methods described in the foregoing section can sometimes be used to obtain estimates of F, the fishing mortality and M, the natural mortality. This is possible if pairs of estimates of Z and effort are available for a number of time periods covering a wide range of efforts.

Let $Z(y)$ and $f(y)$ be the total mortality and the effort in year y. The method (Paloheimo, 1958, 1961 and 1980) is based on Eq. 4.2.3: $Z = F+M$, and the assumption that effort and fishing mortality are related in the simple manner of Eq. 4.3.0.7: $F(y) = q*f(y)$, where q is the catchability coefficient. Inserting Eq. 4.3.0.7 into Eq. 4.2.3 gives:

$$Z(y) = M + q*f(y) \qquad (4.6.1)$$

With Z as the dependent variable and f as the independent variable Eq. 4.6.1 becomes a linear regression with slope q and the natural mortality, M, as the intercept. The fishing mortality for the time period y, $F(y)$, is derived from:

$$F(y) = Z(y) - M$$

Example 17: Estimation of M and q of a tropical fish

Table 4.6.1 and Fig. 4.6.1 show an application of Eq. 4.6.1 to <u>Selaroides leptolepis</u> in the Gulf of Thailand (from Boonyubol and Hongskul, 1978). In this case Z is calculated by Eq. 4.5.3.1, i.e. the observations are Lc, the 50% retention length, and $\overline{Lc}$, the mean length of all fish in the catch. Effort is given in units of millions of trawling hours per year. Fig. 4.6.1 shows the plot of Z on effort.

Using linear regression we find:

slope : q = 0.2532
intercept: M = 2.034 per year

Table 4.6.1 Data for estimation of M and q for <u>Selaroides leptolepis</u> in the Gulf of Thailand (from Boonyubol and Hongskul, 1978) K = 1.16 per year, Lc = 10.0 cm, L_∞ = 20.0 cm

year y	effort [a] f(y)	mean length $\overline{Lc}$ cm	$Z = K*\dfrac{L_\infty - \overline{Lc}}{\overline{Lc} - Lc}$ (Eq. 4.5.3.1)
1966	2.08	13.25	2.41
1967	2.08	13.01	2.69
1968	3.50	12.99	2.72
1969	3.60	13.07	2.62
1970	3.80	12.37	3.73
1971	-	- no data	-
1972	7.19	12.30	3.88
1973	9.94	12.01	4.61
1974	6.06	12.60	3.30

a) effort in units of millions of trawling hours

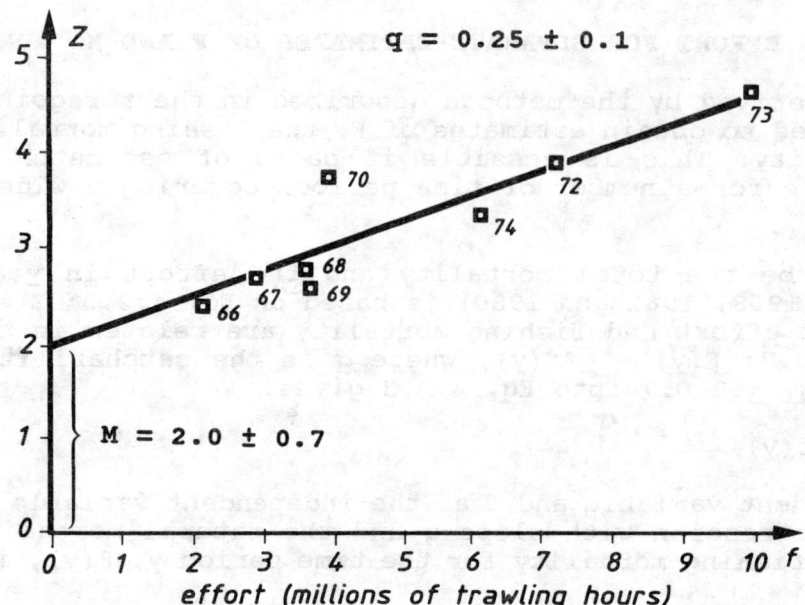

Fig. 4.6.1 Plot of total mortality, Z, on effort, f, for estimation of natural mortality, M, and catchability coefficient, q. Based on data from Table 4.6.1 (from Boonyubol and Hongskul, 1978)

Confidence limits are calculated as described in Section 2.4:

$$sq^2 = \frac{1}{n-2}*[(\frac{sx}{sy})^2 - q^2] = \frac{1}{6}*[(\frac{0.7724}{2.7423})^2 - 0.2532^2] = 0.002533$$

$$sq = 0.0503$$

$$sM^2 = sq^2*(\frac{n-1}{n}*sx^2 + \bar{x}^2) =$$

$$0.002533*(\frac{7}{8}*2.7423^2 + 4.7813^2) = 0.07457 \quad sM = 0.2731$$

The 95% confidence limits for q and M are

q: $[q - sq*t_{n-2}, q + sq*t_{n-2}] =$

$[0.253 - 0.0503*2.45, 0.253 + 0.0503*2.45] = [0.13, 0.38]$

M: $[M - sM*t_{n-2}, M + sM*t_{n-2}] =$

$[2.03 - 0.2731*2.45, 2.03 + 0.2731*2.45] = [1.36, 2.70]$

The estimates of M and q obtained this way are not precise. For example, they are based on the assumption that effort is proportional to fishing mortality, i.e, that q is constant - an assumption which can always be questioned due to an increasing trend in fishing efficiency. One trawling hour in 1970 may not create the same fishing mortality as a trawling hour in 1980 because fishing vessels in 1980 were often better equipped (more efficient gears, larger engines, etc.) than the vessels in 1970. If this is the case in the example of Fig. 4.6.1 the more recent observations should be moved to the right hand side to compensate for the increased efficiency. This would correspond to expressing the effort in units of 1966 trawling hours per year. The effect on the regression analysis would be a smaller slope and a larger intercept, i.e. the estimate of M would be larger and that of q smaller.

Note also that the application of the method is only possible if there is a clear change in effort during the period considered (cf. Exercise 4.6). If effort (the independent variable) remains constant during the period we should not be able to fit a line to the points.

In Section 4.7 we shall continue the discussion of natural mortality, and two alternative simple methods for estimation of M will be suggested. All simple methods for estimation of natural mortality are questionable and it is recommended always to assess the reasonability of any estimate by comparing results of alternative methods and estimates of M for similar species.

(See **Exercise(s)** in Part 2).

4.7 NATURAL MORTALITY

One method of estimating the natural mortality coefficient, M, was introduced in Section 4.6. Many other procedures have been suggested in the literature. Most of them, perhaps all, rank no higher than "guesstimates" or "qualified" guesses. Some of these methods are widely used and others may be so in the future. They therefore merit discussion.

Recall that the natural mortality is the mortality created by all other causes than fishing, e.g., predation including cannibalism, diseases, spawning stress, starvation, and old age. Predation and starvation mortalities and several others are linked to the ambient ecosystem. The same species may have different natural mortality rates in different areas dependent on the density of predators and competitors whose abundance is influenced by fishing activities.

As direct measurements of M are often impossible to obtain, it has been attempted to identify quantities which can be assumed proportional to M and which are easier to measure (or estimate).

The von Bertalanffy curvature parameter, K, has been demonstrated to be linked to the longevity of the fish (Beverton and Holt, 1959) and longevity is related to mortality (Tanaka, 1960; Holt, 1965 and Saville, 1977) (see Eq. 4.5.2.1). Other methods involving longevity are treated in Section 4.7.1.

As a rough generalization, fish species with a high K-value have a high M-value, and species with a low K-value have a low natural mortality. A slow growing species (low K) simply cannot bear high natural mortality - if it did it would soon become extinct. Beverton and Holt (1959) found that values of the ratio M/K mostly lie in the range of 1.5 to 2.5.

Natural mortality must also be linked to L_∞ or the maximum weight of the species, W_∞, since large fish have fewer predators than small fish. It has been suggested that M can be predicted from the body size for certain groups of animals (Taylor, 1960, for bivalves and Peterson and Wroblewski, 1984, for pelagic fish).

Rikhter and Efanov (1976) (Section 4.7.3) demonstrated that fish with a high natural mortality mature early in life, compensating for the high M by starting to reproduce earlier. It can also be mentioned that Gunderson and Dygert (1988) found a relationship of M with the ratio of gonad weight to somatic weight. This is reasonable because fish with a high mortality may compensate by producing more eggs.

Further, as most biological processes go faster at higher temperatures (of course within certain limits) one could imagine the natural mortality to be related to the environmental temperature. Pauly's empirical formula (Section 4.7.2) describes M as a function of K, L_∞ and T, the ambient temperature.

4.7.1 Natural mortality and longevity

Intuitively, we would consider longevity as something more closely related to mortality than K, L_∞ or ambient temperature. The basic concept of mortality coefficients was discussed in Section 4.2 (e.g. Fig. 4.2.1). Alagaraja (1984) suggested another way of illustrating the concept of the mortality coefficient. He tentatively defined the natural life span of fish species (or the longevity) as the age at which 99% of a cohort had died if it had been exposed to natural mortality only (i.e. if Z = M). If Tm stands for longevity and M1% stands for the natural mortality corresponding to a 1% survival, then:

$$M1\% = -\ln(0.01)/Tm \qquad (4.7.1.1)$$

Table 4.7.1.1 shows a collection of M-values defined by the 99% life span as well as the corresponding 99.9% life span.

Table 4.7.1.1 M-values for various life spans

Tm	(months)	3	4	5	6	7	8	10	12
M1%	per year	18	14	11	9	8	6.9	5.5	4.6
M0.1%	per year	28	21	17	14	12	10.4	8.3	6.9

Tm	(years)	2	3	4	5	6	8	10	15	20
M1%	per year	2.3	1.5	1.2	0.92	0.77	0.58	0.46	0.31	0.23
M0.1%	per year	3.5	2.3	1.7	1.38	1.15	0.86	0.77	0.69	0.63

Table 4.7.1.1 can be used for a first rough evaluation of a mortality estimate in conjunction with a modal progression analysis (in particular for tropical species). If, for example, from the modal progression analysis (cf. Section 3.4.2) you find significant numbers (more than 1%) of three year old fish in the catch samples you should not accept a total mortality Z larger than 1.5. If you find very few (less than 1 per mille) you may accept a total mortality up to 2.3.

As longevity is usually as difficult to observe as the natural mortality the relationship between mortalities and life span does not provide any easier way to estimate M, but it presents the concepts in a way which may be easier to grasp. How easy it is to estimate the longevity of a certain species depends on how difficult it is to determine the age of the oldest specimens. If the age of the oldest specimens can be read from the hard parts and if the stock is unexploited it is relatively easy to get an estimate of Tm.

The above aspects were discussed by Hoenig (1983), who also developed a model for estimation of Z from observations on longevity. This model is basically the same as the Alagaraja (1984) model, but applies to Z rather than to M. Hoenig gave the following equation based on observations of Z and Tm and linear regression analysis:

$$\ln Z = a + b*\ln Tm \quad \text{with}$$

a = 1.46 and b = -1.01 for fish
(based on 84 stocks of 53 species)   and

a = 1.23 and b = -0.832 for molluscs
(based on 28 stocks of 13 species)

The above results were based mainly on data from unexploited stocks. Hoenig suggests Tm to be estimated from the mean age of "the oldest specimens" in a sample. This approach, however, is somewhat problematic as the estimate of Tm then depends on the sample size and on the proportion of the sample that is considered to represent "the oldest specimens".

Finally it should be emphasized that the usual assumption, that natural mortality remains constant for all age (size) groups within a species, is highly unrealistic for certain age (size) groups. Naturally, a small fish is exposed to a larger predation mortality than a big fish, simply because small animals have more predators than large animals. This difference in predation mortality (which is perhaps the dominating source of natural mortality for small fish) may be quite big.

When estimating predation mortalities from multispecies cohort analysis combined with work on stomach contents, the ICES Multispecies Working Group (ICES, 1986; Gislason and Sparre, 1987), found a factor of about 10 between the mortalities of 0-group fish and 2-group fish in the North Sea. (see also Section 10.2.)

4.7.2 Pauly's empirical formula

Pauly (1980b) made a regression analysis of M (per year) on K (per year), L_∞ (cm) and T (average annual temperature at the surface in degrees centigrade), based on data from 175 different fish stocks, and estimated the empirical linear relationship:

$$\ln M = -0.0152 - 0.279*\ln L_\infty + 0.6543*\ln K + 0.463*\ln T \quad (4.7.2.1)$$

Table 4.7.2.1 shows values of M (per year) calculated by "Pauly's formula" for various combinations of L_∞, K and T. It should be kept in mind that Eq. 4.7.2.1 gives an estimate of M belonging to the category of "qualified guesses".

Table 4.7.2.1 Natural mortality (per year) calculated by Pauly's formula for various combinations of L_∞, K and T

L_∞	K	T = 5 C°				T = 25 C°			
		0.1	0.5	1.0	2.0	0.1	0.5	1.0	2.0
10		0.24	0.7	1.1	1.7	0.51	1.5	2.3	3.6
80		0.14	0.38	0.6	1.0	0.29	0.8	1.3	2.0
200		0.10	0.30	0.47	0.7	0.22	0.6	1.0	1.6

Eq. 4.7.2.1 may be correct for the "average fish", but may be way off the mark for any particular fish stock. The formula indicates that:

1. Small fish have high natural mortalities
2. Fast growing species have high natural mortalities
3. The warmer the ambient water the higher the natural mortality

The implication is that any other aspect which may influence the natural mortality is considered "random noise around the regression line" (Eq. 4.7.2.1). Other aspects are, for example, the behaviour (schooling, pelagic/demersal), the reproduction physiology, the ecosystem (abundance of predators). Pauly (1983) suggests to account for schooling in Eq. 4.7.2.1 by multiplication by 0.8 so that for schooling species the estimate becomes 20% lower:

$$M = 0.8*\exp[-0.0152 - 0.279*\ln L_\infty + 0.6543*\ln K + 0.463*\ln T] \quad (4.7.2.2)$$

It should also be kept in mind that the quality of the input data to the regression analysis from which Eq. 4.7.2.1 was derived, can be questioned. To arrive at Eq. 4.7.2.1 Pauly needed "observations" of M, which as discussed in the foregoing are difficult to obtain. Any M value which was used for the estimation of Eq. 4.7.2.1 can be questioned. Therefore it is recommended only to use one decimal when presenting the result of Eq. 4.7.2.1. For some species the result seems to be twice or half of what it should be. However, when no other information on M is available as is most often the case one should not hesitate to apply Eq. 4.7.2.1 or the Rikhter and Efanov formula discussed in Section 4.7.3. Usually, it is only for unexploited stocks (M = Z) that we are able to estimate M.

Note that Pauly's formula is based on data for fish stocks only, and that the result depends on how you measure the length of the fish (total length, fork length, standard length, etc.). As Pauly's equation gives only a rough estimate of M we can ignore this detail. Do not use Pauly's formula for crustaceans, molluscs or cephalopods or any other invertebrates, as the formula does not cover these groups.

4.7.3 Rikhter and Efanov's formula

Beverton and Holt (1959) investigated a relationship between longevity, Tm, and the Lm/L_∞ ratio, where Lm is the length at first spawning. Holt (1962) noted that it was generally accepted that the Lm/L_∞ ratio was about 2/3 (average 0.64 with minimum and maximum values of 0.3 and 0.9).

Continuing these investigations in comparative dynamics for high latitude stocks, Rikhter and Efanov (1976) showed a close association between M and Tm50% the age when 50% of the population is mature (also called "the age of massive maturation"):

$$M = 1.521/(Tm50\%^{0.720}) - 0.155 \text{ per year} \qquad (4.7.3.1)$$

They also suggested that Tm50% should be equal to the "optimum age" defined as the age at which the biomass of a cohort is maximal.

Eq. 4.7.3.1 gives these values for Tm50%:

Tm50%	4 months	8 months	1 year	5 years	10 years
M	3.2	1.9	1.4	0.3	0.1

5 VIRTUAL POPULATION METHODS

This chapter deals in principle with the same problems as Chapter 4. However, the analysis and the data requirements are more detailed. The methods of Chapter 4 could be applied to data sets originating from small samples of the commercial catch or from research vessel catches, while the methods of Chapter 5 require estimates of the total numbers caught by commercial fishing.

The methods (or models) described in this chapter are closely linked with those described in Chapter 8. Chapter 5 deals with methods that can be used to analyse the effect that a fishery has had on a particular year class of a stock and Chapter 8 deals with methods that can be used to predict the effect of different levels of fishing effort in the future. The methods in the second category are usually based on the findings of those of the first category. The methods that look at the past, using "historic" data are called "virtual population analysis (VPA)" or "cohort analysis", while those methods dealing with the future, are called "predictive methods" or "Thompson and Bell methods".

VPA and cohort analysis were first developed as age-based methods. However, in recent years also length-based methods have become available, which are of particular interest to tropical fisheries. The age-based methods are discussed in Sections 5.1 and 5.2, while the length-based methods are dealt with in Section 5.3.

Multispecies versions of VPA have also been developed, but they fall outside the scope of this manual. An overview of these models is given by Sparre, 1991.

Information is required on how much was fished, in terms of numbers of fish. The total landings must be distributed over age groups (age-based methods) or length groups (length-based methods). The totals are obtained by raising the age or length distributions of random samples of the landings, using information on the total landings in tons. Tables of the total landings in numbers, by age or length, per year or month, have to be prepared, before we can begin an analysis.

5.1 VIRTUAL POPULATION ANALYSIS (VPA)

Virtual population analysis or VPA is basically an analysis of the catches of commercial fisheries, obtained through fishery statistics, combined with detailed information on the contribution of each cohort to the catch, which is usually obtained through sampling programmes and age readings. The word "virtual", introduced by Fry (1949) is based on the analogy with the "virtual image", known from physics. A "virtual population" is not the real population, but it is the only one that is seen.

The idea behind the method is to analyse that what can be seen, the catch, in order to calculate the population that must have been in the water to produce this catch (see Fig. 5.1.1).

The total landings from a cohort in its lifetime is the first estimate of the numbers of recruits from that cohort. It is however, an under-estimate because some fish must have died from natural causes. Given an estimate of M we can do a backwards calculation and find out how many fish belonging to the cohort were alive year by year and ultimately, how many recruits there were. At the same time we learn the values of the fishing mortality coefficient F, because we have calculated the numbers alive and know from the beginning how many of them were caught in any particular year.

VPA therefore looks at a population in an historic perspective. The advantage of doing a VPA is that once the history is known it becomes easier to predict the future catches, which is usually one of the most important tasks of fishery scientists.

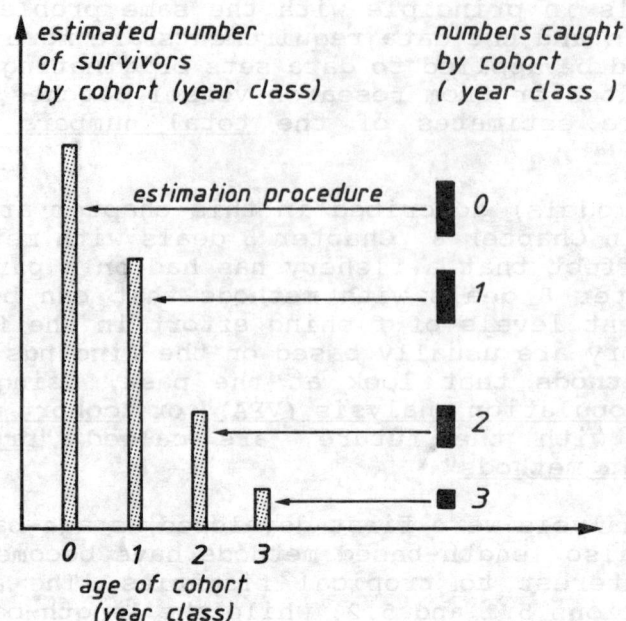

Fig. 5.1.1 The basic features of VPA. For further explanation, see text

A complete review of the development of VPA methods is given by Megrey (1989). The method originated in the USSR, where Derzhavin (1922) was probably the first to combine age data with catch statistics. The method was rediscovered by Fry (1949) and subsequently modified by many authors, including Gulland (1965) and Pope (1972). The modification made by Pope is usually referred to as "Pope's cohort analysis". It will be discussed separately in Section 5.2.

Practical reviews of VPA methods are, among others, given by Pauly (1984) and Jones (1984).

The easiest way to explain the concepts of VPA is to follow an example based on real data.

Example 18: Virtual population analysis (VPA), North Sea whiting

The data given in Table 4.4.3.1 are of the kind required for virtual population analysis, i.e. the total number caught by age group by the entire commercial fishery. To explain the concepts of VPA, consider the 1974 cohort of whiting, the underlined figures given in Table 4.4.3.1. The annotation used is the same as in Section 4.4.3. $C(y,t,t+1)$ = number caught in year y of age between t and t+1 years (in millions).

The numbers caught (in units of millions) were:

    C(1974,0,1) = 599 , number caught between age 0 and age 1
    C(1975,1,2) = 860 , number caught between age 1 and age 2
    C(1976,2,3) = 1071 , number caught between age 2 and age 3
    C(1977,3,4) = 269 , number caught between age 3 and age 4
    C(1978,4,5) =   69 , number caught between age 4 and age 5
    C(1979,5,6) =   25 , number caught between age 5 and age 6
    C(1980,6,7) =   8 , number caught between age 6 and age 7

We start our calculations from the bottom, i.e. with the number caught between age 6 and age 7, $C(1980,6,7) = 8$ million fish. Suppose that we know that the natural mortality was $M = 0.2$ per year for all age groups. Then if we would know also the fishing mortality for the age group between 6 and 7 years, the so-called 6-group, we could calculate how many fish there must have been in the sea on 1 January 1980, N(1980,6) to account for a catch of 8 million whiting in 1980, by using the catch equation (Eq. 4.2.7):

$$C(1980,6,7) = N(1980,6)*\frac{F}{Z}*[1 - \exp(-Z*(7-6))]$$

If we make an initial guess on F(1980,6,7) = 0.5 per year, then Z = 0.5+0.2 = 0.7. The catch equation then becomes:

$$8 = N(1980,6)*\frac{0.5}{0.7}*[1 - \exp(-0.7*(7-6))] = N(1980,6)*0.36$$

thus $N(1980,6) = \frac{8}{0.36} = 22.2$ million

Now, knowing that the number of survivors on 1 January, N(1980,6), which is equal to the number at the end of 1979, we can calculate how many whiting there must have been in the sea on 1 January 1979 to account for the catch (in numbers) during 1979: C(1979,5,6) = 25 million.

However, it is not necessary to guess the value of F again, because we can now also calculate which fishing mortality corresponds to this catch. For the calculation of F, Eq. 4.2.7 is used again, but now in conjunction with the exponential decay model (Eq. 4.2.6). (In this example, in both equations the term (t2-t1) = (t+1-t) has been omitted, because it is equal to 1.)

$$C(1979,5,6) = N(1979,5)*\frac{F}{Z}*[1 - \exp(-Z)] \quad (5.1.1)$$

and

$$N(1980,6) = N(1979,5)*\exp(-Z)$$

which is equivalent to:

$$N(1979,5) = N(1980,6)*\exp(Z) \quad (5.1.2)$$

Inserting the value for N(1980,6) = 22.2 million, calculated above gives:

$$N(1979,5) = 22.2*\exp(Z)$$

Inserting this result and the number caught, C(1979,5,6) = 25 million into Eq. 5.1.1 gives:

$$25 = 22.2*\exp(Z)*\frac{F}{Z}*(1 - \exp(-Z))$$

which after multiplication and rearranging is equivalent to:

$$\frac{25}{22.2} = \frac{F}{Z}*[\exp(Z) - 1]$$

M is assumed to be 0.2, then inserting Z = F+M = F+0.2 gives:

$$1.126 = \frac{F}{F+0.2}*[\exp(F+0.2) - 1]$$

We have thus obtained an equation with F as the only unknown variable. Solving it gives an estimate of F. The equation above, however, is not the type of equation which can be solved by algebraic manipulations. It must be solved by some trial and error method. We shall later discuss how this minor technical problem can be circumvented, but for the time being we forget about it and note that F = 0.696 is the solution, i.e.:

$$1.126 = \frac{0.696}{0.696+0.2} * [\exp(0.696+0.2) - 1]$$

From F = 0.696 and M = 0.2 we can derive

Z(1979,5,6) = M + F(1979,5,6) = 0.2+0.696 = 0.896

With the estimate of Z(1979,5,6) = 0.896 the number of age group 5 fish on 1 January 1979 is easily found by means of the decay model (Eq. 5.1.2):

N(1979,5) = N(1980,6)*exp(Z(1979,5,6))   or

N(1979,5) = 22.2*exp(0.896) = 54.4 million

The results of the calculations made so far may be summarized as follows:

age group t	year y	number caught during year y C(y,t,t+1)	fishing mort. during year y F(y,t,t+1)	no. surviving on 1 Jan. year y N(y,t)
0	1974	599		
1	1975	860	⇑	⇑
2	1976	1071		
3	1977	269	⇑	⇑
4	1978	69		
5	1979	25	0.70	54.4
6	1980	8	0.50*)	22.2

*) initial guess on terminal F

The next pair N(1978,4) and F(1978,4,5) can be calculated exactly as those for the year 1979. In this way we can work backwards in time estimating numbers of survivors and fishing mortalities for each age group (as indicated by the arrows).

Note that contrary to the catch curve methods, we do not assume F (and Z) to remain constant. Each age group may have a different F-value. This method thus provides a more detailed analysis of the population than any of the other methods presented so far. The two VPA equations derived above read in a general form:

$$\frac{C(y,t,t+1)}{N(y+1,t+1)} = \frac{F(y,t,t+1)}{M+F(y,t,t+1)} * \left[\exp[F(y,t,t+1)+M] - 1\right] \qquad (5.1.3)$$

$$N(y,t) = N(y+1,t+1) * \exp[F(y,t,t+1)+M] \qquad (5.1.4)$$

For the year 1978 in the example we get:

$$\frac{C(1978,4,5)}{N(1979,5)} = \frac{69}{54.4} = 1.268, \text{ while } M = 0.2$$

Inserting these values into Eq. 5.1.3, we get by trial and error:

$$F(y,t,t+1) = F(1978,4,5) = 0.757$$

and by inserting this F-value and the number of survivors at 1 January 1979, in Eq. 5.1.4 we get the number of survivors at 1 January 1978:

$$N(1978,4) = N(1979,5)*\exp[F(1979,4,5)+M] = 54.4*\exp[0.757+0.2] = 141.9$$

By repeating this procedure for the years 1977 to 1974 the estimates of fishing mortalities and stock numbers are obtained as shown in Table 5.1.1.

Table 5.1.1 Results of VPA for the 1974 whiting cohort.
(Data from Table 4.4.3.1, numbers in millions)

age group t	year y	number caught during year y $C(y,t,t+1)$	fishing mort. during year y $F(y,t,t+1)$	number of survivors on 1 Jan. of year y $N(y,t)$
0	1974	599	0.16	4390
1	1975	860	0.37	3054
2	1976	1071	1.11	1729
3	1977	269	0.99	465
4	1978	69	0.76	142
5	1979	25	0.70	54.4
6	1980	8	0.50 *)	22.2

*) assumed to be known in advance.

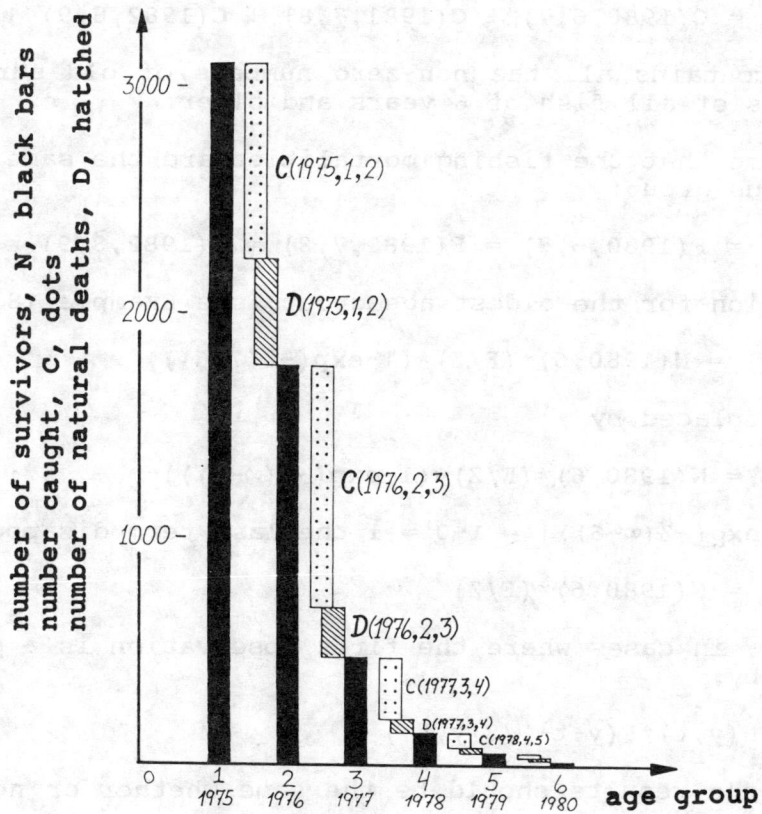

Fig. 5.1.2 Number of survivors, N (black bars), number caught $C = F*\bar{N}$ and number of natural deaths, $D = M*\bar{N}$, as derived by VPA in Table 5.1.1 for North Sea whiting

Fig. 5.1.2 illustrates the cohort dynamics as described by VPA for North Sea whiting (Table 5.1.1). In this case M is relatively small compared to F, which can be seen by comparing the number caught (Eq. 4.2.8 with t2-t1 = 1):

$$C(y,t,t+1) = F(y,t,t+1) * \bar{N}(y,t,t+1) \qquad (5.1.5)$$

to the number of natural deaths:

$$D(y,t,t+1) = M(y,t,t+1) * \bar{N}(y,t,t+1) \qquad (5.1.6)$$

Fig. 5.1.2 presents the results as the number of survivors, N, but we could as well have chosen the fishing mortality, F, as the basic result, because:

F and C determine N and
N and C determine F

so that there is a one-to-one correspondence between N and F when C and M are known.

VPA using a plus-group

In Example 18 we started with the calculation of N(1980,6), but we did not say anything about the fish older than 6 years. This approach is correct, because it is not necessary to account for the older age groups. However, there may be many fish older than 6 years and we may want to account for all these older fish, which are difficult to separate into age groups, by combining them in one so-called plus-group. If we account for these older fish, the formulas must be modified accordingly.

Let us return to the example and replace C(1980,6,7) by the plus-group:

$$C(1980,6+) = C(1980,6,7) + C(1981,7,8) + C(1982,8,9) + ...$$

where the sum contains all the non-zero numbers of old survivors, or the catch in numbers of all fish of 6 years and older.

We further assume that the fishing mortalities are the same for all components of the plus-group:

$$F(1980,6+) = F(1980,6,7) = F(1981,7,8) = F(1982,8,9) = ...$$

The catch equation for the oldest age group (see Example 18):

$$C(1980,6,7) = N(1980,6)*(F/Z)*(1-\exp(-Z(7-6)))$$

should now be replaced by

$$C(1980,6+) = N(1980,6)*(F/Z)*(1-\exp(-Z(\infty-6)))$$

and because $[1-\exp(-Z(\infty-6))] = 1-0 = 1$ the last term disappears, so that

$$C(1980,6+) = N(1980,6)*(F/Z)$$

Thus in general, in cases where the first observation is a plus-group, the VPA is started by:

$$C(y,t+) = N(y,t)*F(y,t+)/Z(y,t+)$$

Theoretically, the results should be the same whether or not the last age group is a plus-group.

The biomass concept

The biomass concept associated with Table 5.1.1 is rather straightforward when we consider the weight of the cohort at a particular time. For example, the weight of the cohort in year 1979 is $N(1979,5)*w(5)$, where $w(5)$ is the body weight of a 5 years old whiting. A biomass concept which reflects the cohort during its entire life span is more difficult to grasp.

The average biomass on 1 of January during the first 6 years of life of the cohort is:

$$\frac{N(1974,0)*w(0) + N(1975,1)*w(1) + \ldots + N(1974,6)*w(6)}{6}$$

The average annual biomass could be defined as:

$$\frac{\overline{N}(1974,0)*\overline{w}(0) + \overline{N}(1975,1)*\overline{w}(1) + \ldots + \overline{N}(1974,6)*\overline{w}(6)}{6}$$

where $\overline{N}$ is defined by Eq. 4.2.9 with $t_2-t_1 = 1$ and $\overline{w}$ is the average annual body weight. The two biomass concepts are different, and it is not obvious how they should be used. The same sort of problem emerges when trying to define the average number of survivors. We shall come back to the biomass concept in the following sections.

Basic features of VPA

From observations on the numbers caught in each age group the VPA estimates how many fish there must have been in the sea to account for that catch, under the assumption that natural mortality is known (see Fig. 5.1.1). If the catch constitutes a small fraction of the stock (i.e. if F is small) the estimation of the stock size becomes more uncertain. Thus, the higher the fishing mortality the more dependable is the VPA.

Natural mortality, M, is assumed to be known from investigations independent of the VPA, but is actually unknown in most cases. The reliability of VPA is also dependent on the size of M relative to F. Often the estimate of M is rather a "guesstimate" (qualified guess), but if M is small compared to F it may not matter so much that M is not well estimated. What a "guesstimate of M" means was discussed in Section 4.7.

A set of equations can only have a unique solution, when the number of equations equals the number of unknown variables. If there are more unknown variables than equations, there will be infinitely many solutions.

The whole set of VPA equations consists of pairs of Eqs. 5.1.3 and 5.1.4 for each age group. There are apparently three unknown variables in each set of two equations, viz. $N(y,t)$, $N(y+1,t+1)$ and $F(y,t,t+1)$. However, in all cases except in the first set with the oldest age group, $N(y+1,t+1)$ is known from the solution of the preceding set of equations and we do end up with two unknown variables in two equations and therefore with a unique solution.

The problem with the first set of equations pertaining to the oldest age group can be solved by making a plausible assumption and formulating it as an additional equation. We can then obtain a solution that is conditioned on this assumption. The solution in the case of a VPA is to assume a value for the F of the oldest age group, the so-called "<u>terminal F</u>".

For example, we might assume that the terminal F is equal to the F of the second oldest age group, so the additional equation would then be:

$F7 = F6$ (assuming that 7 is the oldest age group)

We then have four equations, two sets of Eqs. 5.1.3 and 5.1.4, with four unknown variables, viz. F7, N7, N6 and N5.

If there are more equations than unknown variables there is (usually) no solution. In that case we use regression analysis to find the best "fit" to the data to find a solution and we may calculate confidence limits. In regression analysis the concept of "unknown variables" is replaced by the concept of "parameters".

To calculate confidence intervals for estimates of parameters the number of observations must be larger than the number of parameters. The number of parameters in VPA (the N's and the F for the oldest age group) equals the number of observations (the C's) plus one. Therefore it is not possible to calculate confidence limits for the estimates of the N's (or the F's of the other age groups).

The data used in Example 18 to illustrate VPA were obtained from direct age reading (otoliths). However, input data could have been derived from time series of length-frequency data which were resolved into cohort components by for example, the Bhattacharya method (Section 3.4). This aspect is further discussed at the end of Section 5.3.

The VPA is a method to analyse historical data for estimation of population parameters. The ultimate use of such parameters is to determine the optimum fishing strategy, i.e. the array of F-at-age, or the so-called "<u>fishing pattern</u>", which in the long term gives the highest yield from the stock in question. To assess alternative (future) fishing strategies we require a counterpart to the VPA, namely a model which can predict the stock and the catch for various assumptions on the future fishing pattern. The "Thompson and Bell model" (Section 8.6) is the predictive version of the VPA.

Computer programs

Mesnil (1988) presents a package of microcomputer programs, ANACO, ("ANAlysis of COhorts" or "L'ANAlyse des COhortes") which can perform the VPA calculations as described above. The ANACO package also offers a number of additional options, for example sensitivity analysis. Also the "COMPLEAT ELEFAN" package (Gayanilo, Soriano and Pauly, 1988) and FiSAT contain routines for VPA analysis.

5.2 AGE-BASED COHORT ANALYSIS (POPE'S COHORT ANALYSIS)

As derived from the catch equation, the VPA implied the solution of Eq. 5.1.3 by some numerical techniques (some trial and error method). This is a minor technical problem when one has access to a computer. However, the problem can be circumvented in an easy way, so that VPA can also be carried out on a pocket calculator. The version of VPA suitable for pocket calculators is the "<u>cohort analysis</u>" developed by Pope (1972), reviewed in Jones (1984) and Pauly (1984).

Cohort analysis is conceptually identical to VPA, but the calculation technique is simpler. It is based on an approximation, illustrated in Fig. 5.2.1, which shows the number of survivors of a cohort during one year. The catch is taken continuously during the year, but in cohort analysis the assumption is made that <u>all fish are caught on one single day</u>. This day is chosen to be 1 July, i.e. when one half of the year has elapsed.

Consequently in the first half year the fish suffer only natural mortality so the number of survivors on 1 July becomes:

$$N(y, t+0.5) = N(y,t)*\exp(-M/2)$$

Then, instantaneously, the catch is taken and the number of survivors becomes:

$$N(y,t)*\exp(-M/2) - C(y,t,t+1)$$

This number of survivors then suffers further only natural mortality in the second half year and finally the number of survivors at the end of the year is:

$$N(y+1,t+1) = \left[N(y,t)*\exp(-M/2) - C(y,t,t+1)\right]*\exp(-M/2)$$

For convenience of calculation this equation is rearranged as:

$$N(y,t) = \left[N(y+1,t+1)*\exp(M/2) + C(y,t,t+1)\right]*\exp(M/2) \qquad (5.2.1)$$

Note that the F that caused computational problems in the VPA equation does not occur here. Again we shall demonstrate the method on the basis of the same data set, the North Sea whiting cohort of 1974.

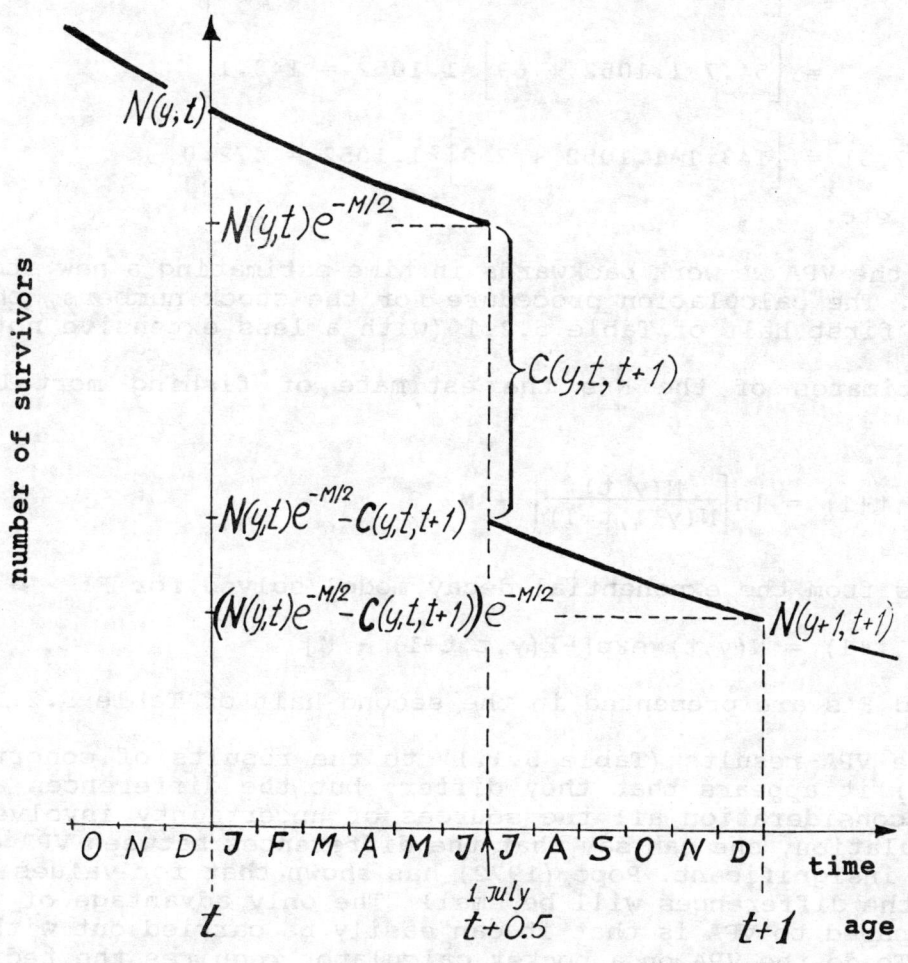

Fig. 5.2.1 Illustration of the approximation behind Pope's cohort analysis (for further explanation, see text)

Example 19: Pope's cohort analysis, North Sea whiting

To apply Eq. 5.2.1 to the whiting example we start the same way as for the VPA by assuming the F for the oldest age group (the so-called "terminal F") to be known, $F(1980,6,7) = 0.5$, while $M = 0.2$, and then start by calculating $N(1980,6)$ from the catch equation:

$$N(1980,6) = \frac{C(1980,6,7)}{(F/Z)*[1 - \exp(-Z)]} = 22.2$$

Then Eq. 5.2.1 is applied to calculate $N(1979,5)$:

$$N(1979,5) = \left[N(1980,6)*\exp(M/2) + C(1979,5,6)\right]*\exp(M/2)$$

$$= \left[22.2*1.1052 + 25\right]*1.1052 = 54.7$$

and we continue in the same way:

$$N(1978,4) = \left[N(1979,5)*\exp(M/2) + C(1978,4,5)\right]*\exp(M/2)$$

$$= \left[54.7*1.1052 + 69\right]*1.1052 = 143.1$$

$$N(1977,3) = \left[143.1*1.1052 + 269\right]*1.1052 = 472.0$$

..... etc. ....

Thus, as for the VPA we work backwards in time estimating a new stock number at each step. The calculation procedure for the stock numbers, the N's, is given in the first half of Table 5.2.1 (with a less extensive notation).

From the estimates of the N's the estimate of fishing mortalities are obtained from:

$$F(y,t,t+1) = \ln\left[\frac{N(y,t)}{N(y+1,t+1)}\right] - M \qquad (5.2.2)$$

which follows from the exponential decay model solved for F:

$$N(y+1,t+1) = N(y,t)*\exp[-F(y,t,t+1) - M]$$

The estimated F's are presented in the second half of Table 5.2.1.

Comparing the VPA-results (Table 5.1.1) to the results of cohort analysis (Table 5.2.1) it appears that they differ, but the differences are small. Taking into consideration all the sources of uncertainty involved in this kind of calculation, one can say that the differences between VPA and cohort analysis are insignificant. Pope (1972) has shown that for values of $F < 1.2$ and $M < 0.3$ the differences will be small. The only advantage of the cohort analysis compared to VPA is that it can easily be carried out with a pocket calculator. To do the VPA on a pocket calculator requires the tedious trial and error way of calculating F (unless the calculator is fully programmable).

Table 5.2.1 The computation procedure of Pope's age-based cohort analysis, illustrated by the 1974 cohort of North Sea whiting (from Table 4.4.3.1)

POPE'S AGE-BASED COHORT ANALYSIS

$M = 0.2$ per year

NATURAL MORTALITY FACTOR:
$G = \exp(M/2) = \exp(0.2/2) = 1.1052$

GUESS ON TERMINAL F: $F6 = 0.5$

STOCK NUMBERS: $N(1980,6) = \dfrac{C(1980,6,7)}{[F6/Z6]*[1 - \exp(-Z6)]} =$

$N6 = \dfrac{8}{[0.5/(0.5+0.2)]*[1 - \exp\{-(0.5+0.2)\}]} = 22.2$

$N(1979,5) = N5 = (N6*G + C5,6)*G =$

$N5 = (N6*G + C5)*G = (22.2*G + 25)*G \quad = \quad 54.7$

$N4 = (N5*G + C4)*G = (54.7*G + 69)*G \quad = \quad 143.1$

$N3 = (N4*G + C3)*G = (143.1*G + 269)*G \quad = \quad 472.1$

$N2 = (N3*G + C2)*G = (472.1*G + 1071)*G = 1760.3$

$N1 = (N2*G + C1)*G = (1760.3*G + 860)*G = 3100.4$

$N0 = (N1*G + C0)*G = (3100.4*G + 599)*G = 4448.9$

FISHING MORTALITIES:

$F6 =$ (initial guess of terminal F) $= 0.50$

$F(1979,5,6) = F5 = \ln\left[\dfrac{N(1979,5)}{N(1980,6)}\right] - M =$

$F5 = \ln(N5/N6) - M = \ln(54.7/22.2) - 0.2 \quad = 0.70$

$F4 = \ln(N4/N5) - M = \ln(143.1/54.7) - 0.2 \quad = 0.76$

$F3 = \ln(N3/N4) - M = \ln(472.1/143.1) - 0.2 \quad = 0.99$

$F2 = \ln(N2/N3) - M = \ln(1760.3/472.1) - 0.2 = 1.12$

$F1 = \ln(N1/N2) - M = \ln(3100.4/1760.3) - 0.2 = 0.37$

$F0 = \ln(N0/N1) - M = \ln(4448.9/3100.4) - 0.2 = 0.16$

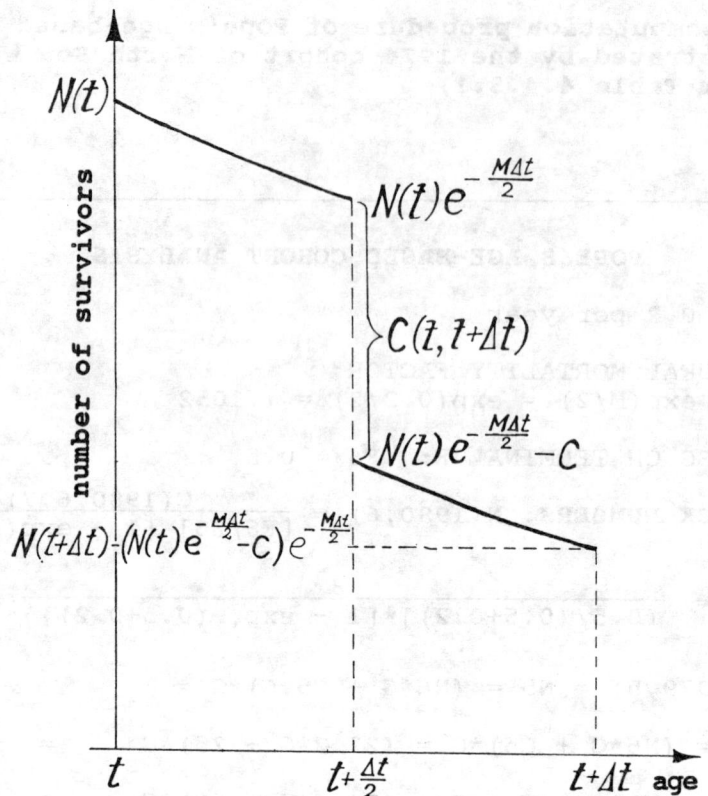

Fig. 5.2.2 Derivation of the formulas for Pope's cohort analysis in the case of variable time intervals (Eqs. 5.2.3 and 5.2.4)

Eq. 5.2.1 was derived for a time period of one year. As was the case with the catch curve (cf. Section 4.4.4) we may consider the catch during any time period, from t to t+Δt. In that case Eq. 5.2.1 should be replaced by the more general expression:

$$N(t) = \left[N(t+\Delta t)*\exp(M*\Delta t/2) + C(t,t+\Delta t)\right]*\exp(M*\Delta t/2) \qquad (5.2.3)$$

The derivation of Eq. 5.2.3 is similar to that of Eq. 5.2.1 and is shown in Fig. 5.2.2. The parallel to Eq. 5.2.2 is:

$$F(t,t+\Delta t) = \frac{1}{\Delta t}*\ln\left[\frac{N(t)}{N(t+\Delta t)}\right] - M \qquad (5.2.4)$$

The year index, y, has disappeared in Eqs. 5.2.3 and 5.2.4, the main reason being that they usually are applied under the assumption of a constant parameter system (cf. Section 4.4.4). Further, when Δt varies it will not conform to the year intervals and the notation used for age groups is not suitable any more.

Eqs. 5.2.3 and 5.2.4 applied under the assumption of constant parameters are typical applications of cohort analysis for tropical fish stocks. They could be applied to all cohorts during one year or to the average annual catch during a sequence of years. We shall return to this in Section 5.3.

(See **Exercise(s)** in Part 2).

5.3 JONES' LENGTH-BASED COHORT ANALYSIS

In this section we deal with the situation when only length composition data for the total fishery are available for one year (or the average length composition for a sequence of years). The approach is basically the same as for the length-converted catch curve (see Section 4.4.5). The name "length-based cohort analysis" is somewhat misleading, as we are not dealing with real cohorts in the present analysis. The real cohort is replaced by a "pseudo-cohort" which is based on the assumption of a constant parameter system (see Section 4.4.1). Thus, it is assumed that the picture presented by all length (or age) classes caught during one year reflects that of a single cohort during its entire life span. We shall come back to this aspect later on. Also this method will be explained based on an example.

Example 20: Jones' length-based cohort analysis, hake, Senegal

Table 5.3.1 shows a data set for the hake fishery off Senegal (CECAF, 1978), which can be used as input for a length-based cohort analysis.

Table 5.3.1 Length composition of the total catch of hake (Merluccius merluccius) off Senegal (from CECAF, 1978), input data for length-based cohort analysis

length group cm L1-L2	number caught ('000) C(L1,L2)
6-12	1823
12-18	14463
18-24	25227
24-30	8134
30-36	3889
36-42	2959
42-48	1871
48-54	653
54-60	322
60-66	228
66-72	181
72-78	96
78-84	16
84-∞	46

As was the case for the catch curve analysis (cf. Section 4.4.5) the length groups can be converted into age intervals by the inverse von Bertalanffy equation (Eq. 3.3.3.2 and Eq. 4.4.5.1 respectively):

$$t(L1) = t_o - \frac{1}{K} * \ln\left[1 - \frac{L1}{L\infty}\right] \qquad (5.3.1)$$

$$\Delta t = t(L2) - t(L1) = \frac{1}{K} * \ln\left[\frac{L\infty - L1}{L\infty - L2}\right] \qquad (5.3.2)$$

For the hake off Senegal (Table 5.3.1) the von Bertalanffy growth parameters and the natural mortality factor have been estimated as:

$K = 0.1$ per year, $L_\infty = 130$ cm and $M = 0.28$ per year.

Putting $t_o = 0$ (compare Section 4.4.5) and applying Eqs. 5.3.1 and 5.3.2 gives the relative ages, $t(L1)$ and Δt as shown in Table 5.3.2, columns B and C respectively.

To convert the cohort analysis equation (Eq. 5.2.3) into a length-based version, only the term $\exp(M*\Delta t/2)$ has to be changed. This is easily done by substituting Δt with Eq. 5.3.2.

$$\exp(\frac{M}{2}*\Delta t) = \exp\left[\frac{M}{2}*\frac{1}{K}*\ln(\frac{L\infty-L1}{L\infty-L2})\right] = \exp\left[\ln(\frac{L\infty-L1}{L\infty-L2})^{M/2K}\right] = \left[\frac{L\infty-L1}{L\infty-L2}\right]^{M/2K} \quad (5.3.3)$$

It is convenient to use a symbol instead of this complicated term, therefore we introduce the symbols:

$N(L1) = N(t(L1))$ = the number of fish that attain length L1
= the number of fish that attain age $t(L1)$
(also called the number of survivors)

$N(L2) = N(t(L1)+\Delta t)$ = the number of fish that attain length L2
= the number of fish that attain age $t(L2)$
$(= t(L1)+\Delta t)$

$C(L1,L2) = C(t,t+\Delta t)$ = the number of fish caught of lengths between L1 and L2
= the number of fish caught of ages between $t(L1)$ and $t(L2)$

$H(L1,L2) = \left[\frac{L\infty-L1}{L\infty-L2}\right]^{M/2K}$ = $\left[\begin{array}{l}\text{the fraction of } N(L1) \text{ which survive} \\ \text{natural deaths during the time period} \\ \text{from } t(L1) \text{ to } t(L1)+\Delta t/2\end{array}\right]^{-1}$

Now Eq. 5.2.3 can be rewritten using these length-based symbols, as:

$$N(L1) = \left[N(L2)*H(L1,L2) + C(L1,L2)\right]*H(L1,L2) \quad (5.3.4)$$

The calculation procedure of Eq. 5.3.4 is similar to that of the age-based cohort analysis (Eq. 5.2.1). We start with the last group and use the length-based form of the catch equation:

$$C(L1,L2) = N(L1)*\frac{F}{Z}*\left[1 - \exp(-Z*\Delta t)\right] \quad (5.3.5)$$

In the case of the hake off Senegal the last group is the hake of length 84 cm and longer:

$$C(84,\infty) = N(84)*\frac{F}{Z}*[1 - \exp(-Z*\Delta t)]$$

In this case Δt refers to all fish longer than 84 cm, so that Δt is very large. Theoretically, the age corresponding to L_∞ is ∞, so theoretically $\Delta t = \infty$, and therefore:

$\exp(-Z*\infty) = 0$

The catch in numbers per length group is known (see Table 5.3.1), thus, approximately:

$$C(84,\infty) = 46 = N(84) * \frac{F}{Z} * (1-0)$$

or

$$N(84) = \frac{C(84,\infty)}{F/Z} = \frac{46}{F/Z}$$

Here again it is necessary to make an initial guess, viz. of F/Z in this case (cf. Table 5.2.1). If F/Z for the last length group is assumed to take the value 0.5 then the number of hake attaining length 84 cm becomes:

$$N(84) = 46/0.5 = 92$$

Column D of Table 5.3.2 gives the values of H(L1,L2) for hake off Senegal with M = 0.28 per year and K = 0.1 per year, i.e. M/2K = 0.28/(2*0.1) = 1.4.

Table 5.3.2 Length groups of the hake (<u>Merluccius merluccius</u>) off Senegal, converted into age intervals, and the factors H(L1,L2). K = 0.1 per year, L_∞ = 130 cm, M = 0.28 per year

A	B	C	D
length group cm L1-L2	relative age t(L1)	Δt	nat.mort. factor H(L1,L2)
6-12	0.473	0.496	1.0719
12-18	0.968	0.522	1.0758
18-24	1.490	0.551	1.0801
24-30	2.041	0.583	1.0850
30-36	2.624	0.619	1.0905
36-42	3.242	0.660	1.0967
42-48	3.902	0.706	1.1039
48-54	4.608	0.760	1.1122
54-60	5.368	0.822	1.1220
60-66	6.190	0.890	1.1337
66-72	7.087	0.984	1.1478
72-78	8.071	1.092	1.1652
78-84	9.163	1.226	1.1873
84-∞	10.389	-	-

<u>column</u> <u>contents</u>

B $\quad t(L1) = 0 - \frac{1}{K} * \ln(1 - \frac{L1}{L_\infty})$ $\qquad$ (Eq. 5.3.1, with t_o = 0)

C $\quad \Delta t = t(L2) - t(L1) = \frac{1}{K} * \ln\left[\frac{L_\infty - L1}{L_\infty - L2}\right]$ $\qquad$ (Eq. 5.3.2)

D $\quad H(L1,L2) = \exp\left(\frac{M*\Delta t}{2}\right) = \left[\frac{L_\infty - L1}{L_\infty - L2}\right]^{M/2K}$ $\qquad$ (Eq. 5.3.3)

Note: In this case H(L1,L2) can be calculated on the basis of either the age-based or the length-based formula. The results are the same, but the length-based approach is much shorter, because it is not necessary to calculate Δt.

The number of hake attaining length 78 cm can be obtained by inserting into Eq. 5.3.4 as N(L2) the value of N(L1) obtained for the larger length group and further the corresponding value for H from Table 5.3.2. In this case N(L2) = N(84) = 92 and H(78,84) = 1.1873.

N(78) = [92*1.1873 + 16]*1.1873 = 148.7

Continuing to move backwards in length (and thus in time) the subsequent stock numbers are calculated:

N(72) = [148.7*1.1652 + 96]*1.1652 = 313.7

N(66) = [313.7*1.1478 + 181]*1.1478 = 621.0

.... etc. ....

The stock numbers for all length groups obtained in this way are given in column D of Table 5.3.3.

Table 5.3.3 **The calculation procedure of Jones' length cohort analysis illustrated by the hake (Merluccius merluccius) off Senegal.**

$K = 0.1$ per year, $L_\infty = 130$ cm, $M = 0.28$ per year
Terminal F/Z assumed to be 0.5000 (indicated by *)

A	B	C	D	E	F	G
length group cm	nat. mort. factor	number caught ('000)	number of survivors	exploitation rate	fishing mortality	total mortality
L1-L2	H(L1,L2)	C(L1,L2)	N(L1)	F/Z	F	Z
6-12	1.0719	1823	98919.3	0.1255	0.04	0.32
12-18	1.0758	14463	84392.7	0.5805	0.39	0.67
18-24	1.0801	25227	59475.8	0.7920	1.07	1.35
24-30	1.0850	8134	27623.0	0.6979	0.65	0.93
30-36	1.0905	3889	15967.8	0.6369	0.49	0.77
36-42	1.0967	2959	9861.5	0.6785	0.59	0.87
42-48	1.1039	1871	5500.5	0.6977	0.65	0.93
48-54	1.1122	653	2818.8	0.5792	0.39	0.67
54-60	1.1220	322	1691.5	0.5072	0.29	0.57
60-66	1.1337	228	1056.6	0.5234	0.31	0.59
66-72	1.1478	181	621.0	0.5890	0.40	0.68
72-78	1.1652	96	313.7	0.5817	0.39	0.67
78-84	1.1873	16	148.7	0.2823	0.11	0.39
84-∞	-	46	92.0**	0.5000*	0.28	0.56

column	contents
B	$H(L1,L2) = \left[\dfrac{L\infty-L1}{L\infty-L2}\right]^{M/2K} = \left[\dfrac{130-L1}{130-L2}\right]^{1.4}$
D	$N(L1) = \left[N(L2)*H(L1,L2) + C(L1,L2)\right]*H(L1,L2)$
**	$N(84) = C(84,\infty)/(F/Z) = 46/0.5 = 92$
E	$F/Z = C(L1,L2)/[N(L1)-N(L2)]$
F	$F = M*(F/Z)/(1-F/Z)$
G	$Z = F+M$

To estimate F we could use Eq. 5.2.4, but it is more convenient to calculate F by:

$$F(L1,L2) = M * \frac{F(L1,L2)/Z(L1,L2)}{1-[F(L1,L2)/Z(L1,L2)]} \qquad (5.3.6)$$

where the exploitation rate F/Z is derived from:

$$F(L1,L2)/Z(L1,L2) = \frac{C(L1,L2)}{N(L1) - N(L2)} \qquad (5.3.7)$$

For example, writing for simplicity F/Z for F(L1,L2)/Z(L1,L2), for the length group 72-78 cm we get:

$$F/Z = \frac{C(72,78)}{N(72)-N(78)} = \frac{96}{313.7-148.7} = 0.5817 \quad \text{and}$$

$$F = M * \frac{F/Z}{1-F/Z} = 0.28 * \frac{0.5817}{1-0.5817} = 0.39$$

The complete results of this calculation procedure for length cohort analysis are given in Table 5.3.3.

Mean number and biomass

We now want to calculate the mean number of fish in the sea and their biomass. Summing the column of N(L1) values in Table 5.3.3 would not give the right number because changing the class interval would give a different sum: The N(L1) values are simply the number alive at any length L1. The procedure is to find the mean number in each class interval and weight it by the time, Δt, spent in that class interval. The same problem was dealt with by Eqs. 4.2.6 and 4.2.9. Some manipulation of either of these equations leads to the result:

$$\bar{N}(L1,L2) * \Delta t = [N(L1)-N(L2)]/Z \qquad (5.3.8)$$

which is the <u>annual</u> mean number in each length class.

$N(L2) = N(L_\infty) = 0$ may be assumed for the last group. The total mean number of fish in the sea of lengths above the first L1 (here 6 cm) becomes in general:

$$\sum_i \bar{N}(L_i,L_{i+1}) * \Delta t \qquad (5.3.9)$$

Correspondingly, we find the annual mean biomass in each length group by multiplying by the mean weight, $\bar{w}(L1,L2)$, in the length group:

$$\bar{B}(L1,L2) * \Delta t = \bar{N}(L1,L2) * \Delta t * \bar{w}(L1,L2) \qquad (5.3.10)$$

The body weight is calculated from

$$\bar{w}(L1,L2) = q * [(L1+L2)/2]^b \qquad (5.3.11)$$

where q and b are the constants of the length-weight relationship described in Section 2.6. The body weight of the last group may be calculated as $\bar{w}(L1,L_\infty)$.

The general sum

$$\sum_i \bar{B}(L_i, L_{i+1}) * \Delta t \tag{5.3.12}$$

is an estimate of the average biomass during the life span of a cohort, or of all cohorts during a year, and is independent of the length class interval.

The body weight may also be used to estimate the yield, i.e. the weight of the catch. The weight of the catch belonging to length group i becomes

$$\sum_i C(L_i, L_{i+1}) * \bar{w}(L_i, L_{i+1}) \tag{5.3.13}$$

Table 5.3.4 shows the calculation of the yield and the average biomass during a year.

Table 5.3.4 The calculation procedure for yield and average biomass in Jones' length-based cohort analysis illustrated by the hake (*Merluccius merluccius*) off Senegal. $q = 0.00001$ kg/cm^3, $b = 3$, $K = 0.1$ per year, $L_\infty = 130$ cm, $M = 0.28$ per year

A	B	C	D	E	F	G	H
length group cm	number caught ('000)	number of survivors	total mort. rate ('000)	mean body weight kg	mean N *Δt ('000)	mean biomass*Δt tonnes	yield tonnes
L1-L2	C	N(L1)	Z	$\bar{w}$(L1,L2)	$\bar{N}$(L1,L2)*Δt	$\bar{B}*\Delta t$	Y(L1,L2)
6-12	1823	98919.3	0.32	0.0073	45369	330.7	13.3
12-18	14463	84392.7	0.67	0.0338	37335	1260.1	488.1
18-24	25227	59475.8	1.35	0.0926	23664	2191.5	2336.3
24-30	8134	27623.0	0.93	0.196	12575	2475.1	1601.0
30-36	3889	15967.8	0.77	0.359	7919	2845.9	1397.6
36-42	2959	9861.5	0.87	0.593	5007	2970.1	1755.3
42-48	1871	5500.5	0.93	0.911	2895	2638.1	1704.9
48-54	653	2818.8	0.67	1.33	1694	2247.1	866.2
54-60	322	1691.5	0.57	1.85	1117	2068.6	596.3
60-66	228	1056.6	0.59	2.50	741	1852.8	570.1
66-72	181	621.0	0.68	3.29	451.1	1481.9	594.6
72-78	96	313.7	0.67	4.22	246.5	1039.9	405.0
78-84	16	148.7	0.39	5.31	144.9	770.1	85.0
84-∞	46	92.0	0.56	12.25	164.3	2012.7	563.5
					Total	26184.6	12977.2

column	contents
E	$\bar{w}(L1,L2) = q*[(L1+L2)/2]^b$
F	$\bar{N}(L1,L2)*\Delta t = [N(L1)-N(L2)]/Z$
G	$\bar{B}*\Delta t = \bar{w}(L1,L2)*[N(L1)-N(L2)]/Z$
H	$Y(L1,L2) = \bar{w}(L1,L2) * C(L1,L2)$

Basic features of length-based cohort analysis

The length cohort analysis defined by Eqs. 5.3.3 to 5.3.7 (see Table 5.3.3) is called "Jones' length-based cohort analysis". (Jones, 1976, and Jones and van Zalinge, 1981, reviewed in Jones, 1984 and Pauly, 1984). As already mentioned, the method is usually applied to "pseudo-cohorts", i.e. we assume a constant parameter (equilibrium) system. In order to simulate an equilibrium condition it is essential that the data pertain to a relatively long time period (say a year or several years), preferably a number of full years.

A length-frequency sample collected during a relatively short time period is not applicable. The September sample of shrimps shown in Fig. 3.4.2.5 consisting of only one cohort is such an example. The method assumes that a larger specimen is also older, but in this case we assumed that all shrimps, irrespective of length, are of the same age. The descending slope of that sample has something to do with the variation in individual growth rates - it is not related to mortality.

It is possible to apply Jones' length-based cohort analysis method to a real cohort, but that implies that we are able to follow a cohort through time, i.e. that we know its age. If that is the case, however, we might as well use Pope's age-based cohort analysis, as this does not present problems in connection with the conversion of length into age.

As Jones' length-based cohort analysis is based on Pope's age-based cohort analysis (Section 5.2) it has the same limitations. The approximation to VPA is valid for values of $F*\Delta t$ up to 1.2 and of $M*\Delta t$ up to 0.3 (Pope, 1972).

A different version of length-based VPA (i.e. a method not requiring limitations of the values of $F*\Delta t$ and $M*\Delta t$) was presented in Sparre (1979).

Computer programs

The program "LCOHOR" in the LFSA package of microcomputer programs (Sparre, 1987) can execute Jones' length-based cohort analysis as described above. Also the "COMPLEAT ELEFAN" package (Gayanilo, Soriano and Pauly, 1988) and FiSAT contain routines for a length-based analysis similar to LCOHOR.

(See **Exercise(s)** in Part 2).

6 GEAR SELECTIVITY

In several methods discussed in Chapter 4 it appeared that the complete length ranges (or age ranges) of fish or shellfish are not always under full exploitation. Most fishing gears, for example trawl gears, are selective for the larger sizes, while some gears (gill nets) are selective for a certain length range only, thus excluding the capture of very small and very large fish. This property of fishing gear is called "gear selectivity". It needs to be taken into account when we want to estimate the real size (or age) composition of the fish in the fishing area. At the same time, it is an important tool for fisheries managers who, by regulating the minimum mesh sizes of a fishing fleet, can more or less determine the minimum sizes of the target species of certain fisheries. Gear selectivity is strongly related to the estimation of the total mortality, Z, the analysis of trawl survey data vis-à-vis commercial fisheries and predictions of future yields (Thompson and Bell, see Chapter 8). A recent special issue of Fisheries Research provides a useful overview of fishing gear selectivity (MacLennan, 1992).

Since it is conceptually easier, we will first discuss the selectivity of trawl gear and then that of gill nets and similar gears.

6.1 ESTIMATION OF TRAWL NET SELECTION

A full description of a bottom trawl net is given in Section 13.1. The fine-meshed end of the net where the catch is collected is called the codend. It appears that the "mesh size" of the codend determines, to a large extent, the selectivity of trawl gear.

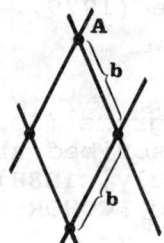

The "mesh size" is usually defined as the length of the "stretched" whole mesh. The mesh size of the netting shown here is 2*b, where b is the length between two knots.

For a detailed discussion of definitions of mesh size and mesh measuring techniques, see FAO (1978b).

It is possible to determine the amount and sizes of fish that escape through the meshes of the codend by covering the codend with a much larger bag with much finer meshes. The idea behind the experiment is illustrated in Fig. 6.1.1. The selectivity of the gear can then be determined by comparing the sizes of the fish in the codend with those of the fish in the cover. The "covered codend method" has been described, among others, by Pope et al., 1975, and Jones, 1976.

Example 21:   Covered codend experiment, Nemipterus japonicus, South China Sea

The experiment deals with the threadfin bream, Nemipterus japonicus, which is caught with a trawl net with a codend mesh size of 4 cm and a cover of much smaller meshes. The typical catch of one trawl haul is given in Table 6.1.1 in the form of two length-frequency tables for the codend and the cover respectively (columns B and C). The fraction of the total catch which was retained in the codend can then be calculated. It is presented as the fraction (e.g. 1/7 = 0.14) retained of each length group. When the fraction retained is plotted against the mid-length of the corresponding length group, it appears that the points are following a sigmoid curve, which reaches 1.00 (100% retention) at a certain length and which approaches 0.00 (0% retention) at a certain small length. This sigmoid curve is called the "gear selection ogive". It resembles a cumulative normal distribution.

The easiest mathematical expression to describe the gear selection ogive is the so-called "logistic curve":

$$S_L = \frac{1}{1 + \exp(S1 - S2*L)} \qquad (6.1.1)$$

where $S_L = \dfrac{\text{number of fish of length L in the codend}}{\text{number of fish of length L in the codend and in the cover}}$

and L is the length interval midpoint (mid-length). S1 and S2 are constants (Paloheimo and Cadima, 1964; Kimura, 1977, and Hoydal et al., 1982).

Eq. 6.1.1 can be rewritten as

$$\ln(1/S_L - 1) = S1 - S2*L \qquad (6.1.2)$$

which represents a straight line, where S1 = a and S2 = b. So the observations of the fractions retained (column E) can be used to determine the logistic curve that fits to the observations. The estimated logistic curve ($S_L$est) can then be used to calculate the fractions that correspond to the curve (column H in Table 6.1.1).

It is seen that if S_L = 0 or if S_L = 1 the expression in Eq. 6.1.2 is not defined.

By applying a few algebraic manipulations it follows that there is a one-to-one correspondence between S1 and S2 and L25%, L50% and L75%, the lengths at which respectively 25%, 50% and 75% of the fish are retained in the codend. The length range from L25% to L75%, which is symmetrical around L50%, is called the "selection range" (see Fig. 6.4.3.1).

The formulas for calculating L25%, L50% and L75% are

$$L25\% = (S1 - \ln 3)/S2 \qquad (6.1.3)$$

$$L50\% = S1/S2 \qquad (6.1.4)$$

$$L75\% = (S1 + \ln 3)/S2 \qquad (6.1.5)$$

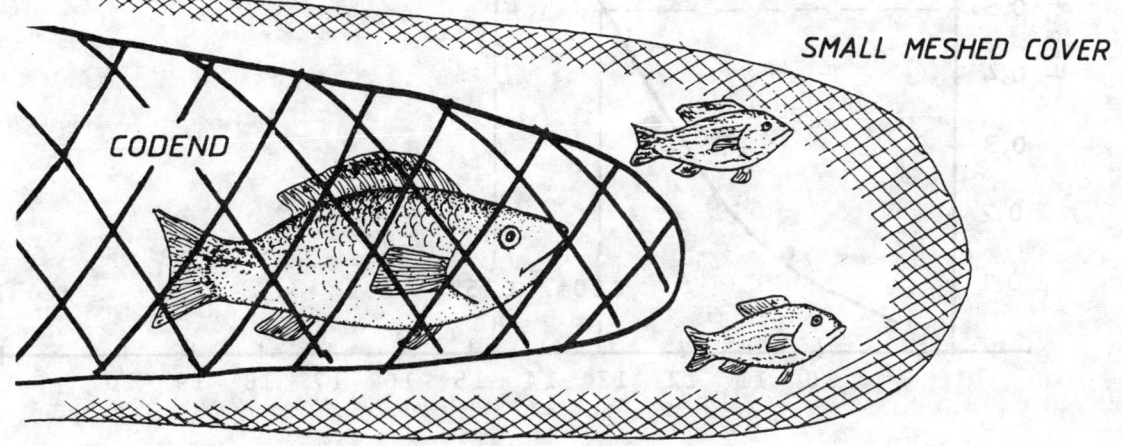

Fig. 6.1.1 Covered codend experiment

Table 6.1.1 Estimation of gear selection ogive for Nemipterus japonicus from a covered codend experiment (from Jones, 1976, cf. Fig. 6.1.2)

A	B	C	D	E	F	G	H
length interval L1-L2	number in codend	number in cover	total number	fraction retained $S_L$obs	$\ln(1/S_L-1)$ (y)	midlength (L1+L2)/2 (x)	fraction retained $S_L$est
9-10	0	1	1	0	-	-	-
10-11	1	6	7	0.14	1.82	10.5	0.13
11-12	2	7	9	0.22	1.27	11.5	0.23
12-13	2	4	6	0.33	0.71	12.5	0.38
13-14	7	5	12	0.58	-0.32	13.5	0.56
14-15	30	13	43	0.70	-0.85	14.5	0.72
15-16	61	8	69	0.88	-1.99	15.5	0.84
16-17	27	3	30	0.90	-2.20	16.5	0.91
17-18	7	0	7	1.00	-	17.5	0.96
18-19	4	1	5	0.80	-	18.5	0.98

intercept = a = S1 = 9.4875   -slope = -b = S2 = 0.7193
L50% = S1/S2 = 13.2 cm   L75% = (S1 + ln 3)/S2 = 14.7 cm
$S_L$est = $1/(1 + \exp(9.4875 - 0.7193*L))$
(used to calculate the curve in Fig. 6.1.2)

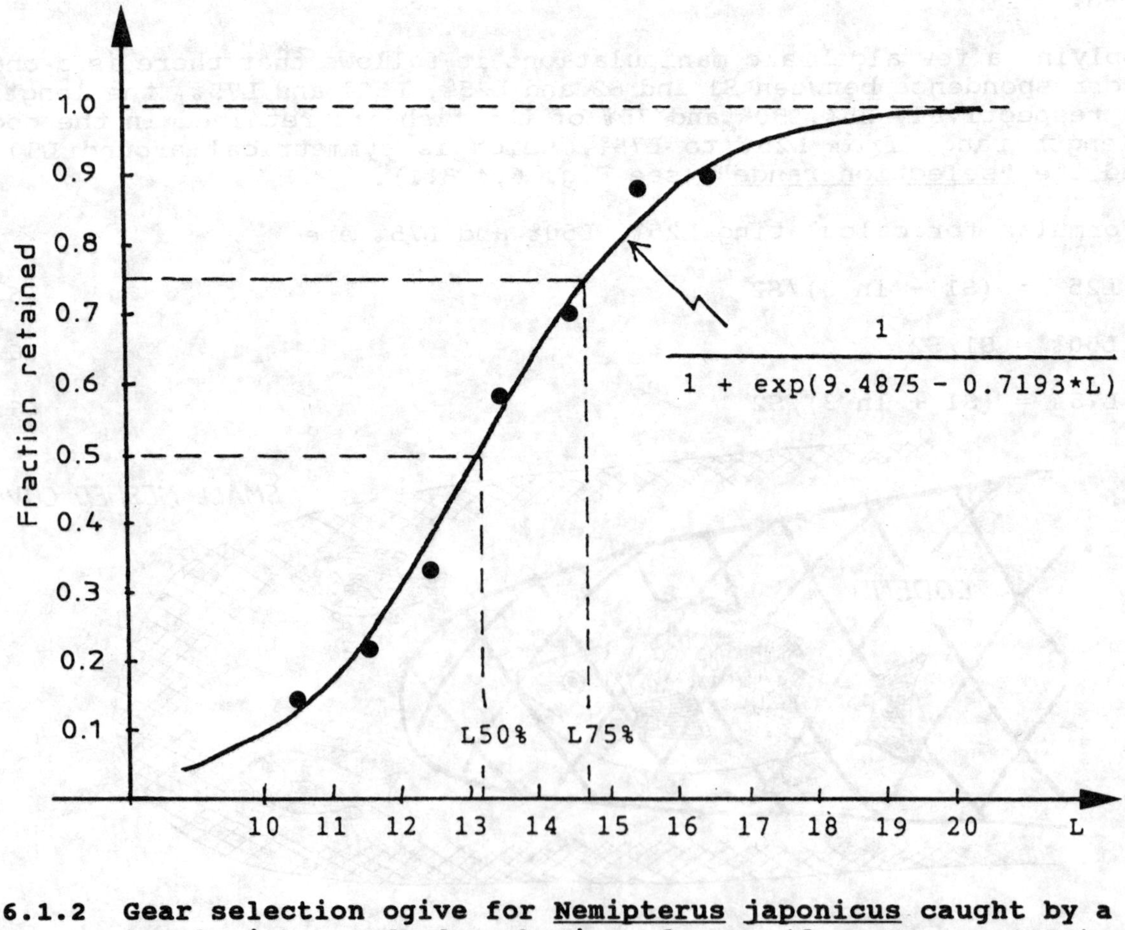

Fig. 6.1.2 Gear selection ogive for Nemipterus japonicus caught by a trawl with a codend mesh size of 4 cm (from Jones, 1976)

S1 and S2 can be derived from L75% and L50% using the following formulas:

$$S1 = L50\% * \ln(3)/(L75\% - L50\%) \qquad (6.1.6)$$

$$S2 = \ln(3)/(L75\% - L50\%) = S1/L50\% \qquad (6.1.7)$$

The regression analysis is done over a length range between zero (0) and full (1) retention, thus excluding those length intervals where no or full retention was obtained and all values beyond those points, even if they are between 0 and 1.

For <u>Nemipterus japonicus</u> the regression gives the following results, as presented in Table 6.1.1 and Fig. 6.1.2:

$$a = S1 = 9.4875 \text{ and } -b = S2 = 0.7193,$$

which gives

$$S_L\text{est} = 1/(1 + \exp(9.4875 - 0.7193*L))$$
$$L25\% = (9.4875 - \ln 3)/0.7193 = 11.7 \text{ cm}$$
$$L50\% = 9.4875/0.7193 = 13.2 \text{ cm}$$
$$L75\% = (9.4875 + \ln 3)/0.7193 = 14.7 \text{ cm}$$

As the probability that a fish will escape through a mesh depends on its shape, and in particular on its body depth compared to the mesh size, it is natural to assume proportionality between d50% (the body depth at which 50% of the fish are retained) and the mesh size:

$$d50\% = A*(\text{mesh size}) \qquad (6.1.8)$$

where A is a constant. As body depth is approximately proportional to body length (cf. Section 2.4) Eq. 6.1.8 implies that a similar expression holds for the length of a fish:

$$L50\% = SF*(\text{mesh size}) \qquad (6.1.9)$$

where SF is called the "<u>selection factor</u>".

In the case of our example (Table 6.1.1) we found L50% = 13.2 cm for a mesh size of 4 cm. Thus, the selection factor is

$$SF = 13.2/4 = 3.3$$

This selection factor can now be used to determine L50% for different mesh sizes, for instance, L50% of <u>Nemipterus japonicus</u> when using meshes of 3 cm would be:

$$L50\% = 3.3*3 = 9.9 \text{ cm}.$$

Further applications of L50% and SF will be discussed in Chapter 8.

6.2 ESTIMATION OF GILL NET SELECTION

6.2.1 Symmetrical selection curves

Gill nets are usually long rectangular nets where the upper edge, the head rope has floats while the foot rope has sinkers. Often gill nets (drifting and set nets) are in the form of gangs of nets with different mesh sizes. For further descriptions of gill nets, see FAO (1978b), Nédélec (1982) or Karlsen and Bjarnason (1986).

The selection properties of gill nets are reviewed in Hamley (1975). Discussions on gill net selectivity can be found in, for example, Baranov (1948); McCombie and Fry (1960), Gulland and Harding (1961), Regier and Robson (1966), Hamley and Regier (1973) and Jensen (1986).

Gill nets are "passive gears", i.e. the fish have to swim into the net to get caught. Theoretically, this implies that fish which move fast, have a larger probability of encounter with the gear than slow moving fish. Further it is known that large fish move faster than small fish of the same species. The swimming speed can be approximated by a constant times a power function of length:

$A*L^B$, where A and B are constants (Yates, 1983).

Rudstam, Magnuson and Tonn (1984) included swimming speed (with B = 0.8 for the cisco, Coregonus artedii, from Wisconsin, USA) into a model for gill net selection. They considered the selection as the product of two probabilities:

(selection) = (probability of encounter)*(probability of being caught given encounter)

We shall, however, only deal with the last factor, the probability of being caught given encounter.

For simple gill nets the selection curve has (unlike trawl selection) a descending slope on the right-hand side. Small fish can pass through the meshes as was the case for trawl nets, but large fish may also avoid being caught in a gill net, because their heads are so large that they cannot be "gilled". This is the simple theory behind gill net selection. The picture becomes somewhat more complicated when other ways by which the fish can get stuck in a gill net are also considered. Baranov (1914) recognized three ways (see also Fig. 6.2.1.1):

a. Gilled, where the mesh is around the fish just behind the gill cover

b. Wedged, where the mesh is around the body as far as the dorsal fin

c. Entangled, where the fish is held in the net by teeth, maxillaries, fins or other projections, without necessarily penetrating the mesh.

For the first two ways of getting stuck in a gill net it has been suggested (Holt, 1963) that the selection curves are bell-shaped and that they can be described by the normal distribution (cf. Section 2.2 and Fig. 6.2.1.2).

Thus for gilling or wedging we use the model:

$$S_L = \exp\left[-\frac{(L-Lm)^2}{2*s^2}\right] \qquad (6.2.1.1)$$

where Lm is the "optimum length for being caught" and s is the standard deviation of the normal distribution. The factor "$n*dL/(s*\sqrt{2\pi})$" which appears in the expression for a normal distribution (Eq. 2.2.1) is not used here. By leaving out this factor, S_L becomes a fraction, i.e. $0 < S_L <= 1$.

Holt (1963) suggested an experiment to estimate Lm and s by using two gill nets with different mesh sizes, ma and mb. The two mesh sizes must be such that their selection curves overlap. The two nets are set to fish in the same area at the same time and the observations are numbers caught by length group. The assumptions behind this method are:

1. The optimum length Lm (the top of the bell-shaped selection curve) is proportional to the mesh size (Lm = SF*m, where SF is the selection factor, cf. Section 6.1)

2. The two selection curves have the same standard deviation

3. The two gears have the same fishing power. This includes that when set, they must have the same length and height.

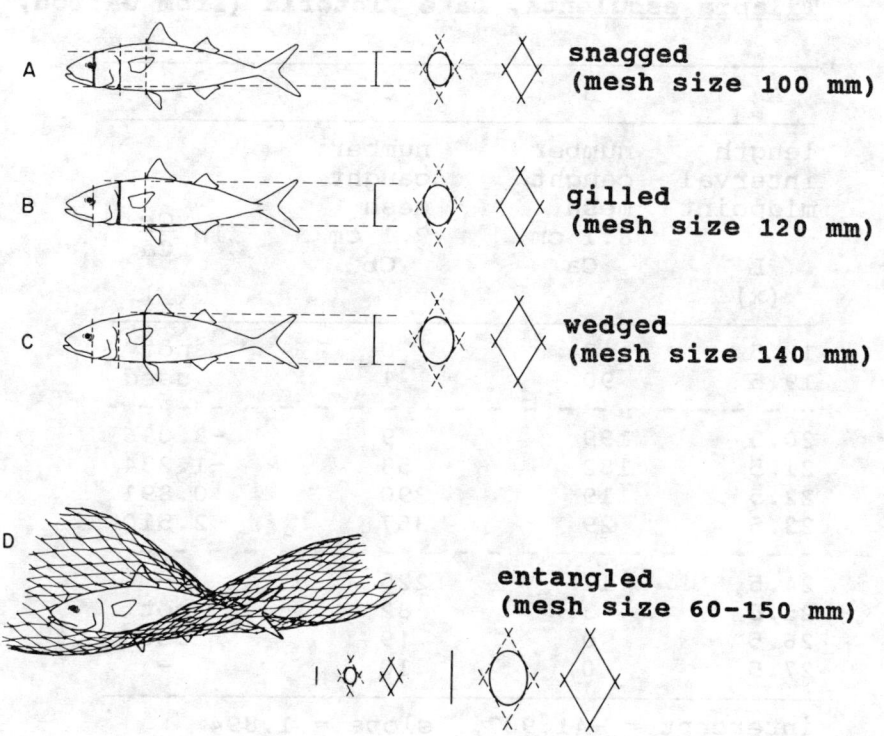

Fig. 6.2.1.1 Gill net selection (from Karlsen and Bjarnason, 1986)

Example 22: Estimation of gill net selection curves, Tilapia, Lake Victoria

Table 6.2.1.1 shows an example of such an experiment on <u>Tilapia esculenta</u> from Lake Victoria. Column B contains the numbers caught, Ca, by length group for the smaller meshed gear (ma = 8.1 cm) and column C contains the corresponding numbers, Cb, for the larger meshed net (mb = 9.1 cm). The parameters to be estimated are:

Lma: Optimum length for the smaller meshed net

Lmb: Optimum length for the larger meshed net

s: The common standard deviation

Input data for the analysis are the numbers caught by length group for each gear, Ca and Cb, and the two mesh sizes ma and mb. The mathematical derivations are lengthy and are by-passed here.

Step 1: Calculate the log ratios

$$y = \ln(Cb/Ca)$$

for each length group (column D of Table 6.2.1.1). Only the lengths where the frequencies overlap can be used.

Step 2: Do a regression analysis of the log ratios (y = ln(Cb/Ca), column D) against the interval midpoint (x = L, column A), and determine a and b:

$$\ln(Cb/Ca) = a + b*L \qquad (6.2.1.2)$$

Table 6.2.1.1 Estimation of gill net selection curves for *Tilapia esculenta*, Lake Victoria (from Garrod, 1961)

A	B	C	D
length interval midpoint L (x)	number caught mesh 8.1 cm Ca	number caught mesh 9.1 cm Cb	$\ln \frac{Cb}{Ca}$ (y)
18.5	7	0	not
19.5	90	1	used
20.5	199	9	-3.096
21.5	182	53	-1.234
22.5	119	290	0.891
23.5	29	357	2.510
24.5	17	225	-
25.5	3	82	not
26.5	0	19	used
27.5	0	10	-

intercept = -41.907, slope = 1.894

$$Lma = \frac{-2*(-41.907)*8.1}{1.894*(8.1+9.1)} = 20.8 \text{ cm}$$

$$Lmb = 20.8*9.1/8.1 = 23.4 \text{ cm}$$

$$s = \sqrt{\frac{-2*(-41.907)*(9.1-8.1)}{1.894^2*(8.1+9.1)}} = 1.17$$

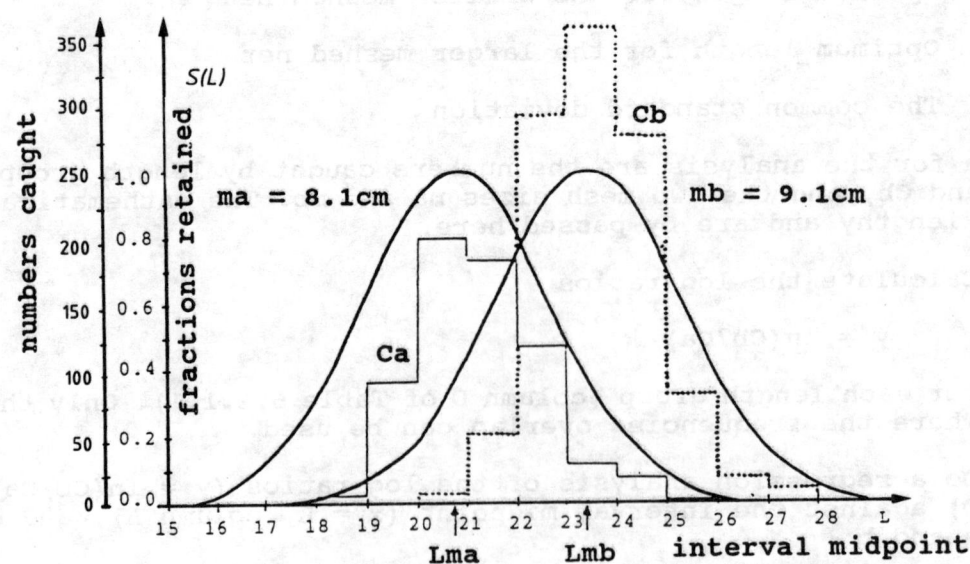

Fig. 6.2.1.2 Selection curves for *Tilapia esculenta* for gill nets of 8.1 and 9.1 cm mesh size from Lake Victoria (from Garrod, 1961)

Step 3: The results are finally obtained by inserting the values for a, b, ma and mb in the following formulas.

The optimum length for the smaller meshed gear is determined by:

$$Lma = -2*\frac{a*ma}{b*(ma+mb)} \qquad (6.2.1.3)$$

and the optimum length for the larger meshed gear by:

$$Lmb = -2*\frac{a*mb}{b*(ma+mb)} = Lma*mb/ma \qquad (6.2.1.4)$$

The common standard deviation is:

$$s = \sqrt{\frac{-2*a*(mb-ma)}{b^2*(ma+mb)}} \qquad (6.2.1.5)$$

The selection factor SF is estimated from:

$$SF = \frac{-2*a}{b*(ma+mb)} \qquad (6.2.1.6)$$

The results are presented in Table 6.2.1.1

Step 4: Calculate the selection curves for each gillnet, using Eq. 6.2.1.1.

Fig. 6.2.1.2 shows the two estimated selection curves for <u>Tilapia esculenta</u>, calculated by means of Eq. 6.2.1.1.

Introducing the selection factor, Eqs. 6.2.1.3 and 6.2.1.4 can be written:

$$Lma = SF*ma \quad \text{and} \quad Lmb = SF*mb$$

In the more complex situation where more than two mesh sizes are used the estimates of an overall selection factor and a common standard deviation can be obtained from the results of an analysis of each pair of successive mesh sizes, under the same assumptions as in the case of two mesh sizes.

Assume that there are n mesh sizes, then there will be n-1 estimates of the intercept, a, and the slope, b (cf. Eq. 6.2.1.2). Thus we have the results:

$$[a(1),b(1)],[a(2),b(2)],\ldots,[a(n-1),b(n-1)]$$

corresponding to the mesh sizes:

$$[m(1),m(2)],[m(2),m(3)],\ldots,[m(n-1),m(n)]$$

Step 5: Estimate the common selection factor by making a regression analysis through the origin, $y(i) = b*x(i)$ with $y(i) = -2*a(i)/b(i)$ as dependent variable and $x(i) = m(i)+m(i+1)$ as independent variable. The slope, b, then becomes the selection factor, SF, based on a rearranged Eq. 6.2.1.6:

$$-2*a(i)/b(i) = SF*[m(i)+m(i+1)], \quad i = 1,2,\ldots,n-1 \qquad (6.2.1.7)$$

SF can be calculated in a similar way as b in Eq. 2.4.13, repeated here for easy reference:

$$b = \Sigma[x(i)*y(i)]/\Sigma x(i)^2 \quad \text{so}$$

$$SF = -2 * \sum_{i=1}^{n-1} [m(i)+m(i+1)]*[a(i)/b(i)] / \sum_{i=1}^{n-1} [m(i)+m(i+1)]^2 \quad (6.2.1.8)$$

Step 6: Estimate the common standard deviation, as the mean value of the individual estimates for each consecutive pair of mesh sizes:

$$s = \sqrt{\frac{1}{n-1} * \sum_{i=1}^{n-1} \frac{-2*a(i)*[m(i+1)-m(i)]}{b(i)^2*[m(i)+m(i+1)]}} \quad (6.2.1.9)$$

Step 7: The optimum length for mesh size i is finally obtained by:

$$Lmi = SF*mi \quad (6.2.1.10)$$

When gill nets of different mesh sizes are used together, the combined selection curve for all mesh sizes may resemble the trawl selection ogive (the logistic curve, Eq. 6.1.1.1) because the fishermen will have ensured the capture of large fish by including large-meshed nets, whereas meshes suitable for the smallest fish will not be used.

So far we have considered selection when the fish are caught by gilling or wedging only, which we described by a bell-shaped selection curve for a particular mesh size. This type of selection is observed to depend on mesh size and a number of other factors such as net construction, visibility and stretchability of the net, net material, shape and behaviour of the fish (Hamley, 1975).

Entangling, however, may be less dependent on mesh size and be more dependent on other factors. The combined selection pattern for gilling, wedging and entangling may deviate from the bell-shaped curve. Since only the fish which are too large to get gilled or wedged are likely to get entangled, the combined selection curve would become a unimodal curve which is skewed to the right-hand side. In extreme cases it would resemble a logistic curve (i.e. selection curve of a trawl). The probability of a fish being entangled is believed to depend on the so-called "hanging ratio", (or "hanging coefficient") which is defined (FAO, 1978b) as:

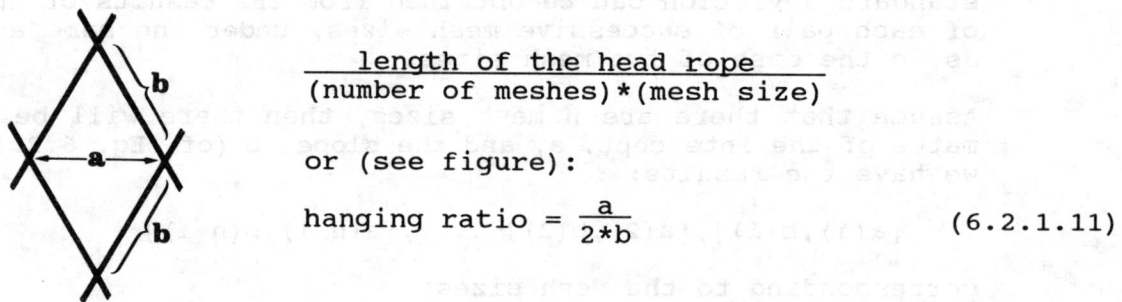

$$\frac{\text{length of the head rope}}{(\text{number of meshes})*(\text{mesh size})}$$

or (see figure):

$$\text{hanging ratio} = \frac{a}{2*b} \quad (6.2.1.11)$$

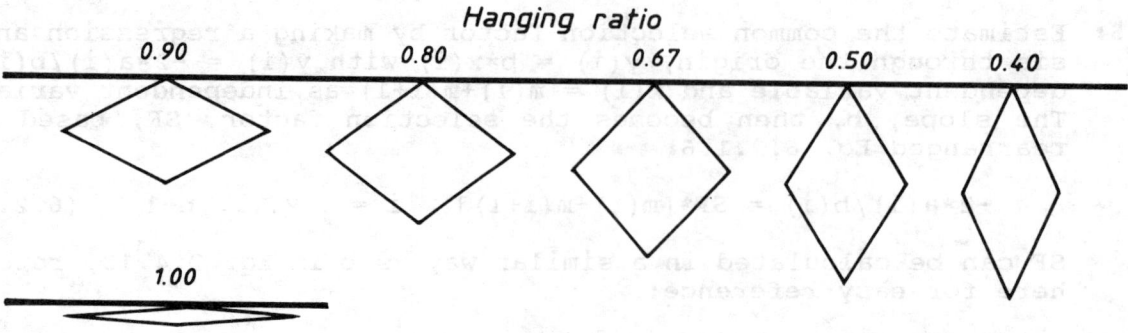

Fig. 6.2.1.3 **Mesh shapes with various hanging ratios for gill nets (from FAO, 1978b)**

Thus, for a square mesh (b = a/√2) we have the hanging ratio √2/2 = 0.707 which represents the maximum opening. Hanging ratios are usually in the range of 0.2 to 0.7 (see Fig. 6.2.1.3). The smaller the hanging ratio the larger the probability of entangling. This is demonstrated by Riedel (1963), who reported catches of <u>Tilapia mossambica</u> with 10 cm mesh gill nets with three different hanging ratios:

hanging ratio	average number caught per day	percent entangled	size range of 95% of catch
0.707	9.3	0	18-23 cm
0.36	29.5	24	13-23 cm
0.24	81.0	80	8-22 cm

6.2.2 The product of two logistic curves

If entangling is an important factor, the normal distribution method described above is not suitable. One way of estimating the selection curve would be to compare the catches of the gill net to a non-selective gear, e.g. to trawl catches. The catches from the non-selective gear would then play the same role as the total combined catch in the cover and the codend of a trawl when operated as described in Section 6.1. The same procedure as used in Table 6.1.1 can be applied. In this case we need a non-symmetrical selection curve of the type shown in Fig. 6.2.2.1. A mathematical expression for this type of curve can be obtained by multiplying two logistic curves (Hoydal et al., 1982). The ascending part of the curve is given by the usual logistic curve (cf. Eq. 6.1.1), it reflects the probability of being gilled or wedged. We call it "SL" where "L" stands for "left-hand side of the selection curve":

$$SL_L = 1/(1 + \exp(S1 - S2*L)) \qquad (6.2.2.1)$$

This type of selection is the dominating one up to length A (see Fig. 6.2.2.1). For lengths larger than B the selection is the combined effect of gilling, wedging and entangling. This part of the curve is modelled by a "<u>reversed logistic curve</u>". We call it "SR" where "R" stands for "right-hand side of the selection curve":

$$SR_L = 1/(1 + \exp(D1 - D2*L) \qquad (6.2.2.2)$$

The parameters in the SR-function, D1 and D2, are negative numbers, whereas the parameters in the SL-function, S1 and S2, are positive numbers. The lengths corresponding to the 50% and 75% de-selection, D50% and D75%, are related to D1 and D2 by the same mathematical expressions as those used for S1, S2, L50% and L75% (cf. Eqs. 6.1.4 to 6.1.7):

$$D50\% = D1/D2 \quad \text{and} \quad D75\% = (D1 + \ln 3)/D2$$

$$D1 = D50\% * \ln(3)/(D75\% - D50\%) \quad \text{and} \quad D2 = D1/D50\%$$

By multiplying the ascending curve, SL, and the descending curve, SR, we obtain the desired type of curve, S. When SL is ascending SR is "neutral" (i.e. approximately equal to 1.0), and when SR is descending SL is "neutral":

$$S_L = SL_L * SR_L = \frac{1}{1 + \exp(S1 - S2*L)} * \frac{1}{1 + \exp(D1 - D2*L)} \qquad (6.2.2.3)$$

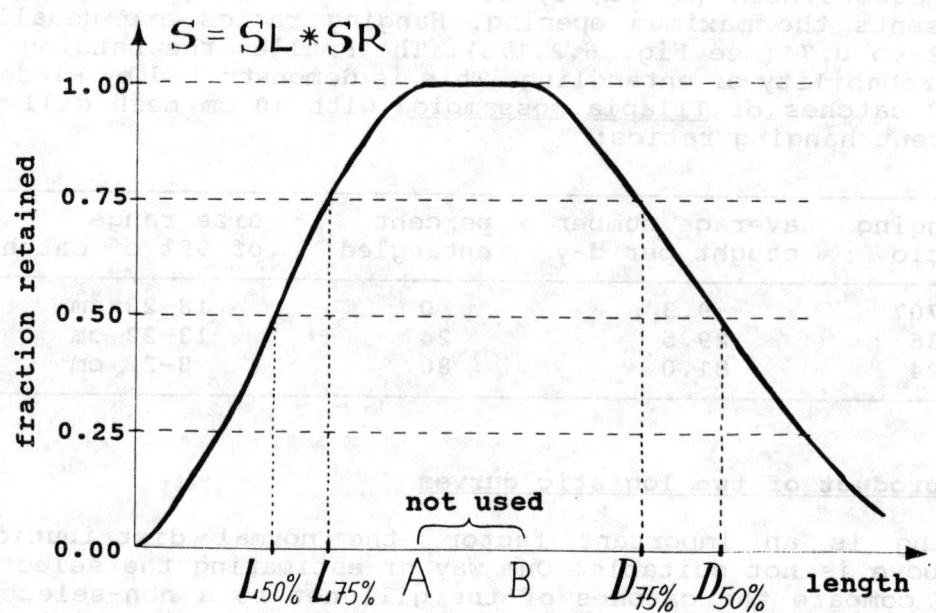

Fig. 6.2.2.1 Asymmetric selection curve

The expression for S_L is supposed to take the maximum value, 1.0, for at least one L-value. Therefore, the expression Eq. 6.2.2.3 (as well as the normal distribution in Eq. 6.2.1.1) should be normalized so that the maximum value equals 1.

In practice this is obtained in the following way. Let S(i) be the point on the selection curve representing length class no. i (estimated by Eq. 6.2.1.1 or Eq. 6.2.2.3) and let MAX {S(j)} designate the maximum value of S(j) among all length classes. We normalize by replacing the value of S(i) by the value:

$$\frac{S(i)}{\underset{j}{MAX}\{S(j)\}}$$

The parameters can be estimated in the same way as the parameters of the trawl selection ogive (cf. Section 6.1). To estimate S1 and S2 we use only the length classes below A (see Fig. 6.2.2.1) and perform the regression analysis:

$$\ln(1/S - 1) = S1 - S2*L \qquad (6.2.2.4)$$

where the dependent variable, $y = \ln(1/S - 1)$, is derived from the comparison with the non-selective gear:

$$S(i) = \frac{Cg(i)/Cn(i)}{\underset{j}{MAX}\{Cg(j)/Cn(j)\}} \qquad (6.2.2.5)$$

where

$$\frac{Cg(i)}{Cn(i)} = \frac{\text{no. of length class i fish caught in gill net}}{\text{no. of length class i fish caught in non-selective gear}}$$

The denominator in Eq. 6.2.2.5, "MAX {Cg(j)/Cn(j)}", is the maximum value of the ratio Cg/Cn among all length groups with non-zero values of Cg and Cn. Thus, S(i) as defined by Eq. 6.2.2.5 takes values between 0 and 1 (including 1).

Eq. 6.2.2.5 is based on the assumption that the numbers Cn are caught by a gear which is non-selective for all those length classes which are caught by the gill net. If this is not the case Eq. 6.2.2.5 should be replaced by:

$$S(i) = \frac{[Cg(i)/Cn(i)]*Sn(i)}{\text{MAX}_j\{(Cg(j)/Cn(j))*Sn(j)\}} \quad (6.2.2.6)$$

where Sn(i) is the selection curve for the "other gear".

The descending (right-hand side) of the selection curve is estimated using the same calculation procedure and the corresponding observations as those used for the left-hand side. Thus we use the data for the length classes above point B (see Fig. 6.2.2.1) and perform the linear regression analysis:

$$\ln(1/S - 1) = D1 - D2*L \quad (6.2.2.7)$$

The method based on the product of two logistic curves is a generalization which includes the trawl selection ogive (Eq. 6.1.1) and the symmetric curve (Eq. 6.2.1.1) as special cases. Assigning the values D1 = $-\infty$ and D2 = 0 makes the factor $1/(1 + \exp(D1 - D2*L))$ in Eq. 6.2.2.3 take the value 1 for all values of L and then Eq. 6.2.2.3 equals Eq. 6.2.1.1. If the curve is symmetrical we estimate parameters so that:

$$L50\% + D50\% = L75\% + D75\%$$

but we do not have to make assumptions beforehand.

The symmetrical curve created by the product of two logistic curves may not be the same as the normal distribution curve. The product of two logistic curves may take the maximum value (1) for a range of values (in Fig. 6.2.2.1 A to B). It may also, however, come very close to the normal distribution as shown by an example in Fig. 6.2.2.2.

Thus, if data for a non-selective gear (or a gear with a known selection curve) are available, there is really no need to use the traditional model, the normal distribution, since the same curve can be obtained as a special case of the logistic product function (Eq. 6.2.2.3). Moreover, the latter is versatile and easy to handle from a computation point of view. Using the product of two logistic curves we are not forced to make the questionable assumption that the selection curves are normally distributed with a common standard deviation.

Finally, it should be mentioned that data collected from gill net catches are difficult to use for the estimation of growth parameters or mortality rates. This is illustrated in Fig. 6.2.2.3 as a hypothetical example. We consider a length-frequency sample representing the population (double line) i.e. a sample from a non-selective gear and a sample representing the gill net catch (shaded line). The population contains four components or cohorts (normal distributions), whereas the gill net sample appears to contain only one component with mean value Lm, which does not coincide with any of the four cohort mean lengths (L0,L1,L2 and L3). Thus, the gill net sample does not give any information which can be used for the separation of cohorts and the estimation of length-at-age.

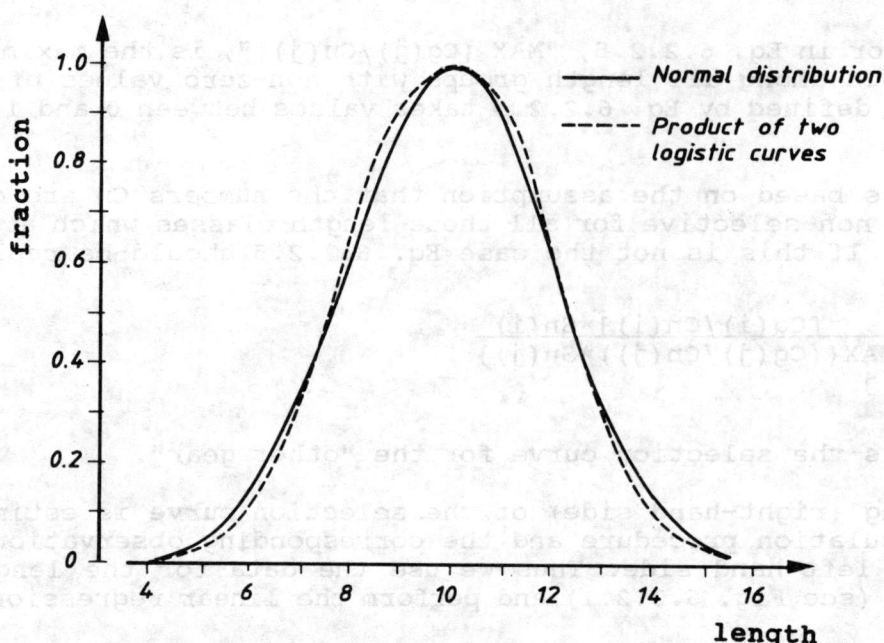

Fig. 6.2.2.2 A product of two logistic curves which is nearly identical to a normal distribution. The parameters are:
Normal distribution: Lm = 10 and s = 2
Product of two logistic curves:
L50% = 7.645   L75% = 8.483
D50% = 2*Lm - L50%   D75% = 2*Lm - L75%

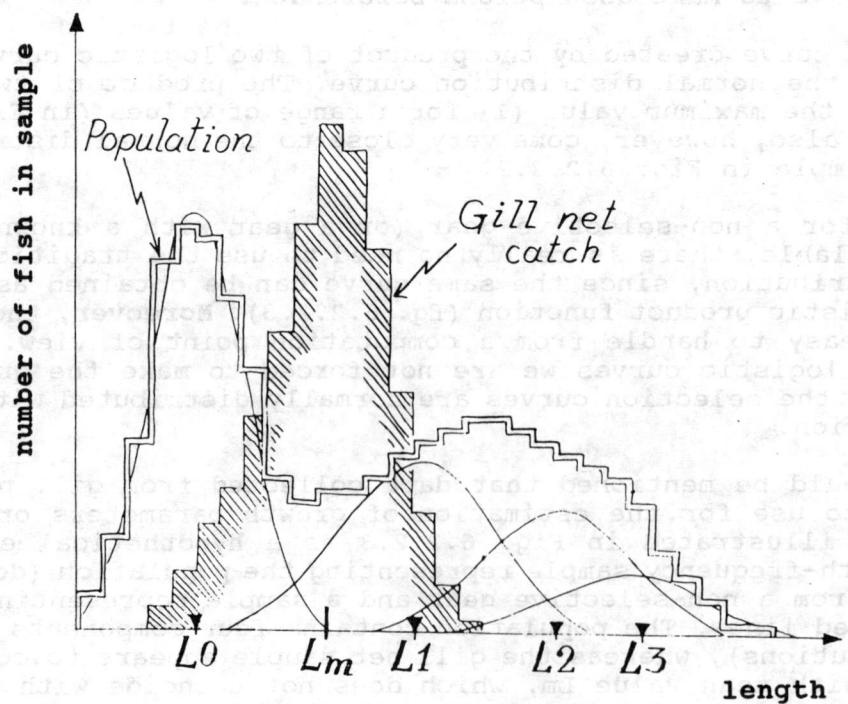

Fig. 6.2.2.3 Hypothetical example to illustrate bias problems when using gill net data for estimation of growth parameters and mortality rates

Also samples collected during a whole year would all give more or less the same picture. Summing a time series over a year would therefore give a curve not very different from the curve for the gill net catch in Fig. 6.2.2.3. Using the descending slope of this curve to estimate total mortality from the length-based catch curve analysis would lead to an over-estimate of Z. Before using any gill net data for estimation of growth parameters or mortality rates, they should be critically examined. The result of such an examination may be either that the data cannot be used at all or that they can be used only after having been adjusted for gear selection as described in Section 6.7.

6.3 DISCUSSION OF SELECTION BY OTHER GEARS

The previous two sections dealt with the selectivity of trawl nets (active gears) and gill nets (passive gears). The scientific literature on selectivity has been mainly concerned with these gears, partly because it is relatively easy to conduct experiments for the estimation of their selection curves. Other types of gear are also more or less selective and the pattern of selection can usually be changed by suitable adjustments of the gear. The gear selection model based on the logistic curve (Eq. 6.1.1) or the model based on the product of two logistic curves (Eq. 6.2.2.3) are believed versatile enough to describe the selection curve of any gear.

Below is a brief discussion of the selection properties of two passive gear types: hooks and traps, and of an active gear type: seine nets.

Much more appears to be known about trawl and gill net selection than about hook fisheries. For hook-and-line fisheries some authors report that the selection curve is bell-shaped, of the gill net type, dependent on hook size, whereas others find a trawl type selection. The idea behind the use of the bell-shaped selection curve is that small fish cannot take a large hook in their mouth and that large fish are not held securely by small hooks.

A discussion of hook selection can be found in Ralston (1982), who observed the trawl type of selection for the Hawaiian deep-sea hand line fishery. From an experiment where four hook sizes were used for catching snappers and groupers he found that small hooks were nearly as efficient as large hooks at catching large fish. For the ascending left-hand selection curve (the small fish) he found a sigmoid curve. Pope et al. (1975) suggested using the gill net type of selection for hook fisheries, although they quote a number of works which show the trawl type of selection.

Pope et al. (1975) suggested the trawl type of selection for traps arguing that traps behave like codends in retaining fish. Munro (1974, 1983) discusses selectivity and other aspects of the operation of portable Antillean fish traps. These 3 m long traps are made of chicken wire-netting with two funnels whose bottoms are the entrance openings. They are used for catching coral reef fish. Munro (1974) developed a model for the catchability of traps as a function of the soak time (the time the trap is left on the fishing ground).

Trap selectivity is complicated because it relies on the fish to move actively into the trap. For small traps (as the Antillean trap mentioned) only one specimen of a territorial fish species can be expected to be caught, but the trapped fish may have been replaced by another before the trap is pulled. As the other extreme we have the hunters, like the jacks (Carangidae) in a coral reef fishery which cover a lot of ground in a short time and therefore become over-represented in trap catches. If a large predator is caught in the trap it may keep potential prey fish from entering the trap. If prey species are already in the trap they may act as live-bait and attract the large predators, which in turn may eat the prey before the trap is pulled.

Trap catches depend on the duration of the soak (Munro, 1974). There is always a chance that a trapped fish finds the entrance opening and leaves through it. There are considerable differences between species. Some species leave the trap with great ease (Munro, 1983). Thus, trap selectivity may not only be a function of the mesh size used in the trap. For example, the size of the entrance opening and the soak time may be of importance. The species composition on the ground where the trap is placed also may influence the selectivity. For the escapement through the meshes, however, it appears reasonable to assume the trawl selection type of ogive. When considering the average of a large number of trap catches some of the above mentioned complications may disappear, i.e. they may turn out to be "random noise" around the selection curve.

In principle a seine net should work like a trawl as far as selection is concerned. However, it is more difficult to deal with the seines because this type of gear is usually used to catch schooling species, such as sardines, mackerels and tunas. These species have a tendency to form schools consisting of fish of the same size. Thus, we should consider a school to be a sampling unit (instead of an individual fish).

6.4 OTHER ASPECTS OF GEAR SELECTIVITY

6.4.1 Knife-edge selection

Fig. 6.4.1.1 shows two selection curves. Curve A has a selection range of 3 cm and curve B, the fat vertical line, has a selection range of 0 cm. Curve B is a so-called "knife-edge selection curve" (Beverton and Holt, 1957). Knife-edge selection should be considered a hypothetical model since it will never describe a real situation. However, knife-edge selection is often used as an approximation to the selection ogive. For lengths below L50% the numbers selected are under-estimated and for lengths above L50% the numbers are over-estimated. These two sources of bias have opposite signs and as the two areas "a" and "b" (see Fig. 6.4.1.1) are the same size they balance out. However, the fish of area "a" will weigh more than those of area "b", since the weight of a fish corresponds to the cube of the length.

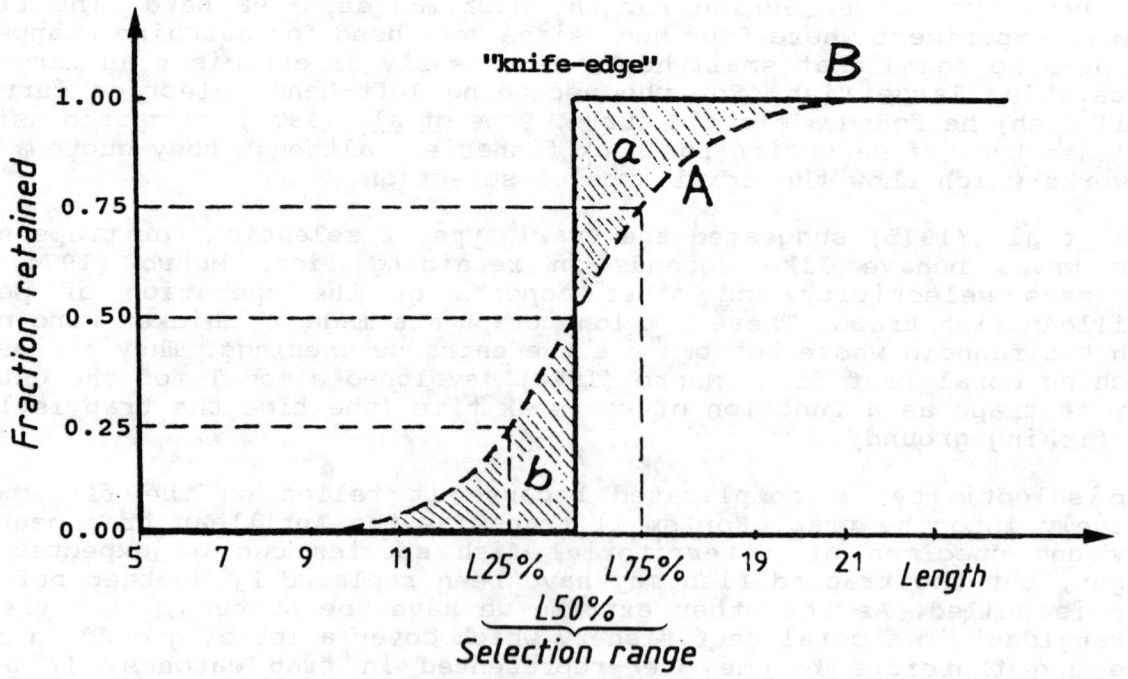

Fig. 6.4.1.1 Trawl selection curve as a function of body length. An illustration of the concepts of knife-edge selection and selection range

6.4.2 Recruitment and selectivity

The recruitment of fish to the fishing area, that is when they move away from the nursery or spawning areas to the fishing grounds, is also size dependent, in the same way as a trawl selection ogive. This means that every size of fish will not be fully represented at the fishing grounds and, thus, when there is a fishery for the size ranges which are not yet fully recruited the probability that a fish is retained by the fishing gear is in fact the product of two probabilities:

1. The probability that the fish is present on (has recruited to) the fishing ground.
2. The probability that the fish is retained by the meshes once it has entered the gear.

Fig. 6.4.2.1 illustrates these points. Curve R is the "<u>recruitment curve</u>", curve G is the "<u>gear selection curve</u>", and curve S is the "<u>resultant curve</u>".

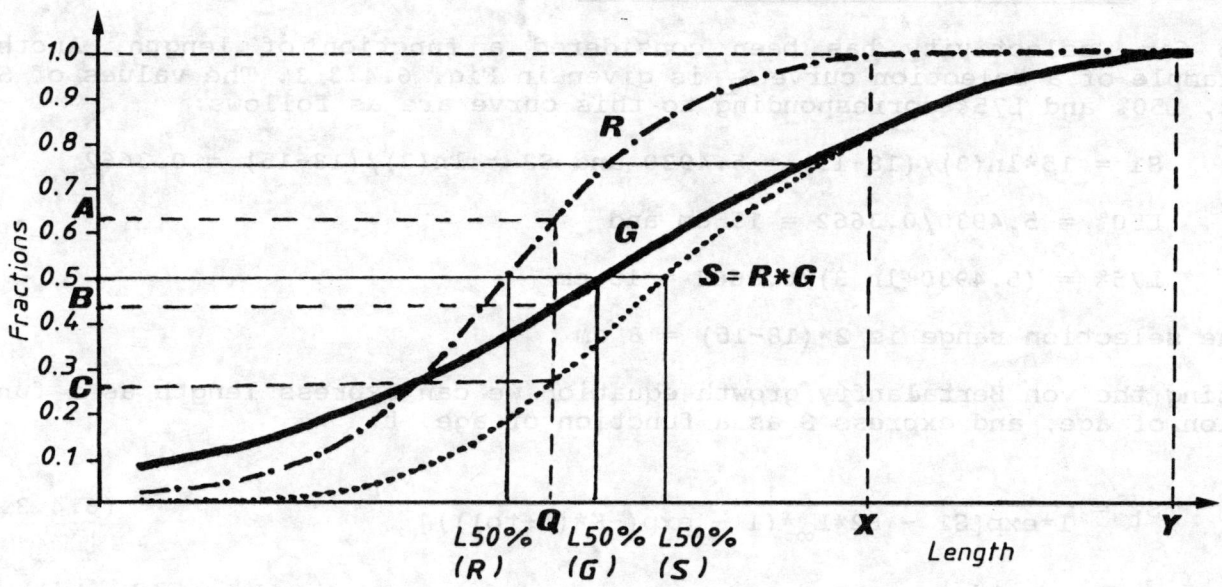

Fig. 6.4.2.1 Curves representing: recruitment to the fishing grounds (R), gear selection (G), and the resultant (S = R*G) (see also text)

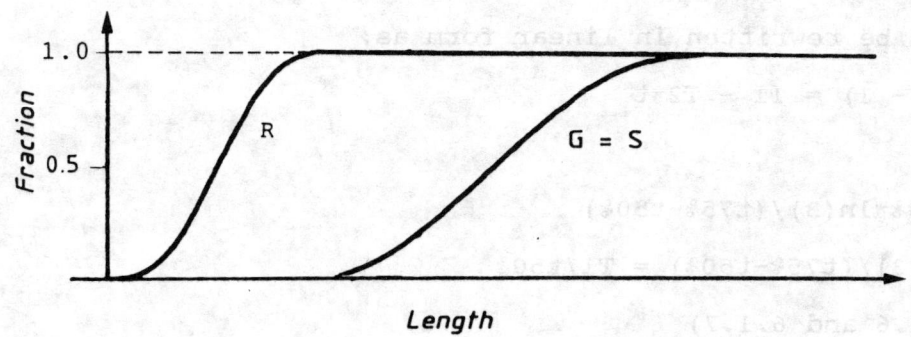

Fig. 6.4.2.2 Example of equality (overlap) between the gear selection curve (G) and the resultant curve (S)

The probability that a certain size of fish will be caught is the product of the probabilities of recruitment and selection. The probability can therefore be described by a "resultant curve", S, where S = R*G (see Fig. 6.4.2.1).

The L50% for the three curves, R, G and S, are different as indicated in Fig. 6.4.2.1. The probability of capture of a fish of length Q is the product of probability A, related to the recruitment curve, and probability B related to the gear selection curve, and the result is probability C.

In this example A*B = C or R(Q)*G(Q = S(Q) or 0.62*0.42 = 0.26. At length X practically all fish have recruited to the fishing grounds while some are still small enough to escape through the meshes. At length Y mesh retention is complete and no fish escape.

When the meshes are so large that there is no overlap of the recruitment curve with the selection curve we can ignore recruitment. The resultant curve is then determined by selection only, see Fig. 6.4.2.2.

6.4.3 Selectivity as a function of age

So far, selectivity has been considered a function of length. Another example of a selection curve S_L is given in Fig. 6.4.3.1. The values of S1, S2, L50% and L75% corresponding to this curve are as follows:

$S1 = 15*\ln(3)/(18-15) = 5.4930$ and $S2 = \ln(3)/(18-15) = 0.3662$

$L50\% = 5.4930/0.3662 = 15$ cm and

$L75\% = (5.4930+\ln 3)/0.3662 = 18$ cm

The selection range is $2*(18-15) = 6$ cm

Using the von Bertalanffy growth equation we can express length as a function of age, and express S as a function of age, t:

$$S_t = \frac{1}{1*\exp[S1 - S2*L_\infty*(1 - \exp(-K*(t-t_o)))]} \quad (6.4.3.1)$$

The graph of this expression is approximately equal to the graph which is the age-based equivalent of Eq. 6.1.1, and defined by:

$$S_t = \frac{1}{1 + \exp(T1 - T2*t)} \quad (6.4.3.2)$$

and which can be rewritten in linear form as:

$$\ln(1/S_t - 1) = T1 - T2*t \quad (6.4.3.3)$$

where:

$T1 = t50\%*\ln(3)/(t75\%-t50\%) \quad (6.4.3.4)$

$T2 = \ln(3)/(t75\%-t50\%) = T1/t50\% \quad (6.4.3.5)$

(cf. Eqs. 6.1.6 and 6.1.7)

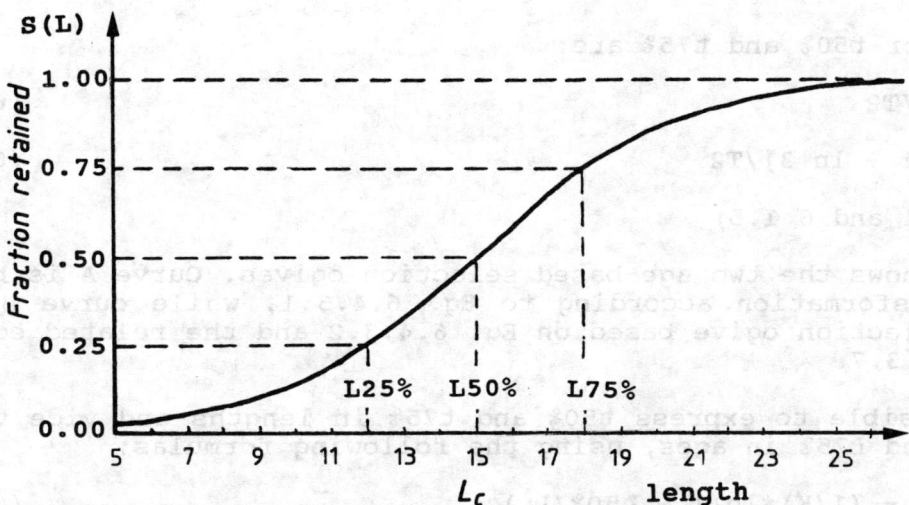

Fig. 6.4.3.1 The selection curve referred to in the text

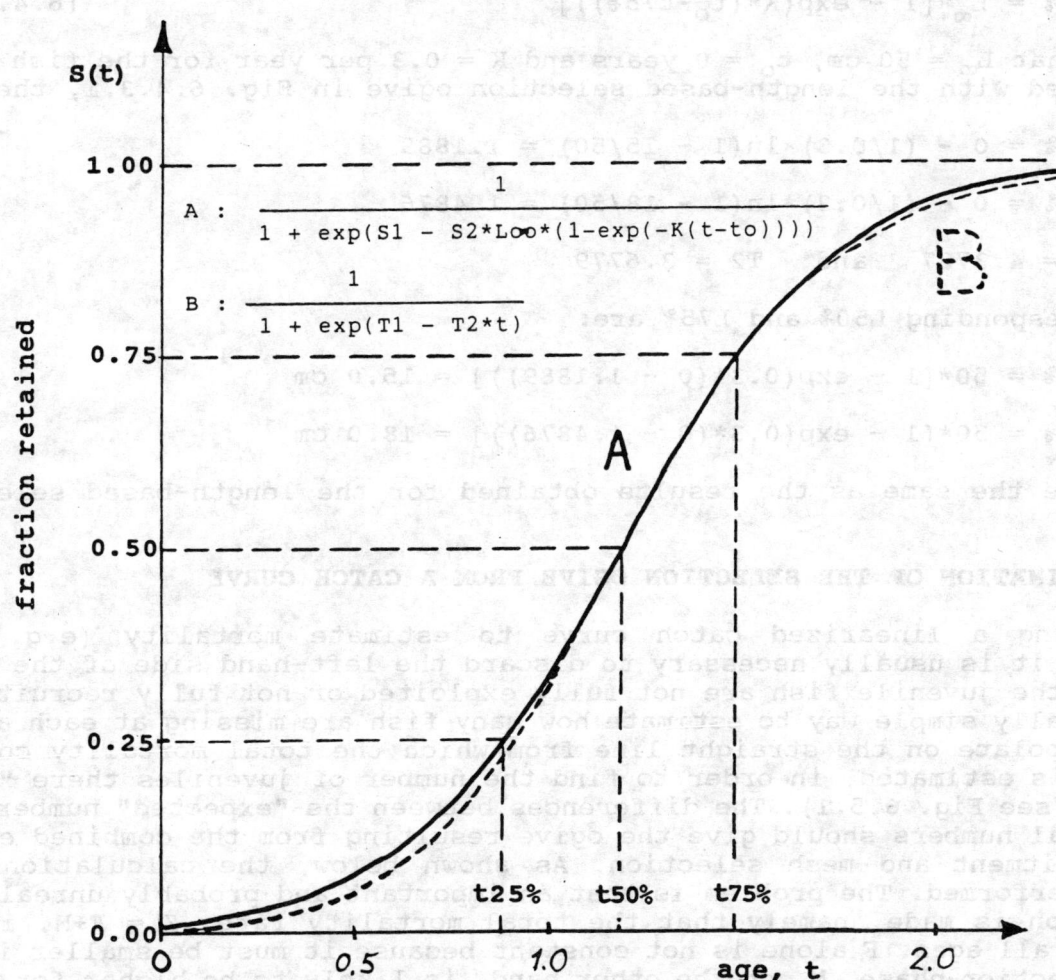

Fig. 6.4.3.2 A: Exact age-transformed selection ogive (Eq. 6.4.3.1)
B: Approximated selection ogive (Eq. 6.4.3.2)

The formulas for t50% and t75% are:

$$t_{50\%} = T1/T2 \qquad (6.4.3.6)$$

$$t_{75\%} = (T1 + \ln 3)/T2 \qquad (6.4.3.7)$$

(cf. Eqs. 6.1.4 and 6.1.5)

Fig. 6.4.3.2 shows the two age-based selection ogives. Curve A is based on the exact transformation according to Eq. 6.4.3.1, while curve B is the approximate selection ogive based on Eq. 6.4.3.2 and the related equations 6.4.3.4 to 6.4.3.7.

It is also possible to express t50% and t75% in lengths and vice versa to express L50% and L75% in ages, using the following formulas:

$$t_{50\%} = t_o - (1/K)*\ln(1 - L_{50\%}/L_\infty) \qquad (6.4.3.8)$$

$$t_{75\%} = t_o - (1/K)*\ln(1 - L_{75\%}/L_\infty) \qquad (6.4.3.9)$$

and

$$L_{50\%} = L_\infty*[1 - \exp(K*(t_o-t_{50\%}))] \qquad (6.4.3.10)$$

$$L_{75\%} = L_\infty*[1 - \exp(K*(t_o-t_{75\%}))] \qquad (6.4.3.11)$$

Assume that L_∞ = 50 cm, t_o = 0 years and K = 0.3 per year for the fish stock associated with the length-based selection ogive in Fig. 6.4.3.1, then:

$$t_{50\%} = 0 - (1/0.3)*\ln(1 - 15/50) = 1.1889$$

$$t_{75\%} = 0 - (1/0.3)*\ln(1 - 18/50) = 1.4876$$

$$T1 = 4.3727 \quad \text{and} \quad T2 = 3.6779$$

The corresponding L50% and L75% are:

$$L_{50\%} = 50*[1 - \exp(0.3*(0 - 1.1889))] = 15.0 \text{ cm}$$

$$L_{75\%} = 50*[1 - \exp(0.3*(0 - 1.4876))] = 18.0 \text{ cm}$$

which are the same as the results obtained for the length-based selection ogive.

6.5 ESTIMATION OF THE SELECTION OGIVE FROM A CATCH CURVE

When using a linearized catch curve to estimate mortality (e.g. Fig. 4.4.5.1) it is usually necessary to discard the left-hand side of the curve because the juvenile fish are not fully exploited or not fully recruited. A conceptually simple way to estimate how many fish are missing at each age is to extrapolate on the straight line from which the total mortality coefficient Z is estimated, in order to find the number of juveniles there "ought to be", (see Fig. 6.5.1). The differences between the "expected" numbers and the actual numbers should give the ogive resulting from the combined effect of recruitment and mesh selection. As shown below, the calculations are easily performed. The problem is that an important and probably unrealistic assumption is made, namely that the total mortality rate, Z = F+M, is the same for all ages. F alone is not constant because it must be smaller in the mesh selection phase. M, on the other hand, is likely to be higher for small fish than for the adults. It is therefore possible that Z remains approximately constant although so far, nobody has shown it to be. Nevertheless, the method has achieved considerable popularity and is therefore mentioned here.

Example 23: Estimation of the resultant selection ogive from a catch curve, hypothetical data

To explain this method (Pauly, 1984a) the example of Table 6.5.1 is used. Columns A-E contain the input data and calculations for a length-converted catch curve analysis (cf. Section 4.4.5). In this case we calculate the total mortality $Z = 1.0$ per year from the growth parameters, $L_\infty = 50$ cm, $K = 0.3$ per year (see Fig. 6.5.1). The result of the regression analysis is:

$$\ln \frac{C}{\Delta t} = 9.208 - 1.0*t$$

In contrast to the example discussed in Section 4.4.5 we now have a use for the intercept ($a = 9.208$).

Under the assumption of constant mortality we expect the values of $\ln(C/\Delta t)$ to be on the regression line, $\ln(C/\Delta t) = a - Z*t$. Thus, the hypothetical true frequency the total population numbers in the sea, CT, is expected to fulfill the equation:

$$\ln(CT/\Delta t) = a - Z*t. \qquad (6.5.1)$$

The idea behind this method is that the number in the sea is proportional to the number caught, i.e.

$$\frac{C}{CT} = \frac{\text{the number in the catch}}{\text{total population number in the sea}}$$

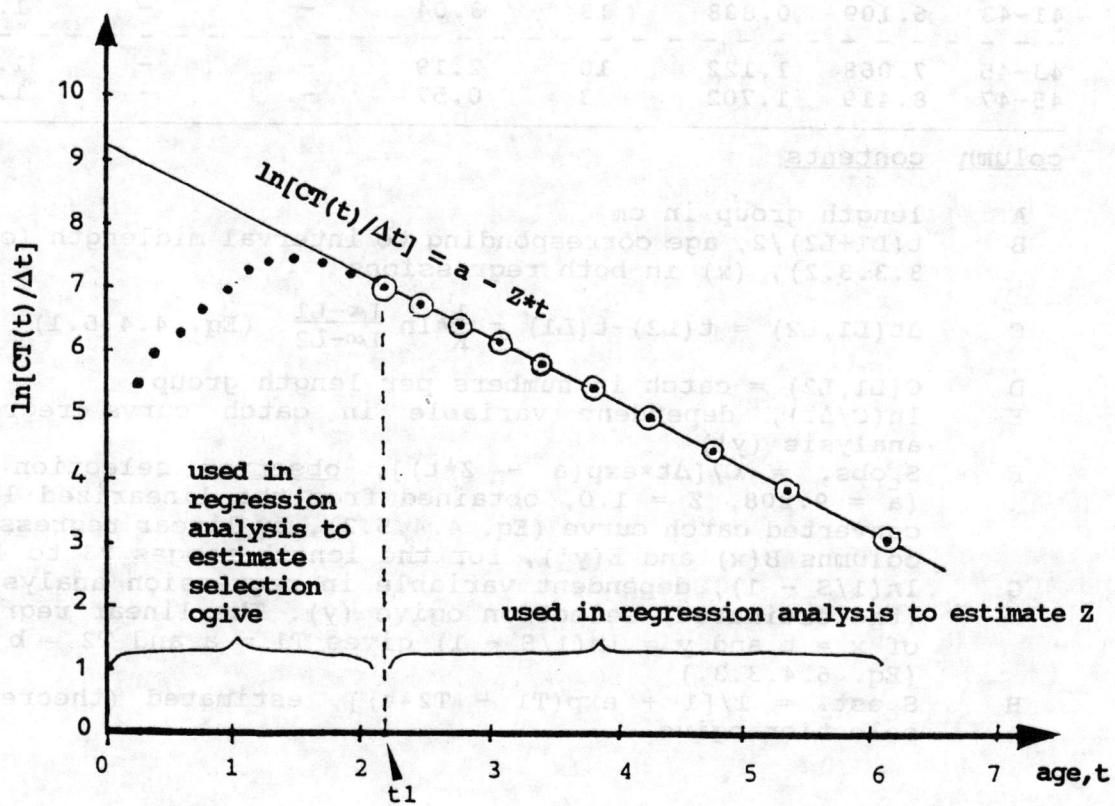

Fig. 6.5.1 Estimation of the selection ogive from a length-converted catch curve analysis based on Table 6.5.1

Table 6.5.1 Example to illustrate estimation of the selection ogive from a catch curve (cf. Fig. 6.5.1). $L_\infty = 50$ cm, $K = 0.3$ per year, $t_o = 0$. (The same data were used in Table 4.5.4.1.)

A	B	C	D	E	F	G	H
L1-L2	t (x)	Δt (L1,L2)	C(L1,L2) (y')	ln (C/Δt) (y)	S_t obs.	ln (1/S-1)	S_t est.
3- 5	0.278	0.145	37	5.54	0.034	3.35	0.03
5- 7	0.426	0.151	56	5.92	0.057	2.81	0.06
7- 9	0.581	0.159	86	6.29	0.097	2.23	0.10
9-11	0.744	0.167	129	6.65	0.163	1.64	0.16
11-13	0.915	0.176	188	6.97	0.267	1.01	0.27
13-15	1.095	0.186	258	7.23	0.416	0.42	0.42
15-17	1.286	0.196	319	7.39	0.590	-0.37	0.59
17-19	1.487	0.208	352	7.43	0.750	-1.10	0.75
19-21	1.703	0.222	351	7.37	0.870	-1.90	0.87
21-23	1.933	0.238	324	7.22	0.943	-2.80	0.94
-----	-----	-----	-----	-----	-----	-----	-----
23-25	2.180	0.257	283	7.00	(0.976)	-	0.98
25-27	2.447	0.278	239	6.76	-	-	0.99
27-29	2.734	0.303	196	6.47	-	-	1.00
29-31	3.054	0.334	158	6.16	-	-	1.00
31-33	3.406	0.371	123	5.80	-	-	1.00
33-35	3.798	0.417	93	5.41	-	-	1.00
35-37	4.243	0.477	69	4.97	-	-	1.00
37-39	4.757	0.557	48	4.46	-	-	1.00
39-41	5.365	0.669	31	3.84	-	-	1.00
41-43	6.109	0.838	18	3.04	-	-	1.00
-----	-----	-----	-----	-----	-----	-----	-----
43-45	7.068	1.122	10	2.19	-	-	1.00
45-47	8.419	1.702	3	0.57	-	-	1.00

<u>column</u>   <u>contents</u>

A   length group in cm
B   t(L1+L2)/2, age corresponding to interval midlength (cf. Eq. 3.3.3.2), (x) in both regressions

C $\Delta t(L1,L2) = t(L2) - t(L1) = \frac{1}{K} * \ln \frac{L\infty - L1}{L\infty - L2}$ (Eq. 4.4.5.1)

D   C(L1,L2) = catch in numbers per length group
E   ln(C/Δt), dependent variable in catch curve regression analysis (y')
F $S_t$obs. = C/[Δt*exp(a - Z*t)], <u>observed</u> selection ogive (a = 9.208, Z = 1.0, obtained from the linearized length-converted catch curve (Eq. 4.4.5.3), by linear regression of columns B(x) and E(y'), for the length ranges 23 to 43 cm
G   ln(1/S - 1), dependent variable in regression analysis for (the estimated) selection ogive (y). The linear regression of x = t and y = ln(1/S - 1) gives T1 = a and T2 = b (Eq. 6.4.3.3.)
H $S_t$est. = 1/[1 + exp(T1 - T2*t)], estimated (theoretical) selection ogive

Let t1 be the age corresponding to the first length group which is supposed to be fully represented in the catch and therefore used in the catch curve regression (in the case of Table 6.5.1, we have t1 = 2.180, see Fig. 6.5.1). For ages above t1, CT_t should be approximately equal to the observed frequencies, since the probability of capture is 1, because selection and recruitment are supposed to have finished before that age. For the ages below t1 we expect that the population in the sea is higher than that represented in the catch, i.e:

$$\ln(CT_t/\Delta t) > \ln(C_t/\Delta t)$$

As CT_t is supposed to be proportional to the population number the ratio

$$C_t/CT_t$$

is an estimate of the probability that a fish of age t will be on the fishing ground and be retained if it encounters the gear, i.e. C_t/CT_t can be used as an estimate of the resultant ogive S_t.

CT can be predicted by Eq. 6.5.1, modified into:

$$CT_t = \Delta t * \exp(a - Z*t) \qquad (6.5.2)$$

Thus, the ogive can be estimated by:

$$S_t = \frac{C_t}{CT_t} = \frac{C_t}{\Delta t * \exp(a - Z*t)} \qquad (6.5.3)$$

The fractions retained of the <u>observed</u> selection ogive are presented in column F of Table 6.5.1. In order to obtain the theoretical (estimated) selection ogive the expression for S_t is used in linear form (Eq. 6.4.3.3):

$$\ln(1/S_t - 1) = T1 - T2*t$$

Eq. 6.4.3.3 enables us to estimate the parameters T1 and T2 by linear regres-sion. Columns B (x) and G (y) of Table 6.5.1 contain the inputs for this regression. (Columns C, D and E contain the results of the catch curve analysis used to calculate the dependent variable, y, column G.) Column H contains the estimated selection ogive. Only values of S_t less than 1 (column F) can be used in the expression $\ln(1/S-1)$ (column G). Carrying out the regression analysis we find:

$$a = T1 = 4.396 \quad \text{and} \quad -b = T2 = 3.701$$

which gives, using Eqs. 6.4.3.6, 6.4.3.7, 6.4.3.10 and 6.4.3.11:

```
t50% = T1/T2            = 1.1877 years
t75% = (T1 + ln 3)/T2   = 1.4846 years
L50% = 50*(1 - exp(-0.3*(1.1877 - 0))) = 15.0 cm
L75% = 50*(1 - exp(-0.3*(1.4846 - 0))) = 18.0 cm
```

The example of Table 6.5.1 is a hypothetical one constructed to give the results of Figs. 6.4.3.1 and 6.4.3.2. Because the data are ideal hypothetical data there is a perfect agreement between the observed fractions retained (column F of Table 6.5.1.) and the theoretical fractions retained (column H).

The exercise provides a check on the adequacy of the choice of points used in the regression analysis for the estimation of Z. The conclusion to be drawn from Table 6.5.1 is that the first length group to be used in the estimation of Z should have been 27-29 cm, as this group is the first one under full exploitation. However, because the logistic curve never attains the value 1 the concept of "full exploitation" is determined by the number of decimals in the table. Taking into account that the logistic curve is an approximation to the real selection curve one cannot expect to get a precise estimate for the first length under full exploitation. If we are somewhere in the "near neighbourhood" of 1 the choice of first length group in the catch curve regression is likely to be good enough.

As emphasized in the introduction to this section, the results of the method described should be treated with a certain reservation.

(See **Exercise(s)** in Part 2).

6.6 GEAR SELECTIVITY AND VPA METHODS

6.6.1 Gear selectivity and fishing mortality

Fishing mortality, F, is clearly related to the selection ogive. When $S_L = 0$ fishing mortality must be zero and when $S_L = 1$ fishing mortality is at its highest level. The obvious relationship between fishing mortality and selection is:

$$F_L = Fm*S_L \qquad (6.6.1.1)$$

where Fm is the "maximum fishing mortality". Thus, F, as a function of length has the same shape as S, but it has a different level (see Fig. 6.6.1.1A).

In Eq. 6.6.1.1 we consider F a continuous function of length, L. In practice, however, it is often convenient to replace the continuous function by a step function as shown in Fig. 6.6.1.1B, where F is assumed to remain constant within each length group.

The continuous selection curve S_L may also be approximated by a step function, S(j), in which the value for length group no. j is S((L1+L2)/2), where L1 and L2 are the lower and upper limits of length group no. j. When we use the length group index, j, as argument rather than the length L, we may write a step-function model for total mortality, Z;

$$Z(j) = M + Fm*S(j) \qquad (6.6.1.2)$$

where M is the natural mortality coefficient (here assumed to remain constant for all length groups), S(j) is the step function of the selection ogive and Fm the maximum fishing mortality. If Z, M and Fm are known the selection can be estimated by:

$$S(j) = F(j)/Fm \qquad (6.6.1.3)$$

where $F(j) = Z(j) - M$.

(Pope et al., 1975 and Hoydal et al., 1982).

Fig. 6.6.1.2 shows F(j), Z(j) and S(j) as functions of length. When we work with a step function rather than with a continuous logistic curve the selection is given as an array of S-values and this array can replace the mathematical expression (Eq. 6.6.1.1) or the array may be applied to estimate the parameters of the logistic curve. Actually, an array of S-values is a more versatile way of presenting selection ogives as no assumptions have to be made about the underlying mathematical expression. (cf. the discussion of age/length keys versus growth equations in Section 3.2.1).

- 195 -

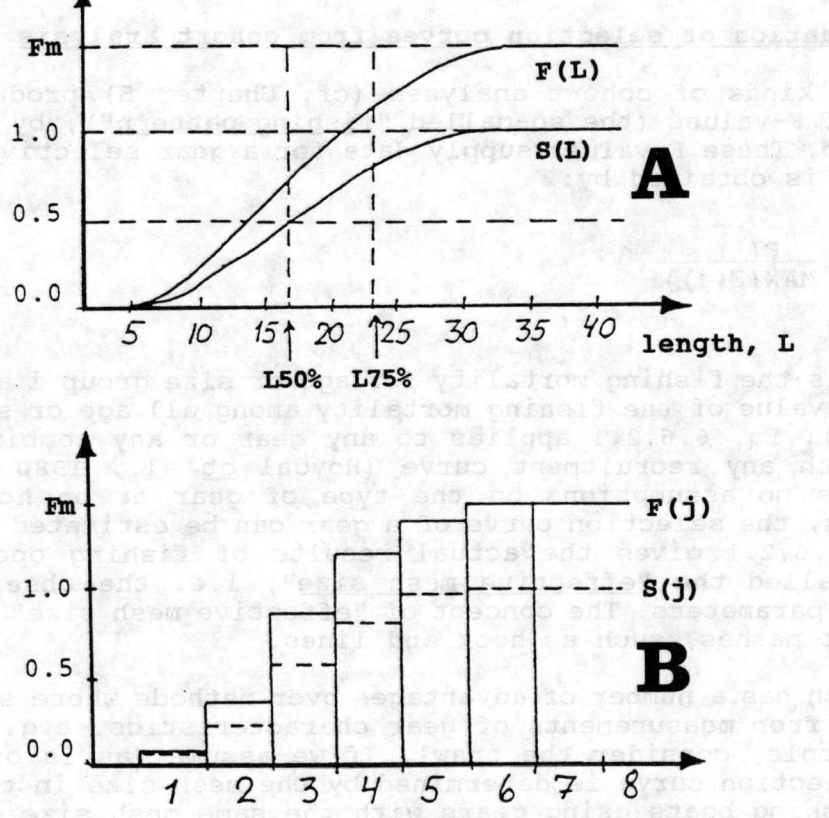

Fig. 6.6.1.1 Relationship between selection ogive and fishing mortality
A: Continuous functions
B: The step-functions corresponding to A

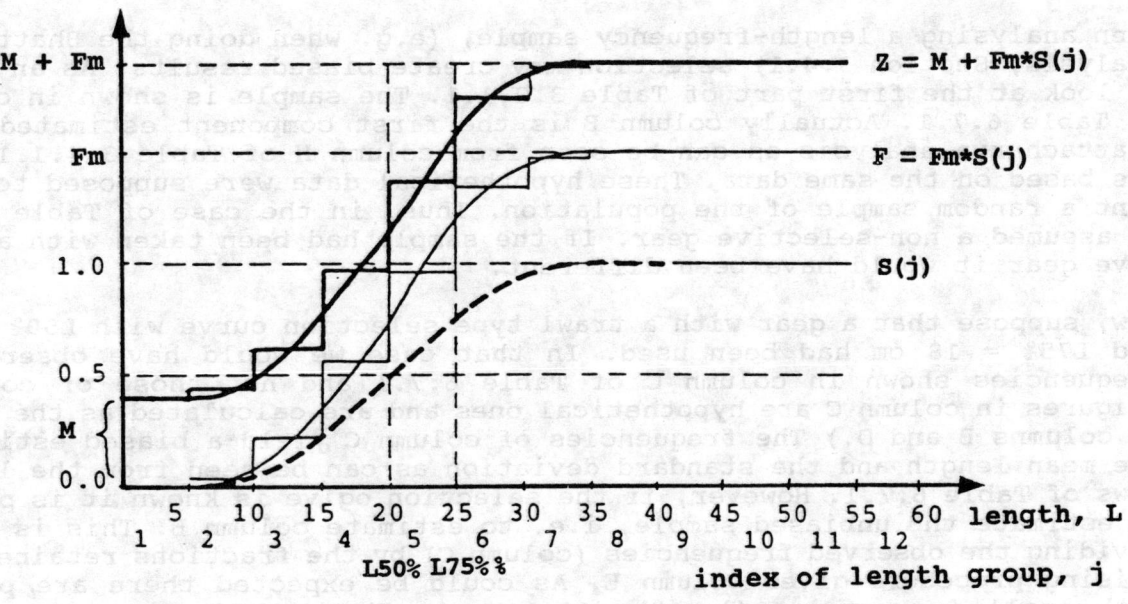

Fig. 6.6.1.2 The relationship between mortality and the selection ogive
(for further explanation, see text)

6.6.2 Estimation of selection curves from cohort analysis

The various kinds of cohort analyses (cf. Chapter 5) produce an array of estimates of F-values (the so-called "fishing pattern"), by age group or by length group. These F-values supply data for a gear selectivity/recruitment curve which is obtained by:

$$S(i) = \frac{F(i)}{\underset{j}{MAX}\{F(j)\}} \qquad (6.6.2.1)$$

where $F(i)$ is the fishing mortality for age or size group i and $MAX\{F(j)\}$ is the maximum value of the fishing mortality among all age or size groups (cf. Eq. 6.6.1.3). Eq. 6.6.2.1 applies to any gear or any combination of gears combined with any recruitment curve (Hoydal et al., 1980 and 1982). The method makes no assumptions on the type of gear or on how the fish are caught. Thus, the selection curve of a gear can be estimated from catch data only. Eq. 6.6.2.1 gives the actual results of fishing operations and is therefore called the "effective mesh size", i.e. the observed selection/recruitment parameters. The concept of "effective mesh size" also applies to gear without meshes, such as hook and lines.

This approach has a number of advantages over methods where selection curves are derived from measurements of gear characteristics, e.g. mesh size. Let us, for example, consider the trawl. If we assume (as is often done) that the gear selection curve is determined by the mesh size in the codend only, then two fishing boats using gears with the same mesh size should have the same gear selection ogive. However, this is likely to be the case only if the two boats operate the gear in exactly the same way. For example, if one boat takes hauls of 5 hours duration and the other boat only uses one hour for a haul the selective properties may be different because of clogging of the net by the catch. Also the towing speed may influence the selectivity. Higher speed may make the meshes more elongated and cause a lower selection factor.

6.7 USING A SELECTION CURVE TO ADJUST LENGTH-FREQUENCY SAMPLES

When analysing a length-frequency sample, (e.g. when doing the Bhattacharya analysis, Section 3.4.1) selection may create biased results. As an example we look at the first part of Table 3.2.1.1. The sample is shown in column B of Table 6.7.1. Actually column B is the first component estimated by the Bhattacharya analysis as can be seen from column H of Table 3.4.1.1, which was based on the same data. These hypothetical data were supposed to represent a random sample of the population. Thus, in the case of Table 3.2.2.1 we assumed a non-selective gear. If the sample had been taken with a selective gear it would have been different.

Now, suppose that a gear with a trawl type selection curve with L50% = 15 cm and L75% = 18 cm had been used. In that case we would have observed the frequencies shown in column C of Table 6.7.1 and not those of column B. (Figures in column C are hypothetical ones and are calculated as the product of columns B and D.) The frequencies of column C yield a biased estimate of the mean length and the standard deviation as can be seen from the last two rows of Table 6.7.1. However, if the selection ogive is known it is possible to estimate the unbiased sample, i.e. to estimate column B. This is done by dividing the observed frequencies (column C) by the fractions retained. This raising procedure gives column E. As could be expected there are problems with small frequencies (lengths 12-14 cm). The method cannot be used to raise a zero frequency and is not dependable for small frequencies.

Table 6.7.1 Example to illustrate the estimation of a random sample from a sample biased by selection (cf. Fig. 6.7.1)

A	B	C	D	E
length interval cm	observed unbiased sample (Table 3.2.1.1)	sample biased by selection	estimated ogive S_L	estimated unbiased sample C/D
12-13	1	0	0.30	0
13-14	4	1	0.37	3
14-15	11	5	0.45	11
15-16	24	13	0.55	24
16-17	38	24	0.63	38
17-18	42	30	0.71	42
18-19	33	26	0.78	33
19-20	20	17	0.84	20
20-21	7	6	0.88	7
21-22	2	2	0.92	2
Total	182	124		180
Mean L	17.3	17.6		17.3
s	1.7	1.6		1.6

<u>Column</u> <u>contents</u>

A length interval in cm
B unbiased random sample of the population (from Table 3.2.1.1)
C sample as it would have been obtained with a trawl net with a selection curve with L50% = 15 cm and L75% = 18 cm
D estimated selection ogive (fraction retained)
   $S_L = 1/(1 + \exp(S1 - S2*L))$ Eq. 6.1.1 where
   $S1 = L50\% * \ln(3)/(L75\% - L50\%)$ (Eq. 6.1.6) and $S2 = S1/L50\%$ (Eq. 6.1.7).
E estimated unbiased samples corrected for selection, frequency of biased sample divided by fraction retained (C/D) (compare with column B).

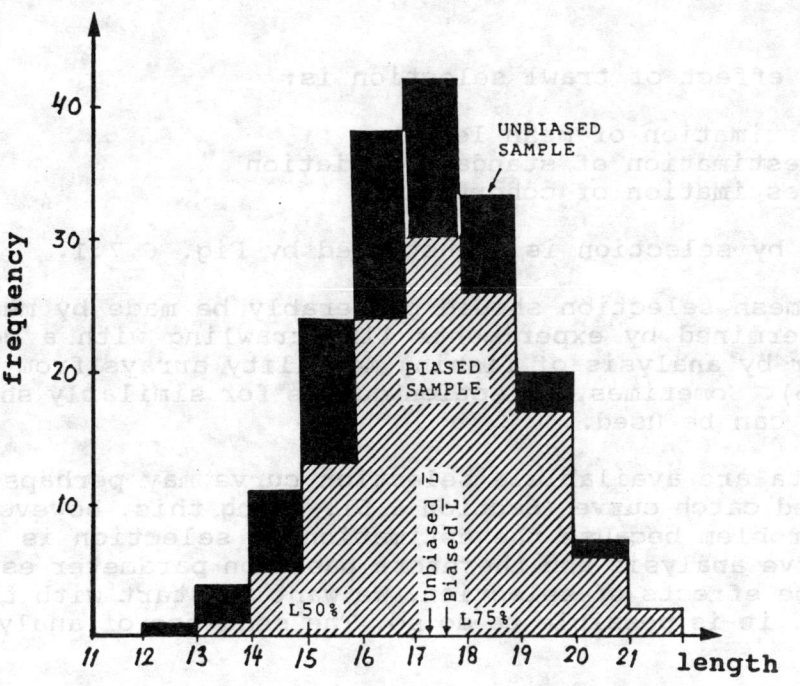

Fig. 6.7.1 Bias created by selection

Table 6.7.2 Example to illustrate the use of a selection curve to adjust a length-frequency sample for selectivity, using the same length-frequency data as presented in Table 6.5.1, with $L_\infty = 49.7$ cm (estimated by the Powell-Wetherall method), $K = 1.0$ per year and $t_o = 0$

A L1-L2	B $t\frac{(L1+L2)}{2}$ (x)	C Δt (L1,L2)	D C (L1,L2) obs	E ln (C/Δt) (y')	F S_t obs	G ln (1/S-1) (y)	H S_t est	I C (L1,L2) est
3-5	0.084	0.044	37	6.74	0.03	3.35	0.03	1121
5-7	0.129	0.046	56	7.11	0.06	2.81	0.06	1000
7-9	0.176	0.048	86	7.49	0.10	2.22	0.10	887
9-11	0.225	0.050	129	7.85	0.17	1.61	0.17	777
11-13	0.276	0.053	188	8.17	0.27	0.99	0.27	689
13-15	0.331	0.056	258	8.44	0.42	0.32	0.43	604
15-17	0.389	0.059	319	8.59	0.60	-0.40	0.61	526
17-19	0.450	0.063	352	8.63	0.76	-1.14	0.77	459
19-21	0.515	0.067	351	8.56	0.88	-1.98	0.88	398
21-23	0.585	0.072	324	8.41	0.95	-2.97	0.95	342
23-25	0.660	0.078	283	8.20	0.98	-3.94	0.98	289
25-27	0.741	0.084	239	7.95	1.00		0.99	241
27-29	0.829	0.092	196	7.66			1.00	197
29-31	0.925	0.102	158	7.35			1.00	158
31-33	1.032	0.113	123	6.99			1.00	123
33-35	1.152	0.128	93	6.59			1.00	93
35-37	1.289	0.146	69	6.16			1.00	69
37-39	1.446	0.171	48	5.64			1.00	48
39-41	1.634	0.207	31	5.01			1.00	31
41-43	1.865	0.261	18	4.23			1.00	18
43-45	2.166	0.355	10	3.34			1.00	10
45-47	2.598	0.554	3	1.69			1.00	3

T1 = 4.428   T2 = 12.492
t50% = T1/T2 = 0.3545 years   t75% = (T1 + ln 3)/T2 = 0.4424 years
L50% = 49.7*(1 - exp(-1*(0.3545 - 0))) = 14.8 cm
L75% = 49.7*(1 - exp(-1*(0.4424 - 0))) = 17.8 cm

In general, the effect of trawl selection is:

1. Over-estimation of mean length
2. Under-estimation of standard deviation
3. Under-estimation of cohort size

The bias caused by selection is illustrated by Fig. 6.7.1.

Correction for mesh selection should preferably be made by means of selection curves determined by experiments like trawling with a covered codend (Section 6.1) or by analysis of fishing mortality arrays from cohort analysis (Section 6.6). Sometimes, selection curves for similarly shaped, closely related species can be used.

When no such data are available a selection curve may perhaps be estimated from a linearized catch curve (Section 6.5). Doing this, however, we encounter a logical problem because the estimation of selection is the last part of the catch curve analysis and therefore based on parameter estimates which are biased by the effects of selection. We want to start with the correction and fortunately, it is possible to do so. The sequence of analyses can be as follows:

Stage 1: Estimate L_∞ by the Powell-Wetherall method

Stage 2: Correct length-frequencies for selection using the value 1.0 for the curvature parameter, K, and the estimate of L_∞ obtained by Stage 1

Stage 3: Separate normally-distributed components by the Bhattacharya method (Section 3.4.1) from the corrected length-frequency distribution

Stage 4: Use the estimated mean lengths of the components in modal progression analysis to estimate the growth parameters K and L_∞ (Section 3.4.2)

Stage 5: Estimate Z using length-converted catch curve analysis with the newly-estimated growth parameters (Section 4.4.5).

This procedure is applicable because the estimation of a selection ogive is not very sensitive to the choice of the curvature parameter, K.

Example 24: Using a selection curve to adjust the length-frequency sample of Table 6.5.1

We apply the Powell-Wetherall method to the data in Table 6.5.1 column D (the numbers caught) and find L_∞ = 49.7 cm (cf. Table 4.5.4.1 which uses the same input data to estimate L_∞ by the Powell-Wetherall method). Redoing the calculations of Table 6.5.1 with K = 1.0 gives the results presented in Table 6.7.2.

The estimated selection ogive $S_t$est in Table 6.7.2 calculated with K = 1.0 per year and L_∞ = 49.7 cm is almost identical to the one presented in Table 6.5.1, calculated with K = 0.3 per year and L_∞ = 50 cm. The values of L50% and L75% are 14.8 cm and 17.8 cm respectively with K = 1.0, while they were 15.0 and 18.0 cm respectively with K = 0.3.

The last column of Table 6.7.2 contains the results of Step 2 of the process, <u>viz</u>. the length-frequencies corrected for selectivity. A comparison with the observed data (column D) shows immediately that large numbers of fish were not accounted for in the sample. The next step (3), the Bhattacharya method will therefore be very different from the one on the original data.

In principle, the above-mentioned method can be applied to any type of gear selection curve, but the narrower the length range selected by the gear the more difficult it is to estimate the length-frequency you would have got with a non-selective gear.

For gill nets or any other gear with a bell-shaped selection curve one has to be careful in the interpretation of length-frequency samples. The mode observed (usually there is only one mode for this type of gear) may have little to do with a cohort, but it may primarily reflect the selection curve of the gear (cf. Section 6.2).

(See **Exercise(s)** in Part 2).

7 SAMPLING

The ideal basis for fish stock assessment is data that fully represent a stock, at least from the moment that it has recruited to the fishery, without any systematic errors or biases. Although it may not be possible in practice to obtain data of such quality, it should be the aim of any programme for the collection of data on a fishery, to obtain samples that fully represent the population under investigation, to know the sources of bias and to find means to correct for those biases.

The full theory of sampling is dealt with in many textbooks, which are widely available, such as Raj (1968), Som (1973) or Cochran (1977).

In the first part of this Chapter, some basic aspects of sampling will be discussed, with emphasis on random sampling. The second part deals with an example of a typical tropical demersal fishery, where many species are landed, partly sorted, for human consumption and partly mixed in the form of by-catch (for fish meal production or other purposes). The aim of this example is to show how complex a sampling scheme may have to be in order to get a representative sample of a particular fishery and how raising factors should be applied to reflect the total fishery for a particular species.

This example was created on the basis of experience in FAO/DANIDA follow-up courses on fish stock assessment, where many data sets on tropical fisheries turned out to be incomplete or biased, due to errors in the design and/or execution of sampling schemes (see Venema et al., 1988).

Good sampling requires large and long-term investments in terms of manpower and general expenses. Therefore, it is important that sampling programmes are designed in such a way that they provide the data needed for the assessment and management of important species and fisheries, and that these data meet the standards as formulated by international and national working groups.

The sampling principles dealt with in this Chapter also apply to the procedures for sampling catches on board of research vessels, for which additional specific details on deck-sampling are given in Section 13.4.

7.1 SIMPLE RANDOM SAMPLING

Let us go back to the problem of estimating the mean length of a cohort as addressed already in Sections 2.2 and 2.3. An estimate is said to be "unbiased" if replicate estimates deviate from the true value in a random manner only. The "true value" is the parameter value we would get from measuring all specimens in the total population (see Section 2.3). An estimate is "biased" if it deviates from the true value in a systematic manner. With an unbiased estimate we can approach the true value as closely as we want by increasing the sample size. With a biased estimate there will always be a deviation between the true value and the estimate, and the deviation is independent of the sample size.

To obtain an unbiased estimate of the mean length the sample should be a "random sample", i.e. any fish from the stock considered should have exactly the same probability of being sampled. Assuming that it is possible to obtain a random sample (this is usually very difficult in practice), how many fish, n, would we need in the sample to obtain a pre-specified accuracy?

Suppose that we require an estimate of the mean length not deviating more than 7% from the true mean length and that we want to be 95% certain of this. We then would require the upper and lower 95% confidence limits not to deviate more than 7% from the estimated mean, $\bar{x}$. So the deviation $t_{n-1}*s/\sqrt{n}$ should have a maximum value of $0.07*\bar{x}$, or

$$t_{n-1}*s/\sqrt{n} = 0.07*\bar{x}$$

or more general: $t_{n-1}*s/\sqrt{n} = \epsilon*\bar{x}$

where ϵ stands for "maximum relative error" (in this case $\epsilon = 0.07$).

Solving this equation with respect to n gives:

$$n = \left[\frac{t_{n-1}*s}{\epsilon*\bar{x}}\right]^2 \qquad (7.1.1)$$

In order to apply Eq. 7.1.1 for estimating the required sample size we must already know the standard deviation, s, from previous samples.

We can also solve Eq. 7.1.1 with respect to ϵ, using, for example, the estimates of s = 2.20 and $\bar{x}$ = 15.07 from Table 2.1.2.

$$\epsilon = \frac{t_{n-1}*2.20}{15.07*\sqrt{n}} = 0.146*\frac{t_{n-1}}{\sqrt{n}}$$

In Fig. 7.1.1 the maximum relative error, ϵ, is shown as a function of the sample size, n. Note that we gain relatively little by increasing n when n > 50. An increase from n = 10 to n = 20, however, produces a reduction in relative error from 10.0% to 6.8%.

If unbiased random samples can be obtained, there is no problem in estimating the required sample size for any pre-specified accuracy. However, samples are usually biased in one way or another. If small fish can escape through the meshes of a trawl we get an over-estimate of the mean length of the population of fish (see Fig. 7.1.2). This is one example of bias. If the larger fish can swim faster than the trawl is towed and thereby avoid being caught we have another type of bias.

However, once you are aware of the bias you can often adjust for it, and in the example of mesh selection you can try to estimate how many fish there would have been if all had been retained by the gear (see Chapter 6).

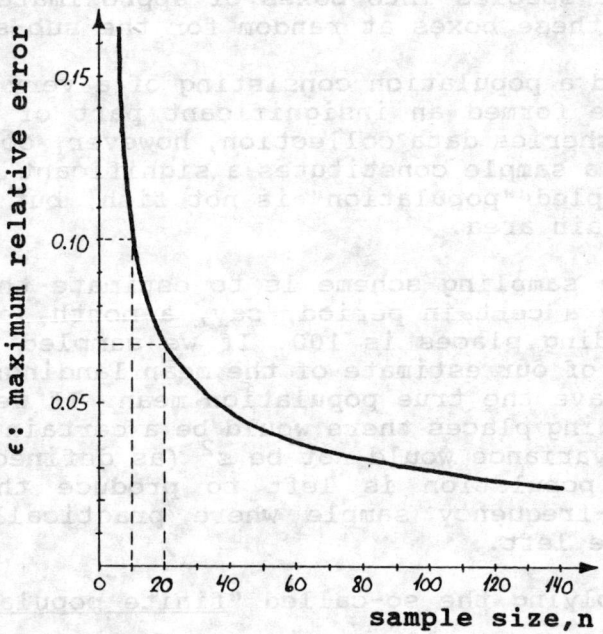

Fig. 7.1.1 Maximum relative error (ϵ) of the estimate of the mean length as a function of sample size (n) (using data from Table 2.1.2)

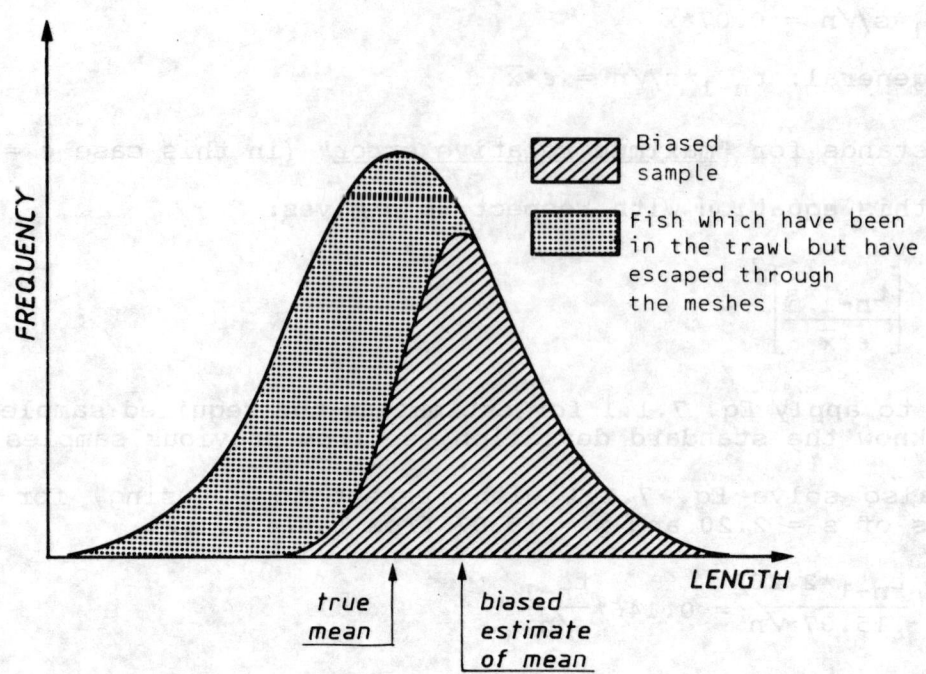

Fig. 7.1.2 Example of bias: bias caused by gear selection

Another type of bias occurs if the spatial distribution of the population of fish is size-dependent. For example, when juvenile fish concentrate in certain nursery grounds and gradually migrate to the fishing grounds this may introduce a bias, if this migration pattern is not well understood. Bias caused by migration will be discussed in Chapter 11.

Another case of possible bias concerns the situation where it would be inefficient to measure the complete catch, for example from a large trawl haul. After sorting the catch by species you would take a sub-sample of, say, 100 fish of one species. If you selected the 100 fish one by one by simple "hand-picking" this would create a biased sample, since there always is a tendency to select the larger specimens. The proper procedure would be to put the catch of that species into boxes of approximately equal weight and then select some of these boxes at random for the sub-sample.

So far we have considered a population consisting of a very large number of fish, so that the sample formed an insignificant part of the population. Some "populations" in fisheries data collection, however, consist of a small number of units, so that a sample constitutes a significant part of it. This is the case when the sampled "population" is not fish, but for example the landing places in a certain area.

Suppose the purpose of a sampling scheme is to estimate the mean landings per landing place during a certain period, say, a month, and suppose that the total number of landing places is 100. If we sampled all 100 landing places then the variance of our estimate of the mean landings would be zero. In that case we simply have the true population mean. If we have personnel for sampling only 50 landing places there would be a certain variance in our estimate. However, this variance would not be s^2 (as defined by Eq. 2.1.2), since only 50% of the population is left to produce the variance, as distinct from a length-frequency sample where practically 100% of the population would still be left.

We allow for this by applying the so-called "<u>finite population correction factor</u>":

$(1 - n/N)$

where N is the population size (N = 100 in the example) and n the sample size. Let Y(i) be the landings at landing place no. i of the sample, i = 1,2,...n, and let $\bar{Y}$ be the mean landings of all landing places, then the estimate of the mean landings is:

$$\bar{Y} = \frac{1}{n} * \sum_{i=1}^{n} Y(i)$$

The variance of the estimate then becomes (see Eqs. 2.3.2 and 2.1.2):

$$VAR(\bar{Y}) = \frac{1 - n/N}{n} * s^2 \qquad (7.1.2)$$

where

$$s^2 = \frac{1}{n-1} * \sum_{i=1}^{n} (Y(i) - \bar{Y})^2 \qquad (7.1.3)$$

The confidence interval of $\bar{Y}$ is (see Eq. 2.3.1):

$$\left[\bar{Y} - t_{n-1}*s*\sqrt{(1/n - 1/N)} \;,\; \bar{Y} + t_{n-1}*s*\sqrt{(1/n - 1/N)} \right]$$

The estimate of the total landings is:

$$Y = N*\bar{Y} \qquad (7.1.4)$$

and its variance is

$$VAR(Y) = N^2 * VAR(\bar{Y}) \qquad (7.1.5)$$

Y is usually the quantity we are interested in. Eq. 7.1.5 follows from the general rule for a random variable, Eq. 2.3.3, where N is a constant.

Note that Eq. 7.1.2 is general in the sense that it also applies to large (in practice infinite) populations as the finite population correction factor, (1-n/N), becomes 1.0 when N is infinite.

Often Eq. 7.1.4 is expressed as:

$$Y = \frac{N}{n} * \sum_{i=1}^{n} Y(i) \qquad (7.1.6)$$

and we say that the sample has been raised to the total, Y, by application of the "raising factor" N/n.

7.2 STRATIFIED RANDOM SAMPLING

Consider again the problem of estimating the total landings, Y, from the 100 landing places during a certain month, as dealt with in Section 7.1. Assume that a sampling programme has been conducted in the previous years, based on which the 100 landing places have been divided into three categories as shown in Table 7.2.1. Such a division of the total population is called a "stratification" and the categories (large, medium, small) are called "strata". Table 7.2.2 shows a numerical example (from Gulland, 1966) corresponding to Table 7.2.1. To obtain an estimate of the standard deviation within each stratum, s(j), j = 1,2,3,... a survey covering all landing places was carried out during one month.

The standard deviation, $s(j)$, in Tables 7.2.1 and 7.2.2 is the square root of the corresponding variance:

$$s(j)^2 = \frac{1}{N(j)-1} * \sum_{i=1}^{N(j)}[Y(j,i)-\overline{Y}(j)]^2 \quad (7.2.1)$$

where the stratum mean is:

$$\overline{Y}(j) = \frac{1}{N(j)} * \sum_{i=1}^{N(j)} Y(j,i) \quad (7.2.2)$$

Note that Eqs. 7.2.1 and 7.2.2 produce the true parameters for the particular month in which the data were collected.

Often we are not in a position to sample all landing places. Let us assume that we only have manpower etc. available for a sample of n (n < 100) landing places. The sample size can be written as the sum:

$$n = n(1)+n(2)+n(3)$$

where n(i) is the number of landing places sampled in stratum no. i. To "design a sampling scheme" means to decide which sample sizes, n(1), n(2) and n(3), should be applied to each of the three strata.

One could now ask why we should complicate the sampling by the introduction of strata. The answer is that we (nearly) always obtain a more precise estimate of the population mean from stratified sampling than from simple random sampling. How much we gain depends on the choice of strata. If the observations within a stratum are approximately of the same size (as is the case in Table 7.2.2) we are likely to gain from the stratification. The less variation there is within a stratum, the more is gained by stratification. On the other hand, for practical reasons there is a limit to the number of strata that can be selected.

The basic rules of stratified sampling are that the sample size per stratum, n(j), should be large when:

1) The stratum is large (if N(j) is large)
2) The standard deviation s(j) is large

To these rules we can add for economic reasons that samples should be large when:

3) Sampling is inexpensive

Mathematically the first two conditions can be expressed as

$$n(j) \text{ proportional to } N(j)*s(j) \quad (7.2.3)$$

or

$$n(j) = n * \frac{N(j)*s(j)}{\sum N(j)*s(j)} \quad (7.2.4)$$

where n is the total sample size. This formula is called the "optimum stratified sampling equation" (or "Neyman allocation").

The first two conditions will be dealt with below in Example 25 and the third condition in Example 26.

Table 7.2.1 An example of stratification (from Gulland, 1966)

landing place category or stratum		number of landing places by stratum	average landings by stratum	standard deviation within each stratum s(j)
1	large	N(1)	$\overline{Y}(1)$	s(1)
2	medium	N(2)	$\overline{Y}(2)$	s(2)
3	small	N(3)	$\overline{Y}(3)$	s(3)

Table 7.2.2 Numerical example of a stratification based on samples from one month (from Gulland, 1966)

Y(j,i) = landings at landing place no. i in stratum j

LARGE LANDING PLACES: Y(1,i)

N(1) = 10   45   59   87   41   71   25   9   69   10   7
$\overline{Y}(1)$ = 42.3
s(1) = 28.91

MEDIUM LANDING PLACES: Y(2,i)

N(2) = 30   17   13   19   26   1   8   27   11   12   26
$\overline{Y}(2)$ = 15.2   5   8   10   16   16   4   16   16   13   29
s(2) = 8.57   14   25   29   27   20   25   2   7   3   12

SMALL LANDING PLACES: Y(3,i)

N(3) = 60   2   6   7   0   1   2   1   5   4   7
$\overline{y}(3)$ = 4.2   8   9   3   2   5   4   2   0   2   8
s(3) = 2.81   5   3   8   9   8   9   1   6   5   3
              3   4   7   5   5   3   2   4   6   1
              6   2   5   1   0   3   8   0   4   3
              3   5   5   0   7   0   9   7   9   0

Example 25: Stratified random sampling

Let us assume that we have funds available to collect 100 observations, that there are two strata and that their sizes, N(j), and standard deviations, s(j), are:

	Stratum 1	Stratum 2
N(j)	1000	2000
s(j)	50	10

Then, the 100 observations should be allocated to the two strata according to Eq. 7.2.4 as follows:

Stratum 1: $\dfrac{1000*50}{1000*50 + 2000*10} * 100 = 71.4$ say 71

Stratum 2: $\dfrac{2000*10}{1000*50 + 2000*10} * 100 = 28.6$ say 29

Usually the budget available for the execution of a sampling programme is limited. The cost of taking a sample will differ from stratum to stratum and it is therefore possible also to take the cost of sampling into account, when designing a sampling scheme.

Let $c(j)$ be the cost of taking a sample unit in stratum j and let C_0 be the additional fixed cost of the whole sampling programme. The total cost, C, for m strata then becomes:

$$C = C_0 + \sum_{j=1}^{m} c(j)*n(j) \qquad (7.2.5)$$

It can be shown that for an optimum allocation the sample size should be proportional to $1/\sqrt{c(j)}$.

An optimum use of the available resources is then achieved (see Eq. 7.2.3) when the sample size is proportional to

$N(j)*s(j)/\sqrt{c(j)}$, or if

$$n(j) = n*\frac{N(j)*s(j)/\sqrt{c(j)}}{\Sigma\ N(j)*s(j)*\sqrt{c(j)}} \qquad (7.2.6)$$

where n is the total sample size. Using the criterion Eq. 7.2.6, we minimize the variance of the estimate of the total landings, Y.

If the total budget available is C and the fixed cost is C_0 then the total number of sample units in all strata is given by:

$$n = [C-C_0] * \frac{\Sigma\ N(j)*s(j)/\sqrt{c(j)}}{\Sigma\ N(j)*s(j)*\sqrt{c(j)}} \qquad (7.2.7)$$

and the number of units sampled in stratum j is given by:

$$n(j) = [C-C_0] * \frac{N(j)*s(j)/\sqrt{c(j)}}{\Sigma\ N(j)*s(j)*\sqrt{c(j)}} \qquad (7.2.8)$$

Example 26: Stratified random sampling, considering costs

Let the total budget for the sampling programme given in Example 25 be 1800 (money units) while the fixed cost is 444 and the prices per sample unit are:

    Stratum 1: 16
    Stratum 2:   9

Then the total number of samples, n, that can be taken with the available budget is determined by Eq. 7.2.7:

$$n = (1800-444)*\frac{1000*50/\sqrt{16} + 2000*10/\sqrt{9}}{1000*50*\sqrt{16} + 2000*10*\sqrt{9}} = 99.96, \text{ say } 100$$

These 100 sample units should then be allocated to stratum 1 and 2 respectively as follows:

$$n(1) = (1800-444)*\frac{1000*50/\sqrt{16}}{1000*50*\sqrt{16} + 2000*10*\sqrt{9}} = 65.19, \text{ say } 65$$

$$n(2) = (1800-444)*\frac{2000*10/\sqrt{9}}{1000*50*\sqrt{16} + 2000*10*\sqrt{9}} = 34.77, \text{ say } 35$$

Note that now 35 sample units are allocated to the inexpensive stratum 2, compared to 29 in Example 25, where the price per sample unit was not taken into account (or assumed to be the same for both strata).

Hereby we conclude the theory for optimum allocation and turn to the question of how the estimates of mean values, the totals and their variances can be calculated. Although the theory given below is general, you may again think of the example given in Table 7.2.2. Having determined the sample sizes, n(j), we obtain an estimate of the mean landings of each stratum, $\bar{Y}(j)$, by:

$$\bar{Y}(j) = \frac{1}{n(j)} * \sum_{i=1}^{n(j)} Y(j,i) \tag{7.2.9}$$

Let m be the total number of strata (j = 1,2,...,m), then the estimate of the total population mean i.e. the mean landings in all strata, is:

$$\bar{Y} = \sum_{j=1}^{m} \frac{N(j)}{N} * \bar{Y}(j) \tag{7.2.10}$$

and finally, the estimate of the total landings, Y, (see Eq. 7.1.4) is:

$$Y = N*\bar{Y}$$

Let the symbol "VARst" stand for the variance of an estimate obtained from stratified ("st") random sampling. The estimate of the variance of the total population mean, $\bar{Y}$, is:

$$VARst(\bar{Y}) = \sum_{j=1}^{m} \left[\frac{N(j)}{N}\right]^2 * VAR(\bar{Y}(j)) \tag{7.2.11}$$

where the estimate of the variance of the estimate of the mean for each stratum Y, $VAR(\bar{Y}(j))$, is defined by Eq. 7.1.2:

$$VAR(\bar{Y}(j)) = \frac{1 - n(j)/N(j)}{n(j)} * s(j)^2 \tag{7.2.12}$$

Note the finite population correction factor: $1 - n(j)/N(j)$

Eq. 7.2.11 follows from the general rules for the variance of a sum of independent random variables given in Eqs. 2.3.3 and 2.3.4.

The variance within stratum j as defined by Eq. 7.1.3 is:

$$s(j)^2 = \frac{1}{n(j)-1} * \sum_{i=1}^{n(j)} [Y(j,i)-\bar{Y}(j)]^2 \tag{7.2.13}$$

Inserting Eq. 7.2.12 into Eq. 7.2.11 the latter may also be written:

$$VARst(\bar{Y}) = \frac{1}{N^2} * \sum_{j=1}^{m} \frac{N(j)*[N(j)-n(j)]}{n(j)} * s(j)^2 \tag{7.2.14}$$

Eq. 7.2.14 is more convenient from a calculation point of view (see Exercise 7.2).

The variance of the total landings in stratum j is:

$$VAR(Y(j)) = N(j)^2 * VAR(\overline{Y}(j)) \quad (7.2.15)$$

and the variance of the total catch (all strata) is:

$$VARst(Y) = N^2 * VARst(\overline{Y}) \quad (7.2.16)$$

In the examples given above the strata consisted of different sizes of landing places in terms of landed weight. Stratifications can also be based on other criteria, for example:

Gear types
Boat types
Fishing seasons
Fishing grounds
Species or species groups
Commercial size categories

Usually one can take advantage of whatever stratification is available. Sometimes we are forced to stratify the sampling, as might be the case when fish are landed already sorted into commercial size categories. Sometimes we will have to do extra work to obtain the stratification. This may apply to a stratification by boat type. If we sample from an auction hall at the time the fish are sold, we may have to trace the boat which caught the sample. If we are to invest extra resources (time, funds and manpower) in obtaining a stratification the increased precision obtained from that stratification should be considered relative to the cost.

7.3 PROPORTIONAL SAMPLING

In some cases (e.g. when starting up) we may not know the variance within strata but only the stratum size, $N(j)$. In that case it is recommended to use "<u>proportional sampling</u>", i.e. to allocate samples in proportion to the stratum size. Applying proportional sampling to the 100 observations of Example 25 given above, the allocation to the two strata should be in the proportions:

Stratum 1: $\frac{1000}{1000+2000} * 100 = 33.3$ say 33

Stratum 2: $\frac{2000}{1000+2000} * 100 = 66.7$ say 67

Compare with the values obtained by optimum allocation in Example 25, respectively, 71 for stratum 1 and 29 for stratum 2.

Note that proportional sampling is only identical to optimum stratified sampling in the exceptional case that the variances of all strata are equal.

(See **Exercise(s)** in Part 2).

7.4 SAMPLING COMMERCIAL CATCHES

In order to be able to carry out assessments of exploited stocks we must have adequate data for each species under investigation. We must know the total weight of the catch and the length and/or age composition of the catch of each stock. In order to obtain this kind of data it will be necessary to sample the landings of the commercial fisheries, according to a predetermined scheme. Such a sampling scheme should take into account the following factors:

1) the total area of distribution of the stock of a species and

2) all the fishing activities taking place in that area, which are catching that particular species. These may include different types of boats (fleets) and different gears. The fleets may either be completely national, or include also those of other countries exploiting the same stock.

Since stocks of fish may occupy areas across international boundaries, the data collection should also be set up in such a way that merging of data collected by different countries is possible. In such cases it is essential that agreements on data requirements are made between countries, for example, through international fisheries bodies, and that these agreements are put into practice through the creation of international working groups. The same criterion applies in the case of large countries where several institutes or sub-institutes are covering the landings of a particular stock.

The data collected should be verified and entered in a computerised data base, which should be accessable to all scientists who have an interest in the data and who are authorised to use them. Data collected through such sampling schemes should never be considered as the private property of the scientists who were responsible for their collection.

Because of the complexity of many fisheries, in particular in the tropics, where many different gears are used to catch a mixture of species and where the landings are often spread over many different sites, it is important to design a sampling scheme that is based on an in-depth knowledge of the fisheries on a particular species. When more than one species has been selected for sampling, it may be possible to combine sampling schemes, if there is sufficient overlap in landing sites, fleets etc.

Sampling schemes, once set up, and in particular when international agreements are involved, should have secured funding over very long periods, in some cases practically forever.

Sampling for stock assessment purposes is very closely connected to sampling or total enumeration systems set up by fishery statisticians. While the fishery statistician is mainly interested in obtaining an estimate of the catches of all important species, usually by type of gear and boat, the fishery scientist will usually concentrate on a smaller number of species. Biological sampling is time-consuming and it is therefore not possible in most cases to sample all the landings in a particular place. However, also for stock assessment purposes it is necessary to estimate how much fish of a particular species has been landed by the vessels that were not sampled, both in the landing places where biological sampling did take place and in all other places. The above makes it obvious that the work of the fishery statisticians is extremely important for the biologists, since the general statistics will be needed to determine the so-called "<u>raising factors</u>", which consist of the ratio between the total number of units and the sampled number of units. Raising factors are therefore used to raise the data (e.g. length-frequencies) obtained from the sample to the level of the total catches. When the samples are small relative to the total landings, the raising factor may become very large and if the samples are biased for one reason or another this will be reflected at a larger scale in the totals.

Therefore, it is very important that the samples are random samples and that they represent a reasonable proportion of the total catch. Rather than thinking in terms of measuring hundreds of fish, plans should be made to measure many thousands of fish. Only then the data may provide a reliable basis for stock assessments. In the example of whiting given in Table 4.4.3.1, the total catches of whiting of all year classes (cohorts) were estimated to be 2,021,800,000 fish, while the number of whiting otoliths sampled in all the countries exploiting this resource and used by the international working group was approximately 10,000. The overall raising factor from samples to total catch was therefore approximately 200,000.

In Table 5.1.1, the total estimated number of survivors from the 1974-cohort is 9,856,600,000 whiting, or about 3.4 times more than the numbers caught, so the number actually sampled is only a very small fraction of the real population in the sea, despite enormous efforts.

The basis for all sampling schemes is a decision on the species to be sampled. This decision will usually be determined by the ecomomic importance of the species (or species groups) and in the case of resources being exploited by more than one country, on the requirements of international working groups.

Assuming that the general fishing pattern of the commercial fisheries is known for each of the selected species, it will be possible to determine which landing places and fleets should be sampled each season.

On that basis, and on that of the availability of funds and personnel, the system can be designed and tested.

The complexity of sampling schemes and the various raising factors involved are illustrated in the following theoretical example, which has many features of a tropical fishery for shrimp and demersal fish.

Example 27: Sampling scheme for a tropical demersal fishery

The sampling scheme illustrated in the example has two major objectives:

1) To determine the total catch in weight of one species "s" (Section 7.5) and

2) To determine the length composition of the total catch of species "s" (Section 7.6).

It is assumed that all specimens of species "s" caught originate from one stock and that all the landings from this stock are covered by the sampling scheme.

The physical features of the example are shown in Fig. 7.4.1. We assume the stock to be confined to one fishing ground exploited only by the boats from three landing places denoted by the index h and labelled, I, II and III. There are three different types of boats, for example:

a) large trawlers,
b) gill netters and
c) "baby"-trawlers.

Each landing place is the home port to a number of boats of each type. A batch of similar boats is called a fleet (index f). Because locations of landing places are factors of practical importance in a sampling scheme, each fleet is also distinguished according to its landing place, so that in this case we actually operate with nine sets of boats, which are distinguished by combinations of f and h, for example, aI, bII, bIII etc. In Fig. 7.4.1 the fleets have been assigned to their respective landing places, while in Table 7.4.1 they have been arranged by type of boat in order to facilitate the calculation of catch per unit of effort (CPUE).

Table 7.4.1 The procedure for estimating total catch in weight for one species, s, during one period, t. (compare Fig. 7.4.1)

TOTAL CATCHES OF SPECIES "s" DURING TIME PERIOD "t" IN WEIGHT UNITS

fleet f	landing place h	effort E (t,f,h)	observed catch W (s,t,f,h)	$\overline{CPUE}$ by fleet $\Sigma W/\Sigma E$	estimated catch $\overline{CPUE}*E$	total catch by fleet and landing place W(s,t,f,h)	total catch by fleet W(s,t,f,*)
a	I	15	57		-	57	
a	II	20	83	$\frac{57+83}{15+20} = 4$	-	83	172
a	III	8	-		4*8 = 32	32	
b	I	25	55		-	55	
b	II	55	105	$\frac{55+105}{25+55} = 2$	-	105	184
b	III	12	-		2*12 = 24	24	
c	I	20	-		1*20 = 20	20	
c	II	25	25	$\frac{25}{25} = 1$	-	25	75
c	III	30	-		1*30 = 30	30	

estimated grand total catch: W(s,t,*,*) = 431

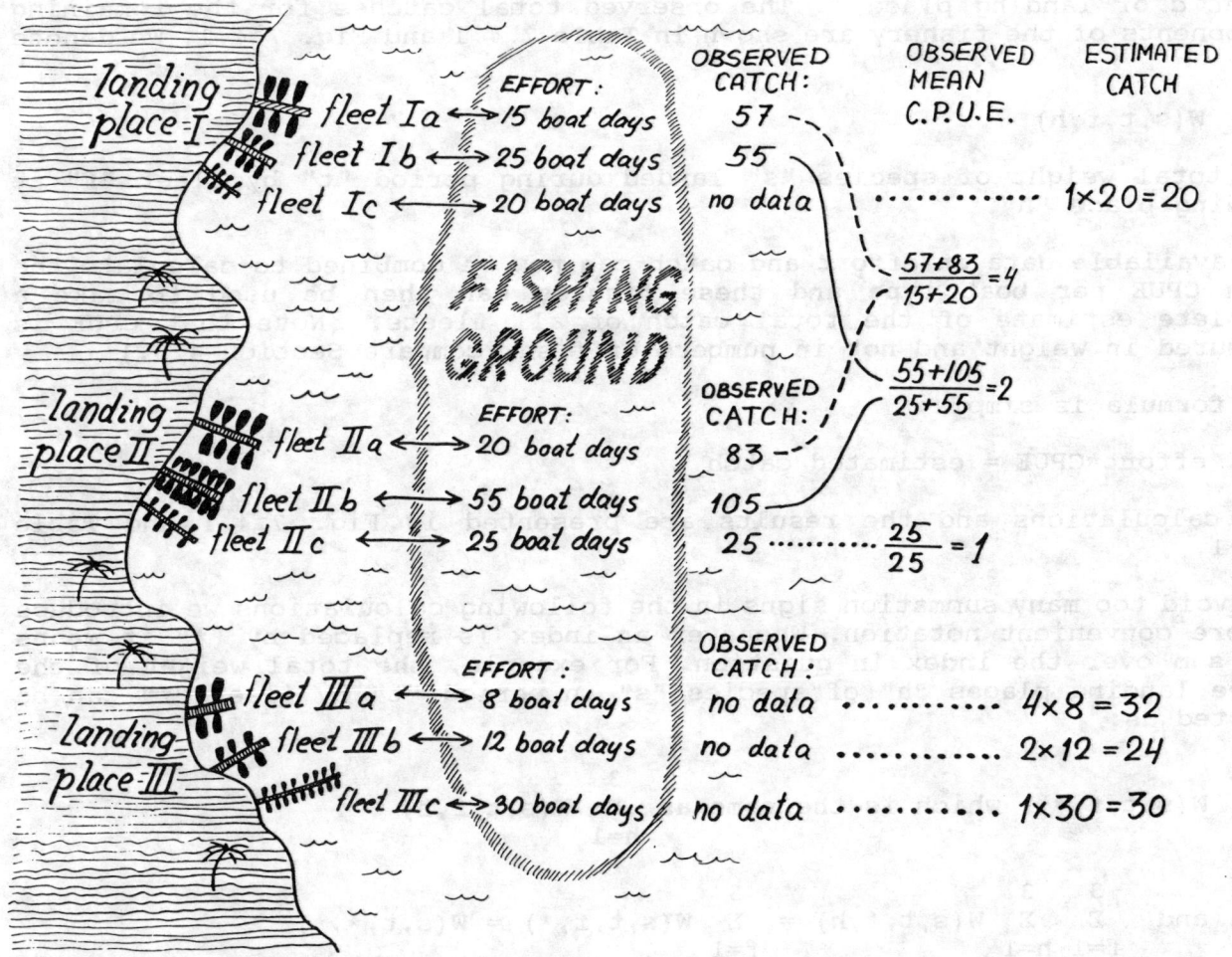

Fig. 7.4.1 Hypothetical example to illustrate the estimation of the total catch in weight of species, s, during a time period, t, say a quarter of a year

The next important assumption we make is that data are available on the total fishing effort, E, of each fleet in each landing place. In the present example effort has been measured as the number of boat-days during the period of time considered. (Note that a number of alternative measures of effort could have been used. In the worst case only the number of boats in each fleet is known. In that case one must assume an average number of effort units per boat per time unit. If we know, for example, the average number of boat-days per fleet per month for one landing place, then this figure can be applied also to landing places for which this information is not available.)

The number of effort units expended during time period t by fleet f from landing place h is denoted:

$$E(t,f,h) = \text{effort}$$

The observed (or estimated) values of the effort $E(t,f,h)$ of the nine sets of boats are shown in Table 7.4.1 and Fig. 7.4.1. For the moment we confine the description of the sampling scheme to a single time period, for example the second quarter of 1978, which makes the index "t" a constant. Later, data from different time periods will be combined and then t becomes a variable index.

7.5 ESTIMATION OF THE TOTAL CATCH IN WEIGHT OF SPECIES S

Assume that due to shortage of funds and personnel the total catches have not been recorded for 1) any of the fleets of landing place III, 2) for fleet c of landing place I. The observed total catches for the remaining components of the fishery are shown in Table 7.4.1 and Fig. 7.4.1. We denote by

$$W(s,t,f,h)$$

the total weight of species "s" landed during period "t" by fleet "f" at landing place "h".

The available data on effort and catch can now be combined to calculate the mean CPUE per boat type and these figures can then be used to make a complete estimate of the total catch of all fleets. (Note that CPUE is measured in weight and not in numbers of fish, compare Section 4.3.)

The formula is simple:

$$\text{effort} * \text{CPUE} = \text{estimated catch}$$

the calculations and the results are presented in Fig. 7.4.1 and Table 7.4.1.

To avoid too many summation signs in the following calculations we introduce a more convenient notation. Whenever an index is replaced by "*" it means the sum over the index in question. For example, the total weight of the three landing places "h" of species "s" in period "t" by fleet "f" can be denoted as:

$$W(s,t,f,*) \quad \text{which is the same as} \quad \sum_{h=1}^{3} W(s,t,f,h)$$

$$\text{and} \quad \sum_{f=1}^{3}\sum_{h=1}^{3} W(s,t,f,h) = \sum_{f=1}^{3} W(s,t,f,*) = W(s,t,*,*)$$

which is the total catch of species s during time period t landed by all fleets to all landing places (Table 7.4.1).

7.6 ESTIMATION OF THE LENGTH COMPOSITION OF SPECIES S IN THE TOTAL CATCH

We now turn to the main objective, the derivation of the length composition of species s, i.e. the derivation of inputs to the age-based or length-based cohort analyses. First some general features of sampling species s for length composition. Like in most sampling procedures we have to raise the outcome of the sampling to the total catch of the boat and eventually to the total catch of all the fleets. Therefore it is useful to introduce the suffix "m" for quantities that have been sampled. Another useful thing is to distinguish quantities of the sample from those of the total sampled catch. This is done by using capital letters for the total catch and small letters for the sample, e.g.

Wm = the <u>total weight</u> of the <u>sampled catch</u>
wm = the <u>weight</u> of the <u>sample</u>
Cm = the total number of fish in the sampled catch
cm = the number of fish in the sample.

A sampling procedure usually starts with sorting of the catch by species and weighing or estimating the total weight caught of each species. The next step is to select a sample at random, to weigh the sample, and after that to measure the length of all the fish in the sample.

In some cases it will be necessary to take sub-samples before it is possible to measure the length of a particular species. This is the case with the by-catch category as will be demonstrated below.

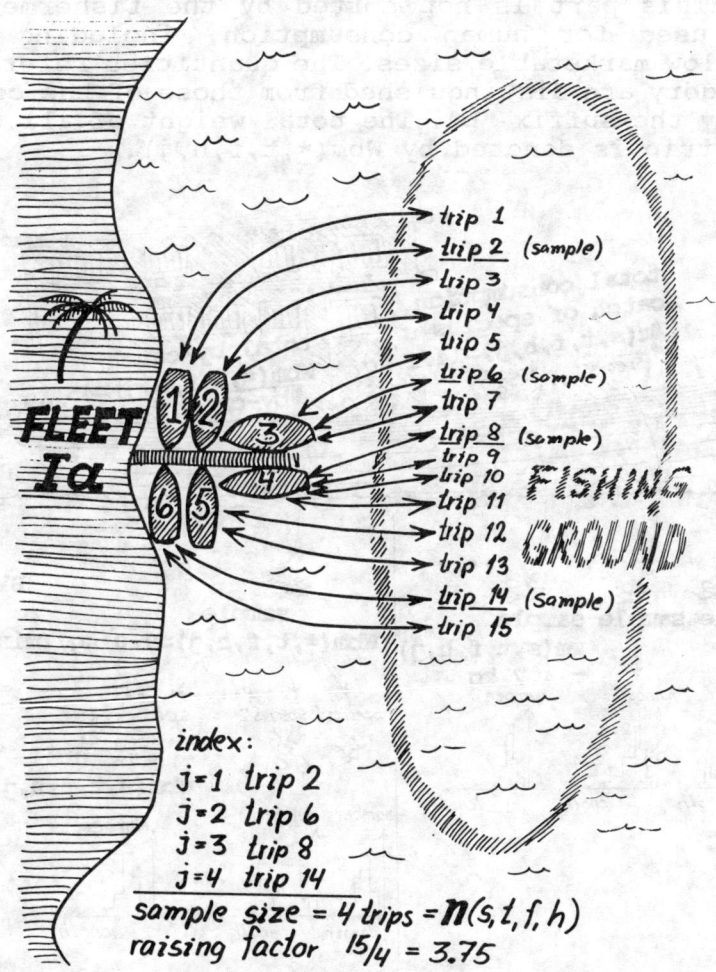

Fig. 7.6.1 Illustration of sampling for the estimation of the length composition of the catch of one fleet from one landing place

In the present example the basic sample for estimation of length compositions is associated with a "trip". Each fleet from each landing place makes a number of trips during the time period considered. In Fig. 7.6.1 fleet a in landing place I is considered as an example. During period ,t, 15 trips were carried out and at the end of four of those 15 trips samples were collected on the jetty when the catch was landed. (Note, usually, it is impossible to take samples from all trips, in this case only four out of the 15 trips were sampled. Care should be taken that these are selected at random, see the discussion in Section 7.1.)

Let the number of samples be denoted:

$$n(t,f,h)$$

In the example of Fig. 7.6.1: $n(t,a,I) = 4$. Each <u>sample</u> is given a number, or an index, j, for our internal book-keeping:

$$j = 1,2,...,n(t,f,h) \quad \text{(see Fig. 7.6.1)}$$

The often complex situation of sampling a single vessel after a trip is illustrated in Fig. 7.6.2 In this case the catch is assumed to consist of two major categories:

1. FISH FOR DIRECT HUMAN CONSUMPTION or CONSUMPTION FISH. This part is sorted into species (or species groups) and contains the marketable sizes

2. BY-CATCH. This part is not sorted by the fishermen. It contains fish not used for human consumption, including valuable fish species below marketable sizes. The quantities referring to the by-catch category are distinguished from those of the consumption fish category by the suffix "b". The total weight of all the by-catch of a sampled trip is denoted by Wbm(*,t,f,h,j).

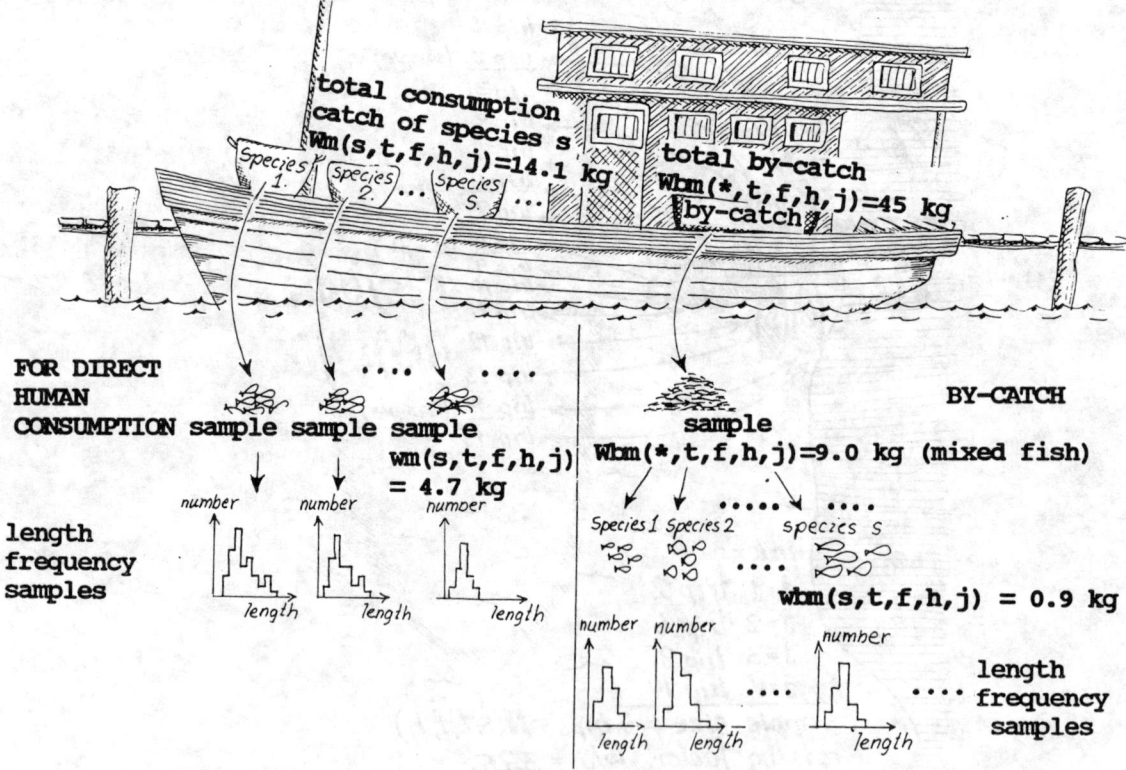

Fig. 7.6.2 Sampling from a single trip (for further explanation, see text)

Consumption fish:

  Total weight of sampled catch: Wm = 14.1 kg
  Total weight of sample: wm = 4.7 kg
  Raising factor: Wm/wm = 14.1/4.7 = 3

By-catch:

  Total weight of sampled by-catch: Wbm = 45 kg
  Total weight of sample (all species): wbm = 9 kg
  Raising factor: Wbm/wbm = 45/9 = 5

Consumption fish (species s)

length-frequency sample
cm(s,t,f,h,j,i)
sample size = 37

raised length-frequency sample
cm*Wm/wm = cm*3
37*3 = 111

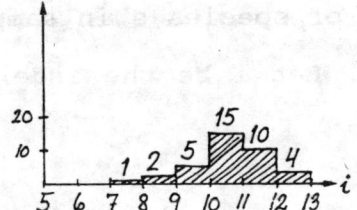
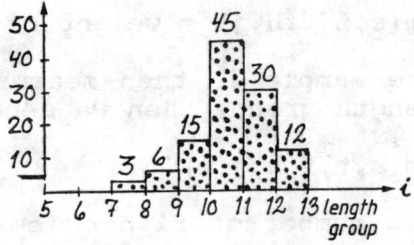

By-catch (species s)

length-frequency sample
cbm(s,t,f,h,j,i)
sample size = 24

raised length-frequency sample
cbm*Wbm/wbm = cbm*5
24*5 = 120

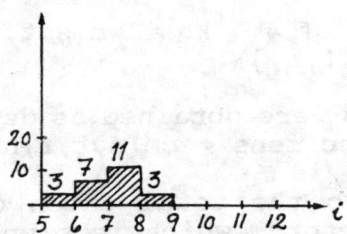
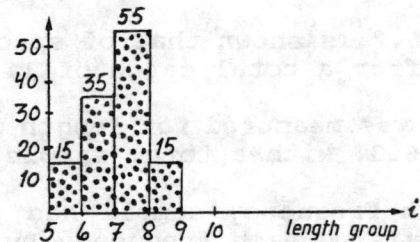

Total catch (species s)

estimated length-
frequency of species s
in the total catch of
the sampled trip
Cm(s,t,f,h,j,i)

combined length-frequency sample
Cm = cm*3 + cbm*5 = 111+120 = 231

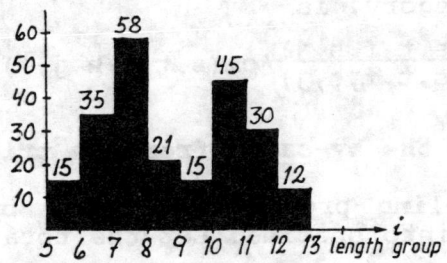

Fig. 7.6.3 Combining length-frequency samples of species s encountered in the consumption fish and by-catch categories of the landings of a single trip (see also Fig. 7.6.2)

Of the species selected for a sampling programme, samples have to be taken from both categories, since from a biological point of view they are of equal importance. We will first deal with the sampling of the category consumption fish, then with the by-catch and then combine the two sets of data. The whole procedure has been illustrated in Figs. 7.6.2, 7.6.3 and 7.6.4.

Sampling the catch for human consumption from one trip

The general sampling procedure for length-frequency data is given below, based on the example of the consumption category of species s:

1) The weight of the total sampled catch of species s, of the consumption fish category, is recorded:

    $Wm(s,t,f,h,j)$ = total weight of consumption catch of species s of sampled trip j.

2) A random sample is taken and the weight of the sample is recorded as follows:

    $wm(s,t,f,h,j)$ = weight of all specimens of species s in sample j.

3) The sample is then measured for length. Let i be the index of a length group, then we denote by

    $cm(s,t,f,h,j,i)$

    the number of fish of species s in length group i, of sample j, from landing place h, caught by fleet f, in time period t.

    The total number of fish of all length groups in the sample is denoted by

    $Cm(s,t,f,h,j,*)$

In Fig. 7.6.2 is shown that of species s a sample of 4.7 kg = $wm(s,t,f,h,j)$ was taken from a total catch of 14.1 kg = $Wm(s,t,f,h,j)$.

The sample was measured for length and frequencies were obtained as depicted in Fig. 7.6.3, with a total sample size of 37 specimens = $cm(s,t,f,h,j,*)$.

This length-frequency sample has to be raised to the total catch of the boat, by raising each frequency by a "raising factor", which is simply the total weight of the catch of species s divided by the weight of the sample

$$\frac{Wm}{wm}$$

In the case of species s the total estimated number caught, in the consumption category, is

$$\frac{Wm(s,t,f,h,j)}{wm(s,t,f,h,j)} * cm(s,t,f,h,j,*) = \frac{14.1}{4.7} * 37 = 111$$

Sampling the by-catch from one trip

The sampling procedure for the by-catch includes an extra step, namely sorting into species. Let the total weight of the by-catch be

   $Wbm(*,t,f,h,j)$ kg

(Index "s" is replaced by "*" because the by-catch is not sorted into species.)

From the Wbm kg a sample is taken of

   wbm(*,t,f,h,j) kg

This sample is then separated into species (see Fig. 7.6.2). We are only interested in species s. The weight of species s from the by-catch sample is

   wbm(s,t,f,h,j) kg

(Note that this is not the weight to be used to raise the sample to the total catch.)

These fish are measured and the lengths are recorded (see Fig. 7.6.3). The number of fish in length group i is denoted as:

   cbm(s,t,f,h,j,i)

and the total number of species s in the sample is

   cbm(s,t,f,h,j,*)

In this case the number is 24 (see Fig. 7.6.3).

These numbers are raised to account for the total by-catch of the boat trip by a raising factor, consisting of the total weight of the by-catch and the weight of the total sample of by-catch (and <u>not</u> the weight of species s only):

$$\frac{Wbm}{wbm}$$

$$\frac{Wbm(*,t,f,h,j)}{wbm(*,t,f,h,j)} * cbm(s,t,f,h,j,i)$$

gives the length-frequencies of species s that represent the entire by-catch.

In the example (see Fig. 7.6.2) Wbm(*,t,f,h,j) = 45 kg and the sample weight wbm(*,t,f,h,j) = 9 kg so that the raising factor becomes Wbm/wbm = 45/9 = 5, and the total number of specimens of species s in the by-catch category is 5*24 = 120.

Combining consumption fish sample and by-catch sample from one trip

We now have to combine the raised length-frequencies of the two categories, in order to obtain a complete picture of the length-frequencies of species s in the catch of the sampled trip. This is done by a simple summation of the two raised frequencies.

An estimate of the <u>total catch in numbers by length group</u>, C, is obtained by simple addition of consumption and by-catch estimates:

$$\frac{Wm}{wm} * cm(s,t,f,h,j,i) + \frac{Wbm}{wbm} * cbm(s,t,f,h,j,i) = Cm(s,t,f,h,j,i)$$

(see Fig. 7.6.3).

For example, the estimated total number for length groups 8-9 cm becomes

$$\frac{14.7}{4.7} * 2 + \frac{45}{9} * 3 = 3*2 + 5*3 = 21$$

Summation of samples from several trips

Again a simple summation will do. The estimated number caught by length group in all $n(t,f,h)$ sampled trips is:

$$\sum_{j=1}^{n(t,f,h)} Cm(s,t,f,h,j,i) = Cm(s,t,f,h,*,i)$$

In Fig. 7.6.4, for example, the estimated total number of all four sampled trips for length group 8-9 cm becomes:

21+22+20+26 = 89

Raising the sampled trips to the total catch of the fleet in one landing place

The total length distribution of the sampled trips $Cm(s,t,f,h,*,i)$ can be raised to account for the entire catch in period t, by fleet f in landing place h, by using a raising factor based on the number of trips:

$$\frac{\text{total number of trips}}{\text{number of trips sampled}} * Cm(s,t,f,h,*,i) = CR(s,t,f,h,*,i)$$

where the suffix "R" stands for "raised". R has only been used here to indicate raising procedures that include quantities that have not been sampled (in this case 11 out of the 15 trips). In the example (Fig. 7.6.1) the raising factor is 15/4 = 3.75 and the result is shown in Table 7.6.1. The estimated total number of specimens in length group 8-9 cm for all trips by fleet f in landing place h becomes:

89*3.75 = 333.75

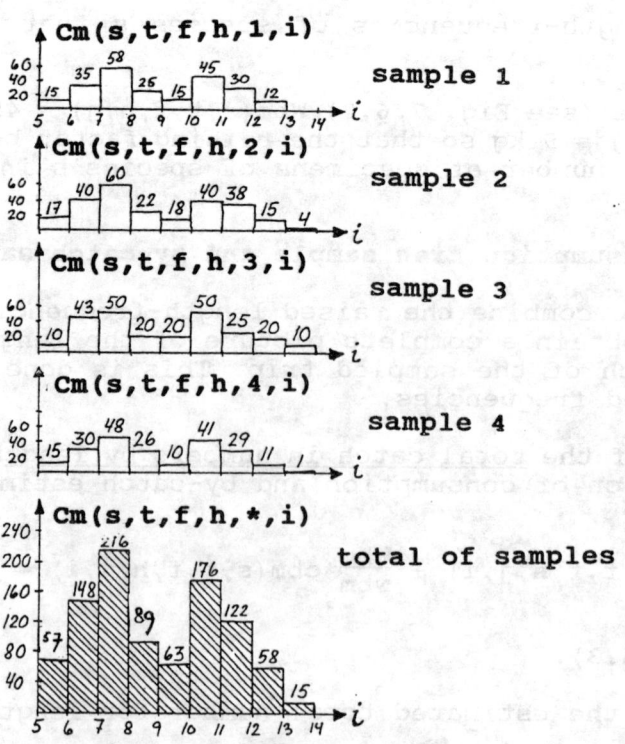

Fig. 7.6.4   Adding up length compositions of species s of sampled trips. The sample used in the example of Fig. 7.6.3 appears as sample no. 1

Table 7.6.1 Raising the sampled trips of fleet f, in landing place h to the total catch of all trips that fleet (see Figs. 7.6.1 and 7.6.4)

length group (i)	total sampled trips (from Fig. 7.6.4) $Cm(s,t,f,h,*,i)$	raised to total number of trips in the sampled landing place $CR(s,t,f,h,*,i)$
5-6	57	213.75
6-7	148	555.00
7-8	216	810.00
8-9	89	333.75
9-10	63	236.25
10-11	176	660.00
11-12	122	457.50
12-13	58	217.50
13-14	15	56.25
total	944 $= Cm(s,t,f,h,*,*)$	3540.00 $= CR(s,t,f,h,*,*)$

This raising procedure is reasonable only if a "trip" is a well-defined unit of effort. If some trips are of a duration of, say, one fishing day and others are of a duration of five fishing days, it is better to use a "fishing day" as unit of effort. Either unit, "trip" or "fishing day" makes sense only if the boats of a fleet are fairly similar, in the sense that they have the same "fishing power". Other possible units of effort are "number of man days", "number of trawling hours", "number of gill net sets", etc.

Summation of sampled landing places for one fleet and raising to all landing places

The total length distribution of all <u>sampled</u> landing places (see Fig. 7.4.1) is obtained by simple addition:

$$\sum_h CR(s,t,f,h,*,i) = CR(s,t,f,*,*,i)$$

This figure can be raised to the total for all landing places by applying a raising factor based on the effort expended in all the landing places:

$$\frac{\text{total effort of all landing places}}{\text{effort of sampled landing places}} * CR(s,t,f,*,*,i) = CRR(s,t,f,*,*,i)$$

Summation of fleets

This too is a simple summation. The total length distribution of species s caught during time period t is:

$$\sum_f CRR(s,t,f,*,*,i) = CRR(s,t,*,*,*,i)$$

We have now obtained a complete picture of the length-frequency distribution of all the landings of species s in one quarter of the year, t. Depending on the type of analysis required we may stop the processing of catch data at this level. In that case the results after one year are the four quarterly length-frequencies as shown in Fig. 7.6.5A. These may be resolved into cohort components by, for example, the Bhattacharya method (Section 3.4) as shown in Fig. 7.6.5B. The numbers caught from the same cohort in different

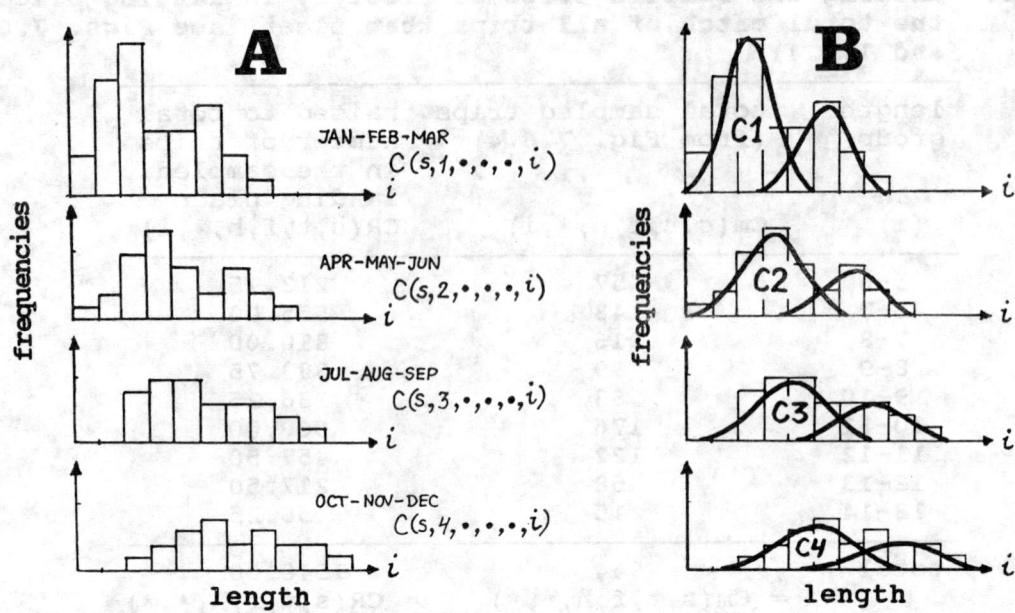

Fig. 7.6.5 A: Total length-frequencies by quarter year.
B: Length-frequencies resolved into normally distributed cohort components (input to Pope's cohort analysis)

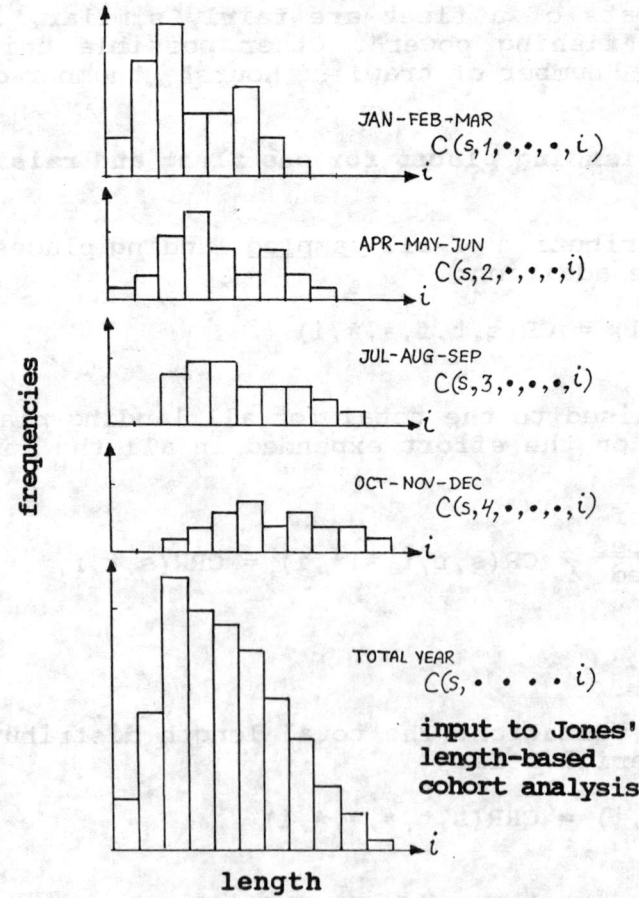

Fig. 7.6.6 Summation of total length compositions of all time periods (total length composition as input for Jones' length-based cohort analysis)

quarters of the year (e.g. the numbers C1,C2,C3 and C4 in Fig. 7.6.5B) form the inputs to Pope's cohort analysis (compare Section 5.2).

Alternatively, we may choose to apply Jones' length-based cohort analysis (compare Section 5.3). In this case we do not separate the cohorts, but proceed as follows:

Summation of time periods

This is the final step, which gives the length-frequencies of species s for the whole year. The summation is simple:

$$\sum_t CRR(s,t,*,*,*,i) = CRR(s,*,*,*,*,i)$$

Fig. 7.6.6 shows an example in which quarterly length compositions are summed to an annual length composition. These final C-values can be used as inputs to Jones' length-based cohort analysis. We may also use the average values for a range of years (see Section 5.3).

Data analysis

The resolution of length-frequency samples into normally distributed components, as shown in Fig. 7.6.5, becomes more problematic the longer the sampling period is. The quarterly samples show a certain cohort structure, whereas in the length-frequency distribution for the whole year the cohorts cannot be distinguished (see Fig. 7.6.6). This example illustrates that for age-based cohort analyses we must work with relatively short time periods, otherwise we are unable to identify the cohorts. Pope's age-based cohort analysis deals with the numbers caught per cohort. The slope of the various length-frequency distributions is not used, since they do not represent the total mortality.

On the other hand, for a Jones' length-based cohort analysis we are interested in the right hand slope of the combined length-frequency distribution, because this slope is a reflection of total mortality. Therefore the combined length-frequencies should represent a long period, so that the individual slopes of length-frequencies of single cohorts are levelled out.

Finally, it is emphasized that the procedure explained above is an example which may not fit to all fisheries. Especially the definition of the effort unit ("trip") may be inappropriate for many fisheries. Also the assumption that the samples are taken from the boats at the time the catch is unloaded may not fit to all cases. In the example it has been assumed that the total catch has been landed, in the form of consumption fish and unsorted by-catch. In this case there were no "discards".

Discards are fish caught but not landed, that is, thrown back into the sea. Discards are believed not to survive the encounter with the fishing gear. From a biological point of view, discards are as important as landings, as the important biological point is that the fish were killed by fishery. Some fisheries, notably shrimp trawling, discard up to 90% and sometimes even more of the weight caught. The discards may well contain good fin fish for human consumption, but which compared to shrimps are of relatively low value. Actually, one should carefully distinguish between "landings" and "catches", the latter including both landings and discards. Discards are difficult or expensive to sample, as reliable estimates would require observers to be placed on board of the commercial vessels. However, if discards are important quantities, attempts to sample them should be made.

Computer programs

The FiSAT and LFSA packages of microcomputer programs contain programs for data manipulations as described in this chapter, i.e. various types of summations and raising procedures for length-frequency samples.

8 PREDICTION MODELS

Future yields and stock biomass levels can be predicted by means of mathematical models which are similar to the ones behind the virtual population analysis, VPA, and the cohort analysis (see Chapter 5). The mathematical formulas used for VPA and cohort analysis, which analyse the history of a fishery, can be transformed in such a way that the knowledge of the past can be used to predict future yields and biomass at different levels of fishing effort. In other words, these models can be used to forecast the effects of development and management measures, such as increases or reductions of fishing fleets, changes in minimum mesh sizes, closed seasons, closed areas, etc. Therefore these models form a direct link between fish stock assessment and fishery resource management.

The prediction models can also incorporate aspects of prices and value of the catch, which make them suitable as a basis for bio-economic analyses, where biological and economic inputs are used to predict future yields, biomass levels and value of the catch under all kinds of assumptions. This chapter contains only a very basic introduction to bio-economic aspects, and for further studies the reader is referred to Sparre and Willmann (1992).

The first prediction models were already developed in the thirties by Thompson and Bell (1934). However, due to the large number of calculations required the so-called "Thompson and Bell model" did not reach a high popularity until the introduction of computers.

In the meantime a simpler model, based on rigorous assumptions, but therefore requiring less calculations was developed by Beverton and Holt (1957). Their "Yield per Recruit" model was widely used, but now it has been replaced by the Thompson and Bell model in regions where VPA and cohort analysis are being applied.

Beverton and Holt's yield per recruit model can be regarded as a special application of the Thompson and Bell model, which means that any general conclusion derived from it also holds for the Thompson and Bell model.

Although it is not likely that the Beverton and Holt model will be much used in the future, it has been incorporated in this chapter for historical and didactical reasons. The yield per recruit model is suited for calculators and therefore for the demonstration of certain aspects of fish stock assessment. The second part of this chapter (Sections 8.6 to 8.8) deals with the Thompson and Bell model, age-based and length-based and with aspects of gear selectivity related to the model.

The final purpose of the use of the predictive models is to provide those responsible for the management of fishery resources with information on the biological and/or economical effect of fishing on the stock. The managers are then expected to take measures that will lead to a level of exploitation of the resources where the maximum yield is obtained, either in the biological or in the economical sense, on a sustainable basis, i.e., without causing damage to the stocks that would affect future yields.

The managers should try to prevent situations where the fishing pressure becomes too high and where stocks are "overfished". An exact prediction of future yields is usually not possible, because stocks are seldom in a "steady state" which is assumed to exist for the application of many models. It has been demonstrated (Sharp and Csirke, 1984) that the abundance of certain stocks, in particular small pelagic species occurring in upwelling areas, usually depends very much on environmental factors which are beyond the control of any human interference. In such cases the predictive value of the models described below is practically nil. However, in the case of some pelagic and most demersal fisheries for fish and shrimp the models have proved to be extremely useful.

Before going deeper into these two models, it is worthwhile to first consider a simpler model and to discuss the concept of overfishing.

The classical model describing a fishery on a particular stock is the one given by Russell (1931). The model is in the form of a "black box" that represents what Ricker (1975) has defined as the "usable stock", the weight of all fish larger than a minimum useful size. The inputs to the usable stock are the weight of the new recruits and the growth of the fish already forming part of the stock. The outputs are the deaths by natural causes and the yield (catch in weight) of the fishery.

In an unfished stock the combined inputs are, on the average, equal to the removal of biomass by natural deaths. When a population is being fished it has an effect on all other factors, viz., there will be a greater rate of recruitment, a faster growth and reduced natural deaths. This is because fishing creates "room" for more new recruits, it removes the large slow-growing fish which are replaced by smaller fast-growing fish, and it removes fish before they can die of old age or other natural causes. A fishery, therefore, has a stimulating effect on the production of fish, as long as the stock is given enough time to adjust to the new situation and as long as the fishing pressure does not become too heavy. When "overfishing" occurs, the growth of individual fish cannot keep pace with the deaths caused by fishing and when it becomes very severe it also affects recruitment. Therefore, there are two types of overfishing, viz., "growth overfishing" and "recruitment overfishing". Growth overfishing occurs when the effort is so high that the total yield decreases with increasing effort. The fish are caught before they can grow to a sufficiently large size to substantially contribute to the biomass. In general, it is reasonable to say that a stock is growth overfished in the biological sense if F exceeds F_{MSY} (cf. Fig. 8.2.3). The term "biological" is used to indicate that only the yield in terms of biomass measured in weight units is considered. Thus, the value of the yield and the cost of fishing are being disregarded here.

In order to understand the concept of recruitment overfishing, it is necessary to first consider the relationship between recruitment and the spawning stock biomass as illustrated in Fig. 8.0.1. As indicated by the question-marks, this relationship is not well understood. The only point that is known for sure is (0,0), i.e., when there is no parent stock there can be no offspring. It is then reasonable to assume that at low levels of the parent stock there is a direct positive linear relationship with the number of offspring or recruitment. Under normal conditions such a direct "linear" relationship is not noted, but when it occurs, it means that the parent stock has come down to a very low level and in that case we speak of recruitment overfishing.

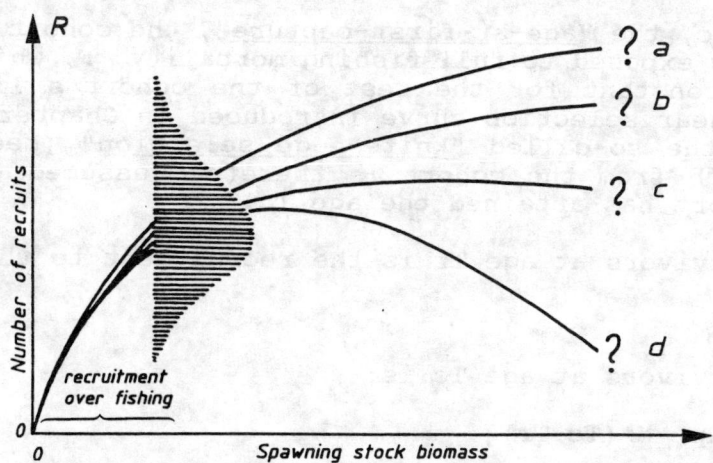

Fig. 8.0.1 The stock-recruitment relationship in connection with the concept of "recruitment overfishing"

8.1 ASSUMPTIONS AND MODELS UNDERLYING THE YIELD PER RECRUIT MODEL OF BEVERTON AND HOLT

The yield per recruit model (Beverton and Holt, 1957) is in principle a "steady state model", i.e., a model describing the state of the stock and the yield in a situation when the fishing pattern has been the same for such a long time that all fish alive have been exposed to it since they recruited. There are extensions to the model dealing with the transitory phase after a change in the fishing pattern, but these are seldom used because models of the Thompson and Bell type (Sections 8.6 to 8.8) provide a simpler description of non-steady state situations.

The rigorous assumptions underlying the Beverton and Holt approach are:

1. Recruitment is constant, yet not specified

2. All fish of a cohort are hatched on the same date

3. Recruitment and selection are "knife-edge" (see Chapter 6)

4. The fishing and natural mortalities are constant from the moment of entry to the exploited phase

5. There is a complete mixing within the stock

6. The length-weight relationship (Eq. 2.6.1) has the exponent 3, i.e. $W = q*L^3$

One of the models behind the Beverton and Holt model is the exponential decay model that was introduced in Section 4.2, and mathematically expressed in Eq. 4.2.2. The definitions and terminology introduced in Section 4.1 are also applicable to the Beverton and Holt model (e.g. $Tc, Tr, R = N(Tr)$).

The assumed life history of a cohort in the Beverton and Holt model, as shown in Fig. 8.1.1 is as follows:

1) At age Tr all fish belonging to a given cohort recruit to the fishing grounds at the same time: "knife-edge recruitment".

2) From age Tr to age Tc the cohort is not exposed to any fishing mortality. (It is assumed that all fish of ages between Tr and Tc escape through the meshes if they enter the gear.) Thus, in that period they suffer only from natural mortality, M, which is assumed to remain constant throughout the life span of the cohort.

3) At age Tc, the "age-at-first-capture", the cohort is assumed to be suddenly exposed to full fishing mortality, F, which is assumed to remain constant for the rest of the cohort's life. The sigmoid shaped gear selection curve introduced in Chapter 6 is approximated by the so-called "knife-edge selection" (see Fig. 6.4.1.1). The catch from the cohort is therefore assumed to be zero until the cohort has attained the age Tc.

The number of survivors at age Tr is the recruitment to the fishery:

$$R = N(Tr) \tag{8.1.1}$$

The number of survivors at age Tc is:

$$N(Tc) = R*\exp[-M*(Tc-Tr)] \tag{8.1.2}$$

The number of survivors at age t, where t > Tc, is:

$$N(t) = N(Tc)*\exp[-(M+F)*(t-Tc)] = R*\exp[-M*(Tc-Tr) - (M+F)*(t-Tc)]$$

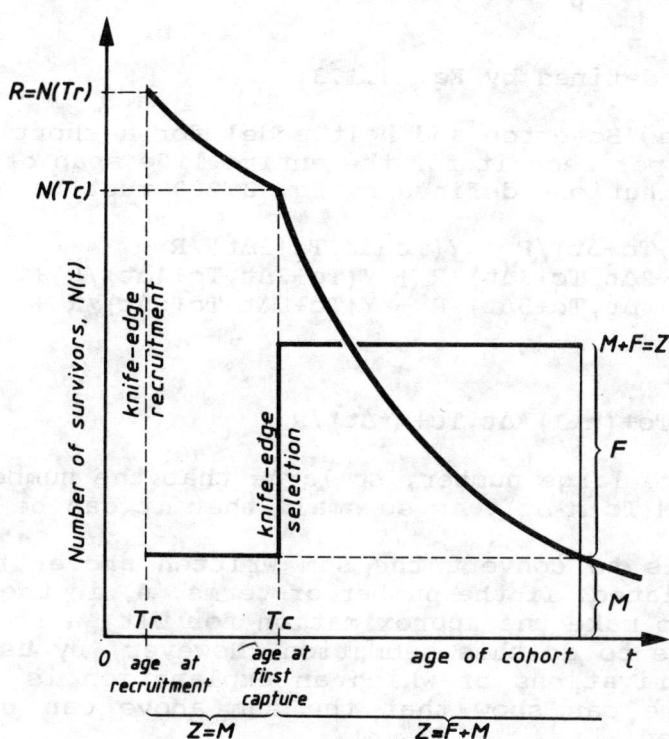

Fig. 8.1.1 The life history of a cohort as assumed in the Beverton and Holt model

The fraction of the total recruitment N(Tr) or R surviving until age t is obtained by dividing both sides of the equation by R and becomes:

$$N(t)/R = \exp[-M*(Tc-Tr) - (M+F)*(t-Tc)] \qquad (8.1.3)$$

This means that Eq. 8.1.3 gives the number of fish at time t "per recruit", i.e. as the fraction of each fish that recruited to the fishery.

The other model underlying the Beverton and Holt model is the "catch equation" in the form of Eq. 4.2.10 as explained in the next section.

8.2 BEVERTON AND HOLT'S YIELD PER RECRUIT MODEL

To derive the mathematical expression for Beverton and Holt's yield per recruit model we take (as usual) a starting point in the catch equation in the form of Eq. 4.2.10:

$$C(t,t+\Delta t) = \Delta t*F*N(t) \qquad (8.2.1)$$

Eq. 8.2.1 gives the number of fish caught from a cohort, in the time period from t to t+Δt when Δt is small. To get the corresponding yield in weight, this number should be multiplied by the individual weight of a fish. If Δt is small, then the body weight of a fish will remain approximately constant during the time period from t to t+Δt, and the yield becomes:

$$Y(t,t+\Delta t) = \Delta t*F*N(t)*w(t) \qquad (8.2.2)$$

where w(t) is the body weight of a t years old fish, as defined by the weight-based von Bertalanffy equation (Eq. 3.1.2.1). To get the yield per recruit for the time period from t to t+Δt Eq. 8.2.2 is divided by the number of recruits, R:

$$\frac{Y(t,t+\Delta t)}{R} = F*\frac{N(t)}{R}*w(t)*\Delta t \qquad (8.2.3)$$

where $N(t)/R$ is defined by Eq. 8.1.3.

Eq. 8.2.3 is the "Beverton and Holt model for a short time period". To get the total yield per recruit for the entire life span of the cohort, Y/R, all the small contributions defined by Eq. 8.2.3 must be added up:

$$\begin{aligned}Y/R = \ &Y(Tc,Tc+\Delta t)/R + Y(Tc+\Delta t,Tc+2\Delta t)/R + \\&Y(Tc+2\Delta t,Tc+3\Delta t)/R + Y(Tc+3\Delta t,Tc+4\Delta t)/R + \\&Y(Tc+4\Delta t,Tc+5\Delta t)/R + Y(Tc+5\Delta t,Tc+6\Delta t)/R + \\&\ldots \\&\ldots \\&+ Y(Tc+(n-1)*\Delta t,Tc+n*\Delta t)/R\end{aligned}$$

where "n" is some large number, so large that the number of fish older than $Tc+n*\Delta t$, i.e., $N(Tc+n*\Delta t)$, is so small that it can be ignored.

The next step is to convert the sum written above into a form which can easily be calculated. If the number of terms, n, in the sum is large (and it must be large to make the approximation for $w(t)$ a reasonable one) it will take a long time to do this summation. However, by using a long series of mathematical derivations of which an explanation is outside the scope of this manual, one can show that the sum above can be written in a more convenient way as:

$$Y/R = F*\exp[-M*(Tc-Tr)]*W_\infty * \left[\frac{1}{Z} - \frac{3S}{Z+K} + \frac{3S^2}{Z+2K} - \frac{S^3}{Z+3K}\right] \qquad (8.2.4)$$

where:

$S = \exp[-K*(Tc-t_o)]$
K = von Bertalanffy growth parameter
t_o = von Bertalanffy growth parameter
Tc = age at first capture
Tr = age at recruitment
W_∞ = asymptotic body weight
F = fishing mortality
M = natural mortality
Z = $F+M$, total mortality

Eq. 8.2.4 is the "Beverton and Holt yield per recruit model" (1957), Y/R model, written in the form suggested by Gulland (1969). Although the equation looks complicated, it is quite easy to handle with a programmable pocket calculator.

Because Beverton and Holt express yields on a "per recruit basis", the yields are relative, i.e. relative to the recruitment. If, say, a recruitment of 100 million fish gives a yield of 10 000 tonnes then 200 million recruits would yield 20 000 tonnes according to the model. This assumption may appear trivial, but it is not, as one could well imagine that the more abundant a species becomes the worse the conditions for the individuals will be, due to, for example, food competition and cannibalism. The results of the model are expressed in units of yield per recruit (grams per recruit). In the example above the yield per recruit becomes:

$$\frac{10\ 000\ 000\ 000}{100\ 000\ 000} = 100 \text{ grams per recruit}$$

The model allows us to calculate Y/R with varying inputs of the different parameters, such as F and Tc and then assess which effect the various input values have on the yield per recruit of the species under investigation. It is important to note here that the two parameters F and Tc are those which can be controlled by fishery managers, because:

1) F is proportional to effort (cf. Eq. 4.3.0.7)

2) Tc is a function of gear selectivity

Therefore, Y/R is considered a function of F and Tc. Most often you will see Y/R plotted against F (or effort).

Fig. 8.2.1 shows the result of a yield assessment with the yield per recruit model. The age-at-entry to the exploited phase, Tc, is kept constant. The independent variable is the effort as expressed by the coefficient F of fishing mortality. The dependent variable is the annual yield in grams per recruit. When the total annual yield is known in a steady state situation, for a given value of F, then the number of recruits can be calculated by dividing the total yield by the yield in grams per recruit.

The "<u>yield per recruit curve</u>" often has a maximum: the "<u>maximum sustainable yield</u> (MSY)". The position of the maximum depends on the age-at-first-capture, Tc, which in turn depends on the mesh size used in the fishery.

A change of mesh size, Tc, leads to a different MSY. Fig. 8.2.2 shows three curves with different Tc's. The highest MSY is reached for the highest value of Tc, at a slightly higher level of effort, F. By combining a range of values of Tc with a range of values of F, the highest maximum sustainable yield, valid for a certain combination of Tc and F, can be determined. The term sustainable means that that yield can be maintained "forever" as long as the conditions do not change. Higher yields may be obtained by a sudden increase in effort, but cannot be maintained, and they will have to be followed by a period of much lower yields. This is always on the assumption that nothing else has changed.

The Y/R model is originally an age-based model, but age is easily converted into length, again using the principles of the conversion of the catch curve (see Sections 4.4.2 and 8.5).

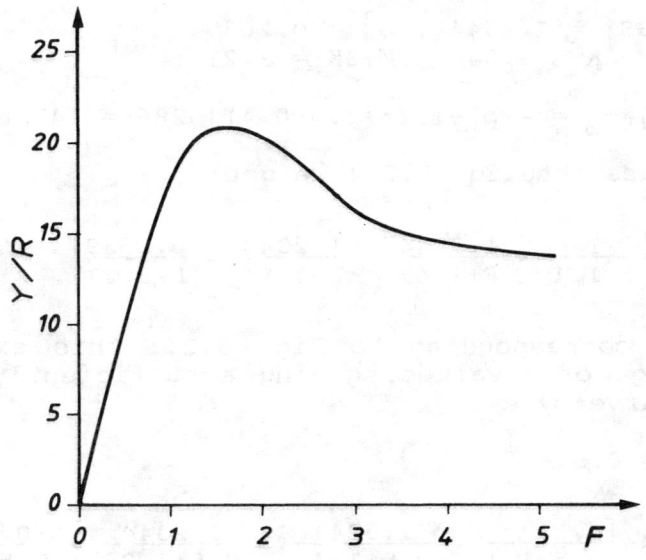

Fig. 8.2.1 Result of a yield assessment with the yield per recruit model

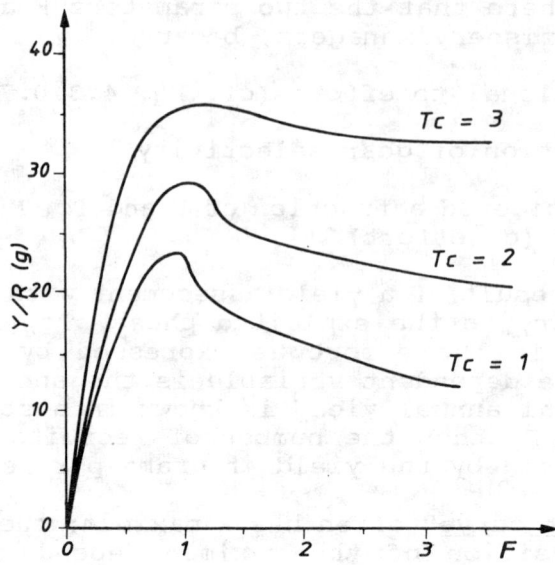

Fig. 8.2.2　Yield per recruit curves with different ages of first capture (Tc)

Example 28: Y/R as a function of F, for a tropical species

As an example we calculate the Y/R for <u>Nemipterus marginatus</u> as a function of F, using the following parameters:

$K = 0.37$ per year　　　$Tc = 1.0$ year　　　$t_o = -0.2$ year
$M = 1.1$ per year　　　$Tr = 0.4$ year　　　$W_\infty = 286$ grams

We start by calculating the terms in Eq. 8.2.4 which are independent of F:

$S = \exp[-K*(Tc-t_o)] = \exp[-0.37*(1.0+0.2)] = 0.6415$

$3S = 1.9244$, $3S^2 = 1.2344$, $S^3 = 0.2639$
$M+K = 1.47$, $M+2K = 1.84$, $M+3K = 2.21$

$\exp[-M*(Tc-Tr)]*W_\infty = \exp[-1.1*(1.0-0.4)]*286 = 147.8$

Inserting these values into Eq. 8.2.4 we get:

$$Y/R = F*147.8*\left[\frac{1}{F+1.1} - \frac{1.9244}{F+1.47} + \frac{1.2344}{F+1.84} - \frac{0.2639}{F+2.21}\right]$$

To produce a graph corresponding to Fig. 8.2.1 this expression must be evaluated for a range of F-values, giving a sufficiently large number of points to fit the curve by eye.

For example, for $F = 0.5$:

$$Y/R = 0.5*147.8*\left[\frac{1}{0.5+1.1} - \frac{1.9244}{0.5+1.47} + \frac{1.2344}{0.5+1.84} - \frac{0.2639}{0.5+2.21}\right]$$

$= 0.5*147.8*0.0785 = 5.8$ grams per recruit

Table 8.2.1 Yield per recruit and average biomass per recruit of <u>Nemipterus marginatus</u> as a function of F. Parameters as indicated in the legend of Fig. 8.2.3

F	Y/R	B/R	B/R as % of Bv	F	Y/R	B/R	B/R as % of Bv
0.0	0	22.4 = Bv	100	1.3	7.66	5.9	26
0.1	1.92	19.2	86	1.5	7.79	5.2	23
0.2	3.33	16.7	75	1.7	7.86	4.6	21
0.3	4.38	14.6	65	1.9	7.90	4.2	19
0.4	5.18	13.0	58	2.1	7.92	3.8	17
0.5	5.79	11.6	52	2.3 F(MSY)	7.93 MSY/R	3.5 MSB/R	15 MSB/Bv
0.6	6.26	10.4	46	2.5	7.92	3.2	14
0.7	6.62	9.5	42	3.0	7.88	2.6	12
0.8	6.91	8.6	38	4.0	7.77	1.9	8
0.9	7.14	7.9	35	5.0	7.66	1.5	7
1.0	7.32	7.3	33	6.0	7.57	1.3	6
1.1	7.46	6.8	30				

Repeating this calculation for F values ranging from F = 0 to F = 6.0 produces the results given in the first and fourth columns of Table 8.2.1, which have been used to produce the graph shown in Fig. 8.2.3.

By testing various F-values it is found that F = 2.3 gives the maximum value of Y/R, the "<u>Maximum Sustainable Yield per Recruit</u>" (MSY/R):

MSY/R = 7.9 grams per recruit

which corresponds to the biologically optimum fishing mortality:

F_{MSY} = 2.3 per year (see Table 8.2.1 and Fig. 8.2.3).

Because the Y/R-model assumes a constant parameter system (cf. Section 4.4.1) the results to be read from the curve only apply after the system has had constant parameters for a while. When F is changed it takes some time before the Y/R becomes the one predicted by the curve. How long this transient period is depends on the longevity of the species in question. From Table 8.2.1 and Fig. 8.2.3 it appears that the F_{MSY}-level is not determined with any great precision. Actually, for F > 1.5 the Y/R remains the same for a wide range of effort.

Curve B in Fig. 8.2.4 is an example of a Y/R-curve which differs in shape from the one in Fig. 8.2.3 (which is reproduced as curve A). Curve B has a pronounced maximum, it has a lower value of F_{MSY} and a higher value of MSY/R compared to curve A. The only difference in the input values of the two curves is the value of the natural mortality rate, M, <u>viz</u>. M = 0.2 per year for curve B and M = 1.1 per year for curve A. The conclusions that can be be drawn from the differences between these two curves are:

1. A lower M produces a lower F_{MSY} and a higher MSY/R, while fishing effort levels above F_{MSY} lead to a severe reduction of the total yield

2. If M is high it is difficult to estimate F_{MSY} by the Y/R curve

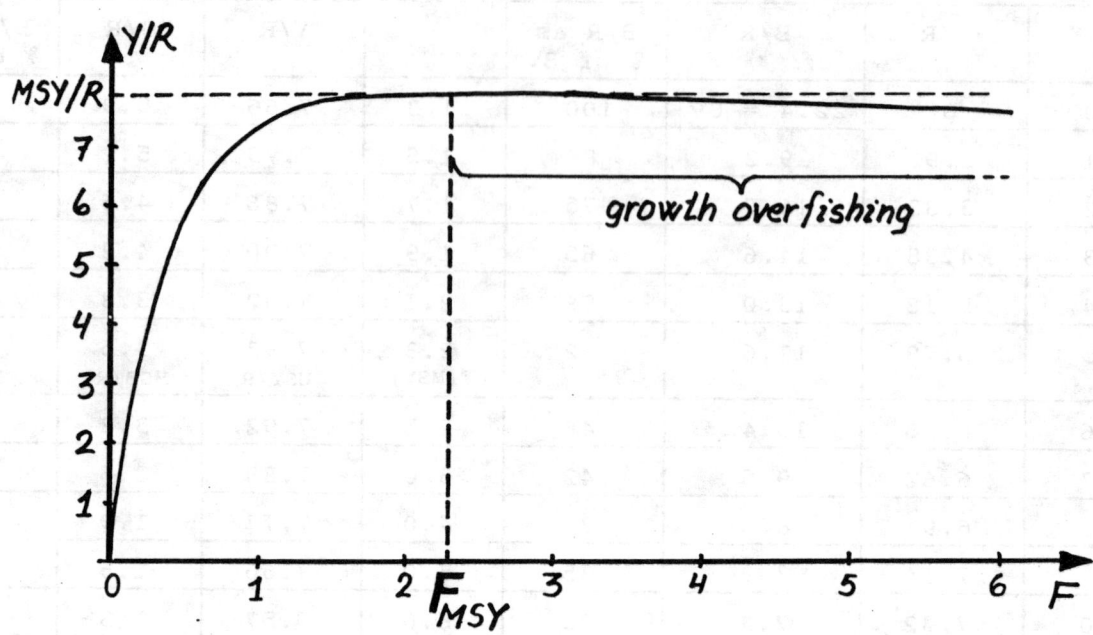

Fig. 8.2.3 Yield per recruit curve of <u>Nemipterus marginatus</u> as a function of F for the parameters:
K = 0.37 per year   Tc = 1.0 year t_o = -0.2 year
M = 1.1 per year   Tr = 0.4 year W_∞ = 286 grams

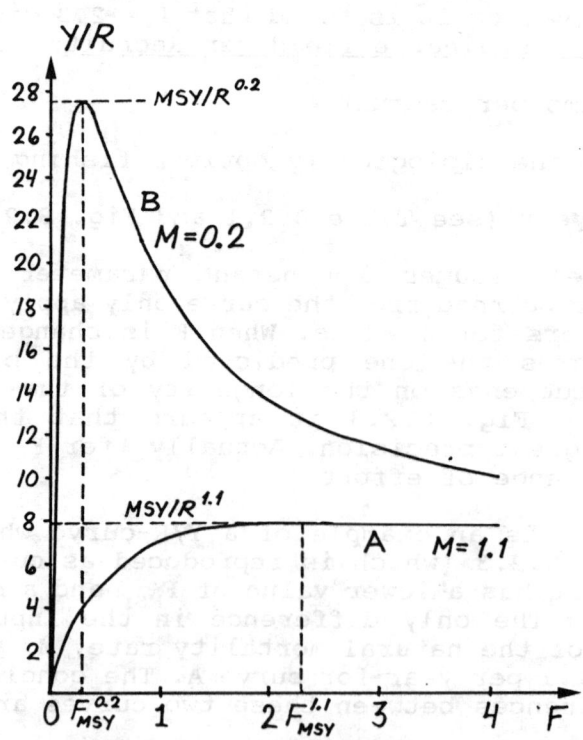

Fig. 8.2.4 Yield per recruit curve as a function of F for the parameters:
A: M = 1.1 per year   B: M = 0.2 per year
K = 0.37 per year   Tc = 1.0 year t_o = -0.2 year
Tr = 0.4 year W_∞ = 286 grams

These conclusions are the logical consequences of the effect of the level of natural mortality, M, on the production of biomass.

If M is high, the fish will soon reach the age where losses due to natural mortality exceed the gain in biomass due to growth. Therefore, F has to be high to catch the fish before they die of natural causes.

If M is low the gain in biomass due to growth will exceed the losses caused by natural mortality for a large part of the cohort's life span. In that case, it pays to let the fish grow to a large size and that means that for a biologically optimum exploitation F should be low.

In some cases (cf. Exercise 8.3) the Y/R-curve does not even have a maximum and an inexperienced management might come to the wrong conclusion that effort should be increased indefinitely. In such cases, which are common in tropical fisheries, it is recommended to look also at the biomass per recruit (B/R) curve, which is introduced in the next section. The two curves provide different information and it is therefore advisable to always plot them together.

The Y/R as a function of Tc, age-at-first-capture, is closely related to the estimation of the optimum mesh size (cf. Fig. 8.2.2, see Exercises 8.3 and 8.4).

8.3 BEVERTON AND HOLT'S BIOMASS PER RECRUIT MODEL

Beverton and Holt's biomass per recruit model expresses the annual average biomass of survivors as a function of fishing mortality (or effort). The average biomass is related to the catch per unit of effort (cf. Section 4.3). Eq. 4.3.0.2 expresses the relationship between CPUE and numbers caught, $CPUE(t) = q*N(t)$, which multiplied by the body weight on both sides gives:

$$CPUE(t)*w(t) = q*N(t)*w(t)$$

or if $N(t)*w(t)$ is replaced by $B(t)$, the symbol for biomass:

$$CPUEw(t) = q*B(t) \qquad (8.3.1)$$

where CPUEw is the "weight of the catch per unit of effort". Thus, the biomass is expected to show the same decline with increasing fishing effort as the CPUEw curve shown in Fig. 8.3.1.

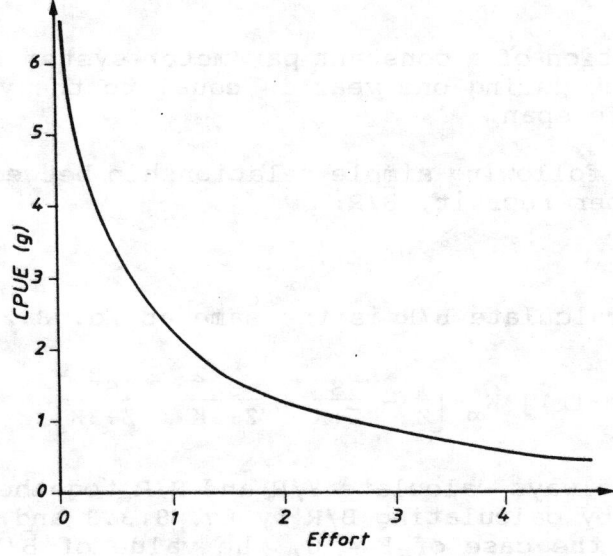

Fig. 8.3.1 Curve of CPUE (in weight) as a function of effort

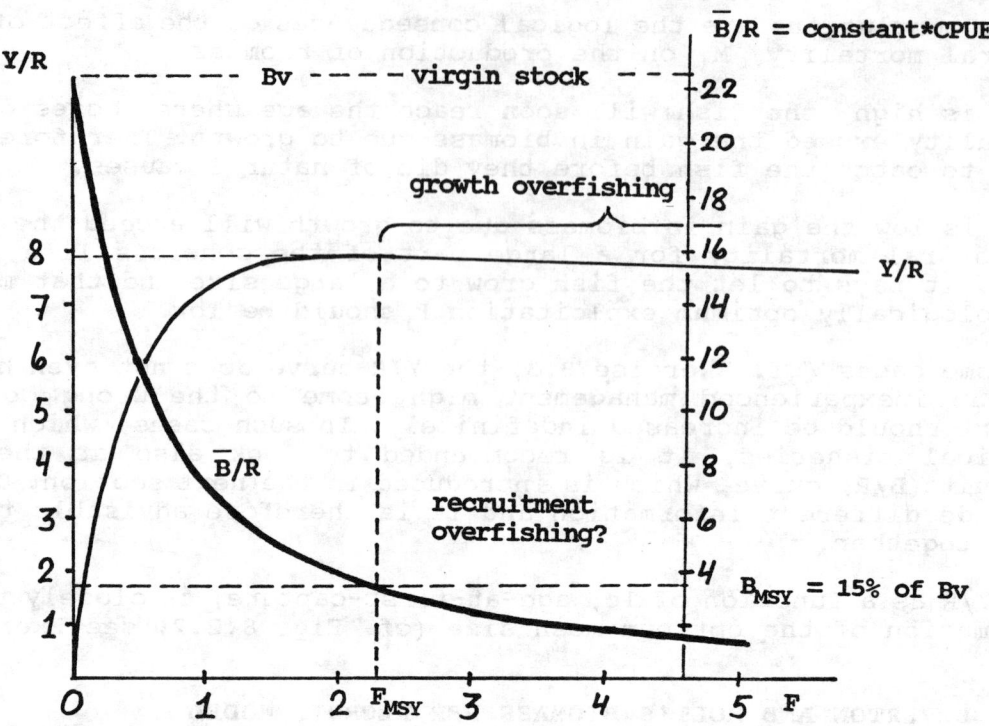

Fig. 8.3.2 Biomass per recruit curve of <u>Nemipterus marginatus</u> corresponding to the Y/R-curve of Fig. 8.2.3 which is repeated here

The catch in numbers per year can be expressed as

$$C = F*\overline{N}$$

(cf. Eq. 4.2.8 with t2-t1 = 1 year). By a similar argument it can be shown that the yield per year is

$$Y = F*\overline{B}$$

where $\overline{B}$ is the average biomass in the sea during a year. It follows that:

$$\frac{\overline{B}}{R} = \frac{Y}{R}*\frac{1}{F}$$

Because of the assumption of a constant parameter system (see Section 4.4.1) the yield from a stock during one year is equal to the yield from a single cohort during its life span.

Therefore we have the following simple relationship between Y/R, (Eq. 8.2.4) and average biomass per recruit, B/R:

$$Y/R = F*\overline{B}/R \tag{8.3.2}$$

The formula used to calculate $\overline{B}/R$ is the same as Eq. 8.2.4, divided by F:

$$\overline{B}/R = \exp[-M*(Tc-Tr)]*W_\infty*\left[\frac{1}{Z} - \frac{3S}{Z+K} + \frac{3S^2}{Z+2K} - \frac{S^3}{Z+3K}\right] \tag{8.3.3}$$

It is recommended to always calculate Y/R and $\overline{B}/R$ together. The easiest way to do so is to start by calculating $\overline{B}/R$ by Eq. 8.3.3 and then use Eq. 8.3.2 to calculate Y/R. In the case of F = 0, the value of $\overline{B}/R$ is the so-called virgin biomass per recruit, Bv/R, the biomass of the unexploited stock.

The average biomass per recruit as defined by Eq. 8.3.2 or Eq. 8.3.3 is the average biomass of the exploited part of the cohort, i.e. the biomass of fish of age Tc or older.

The $\bar{B}/R$ values related to the Y/R values calculated in Section 8.2 are presented in Table 8.2.1, where also $\bar{B}/R$ is given as a percentage of the virgin biomass, Bv. It shows that for Nemipterus marginatus the biomass corresponding to the biologically optimum F level, F_{MSY}, is only 15% of the virgin biomass, Bv. Fig. 8.3.2 shows the "biomass per recruit curve" which is always decreasing with increasing effort. The curve is proportional to the catch per unit of effort on the assumption underlying the model (see Eq. 8.3.1). This means that in any fishery one should expect a decrease in the catch per unit effort and the biomass when effort (e.g., the number of boats) increases. A decrease in the catch per unit effort is, therefore not per se an indication that a stock is overfished. Overfishing occurs when the fishing effort becomes so high that the growth cannot balance the death process.

Another, and sometimes more appropriate, way of using the $\bar{B}/R$-curve is to interpret it as a CPUEw-curve. When managing a fishery, considerations on the possible income per boat are essential and this quantity of course is closely related to CPUEw (see Sparre and Willmann, 1992).

(See **Exercise(s)** in Part 2).

8.4 BEVERTON AND HOLT'S RELATIVE YIELD PER RECRUIT MODEL

For fisheries management purposes, it is important to be able to determine changes in the Y/R for different values of F, for example if F is increased by 20% then the yield will decrease by 15%. The absolute values of Y/R expressed in grams per recruit are not important for this purpose. Therefore, Beverton and Holt (1966) also developed a "relative yield per recruit model" which can provide the kind of information needed for management. This model has the great advantage of requiring fewer parameters, while it is especially suitable for assessing the effect of mesh size regulations. It belongs to the category of length-based models, because it is based on lengths rather than ages.

The "relative Beverton and Holt yield per recruit model" is defined by:

$$(Y/R)' = E*U^{M/K}*\left[1 - \frac{3U}{1+m} + \frac{3U^2}{1+2m} - \frac{U^3}{1+3m}\right] \quad (8.4.1)$$

where

$$m = \frac{1-E}{M/K} = K/Z$$

$U = 1 - L_c/L_\infty$ the fraction of growth to be completed after entry into the exploited phase

$E = F/Z$ the exploitation rate or the fraction of deaths caused by fishing (cf. Section 4.2).

$(Y/R)'$ is considered a function of U and E and the only parameter is M/K. The equation gives a quantity which is proportional to Y/R as defined by Eq. 8.2.4 as can be shown by a number of algebraic manipulations. It can be shown that $(Y/R)' = (Y/R)*\exp[-M*(Tr-t_o)]/W_\infty$.

Note that a separate estimate of K is not required as input and that Eq. 8.4.1 is based on lengths (L_∞ and L_C in U) rather than ages.

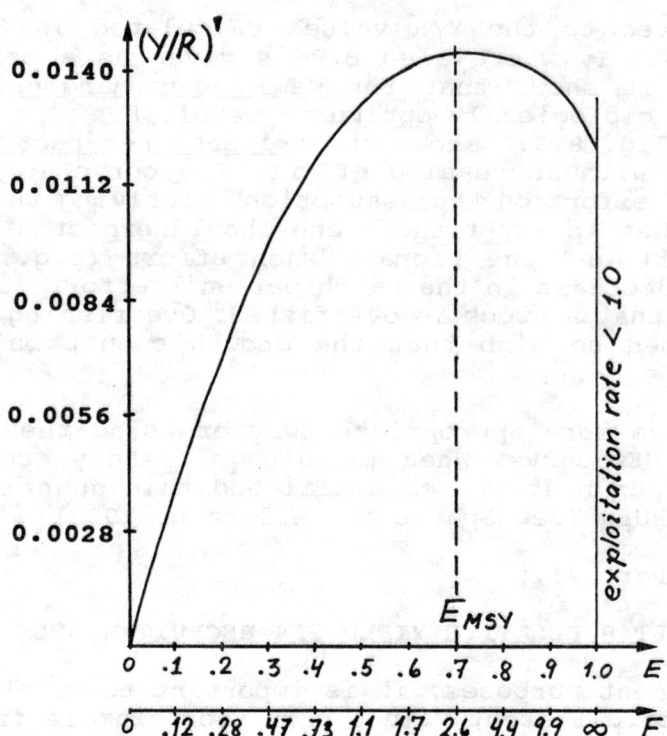

Fig. 8.4.1 Beverton and Holt's relative yield per recruit (Y/R)' curve corresponding to the Y/R-curve of Fig. 8.2.3 (Lc = 10.2 cm)

(Y/R)' can be calculated for given input values of M/K, L_∞ and Lc for values of E ranging from 0 to 1, corresponding to F values ranging from 0 to ∞.

The plot of (Y/R)' against E gives a curve with a maximum value, E_{MSY}, for a given value of Lc. Thus, when Lc, F and Z are known for a certain fishery the actual exploitation rate can be compared with the E_{MSY} level and management measures be proposed as necessary.

Fig. 8.4.1 shows the (Y/R)'-curve corresponding to the Y/R-curve of <u>Nemipterus marginatus</u> (Fig. 8.2.3) in the case of:

$$Lc = L(Tc) = L(1.0) = 28.4*[1 - \exp(-0.37*(1+0.2))] = 10.2 \text{ cm}$$

where L_∞ = 28.4 cm (see Section 3.1.2 and Fig. 3.1.2.1).

$$U = 1 - Lc/L_\infty = 1 - 10.2/28.4 = 0.641$$

As an example we calculate (Y/R)' for E = 0.5:

$$m = \frac{1-0.5}{1.1/0.37} = 0.168$$

$$(Y/R)' = 0.5*0.641^{2.973}*\left[1 - \frac{3*0.641}{1+0.168} + \frac{3*0.641^2}{1+2*0.168} - \frac{0.641^3}{1+3*0.168}\right] = 0.0135$$

(See **Exercise(s)** in Part 2).

8.5 YIELD PER RECRUIT FROM LENGTH DATA

Almost the same algebra as referred to in Section 8.4 transforms the equation for Y/R (Eq. 8.2.4) into a length-based model. The original parameters and variables are F, M, W_∞, K, t_o, Tr and Tc. In the length-transformed model we have L_∞, Lr and Lc instead of t_o, Tr and Tc. The new equation is

$$Y/R = F*A*W_\infty * \left[\frac{1}{Z} - \frac{3U}{Z+K} + \frac{3U^2}{Z+2K} - \frac{U^3}{Z+3K}\right] \qquad (8.5.1)$$

where $U = 1 - Lc/L_\infty$ as in Eq. 8.4.1 and

$$A = \left[\frac{L\infty-Lc}{L\infty-Lr}\right]^{M/K}$$

Recalling that several methods of parameter estimation described in the preceding chapters give Z/K or M/K it may be of interest also to formulate Eq. 8.5.1 in such terms. Division by K outside and multiplication by K inside the brackets, and substituting z for Z/K gives:

$$Y/R = \frac{F}{K}*A*W_\infty * \left[\frac{1}{z} - \frac{3U}{z+1} + \frac{3U^2}{z+2} - \frac{U^3}{z+3}\right] \qquad (8.5.2)$$

This equation contains F/K, M/K (in A) and Z/K (in z), has no reference to age and does not require a separate estimate of K.

Marten (1978), using linear growth instead of the von Bertalanffy model presents a similar length-based Y/R model.

8.6 AGE-BASED THOMPSON AND BELL MODEL

As stated in the introduction to this chapter the first predictive model was developed much earlier than the Beverton and Holt model by Thompson and Bell (1934). The Thompson and Bell model is the exact opposite of the models discussed in Chapter 5, VPA and cohort analysis. It is used to <u>predict</u> the effects of changes in the fishing effort on future yields, while VPA and cohort analysis are used to determine the numbers of fish that must have been present in the sea, to account for a known sustained catch, and the fishing effort that must have been expended on each age or length group to obtain the numbers caught (see Sections 5.1 and 5.2). Therefore, VPA and cohort analysis are called <u>historic</u> or <u>retrospective</u> models, while the Thompson and Bell model is <u>predictive</u>.

The Thompson and Bell method consists of two main stages: 1) Provision of essential and optional inputs and 2) the calculation of outputs in the form of predictions of future yields, biomass levels and even the value of the future yields.

    1) <u>Provision of inputs</u>: The main input is a so-called "<u>reference F-at-age-array</u>", an array of F-values per age group. In principle any F-array could be used as input, but, of course, not just any F-array will produce results which are related to the real situation of a fishery. Therefore, it is customary to use an F-array that has been obtained from an analysis of historical data, in other words from a VPA or a cohort analysis. However, the reference F-array may also originate from other sources as is actually the case in Example 29, given below.

Another important input parameter is the number of recruits, which may also be obtained from VPA or cohort analysis. This input is needed to obtain predictions of yields etc. in absolute quantities. However, if this input is not available the Thompson and Bell model can still be used to provide relative figures as output, for example, in the form of units "per 1000 recruits" (see Example 29).

The model further requires a "<u>weight-at-age-array</u>", the weights of individual fish per age group. For economic analyses the model also requires inputs of value, usually in the form of the price per kg by age group. (For the length-based Thompson and Bell model the same type of input is required per length group.)

2) <u>Outputs</u>: The output of the model is in the form of predictions of the catch in numbers, the total number of deaths, the yield, the mean biomass and the value, all per age group, related to values of F for each age group. New values of F can be obtained by multiplying the reference F-array as a whole by a certain factor, usually called X, or by applying such factors only to a part of the reference F-array. The latter is applied, for example, in the case of a change in the minimum mesh size, or to separate the effect of fleets with different characteristics (e.g. artisanal and industrial) on a particular stock. By carrying out a whole series of calculations with different values for X (F-factors), graphs can be drawn that illustrate clearly the effects of changes in F on the yield, the average biomass and the value of the catch.

The Thompson and Bell model is a very important tool for the fishery scientist to demonstrate the effect that certain management measures, such as changes in the minimum mesh size, decreases or increases of fishing effort, or closed seasons will have on the yield, the biomass and the value of the catch. Since a large number of calculations is required, it is essential to use computers.

An important aspect of the Thompson and Bell model is that it allows for the incorporation of the value of the catch. Therefore, the model has become the basis for the development of so-called bio-economic models, which are extremely useful for the provision of predictions needed for management decisions.

Computer programs

The LFSA package contains programs to carry out relatively simple Thompson and Bell analyses, both length-based and age-based. Similar programs have been incorporated in the FiSAT package. A series of computer programs for bio-economic analysis of fisheries has been developed and published by FAO, the so-called BEAM (**B**io-**E**conomic **A**nalytical **M**odel) programs (BEAM 1 and 2, Coppola <u>et al</u>., 1992, BEAM 3, Cochet and Gilly, 1990 and BEAM 4, Sparre and Willmann, 1992).

Example 29: Age-based Thompson and Bell analysis, tropical shrimp

Input

To illustrate the model we use data from the Kuwait shrimp fishery (from Garcia and van Zalinge, 1982). Columns A to E in Table 8.6.1 contain the input data. In this case the fishing mortalities, the F's, were estimated from catch data and estimates of the biomass obtained by the swept area method (cf. Chapter 13). However, the F-array could also have been estimated by cohort analysis or VPA.

The life span of the shrimp <u>Penaeus semisulcatus</u> is not much over one year so the age groups in Column A of Table 8.6.1 are given in months. The species is recruited to the fishery at the age of one month ($Tr = 1$). Column B gives the average weight per age group. Column C contains a relative value, proportioned to the price per kg of unpeeled tails per age group. Column D contains the fishing mortalities, the "<u>reference F-at-age-array</u>", and Column E the total mortality per year per age group.

In Column F we start with 1000 recruits, which have an age of 1 month at the beginning of the period. In other words, the population or stock number of age group 1 is 1000. All subsequent calculations are relative to 1000 recruits. In case a cohort analysis had been carried out and an estimate of the actual number of recruits obtained, the values obtained per 1000 recruits could be converted into actual yields and stock size (see Section 5.2).

Table 8.6.1 Age-based Thompson and Bell model illustrated by data from the Kuwait shrimp fishery (from Garcia and van Zalinge, 1982). $M = 3.0$ per year for all ages

INPUT					OUTPUT					
A	B	C	D	E	F	G	H	I	J	K
age t months	mean wght. $\overline{w}(t)$ g	value per g $\overline{v}(t)$ money unit	fish. mort. F(t) per year	total mort. Z(t) per year	population*) N(t) number	deaths N(t)− N(t+Δt) number	catch C(t) number	yield Y(t) g	mean biomass B(t) g	value Y*$\overline{v}$ money unit
1=Tr	5.7	0.73	1.20	4.20	1000.0	295.3	84.4	481	4809	351
2	9.3	0.93	1.32	4.32	704.7	213.0	65.1	605	5504	563
3	13.0	1.20	1.32	4.32	491.6	148.6	45.4	590	5367	708
4	17.6	1.45	1.44	4.44	343.0	106.1	34.4	606	5046	878
5	22.0	1.70	1.92	4.92	236.9	79.7	31.1	684	4276	1163
6	26.1	1.90	1.20	4.20	157.2	46.4	13.3	346	3463	658
7	30.3	2.08	1.56	4.56	110.8	35.0	12.0	363	2793	755
8	33.8	2.14	1.20	4.20	75.8	22.4	6.4	216	2161	462
9	37.0	2.18	1.20	4.20	53.4	15.8	4.5	167	1667	363
10	40.3	2.23	1.80	4.80	37.6	12.4	4.7	187	1250	418
11	43.1	2.24	2.76	5.76	25.2	9.6	4.6	199	863	445
12	44.7	2.27	2.52	5.52	15.6	5.8	2.6	117	559	267
13	−	−	−	−	9.9	−	−	−	−	−

Totals 4561 37758 7031
Mean biomass: 37758/12 = 3146.5

*) At beginning of period

Output based on reference F-at-age-array

On the basis of the input figures presented in Columns A to E and the number of recruits at age 1 month (= 1000), the population per age group, expressed in numbers present at the beginning of each month can be calculated (Column F). Also the following can be calculated: the number of deaths per month (Column G), the catch in numbers, equivalent to the number of deaths due to fishing (Column H), the yield in grams (Column I), the mean biomass in grams (Column J), and the value expressed in money units (Column K).

The computation procedures will now be presented step-by-step, using as numerical examples the calculations for the first three age groups.

Step 1: Calculate the population number at the beginning of each period (month):
$N(1) = 1000$, use
$N(t+\Delta t) = N(t)*\exp(-Z*\Delta t)$, where
$\Delta t = 1$ month $= 0.08333$ year, to calculate subsequent numbers
$N(2) = 1000*\exp(-4.20*0.08333) = 704.7$
$N(3) = 704.7*\exp(-4.32*0.08333) = 491.6$

Step 2: Calculate the total number of deaths in each period:
Total number of deaths $D(t) = N(t) - N(t+\Delta t)$
$D(1) = 1000-704.7 = 295.3$
$D(2) = 704.7-491.6 = 213.1$
$D(3) = 491.6-343.0 = 148.6$

Step 3: Calculate the numbers caught in each period:
$C(t) = [N(t) - N(t+\Delta t)]*F(t)/Z(t) = D(t)*F(t)/Z(t)$
$C(1) = 295.3*1.20/4.20 = 84.4$
$C(2) = 213.1*1.32/4.32 = 65.1$
$C(3) = 148.6*1.32/4.32 = 45.4$

Step 4: Calculate the yield (= catch in weight) in each period:
$Y(t) = C(t)*\bar{w}(t)$
$Y(1) = 84.4*5.7 = 481$
$Y(2) = 65.1*9.3 = 605$
$Y(3) = 45.4*13.0 = 590$

Step 5: Calculate the mean biomass in each period:
$\underline{B}(t) = Y(t)/[F(t)*\Delta t]$
$\underline{B}(1) = 481/(1.20*0.08333) = 4810$
$\underline{B}(2) = 605/(1.32*0.08333) = 5500$
$\underline{B}(3) = 591/(1.32*0.08333) = 5373$

Note: This calculation of biomass is derived from Eq. 4.2.8, $C = F*\Delta t*\bar{N}$, which by multiplication by $\bar{w}$ on both sides becomes

$Y = F*\Delta t*\bar{B}$ and $\bar{B} = Y/(F*\Delta t)$

Step 6: Calculate the value of yield in each period:
$V(t) = Y(t)*\bar{v}(t)$
$V(1) = 481*0.73 = 351$
$V(2) = 605*0.93 = 563$
$V(3) = 590*1.20 = 708$

Step 7: Calculate the total yield, the mean biomass over the whole period and the total value (see last row of Table 8.6.1):

Total yield is the sum of all monthly yields.
Total value is the sum of all monthly values.

The approximate average biomass is (cf. Fig. 4.2.3):

$$\bar{B} = \sum_{t=1}^{12} (\bar{B}(t) * \Delta t) / \sum_{t=1}^{12} \Delta t$$

As $\Delta t = 1/12$ and the total period 12 months, in this case

$$\bar{B} = \frac{\Sigma \bar{B}(t)}{12} = \frac{37758}{12} = 3146.5$$

The mean biomass concept in the more complicated case where Δt does not remain constant is discussed in Section 5.3.

The following block of equations summarizes the formulas for the age-based Thompson and Bell model in a general form, including X (F-factor). The index i refers to the age interval $(t_i, t_i + \Delta t)$. The index t_i refers to the start of the interval, while the index $t_i + \Delta t$ refers to the end of it.

$$\begin{aligned}
&\text{age interval: } i = (t_i, t_i + \Delta t) \\
&Z_i = M + X * F_i \\
&N(t_i + \Delta t) = N(t_i) * \exp(-Z_i * \Delta t) \\
&C_i = [N(t_i) - N(t_i + \Delta t)] * X * F_i / Z_i \\
&\bar{w}_i = w(t_i + \Delta t / 2) \\
&Y_i = C_i * \bar{w}_i \\
&\bar{B}_i = Y_i / [F_i * \Delta t * X] \\
&V_i = Y_i * \bar{v}_i
\end{aligned} \qquad (8.6.1)$$

Thompson and Bell using a plus-group

The last age group line in Table 8.6.1, age 13, contains only the number of survivors and none of the other entries. That is because in Example 29 the number of survivors older than 12 months has been considered to be an insignificant number and has therefore been ignored.

In cases where the number is significant there is a way to account for it, even when taking only 12 age groups into account. This is done by treating the age group 12 as a plus group, i.e., replacing the number of deaths between ages 12 and 13, N(12)-N(13), by the total number of deaths after age 12. Since all specimens will eventually die, this number is in Example 29 N(12) = N(12+) = 15.6.

Assuming further that the older age groups have the same mortalities as age group 12, the number of shrimp caught from the plus group becomes

$$C(12+) = \frac{F(12)}{Z(12)} * N(12)$$

$$C(12+) = (2.52/5.52) * 15.6 = 7.1$$

Thus, by leaving out the plus groups in Table 8.6.1 a catch of 7.1-2.6 = 4.5 has been ignored.

If growth has stopped at age 12 then $\overline{w}(12)$ is the maximum body weight and the yield corresponding to C(12+) becomes 44.7*4.5 = 201 g. This yield corresponds to a value of 201*2.27 = 456 money units, which constitutes some 6% of the total a significant amount. Therefore, to be on the safe side with regard to ignoring significant catches, it is better to always treat the last group as a plus-group.

Prediction, output based on different F-arrays

With the output based on the reference F-at-age-array calculated above all the basic data are available to predict the effect of increases and decreases in fishing effort or fishing mortality. New figures for total yield, total mean biomass and total value can be obtained by raising the fishing mortalities in Column D of Table 8.6.1 by a certain percentage. The F-array presented in Table 8.6.1 called the reference F-array is then replaced by a new one, by multiplying the reference F-array, or a part of it with the factor X = (new F)/(reference F).

If, for example, the effort is increased by 20 percent, the new fishing mortalities in Column D would become

$$1.20*1.20 = 1.44, \quad 1.32*1.20 = 1.58, \text{ etc.}$$

By then going again through the whole procedure using the new F's the related total yield, total mean biomass and total value are obtained.

An example of results of such a series of calculations with X = F-factors ranging from 0 to 3.0 is presented in Table 8.6.2. The reference F-array, where X = 1.0, gives a total yield of 4560, a total mean biomass of 3146 and a total value of 7029. (These amounts were obtained with the same input data as used in Table 8.6.1, however there are slight differences with the results presented in that table due to the fact that the computer program used to calculate Table 8.6.2 used the maximum number of digits for all calculations. Such small and, from the stock assessment point of view, insignificant differences can also be found in some other calculations presented in this manual.)

In Fig. 8.6.1 the total yield, total mean biomass and total value figures corresponding to Table 8.6.2 have been plotted against X (F-factor) and the respective curves were drawn. Note that the value curve has a maximum, whereas the yield curve has no maximum in the range of F-Factors (X) considered. When the price per kg varies with the size of the shrimps the two curves will have their maximum at different levels of F.

Table 8.6.2 Yield, value of yield and biomass for various F-levels. The reference F-array is given in column D of Table 8.6.1 (compare Fig. 8.6.1)

F-factor X	total yield	total mean biom.	total value
0	0	5382	0
0.4	2549	4271	4209
0.8	4055	3466	6396
1.0	4561*)	3146*)	7031*)
1.2	4954	2870	7465
1.5	5383	2522	7842
2.0	5814	2075	8025
3.0	6138	1497	7683

*) cf. Table 8.6.1

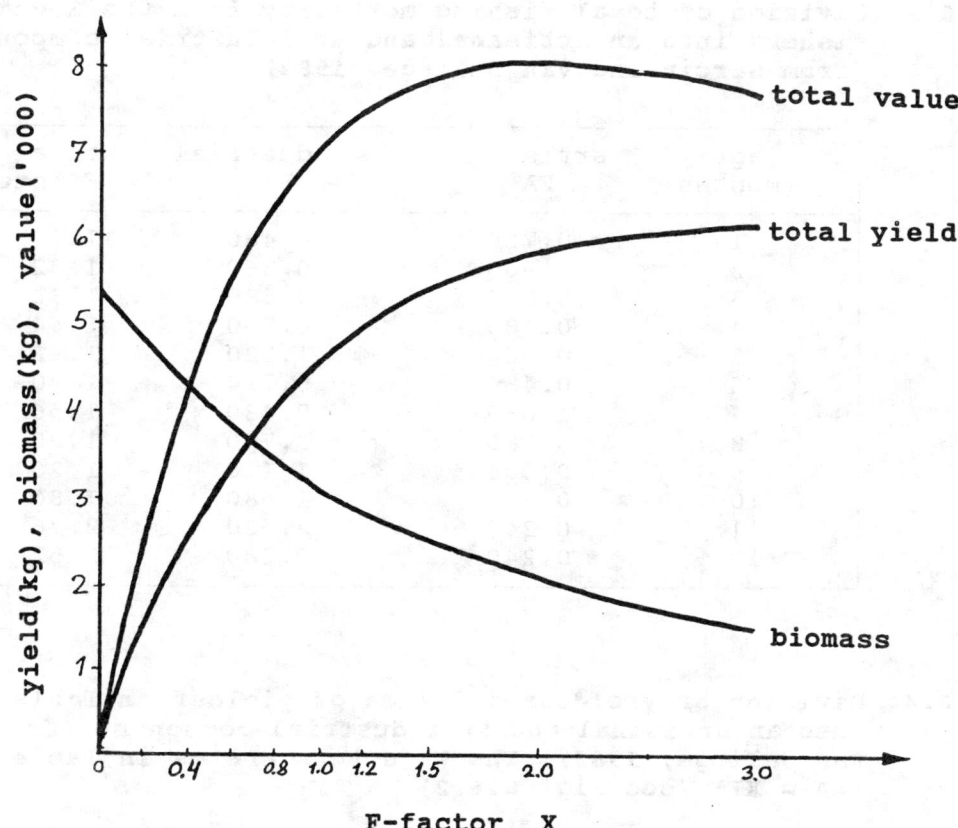

Fig. 8.6.1 Yield, biomass and value of yield per 1000 shrimps calculated by the age-based Thompson and Bell model. Based on data in Table 8.6.2

Recall that biomass is proportional to catch per unit of effort (Sections 4.3 and 8.3). Fig. 8.6.1 illustrates the important conflict between the desire to maximize the total yield from a fishery, by weight or by value, and the need to give the fishermen and boat owners the necessary income. The catch per boat decreases steadily as the effort increases and may in practice become too small to make the fishery profitable, even at effort levels smaller than those corresponding to the maximum on the curve for total value of the yield.

Prediction by fleet

The shrimp fishery in Kuwait waters is composed of an "artisanal fishery" and an "industrial fishery". Table 8.6.3 shows the results of a division of the total fishing mortality given in Table 8.6.1 into an artisanal component and an industrial component (from Garcia and van Zalinge, 1982). Such a division of fishing mortalities caused by different fleets is usually based on the proportions of the numbers of shrimps (or fish) caught by each fleet.

The fishing mortality exerted by one fleet, say, fleet no. i, $F(i)$, is

$$F(i) = F_{total} * C(i) / C_{total} \qquad (8.6.1)$$

where $C(i)$ is the number of shrimps (or fish) caught by fleet no. i, and $F(total)$ and $C(total)$ are the fishing mortality and the numbers caught by all fleets. $F(total)$ may be derived from cohort analysis. The split of the catch (column H in Table 8.6.1) into fleet components is obtained by:

$$C(i) = C_{total} * F(i) / F_{total} \qquad (8.6.2)$$

Table 8.6.3 Division of total fishing mortality from the Kuwait shrimp fishery into an artisanal and an industrial component (from Garcia and van Zalinge, 1982)

age (months)	artisanal FA	industrial FI	total F total
1	0.720	0.480	1.20
2	0.960	0.360	1.32
3	0.840	0.480	1.32
4	0.480	0.960	1.44
5	0.600	1.320	1.92
6	0.480	0.720	1.20
7	1.080	0.480	1.56
8	0.480	0.720	1.20
9	0.084	1.116	1.20
10	0.120	1.680	1.80
11	0.240	2.520	2.76
12	0.240	2.280	2.52

Table 8.6.4 Division of yields and values of yield from Table 8.6.1 into an artisanal and an industrial component (from Garcia and van Zalinge, 1982). The X-factors are as in Table 8.6.2 (XA = XI) (see Fig. 8.6.2)

total yield	total value	artisanal fleet			industrial fleet		
		F-factor XA	yield	value	F-factor XI	yield	value
0	0	0	0	0	0	0	0
2549	4209	0.4	1048	1531	0.4	1501	2678
4055	6396	0.8	1773	2486	0.8	2284	3910
4560	7029	1.0	2048	2815	1.0	2512	4216
4954	7465	1.2	2281	3073	1.2	2673	4392
5383	7842	1.5	2563	3354	1.5	2819	4488
5814	8025	2.0	2903	3627	2.0	2910	4398
6138	7683	3.0	3291	3783	3.0	2847	3900

Table 8.6.5 Assessment of the effect of varying the industrial effort (XI) while the artisanal effort is kept constant (XA = 1.0) (see Fig. 8.6.3)

total yield	total value	artisanal fleet			industrial fleet		
		F-factor XA	yield	value	F-factor XI	yield	value
2479	3603	1.0	2479	3603	0	0	0
3522	5403	1.0	2289	3250	0.4	1234	2154
4270	6598	1.0	2124	2950	0.8	2146	3648
4560	7029	1.0	2048	2815	1.0	2512	4216
4811	7383	1.0	1979	2691	1.2	2832	4692
5120	7783	1.0	1883	2520	1.5	3237	5263
5501	8203	1.0	1740	2271	2.0	3761	5932
5951	8499	1.0	1510	1880	3.0	4441	6619

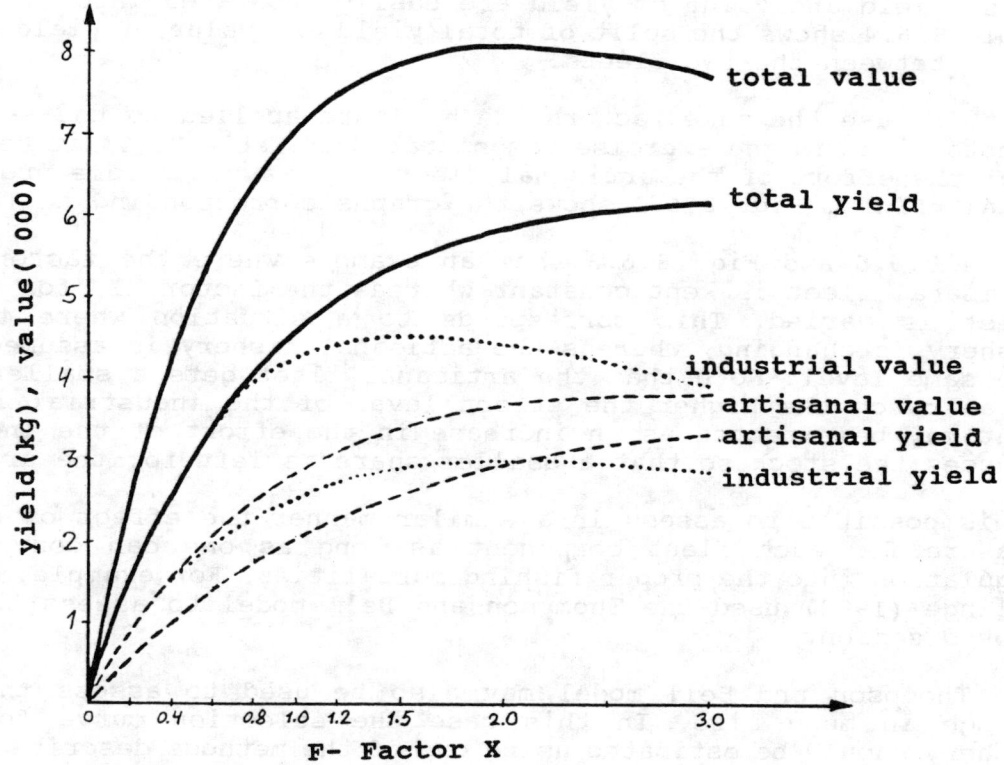

Fig. 8.6.2 Total yield and total value of yield from Fig. 8.6.1 separated into an artisanal and an industrial component (cf. Table 8.6.4)

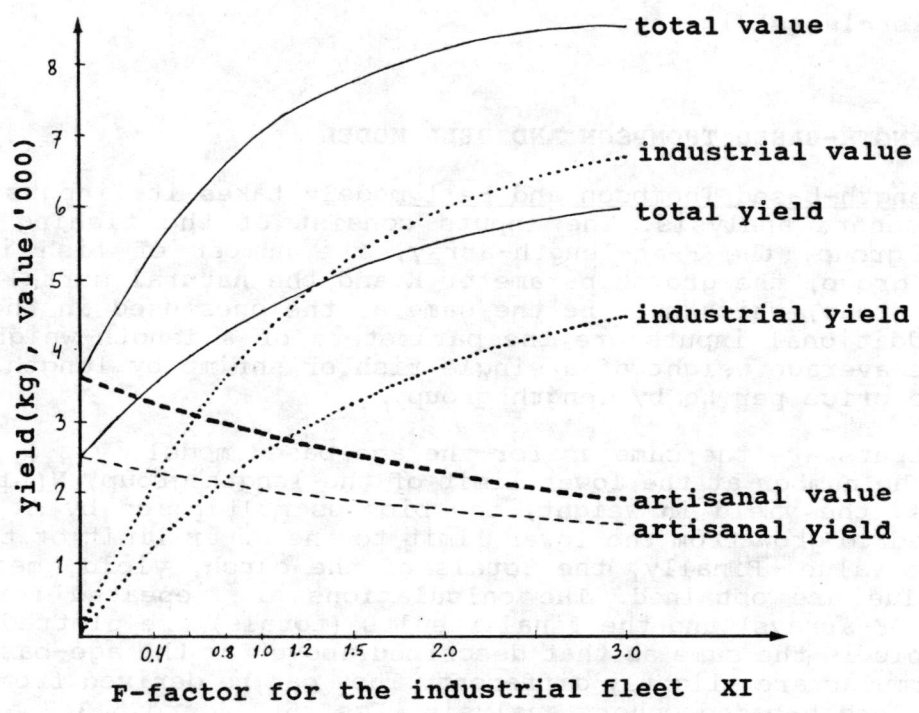

Fig. 8.6.3 Assessment of the effect of changes in the industrial fishery while the artisanal fishery is kept at a constant level (cf. Table 8.6.5)

Thus, yield and value of yield are easily separated into fleet components. Table 8.6.4 shows the split of total yield and value of yield given in Table 8.6.2 between the two fleets.

In this case the same factors, XA = XI are applied to the F-values of both fleets, i.e. in the exercise demonstrated in Table 8.6.4 it has been assumed that the effort of the artisanal fleet is always the same proportion of the total effort. Fig. 8.6.2 shows the graphs corresponding to Table 8.6.4.

Table 8.6.5 and Fig. 8.6.3 show an example where the factor, XA, for the artisanal fleet is kept constant whereas the factor XI, for the industrial fleet is varied. This corresponds to a situation where the industrial fishery is changing, whereas the artisanal fishery is assumed to remain at the same level. Note that the artisanal fleet gets a smaller share of the total catch the higher the effort level of the industrial fleet. This is what may be expected as an increase in the effort of the industrial fleet reduces the stock so that a smaller share is left for the artisanal fleet.

It is possible to assess in a similar manner the effect of any regulatory measure for each fleet component as long as one can convert the effort regulation into the proper fishing mortalities. For example, Garcia and van Zalinge (1982) used the Thompson and Bell model to assess the effect of a closed season.

The Thompson and Bell model may also be used to assess the effect of a change in mesh size. In this case the selection curve for the current fishery should be estimated using one of the methods described in Chapter 6. The method will be discussed in Section 8.8.

The application of the Thompson and Bell model described above (including the mesh assessment described below) is essentially the method applied today to predict catches and to set catch quotas in the ICES area (Northeast Atlantic) and in many other places.

(See **Exercise(s)** in Part 2).

8.7 LENGTH-BASED THOMPSON AND BELL MODEL

The "length-based Thompson and Bell model" takes its inputs from a length-based cohort analysis. The inputs consist of the fishing mortalities by length group, the F-at-length-array, the number of fish in the smallest length group, the growth parameter K and the natural mortality factor H by length group, which must be the same as the ones used in the cohort analysis. Additional inputs are the parameters of a length-weight relationship (or the average weight of a single fish or shrimp by length group) and the average price per kg by length group.

The outputs are the same as for the age-based model, viz., for each length group the number at the lower limit of the length group, $N(L1)$, the catch in numbers, the yield in weight, the biomass multiplied by Δt, i.e. the time required to grow from the lower limit to the upper limit of the length group and the value. Finally, the totals of the catch, yield, mean biomass * Δt and value are obtained. The calculations are repeated for a range of X values (F-arrays) and the final results (totals) are plotted in graphs. The principle is the same as that described above for the age-based models, only the formulas are slightly different. They can be derived from those used for Jones' length-based cohort analysis, Eqs. 5.3.4 and 5.3.7 as follows:

First Eq. 5.3.7 is rearranged:

$$C(L1,L2) = [N(L1)-N(L2)] * \frac{F(L1,L2)}{Z(L1,L2)} \qquad (8.7.1)$$

then it is inserted into into Eq. 5.3.4 which gives:

$$N(L1) = \left[N(L2)*H(L1,L2) + \frac{N(L1)-N(L2)}{Z(L1,L2)}*F(L1,L2) \right] *H(L1,L2)$$

where

$$H(L1,L2) = \left[\frac{L\infty-L1}{L\infty-L2} \right]^{M/2K}$$

which is the same factor as used in Jones' length-based cohort analysis (Eq. 5.3.3).

Solving this equation with respect to N(L2) gives:

$$N(L2) = N(L1) * \frac{1/H(L1,L2) - F(L1,L2)/Z(L1,L2)}{H(L1,L2) - F(L1,L2/Z(L1,L2)} \qquad (8.7.2)$$

In order to calculate the yield (catch in weight) by length group the catch C (in numbers) has to be multiplied by the mean weight of the length group, $\bar{w}(L1,L2)$, which is obtained from Eq. 5.3.11 as follows:

$$\bar{w}(L1,L2) = q*[(L1+L2)/2]^b$$

where q and b are the parameters of the length-weight relationship.

The yield of this length group is then given by

$$Y(L1,L2) = C(L1,L2) * \bar{w}(L1,L2) \qquad (8.7.3)$$

The value of the yield is given by:

$$V(L1,L2) = Y(L1,L2) * \bar{v}(L1,L2) \qquad (8.7.4)$$

where $\bar{v}(L1,L2)$ is the average price per kg of fish between lengths L1 and L2.

During the time $\Delta t(L1,L2)$ that it takes a cohort to grow from L1 to L2, the number of survivors decreases from N(L1) to N(L2). The mean number of survivors of that length group is calculated as follows:

$$\bar{N}(L1,L2) * \Delta t(L1,L2) = [N(L1)-N(L2)]/Z(L1,L2) \qquad (8.7.5)$$

The corresponding mean biomass * Δt is:

$$\bar{B}(L1,L2) * \Delta t(L1,L2) = \bar{N}(L1,L2) * \Delta t(L1,L2) * \bar{w}(L1,L2) \qquad (8.7.6)$$

The annual yield is simply the sum of the yield of all length groups:

$$Y = \Sigma Y_i$$

The annual value is likewise the sum of the value of all length groups:

$$V = \Sigma V_i$$

As discussed in Section 5.3

$$\bar{B} = \Sigma \bar{B}_i * \Delta t_i$$

is an estimate of the average biomass during the life span of a cohort, or of all cohorts during a year. In the age-based method, Section 8.6, it was not necessary to multiply each biomass by Δt because this was constant and equal to 1/12 of a year, or one month, but in this case Δt is variable.

Eqs. 8.7.1 to 8.7.6 have been presented for one specific length class (L1,L2). Like in the age-based version, the following block of equations summarizes the formulas for the length-based Thompson and Bell model in a general form, including X (F-factor). The index i refers here to the length interval (L_i, L_{i+1}). The index L_i refers to the lower limit of that length interval, while the index L_{i+1} refers to the upper limit.

---

length interval: $i = (L_i, L_{i+1})$

$Z_i = M + X*F_i$

$N(L_{i+1}) = N(L_i) * \dfrac{1/H_i - X*F_i/Z_i}{H_i - X*F_i/Z_i}$ where

$H_i = \left[\dfrac{L_\infty - L_i}{L_\infty - L_{i+1}}\right]^{M/2K}$

$C_i = [N(L_i) - N(L_{i+1})] * X * F_i / Z_i$ (8.7.7)

$\bar{w}_i = q * [(L_i + L_{i+1})/2]^b$

$Y_i = C_i * \bar{w}_i$

$V_i = Y_i * \bar{v}_i$

$\bar{N}_i * \Delta t_i = [N(L_i) - N(L_{i+1})]/Z_i$

$\bar{B}_i * \Delta t_i = \bar{N}_i * \Delta t_i * \bar{w}_i$

Basic features of the length-based Thompson and Bell analysis

Since the length-based Thompson and Bell analysis is derived from Jones' length-based cohort analysis (Section 5.3) which in turn is based on Pope's age-based cohort analysis (Section 5.2), the length-based Thompson and Bell method has the same limitations as Pope's age-based cohort analysis. The approximation to VPA in the predictive mode is valid for values of $F*\Delta t$ up to 1.2 and of $M*\Delta t$ up to 0.3 (Pope, 1972). If the F's are high, nonsensical results will come out of the analysis, such as negative stock numbers. If that is the case, smaller length groups and hence, smaller Δt values, are required.

Example 30: Length-based Thompson and Bell analysis, hake, Senegal

As an example of a length-based Thompson and Bell analysis we use the data from Table 5.3.3 for hake (<u>Merluccius merluccius</u>) caught off Senegal. The following input parameters are used (cf. Section 5.3):

L_∞ = 130 cm, K = 0.1 per year, M = 0.28 per year, q = 0.00001 kg/cm^3, b = 3, N (first length group) = N(6) = 98919.3

Using the F-values and the natural mortality factors, H, from Table 5.3.3 and the body weights derived from $\bar{w}_i = q*[(L_i+L_{i+1})/2]^b$, the length-weight relationship, and some (in this case arbitrarily selected) prices per kg for hake, the input may be summarized as in Table 8.7.1.

Table 8.7.1   Input data for length-based Thompson and Bell analysis, hake, Senegal

length group (L_i,L_{i+1})	$F(L_i,L_{i+1})$	$H(L_i,L_{i+1})$	$\bar{w}(L_i,L_{i+1})$	$\bar{v}(L_i,L_{i+1})$
6-12	0.04	1.0719	0.0073	1.0
12-18	0.39	1.0758	0.0338	1.0
18-24	1.07	1.0801	0.0926	1.0
24-30	0.65	1.0850	0.196	1.5
30-36	0.49	1.0905	0.359	1.5
36-42	0.59	1.0967	0.593	2.0
42-48	0.65	1.1039	0.911	2.0
48-54	0.39	1.1122	1.33	2.5
54-60	0.29	1.1220	1.85	2.5
60-66	0.31	1.1337	2.50	2.5
66-72	0.40	1.1478	3.29	3.0
72-78	0.39	1.1652	4.22	3.0
78-84	0.11	1.1873	5.31	3.0
84-∞	0.28	-	12.25	3.0

Using Eqs. 8.7.7 with X = 1 and the input data from Table 8.7.1 we can calculate the numbers in the subsequent length classes, the catch, yield, mean biomass * Δt and the value, as presented in the following example:

$N(12) = N(6)*[1/H(6,12) - F(6,12)/Z(6,12)]/[H(6,12) - F(6,12)/Z(6,12)]$

$\quad\quad\quad = 98919.3*[1/1.0719 - 0.04/0.32]/[1.0719 - 0.04/0.32]$

$\quad\quad\quad = 84400.8$

$C(6,12) = [N(6)-N(12)]*X*F(6,12)/Z(6,12)$

$\quad\quad\quad = [98919.3-84400.8]*1*0.04/0.32 = 1814.8$

$\bar{w}(6,12) = q*(6+12)/2]^b$

$\quad\quad\quad = 0.00001*9^3 = 0.007290$

$Y(6,12) = C(6,12)*\bar{w}(6,12)$

$\quad\quad\quad = 1814.8*0.007290 = 13.23$

$\bar{B}(6,12)*\Delta t(6,12) = [(N(6)-N(12)]/Z(6,12)]*\bar{w}(6,12)$

$\quad\quad\quad = [(98919.3-84400.8)/0.32]*0.007290 = 330.7$

$V(6,12) = Y(6,12)*\bar{v}(6,12) = 13.23*1.0 = 13.23$

These calculations will then continue until the last length group is reached. Since that is a so-called plus group a few additional assumptions have to be made: $N(\infty) = 0$ and $\overline{w}(84,\infty) = \overline{w}(84,90)$. The results are:

$$C(84,\infty) = [N(84)-N(\infty)]*F(84,\infty)/Z(84,\infty)$$

$$= [92-0]*0.28/0.56 = 46$$

$$\overline{w}(84,\infty) = \overline{w}(84,130) = q*[(84+130)/2]^b$$

$$= 0.00001*107^3 = 12.25$$

$$Y(84,\infty) = C(84,\infty) * \overline{w}(84,\infty)$$

$$= 46*12.25 = 563.5$$

$$\overline{B}(84,\infty)*\Delta t(84,\infty) = [(N(84)-N(\infty))/Z(84,\infty)] * \overline{w}(84,\infty)$$

$$= [(92-0)/0.56]*12.25$$

$$= 2012.5$$

$$V(84,\infty) = Y(84,\infty)*\overline{v}(84,\infty)$$

$$= 563.5*3.0 = 1690.5$$

Following these procedures the final result will be like Table 8.7.2. However, it should be noted that there are differences between the results of calculations presented above, calculated by means of a pocket calculator, and those in Table 8.7.2 calculated by using 8 significant digits in all calculations.

Notice that the values of $N(L_i)$ and $F(L_i,L_{i+1})$ are exactly the same as those calculated by Jones' length-based cohort analysis in Table 5.3.3.

The calculations can now be repeated for different values of X.

Table 8.7.2 Output from length-based Thompson and Bell analysis, hake, Senegal, using the F-factor X = 1.0. Weights are in tonnes (cf. Table 5.3.4)

length group (L_i,L_{i+1})	F (L_i,L_{i+1}) X = 1.0	$N(L_i)$	C (L_i,L_{i+1})	yield (L_i,L_{i+1})	mean biomass *Δt $\overline{B}*\Delta t$	value (L_i,L_{i+1})
6-12	0.04	98919.3	1823	13.3	330.7	13.3
12-18	0.39	84392.7	14463	488.1	1260.1	488.1
18-24	1.07	59475.8	25277	2336.3	2191.5	2336.3
24-30	0.65	27623.0	8143	1601.0	2475.2	2401.5
30-36	0.49	15967.8	3889	1397.6	2845.9	2096.4
36-42	0.59	9861.5	2959	1755.2	2970.1	3510.5
42-48	0.65	5500.5	1871	1704.9	2638.4	3409.9
48-54	0.39	2818.8	653	866.2	2247.1	2165.5
54-60	0.29	1691.5	322	596.3	2069.4	1490.8
60-66	0.31	1056.6	228	570.1	1853.8	1710.3
66-72	0.40	621.0	181	594.6	1481.9	1783.8
72-78	0.39	313.7	96	405.0	1040.1	1215.0
78-84	0.11	148.7	16	85.0	772.0	255.1
84-∞	0.28	92.0	46	563.5	2012.6	1690.6
Total			59908	12977.3	26189.0	24567.1

Table 8.7.3 Output from length-based Thompson and Bell analysis, hake, Senegal, using the F-factor X = 2.0. (cf. Table 8.7.2)

length group L_i, L_{i+1}	F (L_i, L_{i+1}) X = 2.0	$N(L_i)$	C (L_i, L_{i+1})	yield (L_i, L_{i+1})	mean biomass $*\Delta t$ $\overline{B}*\Delta t$	value (L_i, L_{i+1})
6-12	0.08	98919.3	3611.6	26.3	327.6	26.3
12-18	0.77	82724.1	26041.2	878.9	1134.4	878.9
18-24	2.13	47271.6	32863.1	3043.4	1427.4	3043.4
24-30	1.29	10092.9	5154.2	1014.5	784.2	1521.7
30-36	0.98	3823.1	1652.3	593.8	604.6	890.7
36-42	1.18	1699.7	881.6	523.0	442.5	1046.0
42-48	1.29	609.2	351.7	320.5	248.0	640.9
48-54	0.77	181.3	74.9	99.3	128.8	248.3
54-60	0.58	79.3	27.4	50.8	88.1	126.9
60-66	0.62	38.5	14.9	37.3	60.6	111.9
66-72	0.80	16.8	8.5	27.9	34.7	83.6
72-78	0.78	5.4	2.8	11.9	15.3	35.8
78-84	0.22	1.5	0.3	1.7	7.5	5.0
84-∞	0.56	0.8	0.5	6.7	12.0	20.2
Total			70685.0	6636.0	5315.9	8679.7

Table 8.7.3 shows the results corresponding to Table 8.7.2 but with the F-factor X = 2.0, i.e. the prediction of catch under the assumption of a doubling of the fishing effort. In this case the effect of doubling the effort would be a dramatic fall in the yield and the value of the yield.

Table 8.7.4 shows the summary results for 16 different F-factors (X). Each row is based on calculations like those illustrated by Tables 8.7.2 and 8.7.3. The total yield, mean biomass and value given in the last row of these two tables can also be found in Table 8.7.4. The two last rows of Table 8.7.4 show the maximum sustainable yield (MSY) and the maximum sustainable economic yield (value) (MSE) together with the corresponding F-factor and stock biomass. When kg prices differ from one length group to another the F-factor giving MSY usually differs from the F-factor giving MSE. Tables 8.7.2 to 8.7.4 were calculated by the program "MIXFISH" in the LFSA package (Sparre, 1987). This program calculates MSY and MSE by using an iterative technique.

The results of Table 8.7.4 have been plotted in Fig. 8.7.1. The graphs clearly show that the present level of fishing effort is well above that giving the maximum sustainable yield and the conclusion to be drawn from this analysis is that the stock is overfished because a reduction in effort would give a higher yield.

In case of economic interaction, where several fleets are exploiting one resource the catches predicted by the length-based Thompson and Bell analysis can be partitioned in exactly the same way as shown in Section 8.6 (cf. Tables 8.6.3 to 8.6.5).

The assumption behind the length-based Thompson and Bell analysis (and behind Jones' length-based cohort analysis) is that the stock remains in a steady state, with all parameters (e.g. recruitment) remaining constant. Thus, we obtain a prediction of the "average long term catches". Deviations from the predicted catches are therefore to be expected in individual years.

(See **Exercise(s)** in Part 2)

Table 8.7.4 Results of the length-based Thompson and Bell analysis, hake, Senegal. MSY = Maximum Sustainable Yield. MSE = Maximum Sustainable Economic Yield (value) (cf. Fig. 8.7.1)

A	B	C	D
F-factor X	total yield	total mean biomass *Δt	total value
0.0	0	571297	0
0.2	18903	268193	48329
0.4	20717	135343	49701
0.6	18360	73209	40925
0.8	15474	42376	31836
1.0*)	12977*)	26189*)	24567*)
1.2	10999	17216	19168
1.4	9470	11976	15236
1.6	8287	8761	12370
1.8	7365	6697	10259
2.0**)	6636**)	5316**)	8680**)
2.2	6053	4357	7480
2.4	5580	3670	6554
2.6	5191	3163	5829
2.8	4868	2780	5253
3.0	4596	2484	4790

MSY = 20919 for F-factor X = 0.343   Biomass at MSY = 163296
MSE = 51544 for F-factor X = 0.301   Biomass at MSE = 188207

*) cf. Table 8.7.2
**) cf. Table 8.7.3

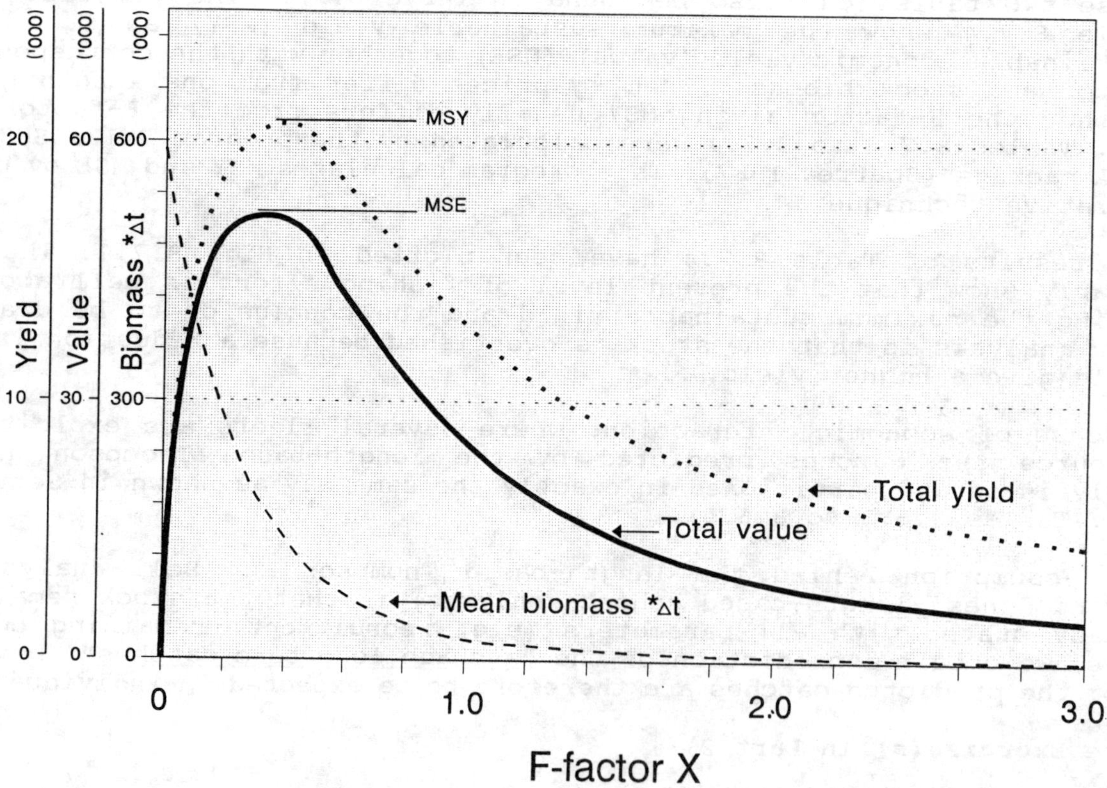

Fig. 8.7.1 Graphic presentation of the results of the length-based Thompson and Bell analysis, hake, Senegal (cf. Table 8.7.4)

8.8 PREDICTION OF THE EFFECTS OF CHANGES OF MESH SIZES USING THE THOMPSON AND BELL METHOD

The regulation of mesh sizes is an important management tool for many fisheries. It is, therefore, important to be in a position to predict the result of a change of mesh size. Since a change of mesh size will cause a change in the fishing pattern, the array of F-values, we can use the formulas presented in Sections 6.6.1 and 6.6.2 to come to a prediction, in other words use the "current" situation to predict a "new" situation.

We may express the current fishing mortality by the age-based or the length-based model (cf. Eq. 6.6.1):

$$F_t \text{current} = Fm * S_t \text{current} \qquad (8.8.1)$$

and

$$F_L \text{current} = Fm * S_L \text{current} \qquad (8.8.2)$$

where Fm is the maximum fishing mortality and $S_t$current or $S_L$current the selection curve for the current gear, for example, if the gear has the trawl type of selection ogive:

$$S_t \text{current} = 1/[1 + \exp(T1 - T2*t)] \qquad (8.8.3)$$

and

$$S_L \text{current} = 1/[1 + \exp(S1 - S2*L)] \qquad (8.8.4)$$

The parameters T1 and T2 are defined by Eqs. 6.4.3.3 and 6.4.3.4 respectively, while S1 and S2 are defined by Eqs. 6.1.6 and 6.1.7 respectively.

The parameters t50% and t75% are the ages at which 50% and 75% of the fish are retained by the gear, respectively. Usually we know the lengths L50% and L75% which correspond to t50% and t75%.

With the known parameters L50% and L75% for the gear currently in use we are in a position to calculate a new age-based or length-based selection curve for new values of L50% and L75% (or t50% and t75%). From the new selection ogive, and the Fm of the current fishery, we can calculate a new array of fishing mortalities, using Eq. 6.6.1.1:

$$F_t \text{new} = Fm * S_t \text{new} \qquad (8.8.5)$$

and

$$F_L \text{new} = Fm * S_L \text{new} \qquad (8.8.6)$$

The new F's are then used as inputs to the Thompson and Bell model, and the results for the alternative F-patterns, F(current) and F(new), can be compared (Hoydal et al., 1980 and 1982). This method is a generalization of the methods suggested by Gulland (1961), Jones (1961) and Kimura (1977).

Computer programs

The program "MIXFISH" in the LFSA package (Sparre, 1987) contains an option for mesh assessment corresponding to the procedure described above. It produces an output table showing the total yield for various combinations of effort and L50%, i.e. a table of the form:

		\multicolumn{5}{c}{Relative Effort}			
		- 20%	- 10%	no change	+ 10%	+ 20%
Relative value of L50%	- 30%					
	- 15%					
	no change			YIELD		
	+ 15%					
	+ 30%					

MIXFISH assumes L75% to be proportional to L50%. MIXFISH allows you to test any combination of L50% and effort, and thereby enables you to determine the optimum combination of L50% and effort. The middle cell (marked "YIELD") corresponds to the current fishing regime.

A similar program has also been incorporated in FiSAT.

9 ESTIMATION OF MAXIMUM SUSTAINABLE YIELD USING SURPLUS PRODUCTION MODELS

In contrast to Chapters 3 to 8, Chapters 9 and 13 do not deal with "analytical models", but with "holistic models" (cf. Fig. 1.8.1), wherein the stock is considered as one big unit of biomass and wherein no attempt is made to model on an age or length base. The "surplus production models" which are discussed in this chapter deal with the entire stock, the entire fishing effort and the total yield obtained from the stock, without entering into any details such as the growth and mortality parameters or the effect of the mesh size on the age of fish capture etc. Surplus production models were introduced by Graham (1935), but they are often referred to as "Schaefer-models" (see below).

The objective of the application of "surplus production models" is to determine the optimum level of effort, that is the effort that produces the maximum yield that can be sustained without affecting the long-term productivity of the stock, the so-called maximum sustainable yield (MSY). The theory behind the surplus production models has been reviewed by many authors, for example, Ricker (1975), Caddy (1980), Gulland (1983) and Pauly (1984).

Because holistic models are much simpler than analytical models, the data requirements are also less demanding. There is, for example, no need to determine cohorts and therefore no need for age determination. This is one of the main reasons for the relative popularity of surplus production models in tropical fish stock assessment. The surplus production models can be applied when reasonable estimates are available of the total yield (by species) and/or of the catch per unit of effort (CPUE) by species and/or the CPUE by species and the related fishing effort over a number of years. The fishing effort must have undergone substantial changes over the period covered.

9.1 THE SCHAEFER AND FOX MODELS

The maximum sustainable yield (MSY) can be estimated from the following input data:

$f(i)$ = effort in year i, i = 1,2,...,n

Y/f = yield (catch in <u>weight</u>) per unit of effort in year i.

Y/f may be derived from the yield, Y(i), of year i for the entire fishery and the corresponding effort, f(i), by

$$Y/f = Y(i)/f(i), \quad i = 1,2,...,n \quad (9.1.1)$$

or by direct observations on the basis of samples from the fishery.

The simplest way of expressing yield per unit of effort, Y/f, as a function of the effort, f, is the linear model suggested by Schaefer (1954):

$$Y(i)/f(i) = a + b*f(i) \quad \text{if } f(i) \leq -a/b \quad (9.1.2)$$

Eq. 9.1.2 is called the "<u>Schaefer model</u>".

The slope, b, must be negative if the catch per unit of effort, Y/f, decreases for increasing effort, f, (see Fig. 9.1.1). The intercept, a, is the Y/f value obtained just after the first boat fishes on the stock for the first time. The intercept therefore must be positive. Thus, -a/b is positive and Y/f is zero for f = -a/b. Since a negative value of catch per unit of effort Y/f is absurd, the model only applies to f-values lower than -a/b.

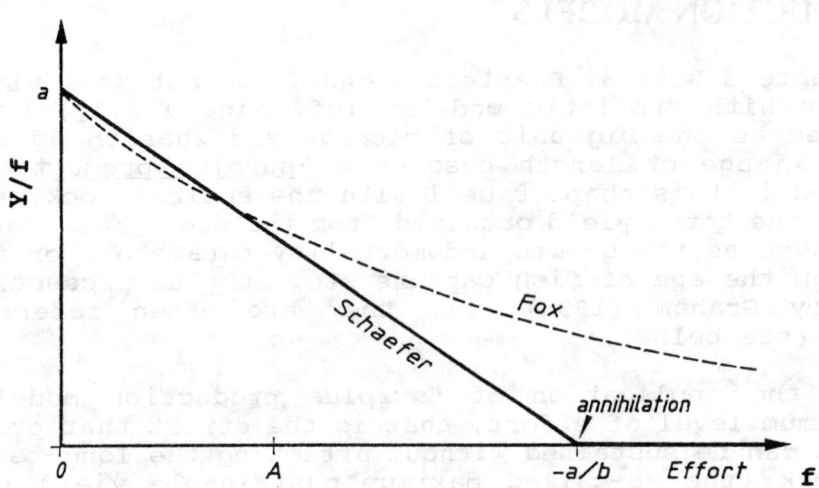

Fig. 9.1.1 Illustration of the different assumptions behind the Schaefer model and the Fox model

An alternative model was introduced by Fox (1970). It gives a curved line when Y/f is plotted directly on effort, f, (see Fig. 9.1.1), but a straight line when the logarithms of Y/f are plotted on effort:

$$\ln (Y(i)/f(i)) = c + d*f(i) \qquad (9.1.3)$$

Eq. 9.1.3 is called the "Fox model", which can also be written:

$$Y(i)/f(i) = \exp(c + d*f(i)) \qquad (9.1.4)$$

Both models conform to the assumption that Y/f declines as effort increases, but they differ in the sense that the Schaefer model implies one effort level for which Y/f equals zero, namely when f = -a/b whereas in the Fox model, Y/f is greater than zero for all values of f.

This can easily be seen in Fig. 9.1.1 where the plot of Y/f on f gives a straight line in case of the Schaefer model and a curved line, which approaches zero only at very high levels of effort, without ever reaching it (asymptotic) in the case of the Fox model.

In Section 8.3 it was demonstrated that CPUEw(t) = q*B(t) (Eq. 8.3.1). Since Y/f is also the catch per unit of effort in weight, we can write

$$Y(i)/f(i) = q*B = a + b*f(i) \quad \text{for the Schaefer model and}$$

$$Y(i)/f(i) = q*B = \exp(c + d*f(i)) \quad \text{for the Fox model}$$

where B is the biomass and q the catchability coefficient (a constant).

Fig. 9.1.1 illustrates another basic feature of the two models. For f close to zero Y/f takes the maximum value and so does the biomass because Y/f = q*B, and q is a constant. The biomass corresponding to f = 0 is called the "virgin stock biomass" or the "unexploited biomass", denoted by "Bv". Thus, replacing Y/f by q*Bv in Eqs. 9.1.2 and 9.1.4 gives:

$$q*Bv = a \qquad \text{or} \quad Bv = a/q \qquad \text{(Schaefer)}$$

$$q*Bv = \exp(c) \qquad \text{or} \quad Bv = \exp(c)/q \qquad \text{(Fox)}$$

The Bv for the two models must be the same. When increasing f from zero to level A (see Fig. 9.1.1) the two curves are approximately equal, but to the right of A the differences become larger. Thus, the choice between the two models becomes important only when relatively large values of f are reached. It cannot be proved that one of the two models is superior to the other. You may choose the one you believe is the most reasonable in each particular case or the one which gives the best fit to the data. However, the Beverton and Holt model is more in agreement with the Fox model, because they have a similar curvilinear relationship between catch per unit effort and effort (cf. Figs. 8.3.1 and 9.1.1) and between the mean biomass per recruit B/R and F (cf. Figs. 8.3.2 and 9.1.1).

Example 31: Schaefer and Fox models, demersal fish, Java Sea

Fig. 9.1.2 shows an example of plots of CPUE, $Y(i)/f(i)$, against effort, $f(i)$, and of the yield, $Y(i)$, against effort. Data are from the demersal fishery off the North coast of Java during the years 1969-1977 (from Dwiponggo, 1979). In this case yield, Y, refers to the annual catch measured in units of 1000 tonnes and Y/f refers to yield per thousand standard vessels each year. Effort is given in units of standard vessels per year.

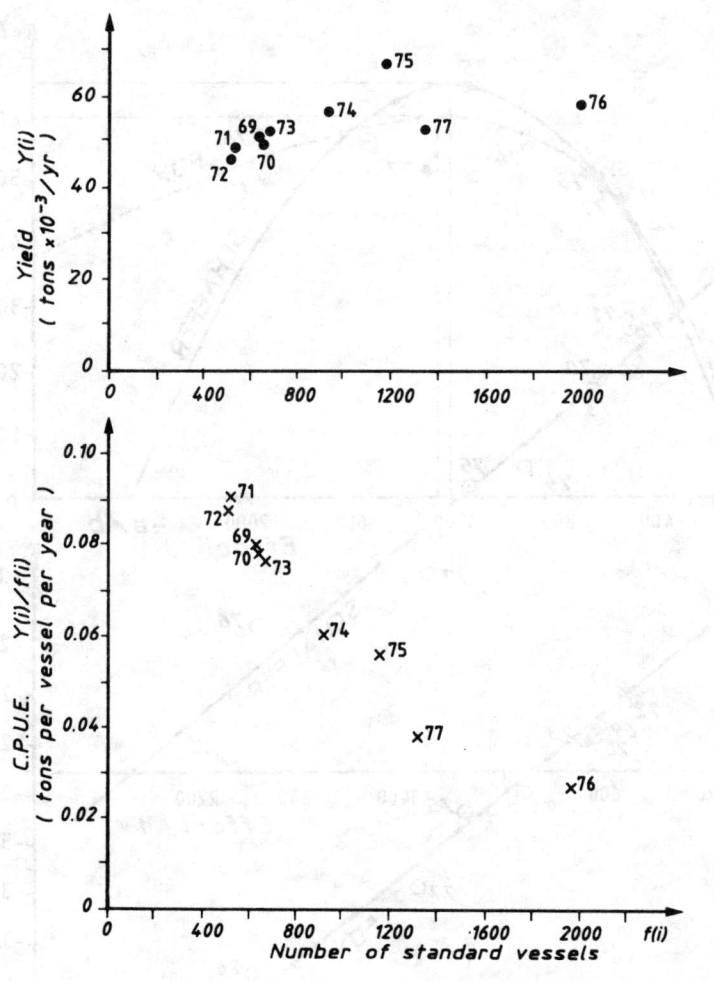

Fig. 9.1.2   Trends of yield, $Y(i)$, and catch per unit of effort, $Y(i)/f(i)$, off the North coast of Java (based on data from Table 9.1.1)

Because the fleet consisted of a number of different boat types, the effort of each boat category has been converted to a standard unit before summing to obtain total effort. Not surprisingly, the trend of Y/f shows a decline for increasing effort. Considering the stock biomass as a limited resource shared by the boats participating in the fishery, we expect a smaller share per boat the more boats enter the fishery.

There is no clear trend in the relationship between yield, Y, and effort, f, in this case. We will now show how the two models can be applied to this type of data.

The estimation procedures for the parameters (Schaefer: a and b, Fox: c and d) will be explained on the basis of the data given in Table 9.1.1 and Fig. 9.1.2. Since we are dealing with a straight line in the case of the Schaefer model and a curve which has been linearized by taking the logarithm in case of the Fox model, the determination of a,b and c,d requires two linear regressions, of $f(i)$ on $Y(i)/f(i)$ and $f(i)$ on $\ln(Y(i)/f(i))$ respectively. The results of the two regressions are presented in Table 9.1.1, including the standard deviations of the slopes and the intercepts. The lines are shown in Fig. 9.1.3. We have thus determined the relationships between catch per unit of effort and effort for both models.

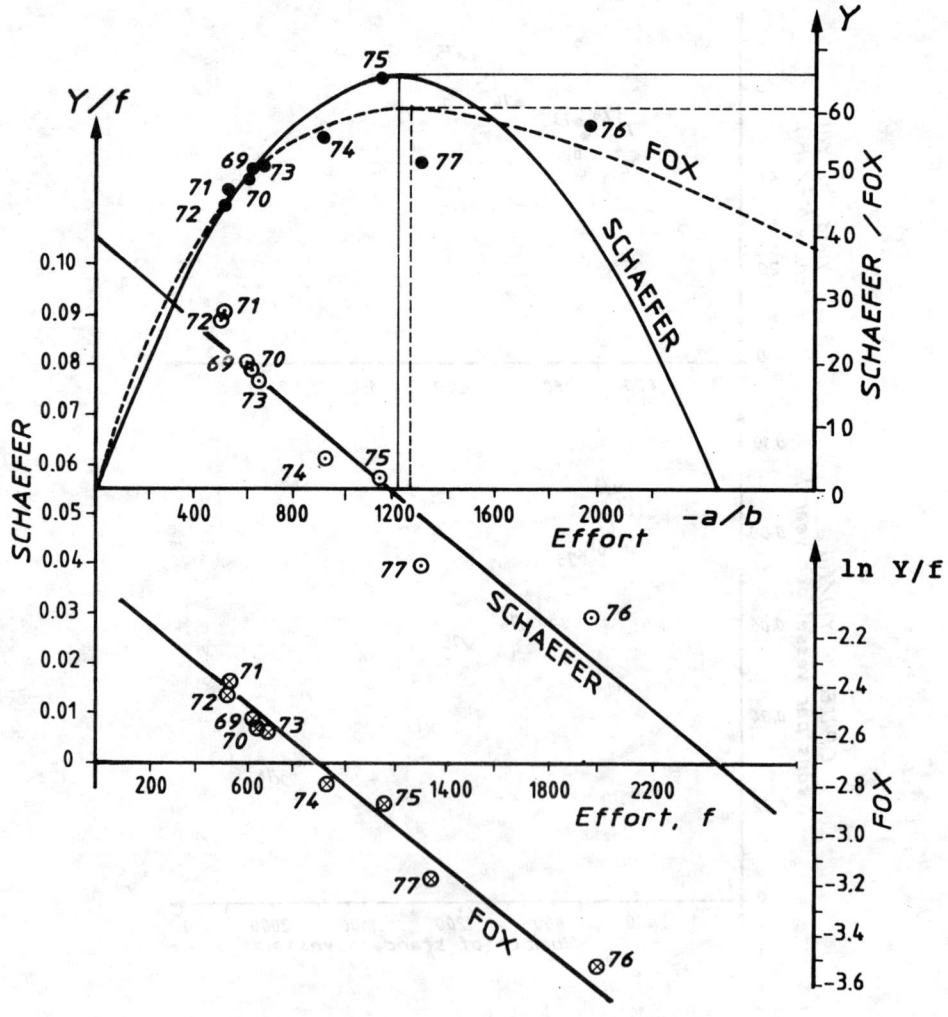

Fig. 9.1.3 **The Schaefer and the Fox models illustrated by the demersal fisheries off the North coast of Java (based on data from Table 9.1.1)**

The objective is, however, to obtain an estimate of the maximum sustainable yield (MSY) and to determine at which level of effort MSY has been or will be reached. To that purpose we have to rewrite Eqs. 9.1.2 and 9.1.4 expressing the yield as a function of effort, by multiplying both sides of the equation by f(i):

Schaefer: $Y(i) = a*f(i) + b*f(i)^2$ if $f(i) < -a/b$ (9.1.5)

or $Y(i) = 0$ if $f(i) = -a/b$

(Note, a value of f(i) > -a/b does not make sense since it would give a negative yield).

Fox: $Y(i) = f(i)*\exp[c + d*f(i)]$ (9.1.6)

Eq. 9.1.5, the Schaefer model, is a parabola (see Fig. 9.1.3), which has its maximum value of Y(i), the MSY level, at an effort level

$f_{MSY} = -0.5*a/b$ (9.1.7)

and the corresponding yield:

$MSY = -0.25*a^2/b$ (9.1.8)

Eq. 9.1.6, the Fox model, is an asymmetric curve with a maximum (the MSY level), with a fairly steep slope on the left side and a much more gradual decline on the right of the maximum (see Fig. 9.1.3).

The MSY and f_{MSY} for the Fox model can be calculated by formulas which are derived from Eq. 9.1.6 by differentiating Y with respect to f and solve dY/df = 0 for f,

$f_{MSY} = -1/d$ (9.1.9)

$MSY = -(1/d)*\exp(c-1)$ (9.1.10)

The results of the calculations of the example are presented at the bottom of Table 9.1.1.

From Table 9.1.1 and Fig. 9.1.3 it can be seen that the two models give slightly different results. According to the Fox model the MSY level is 60900 tons, at an effort level f_{MSY} of 1274 standard vessels, while according to the Schaefer model the MSY level is much higher (65800 tons) and at a lower f_{MSY} of 1235 standard vessels. According to both models the effort level surpassed f_{MSY} in 1976 and the yield was below MSY (see Fig. 9.1.2).

So far we have dealt mainly with the computational procedure, which was straight forward. One could now ask why we should be concerned with complicated models like the Beverton and Holt yield per recruit model when estimates of MSY can be obtained so easily from the surplus production models. One answer is that what we gain in simplicity with the surplus production models has the cost of having to make a number of assumptions about the dynamics of fish stocks, which may be (and nearly always are) impossible to justify. Some of these assumptions are discussed below. The reasoning given below is based on the Schaefer model, but it also applies to the Fox model.

(See **Exercise(s)** in Part 2).

Table 9.1.1 The calculation procedure for estimating the MSY and f_{MSY} by the Schaefer model and by the Fox model using catch and effort data from the trawl fishery off the North coast of Java (Dwiponggo, 1979) [1]

year	yield (thousand tonnes)	effort (no. of standard vessels)	SCHAEFER	FOX
i	Y(i)	f(i) x	Y(i)/f(i) y	ln(Y(i)/f(i)) y
1969	50	623	0.080	-2.523
1970	49	628	0.078	-2.551
1971	47.5	520	0.091	-2.393
1972	45	513	0.088	-2.434
1973	51	661	0.077	-2.562
1974	56	919	0.061	-2.798
1975	66	1158	0.057	-2.865
1976	58	1970	0.029	-3.525
1977	52	1317	0.039	-3.232
mean value		923.22	0.0667	-2.7648
standard deviation		485.14	0.02171	0.3873
intercept, a or c			0.1065	-2.0403
slope, b or d			-0.00004312	-0.0007848

	SCHAEFER	FOX
variance of slope $sb^2 = [(sy/sx)^2 - b^2]/(9-2)$	2.041×10^{-11}	3.087×10^{-9}
standard deviation of slope, sb	0.00000452	0.0000556
confidence limits of slope $b + t_{9-2} \cdot sb \quad t_7 = 2.37$ $b - t_{9-2} \cdot sb$	-0.000032 -0.000053	-0.00065 -0.00092
variance of intercept $sa^2 = sb^2 \cdot [sx^2 \cdot (n-1)/n + \bar{x}^2]$	0.00002167	0.003277
standard deviation of intercept, sa	0.0049	0.0572
confidence limits of intercept $a + t_{9-2} \cdot sa$ $a - t_{9-2} \cdot sa$	0.118 0.095	-1.90 -2.18
MSY Schaefer: $-0.25 \cdot a^2/b$ Fox : $-(1/d) \cdot \exp(c-1)$	65.8	60.9
f_{MSY} Schaefer: $-0.5 \cdot a/b$ Fox : $-1/d$	1235	1274

[1] In the formulas given in this table a and b should be replaced by c and d for the Fox-model

The assumption of an equilibrium situation

To explain the concept of an equilibrium situation we consider a situation where a virgin stock starts to be exploited in, say, 1971 (see Fig. 9.1.4), by, say, 1000 boats. Suppose the "Schaefer line" in Fig. 9.1.4 applies to this stock. According to the Schaefer model the yield in 1971 corresponding to 1000 boats should be x.

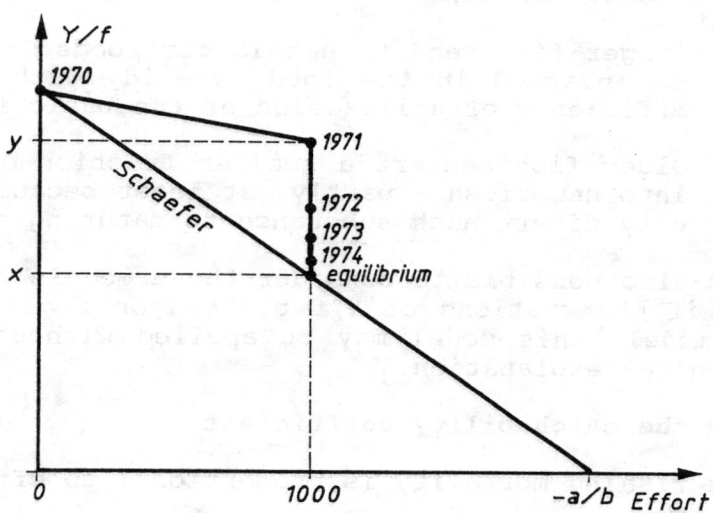

Fig. 9.1.4 Illustration of the concepts of an equilibrium situation and the transition period (for further explanations, see text)

However, it turned out to be y, i.e. a larger value than predicted by the model. This is because when fishing started in 1971 the biomass was still the virgin stock biomass, Bv, and only after a certain period of exploitation the biomass declines.

When fishing continued in 1972 the biomass was reduced due to the removal by fishing in 1971 and the 1972 catch therefore became smaller than that of 1971. Each year the resource is reduced, the reduction being smaller the longer time has elapsed since the introduction of the 1000 boats. Eventually, the system will stabilize at the Y/F-level x. We say that the system has reached an "equilibrium situation" after a "transition period".

For the equilibrium situation the production of biomass per time unit, equals the removal by fishing, the yield per time unit, plus the amount of fish dying of natural causes. This has also been illustrated in Fig. 9.2.1.

The "equilibrium situation" in the surplus production models is comparable to the "stabilized constant parameter system" in the Beverton and Holt models (cf. Section 8.1).

The biological assumptions

The biological reasoning behind the model was adequately formulated by Ricker (1975) as follows:

"1. Near maximum stock density, efficiency of reproduction is reduced, and often the actual number of recruits is less than at smaller densities. In the latter event, reducing the stock will increase recruitment.

2. When food supply is limited, food is less efficiently converted into fish flesh by a large stock than by a smaller one. Each fish of the larger stock gets less food individually; hence a larger fraction is used merely to maintain life, and a smaller fraction for growth.

3. An unfished stock tends to contain more older individuals, relatively, than a fished stock. This makes for decreased production, in at least two ways:

   (a) Larger fish tend to eat larger foods, so an extra step may be inserted in the food pyramid, with consequent loss of efficiency of utilization of the basic food production.

   (b) Older fish convert a smaller fraction of the food they eat into new flesh - partly, at least because mature fish annually divert much substance to maturing eggs and milt."

However, it is also possible to consider these models as purely empirical. For instance, if observations of Y/f plotted on f give a curve complying with the Fox model, this model may be applied without any concern for a possible biological explanation.

Assumptions on the catchability coefficient

We assume that fishing mortality is proportional to effort (Eq. 4.6.1):

$F = q*f$

This assumption in itself is not controversial (if f is a reasonable measure of effort). The problems come when f is measured in, for example, the number of boat days per year over a series of years. In most cases the efficiency of the boats has changed over a long period; often the boats have become larger and better equipped. Thus, 100 boat days in, say, 1978 may create a larger fishing mortality than 100 boat days in 1968. This means that q becomes a function of time or rather, a function of the technical development which is usually a function of time. It has proved very difficult to account for changes in q, caused by increased fishing efficiency and usually it is assumed that q remains constant. Therefore, one should be cautious not to include too long a time series of data in the surplus production analysis. Alternatively, changes of q must be taken into account. In some fisheries on small pelagic species in upwelling areas, for example, the anchoveta in Peru and Chile it occurs that the fish concentrate in small areas because of changes in the environmental conditions. In such cases there is no direct relationship between q, f and F, and surplus production models cannot be applied.

9.2 GULLAND'S FORMULA

In this section we consider the case of poorly investigated stocks. Time series of catch and effort data are not available, but some estimates of overall biomass and the natural mortality have been obtained.

Several empirical formulas have been developed with the objective of providing a first rough estimate of the MSY based on such scanty data. These formulas have found wide application after first estimates were obtained of the standing biomass after one or a series of exploratory bottom trawl surveys and/or acoustic surveys. The first formula was developed by Gulland (1971), a modification was proposed by Cadima (in Troadec (1977) and finally a set of formulas strictly based on the Schaefer and Fox surplus production models was developed by Garcia, Sparre and Csirke (1989).

EQUILIBRIUM CONDITIONS

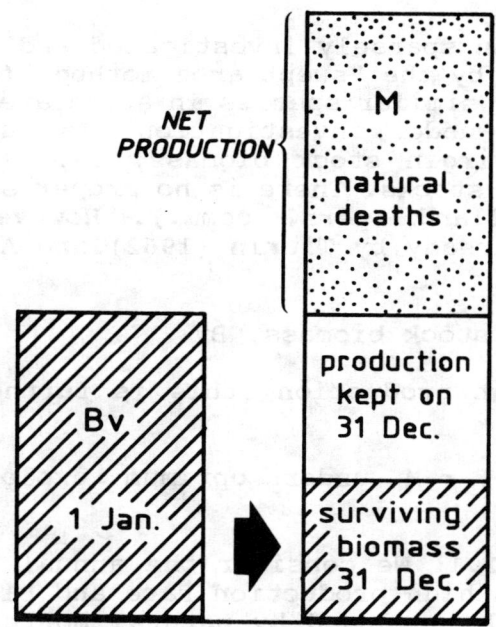

A1: Virgin stock (Bv)
High M
High net production

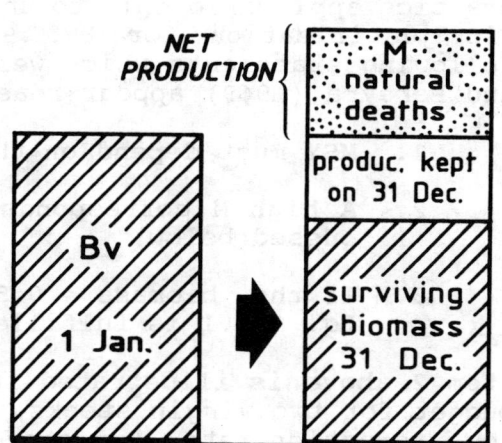

B1: Virgin stock (Bv)
Low M
Low net production

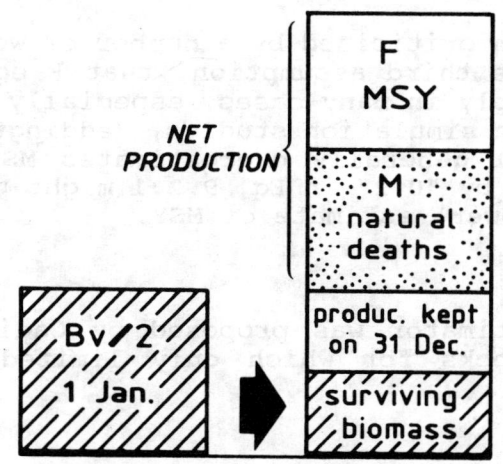

A2: Exploited stock (Bv/2)
High M
High net production

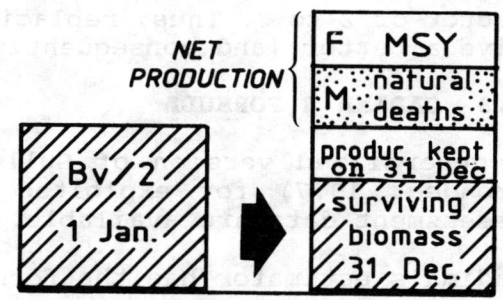

B2: Exploited stock (Bv/2)
Low M
Low net production

Fig. 9.2.1 Illustration of Gulland's formula MSY = 0.5*M*Bv
(for further explanation, see text)

Gulland (1971) suggested the following way of estimating maximum sustainable yield:

$$MSY = 0.5*M*Bv \qquad (9.2.1)$$

where Bv is the virgin stock biomass and M the natural mortality.

This formula has been used especially on sparsely investigated and lightly exploited stocks. Bv is often estimated by the "swept area method" (Chapter 13) and M is often a value estimated for similar species in a sea area which is believed to be similar to the one under investigation. As Gulland's formula requires an estimate of the virgin stock biomass, Bv, it is in practice applicable only to unexploited stocks. There is no proper scientific justification for Eq. 9.2.1 (Gulland, pers. comm.). However, the following statements which were made already by Tiurin (1962) and Alverson and Pereyra (1969) appear reasonable:

1. MSY must depend on the virgin stock biomass, Bv

2. A high M corresponds to a high production (this is further discussed below)

3. If the biomass = 0.5*Bv and F = M under optimum exploitation Eq. 9.2.1 is fulfilled.

Item 2 above is illustrated in Fig. 9.2.1. We consider the annual biomass budget for two virgin stocks, A1 with a high production rate and B1 with a low production rate. For stock A1 the loss caused by natural mortality is large and to maintain the stock the net production must be of equal size. For stock B1 the loss due to natural deaths is small and only a small net production is required to counter-balance this loss. The cases A2 and B2 deal with the same two stocks but after they have been for a while under optimum exploitation and produce the maximum sustainable yield. In A2 a large part of the natural deaths has been replaced by deaths due to fishing. This is also the case for B2, but the potential for replacing natural deaths by deaths due to fishing is smaller in case B2 than in case A2.

Although widely used, Eq. 9.2.1 has been criticized by a number of workers. Caddy and Csirke (1983) showed that the third assumption, that F equals M under optimum exploitation, does not apply in many cases, especially stocks of prey species (e.g. shrimps). Based on simulation studies, Beddington and Cooke (1983) concluded, that Eq. 9.2.1 generally overestimates MSY by a factor of 2 to 3. Thus, replacing "0.5" by "0.2" in Eq. 9.2.1 might perhaps give a better (and consequently much lower) estimate of MSY.

9.3 CADIMA'S FORMULA

A generalized version of Gulland's estimator was proposed by Cadima (in Troadec, 1977) for exploited fish stocks for which only limited stock assessment data are available.

Cadima's estimator has the form:

$$MSY = 0.5*Z*\bar{B} \qquad (9.3.1)$$

where $\bar{B}$ is the average (annual) biomass and Z the total mortality. Since $Z = F+M$ and $Y = F*\bar{B}$, Cadima suggested that in the absence of data on Z, Eq. 9.3.1 could be rewritten:

$$MSY = 0.5*(Y + M*\bar{B}) \qquad (9.3.2)$$

where Y is the total catch in a year and B is the average biomass in the same year.

As most stocks in the world are now already being exploited this equation is quite frequently used in developing and some developed fisheries, where catch and effort time series are not yet available, but where biomass estimates are occasionally obtained from, for instance, trawl- or acoustic surveys.

9.4 MSY ESTIMATORS BASED ON THE SURPLUS PRODUCTION MODEL

Based on the considerations in the foregoing section, Garcia, Sparre and Csirke (1989) suggested two alternative ways to estimate the potential yield of exploited fish stocks which have basically the same foundation and applications as the Gulland and Cadima estimators, but which are consistent with the underlying models. The two estimators have been derived from the Schaefer model and the Fox model.

Both methods assume that the observations:

$\bar{B}$ (average biomass)   and   Y (current yield)

are available for one year only. They also assume that natural mortality, M, is known and that there is a relationship between M and f_{MSY} of the form:

$$f_{MSY} = k*M \qquad (9.4.0.1)$$

where k is a constant.

As $f = Y/\bar{B}$ and $Y/f = \bar{B}$, we can write the surplus production models Eqs. 9.1.2 and 9.1.3 in the form:

Schaefer: $\quad \bar{B} = a + b*(Y/\bar{B}) \qquad (9.4.0.2)$

Fox: $\quad \ln\bar{B} = c + d*(Y/\bar{B}) \qquad (9.4.0.3)$

Suppose observations B1 and Y1 are available and we have a "guesstimate" of M, then combined with the assumption of Eq. 9.4.0.1 we get:

Schaefer: $\quad B1 = a + b*(Y1/B1)$ and $f_{MSY} = k*M = -a/2b \qquad (9.4.0.4)$

Fox: $\quad \ln B1 = c + d*(Y1/B1)$ and $f_{MSY} = k*M = -1/d \qquad (9.4.0.5)$

These equations can be solved for a and b, c and d in the Schaefer and the Fox model, respectively:

Schaefer: $\quad a = \dfrac{2*f_{MSY}*B1^2}{2*f_{MSY}*B1 - Y1} \qquad b = \dfrac{B1^2}{2*f_{MSY}*B1 - Y1} \qquad (9.4.0.6)$

Fox: $\quad c = \ln(B1) + Y1/(B1*f_{MSY}) \qquad d = -1/f_{MSY} \qquad (9.4.0.7)$

Once we have (a,b) or (c,d) we can estimate the MSY by Eqs. 9.1.8 and 9.1.10:

Schaefer: $\quad MSY = -0.25*a^2/b$

Fox: $\quad MSY = -(1/d)*\exp(c-1)$

and draw the yield curves (cf. Eqs. 9.1.5 and 9.1.6).

The MSY expression corresponding to the Schaefer model is found by inserting Eq. 9.4.0.6 into Eq. 9.1.8:

$$MSY = \frac{f^2_{MSY} * \overline{B}^2}{2 * f_{MSY} * \overline{B} - Y} \qquad (9.4.0.8)$$

The yield curve is determined by a and b (Eq. 9.4.0.6). When f_{MSY} is not known (as is most often the case) it may be replaced by k*M. In the special case where k = 1 and f_{MSY} = M we get:

$$MSY = \frac{M^2 * \overline{B}^2}{2 * M * \overline{B} - Y} \qquad (9.4.0.9)$$

If the stock is unfished (i.e. when f = 0, Y = 0 and B = Bv) Eq. 9.4.0.9 becomes Gulland's original formula (Eq. 9.2.1).

If the stock in question responds better to the Fox production model (Eq. 9.1.10), we get the expression:

$$MSY = f_{MSY} * \overline{B} * \exp[Y/(f_{MSY} * \overline{B}) - 1] \qquad (9.4.0.10)$$

The yield curve is determined by c and d (Eq. 9.4.0.7). In the special case where k = 1 and f_{MSY} = M, Eq. 9.4.0.10 becomes:

$$MSY = M * \overline{B} * \exp[Y/(M * \overline{B}) - 1] \qquad (9.4.0.11)$$

When Y = 0, the estimate of MSY comparable to Gulland's estimator becomes

$$MSY = M * \overline{B} * \exp(-1) = 0.37 * M * \overline{B} \qquad (9.4.0.12)$$

Thus, one pair of observations (B1,Y1) and assumptions on M and the relationship between M and f_{MSY} (f_{MSY} = k*M) are sufficient information to get a <u>first rough estimate</u> of the yield curve (Schaefer: a,b or Fox: c,d) from which a <u>first rough estimate</u> of MSY can be derived.

The interested reader is referred to the original paper (Garcia, Sparre and Csirke, 1989) for further details.

9.4.1 Validation of estimates of MSY based on empirical formulas

When working with mathematical models the fishery scientist should check whether the basic assumptions of the models are fulfilled. This applies in particular to the above-mentioned empirical formulas (Garcia, Sparre and Csirke, 1989). For example, the biomass (B) is meant to be the exploited average biomass, and both the catch and the biomass referred to should be comparable and have the same age (or size) structure. For instance, the biomass figure should not include small sizes which are not available to the fishery. This biomass is the annual biomass value and seasonal oscillations caused by change in growth, mortality or recruitment, which are likely to be more important in short-lived species such as shrimps, squids and anchovies, should be taken into account and as far as possible be levelled off to obtain an appropriate annual average of the total biomass.

If seasonal oscillations of biomass are caused by migrations, then the peak biomass representing the real size of the stock should be used. If another country is exploiting the same stock at another season of the year, in another area, the catch of that country should be included in the calculations and the MSY estimate should refer to the whole unit stock. The estimated potential yield could also be validated by comparison with other similar stocks for which better information might be available.

Some of the questions to be asked are: How does an estimate of the density, expressed as MSY/km^2 stand with respect to similar estimates for other stocks of the same species in ecologically similar areas exploited under similar fishing regimes? Does the size structure of the catch provide support for assessment implying that the stock is heavily over-fished (e.g. predominance of juveniles) or under-fished (e.g., predominance of large, old fish)?

It should be noted that a closer look at the length-frequency composition of the catch would give some guidance on relative levels of exploitation and this should be among the first data to be collected in any developing fishery in order to allow some estimation of the total mortality rate for cross-checking with other methods (cf. Chapters 3 and 4).

If the stock in question has been exploited for some time it is likely that a time series of catch data is available, which should also be examined. Even if no detailed effort data are available, the indication that after a period of sustained increase, the total catch has been stable for some time may mean that the MSY has been reached at least for the present regime of exploitation, while if the catch has dropped from a previous high level it may mean that the stock is over-fished and an average of the highest catches experienced in the past may provide an independent approximation to the MSY. In interpreting catch time series as suggested above, one assumes that such variations in catches are caused by changes in fishing effort and not by environmental or socio-economic changes.

9.5 MUNRO AND THOMPSON PLOT

The surplus production models are usually applied to time series of CPUE and effort. Munro and Thompson (1983 and 1983a), however, applied the surplus production model to a set of data from the Jamaican coral reef fishery all collected in the same year but representing different fishing grounds fished at different levels of effort. Fig. 9.5.1 shows a schematic map of Jamaica divided into 9 areas, which except for area B coincide with the parishes of Jamaica. The fishery studied by Munro and Thompson (1983) is a local trap fishery operated from canoes. Coral reef fish are not considered to be very mobile and it was assumed that each area (Fig. 9.5.1) has its own stocks which are independent of the neighbouring stocks (little mixing). The basic assumption is that the "ecological regimes" in the sea areas opposite the various parishes do not differ substantially around the island. Based on that assumption it makes sense to further assume that the relation between the yield and fishing effort in the different areas will follow the same model.

Table 9.5.1 shows the CPUE and effort data collected in the various parishes (Fig. 9.5.1) for the Jamaican shelf fishery on shelf-dwelling species in 1968.

Effort is expressed in units of canoes per km^2 per year to accommodate the assumption that each area has the same relative potentials, i.e. can support the same production per area unit. Thus, if exploited at the same rate (same effort per unit area per year) all areas should have the same yield per unit area per year ($kg/km^2/year$) (cf. column D of Table 9.5.1).

The yield per unit area was based on the area of the shelf and the proximal banks of each parish. For further details the reader is referred to the original papers (Munro and Thompson, 1983 and 1983a).

The relationship between CPUE and effort is here assumed to follow the Fox model (Eq. 9.1.3). Fig. 9.5.2 shows the plot of ln(CPUE) on effort as well as the yield per no. of canoes per km^2. Munro and Thompson had reasons to exclude area F from the regression analysis.

Table 9.5.1 Input data for Munro and Thompson Plot (for further details, see text). From Munro and Thompson (1983a)

parish (stock)	A effort canoes per km^2	B CPUE kg per canoe per year	C = lnB ln(CPUE) Fox model	D = A*B yield kg per km^2 per year
A	1.63	2367	7.769	3858
B	0.38	3279	8.095	1246
C	3.09	1407	7.249	4348
D	5.63	556	6.321	3130
E	4.43	974	6.881	4315
(F)	5.51	1306	7.175	7196
G	4.58	564	6.335	2583
H	4.20	767	6.642	3221
I	1.49	1875	7.536	2794

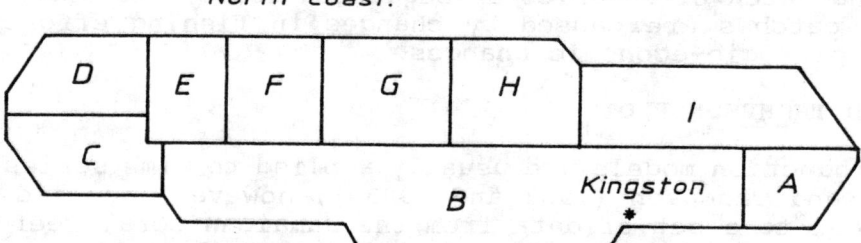

Fig. 9.5.1 Schematic map of Jamaica showing the parishes used for the "Munro and Thompson plot" (Fig. 9.5.2)

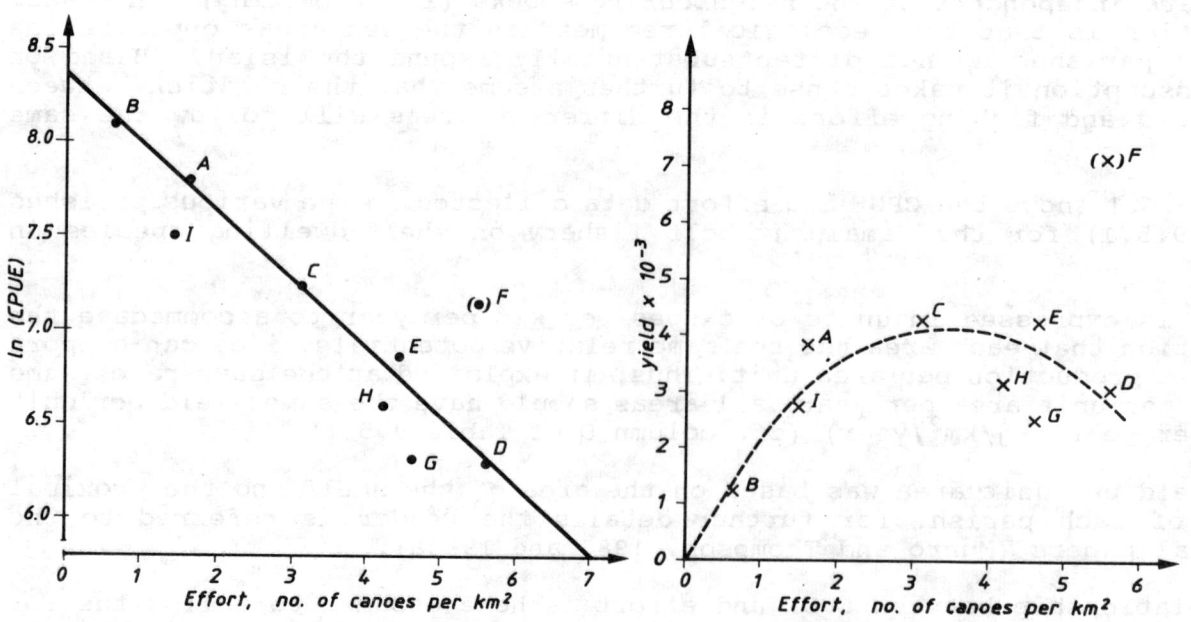

Fig. 9.5.2 Munro and Thompson plot. Based on data in Table 9.5.1 (from Munro and Thompson, 1983)

When applying a Munro and Thompson plot care should be taken to exclude all those fish which can move freely between areas, such as large pelagics.

The Munro and Thompson plot may be useful in situations where only limited data are available from certain parts of a region which have similar fisheries on coral reef stocks or other resources with a similar low mobility.

9.6 STANDARDIZATION OF EFFORT

In Section 4.3 it was suggested that effort is proportional to fishing mortality. This is true, of course, if we define effort as something proportional to fishing mortality, but such a definition has no practical application. In practice we will have to choose a measure for effort which we believe is related to fishing mortality or rather "fishing power". There are many possible choices. For a trawl fishery we may consider:

    number of trawlers
    number of trawler-days
    number of standard trawlers (taking into account the boat type)
    number of standard trawler-days
    ... etc.

For a handline fishery it may be more appropriate to consider the number of fisherman-days or the number of hooks used times the number of days. In this case it may be necessary to take into account that the fishermen on the same boat compete, so that effort is not a linear function of the number of fishermen.

In general, a measure which can be shown to be linearly related to the catch rate is a suitable measure. That is, if it can be shown that two units of effort catch twice as much as one unit of effort when operating under equal conditions the effort measure is a suitable one. For example, the number of fishing hours times engine horsepower may be a suitable measure of effort in some bottom trawl fisheries, whereas in a gill net fishery the boat type and the number of hours are likely to be less important than the number of gill nets set per day. In both of the above mentioned cases the number of fishermen may not be linearly related to the fishing power.

There are already considerable difficulties in defining suitable measures of effort for a single gear as discussed above, but when trying to define the effort for combinations of gears exploiting the same resources some rather really intricate problems are encountered. In tropical fisheries many different gears are used to capture the same resources, therefore several methods for standardization of effort units will be discussed here.

Relative effort

Before we start the discussion on standardization of effort units we notice that:

$$\frac{\text{Yield}}{\text{CPUE}} = \text{effort} \quad \text{as} \quad \text{CPUE} = \text{yield/effort}$$

In the following we shall be using this relation, not for yield and effort, but for quantities proportional to yield, effort and CPUE, so that the final result is a measure proportional to effort. Therefore we call it "<u>relative effort</u>". We also assume that all the effort units defined are suitable ones.

Table 9.6.1 Example (hypothetical) to illustrate summation of effort for different effort units

year y	1. PURSE SEINE Y1(y)	f1(y)	CPUE1(y) Y1/f1	R1(y) = rel. Y/f	2. BEACH SEINE Y2(y)	f2(y)	CPUE2(y) Y2/f2	R2(y) = rel. Y/f
1971	100	10	10.00	1.089	300	800	0.375	1.047
1972	200	21	9.52	1.037	250	669	0.374	1.043
1973	400	43	9.30	1.013	200	543	0.368	1.028
1974	700	81	8.64	0.941	150	430	0.349	0.974
1975	1200	142	8.45	0.920	100	307	0.326	0.909

$\overline{CPUE1}(y1,y2) = 9.18$ $\overline{CPUE2}(y1,y2) = 0.358$

year y	3. POLE AND LINE Y3(y)	f3(y)	CPUE3(y) Y3/f3	R3(y) = rel. Y/f	4. TROLLING Y4(y)	f4(y)	CPUE4(y) Y4/f4	R4(y) = rel. Y/f
1971	40	10000	0.00400	1.11	350	200000	0.00175	1.04
1972	80	20900	0.00383	1.06	344	200000	0.00172	1.02
1973	120	31100	0.00386	1.07	339	200000	0.00170	1.01
1974	80	23200	0.00345	0.96	333	200000	0.00167	0.99
1975	40	13700	0.00292	0.81	320	200000	0.00160	0.95

$\overline{CPUE3}(y1,y2) = 0.00361$ $\overline{CPUE4}(y1,y2) = 0.00169$

Yi(y) = yield of gear i in year y. i = 1,..,4. y = 1971,...,1975.
fi(y) = effort of gear i in year y
CPUEi(y) = Yi(y)/fi(y) = catch per unit of gear i effort in year y
Ri(y) = CPUEi(y)/$\overline{CPUEi}$(y1,y2) = relative CPUE

year	total Y sampled YS(y)	relative CPUE R(y)	total yield YT(y)	relative effort YT(y)/R(y)	normalized rel. effort E(y)
1971	790	1.0524	7900	7507	0.653
1972	874	1.0341	8740	8452	0.735
1973	1059	1.0213	10590	10369	0.902
1974	1263	0.9590	12630	13170	1.145
1975	1660	0.9224	16600	17995	1.565

Mean YT/R   11499

$YS(y) = \sum_{i=1}^{4} Yi(y)$ = sum of yields of gears for which effort is known (yield of sampled gears), per year

$R(y) = \sum_{i=1}^{4} [Ri(y)*Yi(y)/YS(y)]$ = sum of relative CPUE weighted by the yields in year y

YT(y) = total yield of all gears (including gears for which effort is not known)

YT(y)/R(y) = relative effort of year y

$E(y) = \frac{YT(y)/R(y)}{\text{mean YT/R}}$ = normalized relative effort of year y

Example 32: Summation of effort for different effort units

In the example given in Table 9.6.1 we consider the effort of four gears measured in numbers of units per year. As can be seen, the original effort units are not compatible. The yields corresponding to the various efforts are also available from a sampling scheme. The four sampled gears given in the Table are assumed to constitute only one tenth of the total catch of the stock in question. To make the different gear types (effort units) compatible each unit must be converted into CPUE, which then in turn is converted into "relative CPUE" as shown in the Table. The relative catch per unit of effort of gear i in year y is defined as follows:

$$R_i(y) = \frac{CPUE_i(y)}{\overline{CPUE_i}(y1,y2)} \tag{9.6.1}$$

where $\overline{CPUE_i}(y1,y2) = \frac{1}{y2-y1} * \sum_{j=y1}^{y2} CPUE_i(j)$

when a time period over the years y1,y1+1,...,y2 is considered. (In Table 9.6.1, y1 = 1971 and y2 = 1975).

As the relative CPUE of a gear has no dimension you may say that we have obtained compatible units of CPUE by the conversion into the relative CPUE. The relative CPUEs can be summed.

In the hypothetical case that all CPUE observations were proportional to the population size (cf. Section 4.3) the relative CPUEs would become identical for all gears.

However, in reality some gears are less important than others. The purse seine in Table 9.6.1 is the dominating one as far as development in yield is concerned, whereas the pole and line fishery is rather unimportant in terms of yield. This is accounted for by calculating the sum of the relative CPUEs weighted by the corresponding yields. For example for year 1971 (see Table 9.6.1):

$$R(1971) = \frac{100*1.089 + 300*1.047 + 40*1.11 + 350*1.04}{100 + 300 + 40 + 350} = 1.052$$

Dividing the total yield of the species under consideration, YT(y), including the yield not covered by the catch/effort sampling scheme, by the weighted sum of relative CPUEs gives a figure proportional to the total effort, R(y), shown in the column "YT(y)/R(y)" in Table 9.6.1. The last column contains the normalized relative effort, E(y). (This concept is introduced in order not to confound relative effort with absolute effort.)

The first gear in Table 9.6.1, the purse seine, is the important one, in the sense that the trend in the yield of this gear is the same as the trend for the normalized relative effort. The beach seine shows the opposite trend, but because it is a relatively unimportant gear it has less influence on the combined effort. If the purse seine catches had been only 10% of what they were the trend would be changed as the beach seine would then become a relatively more important gear. The E(y) values would then have the same trend as the yields of the beach seines (see Table 9.6.2).

The method described above has been used by the North Sea Round Fish Working Group of ICES (ICES, 1980). This method does not require a direct comparison of the different boat types. It requires only a type of data which is often available. The method can be questioned, and in fact, the results achieved by the ICES working group were not "overwhelmingly convincing" when the ICES working group correlated the normalized relative effort figures with the fishing mortalities obtained from VPA.

Table 9.6.2 Exploring the concept of normalized relative effort for several gears combined (cf. Table 9.6.1)

year	From Table 9.6.1 E(y)	If purse seine catches were only 10% of those given in Table 9.6.1 E(y)
1971	0.653	1.014
1972	0.734	1.018
1973	0.903	1.035
1974	1.146	0.984
1975	1.564	0.949

Relative fishing power

A more direct (and probably more dependable) method to standardize effort is the one suggested by Robson (1966) (discussed in Gulland, 1983). It does, however, require additional data. The method works with the concept of "<u>relative fishing power</u>". With the fishing power of vessel B relative to vessel A we mean:

$$PA(B) = \frac{\text{CPUE of vessel B}}{\text{CPUE of vessel A}} \qquad (9.6.2)$$

when the two boats are fishing under the same conditions (at the same time and in the same area). Vessel A is often called the "<u>standard vessel</u>".

Suppose the boats participating in a certain fishery can be divided into 5 homogeneous groups, so that each group consists of boats with similar fishing powers. Suppose also that the CPUE is in units of catch per time unit (e.g. catch per trawling hour), and further that the following data have been collected:

Boat type	A(standard)	B	C	D	E
Fishing power (PA)	1.0	PA(B)	PA(C)	PA(D)	PA(E)
Number of boats (N)	NA	NB	NC	ND	NE
Average number of fishing days per boat (d)	dA	dB	dC	dD	dE

The total effort would then be estimated by:

total effort =

$1.0*NA*dA + PA(B)*NB*dB + PA(C)*NC*dC + PA(D)*ND*dD + PA(E)*NE*dE$

(9.6.3)

In certain cases it can be assumed that the fishing power is proportional to some characteristics of the boat or gears which are relatively easy to obtain, such as GRT (tonnage) or HP (horsepower) or their product for trawlers and, for example, the number or length of nets for gill netters. As we are usually only interested in the relative effort, the PA's (fishing power) in Eq. 9.6.3 can simply be replaced by the boat/gear characteristics.

9.7 THE DERISO/SCHNUTE DELAY DIFFERENCE MODEL

A family of models which attempts to be a compromise between the surplus production models and the age-structured models has been presented by, for example, Deriso (1980), Ludwig and Walters (1985), Ludwig (1987) and Schnute (1985, 1987). These models, however, have limited use in tropical areas as they were developed for long-lived (slow growing) species, which are not exploited in the first part of their life. They are based on a series of rather strong assumptions which make them inapplicable to certain species. Schnute (1985) states about the models: "Among other things, they reflect the undeniable reality that the population consists of cohorts that get one year older each year". This clearly shows that these models are not intended for stocks where this "reality" can be denied, e.g. shrimps. Actually, the methods give nonsense results for such species (e.g. negative biomass).

The theoretical biological basis for these models does not go beyond the models already introduced (the basic reasoning is similar to that of Pope's cohort analysis), but the mathematical theory applied for the estimation procedure is somewhat more sophisticated than for most of the models presented in this manual. The main difference from the age-structured models given in this manual is that the Ford-Walford equation replaces the von Bertalanffy model for growth in weight (Eq. 3.1.2.1).

The models are highly sophisticated and not quite simple to use. Their description in non-mathematical language would require a very long chapter. The interested reader is therefore referred to the above-mentioned papers by Deriso, Schnute and others.

10 MULTISPECIES/MULTIFLEET PROBLEMS

So far, the models and methods described have dealt mostly with a single stock exploited by one fleet. However, this situation is the exception rather than the rule. In most cases a fleet exploits several stocks and several fleets compete in exploiting the same resources. In this connection we operate with three main types of interaction between components of a multispecies/multifleet system:

1. Biological interaction
2. Economic interaction
3. Technical interaction

Biological interaction is the interaction between fish stocks, and within fish stocks, caused by predation and food competition.

Economic interaction is the competition between fleets, e.g. between an industrial fishery and an artisanal fishery. The more one fleet catches of the limited resource the less will be left for its competitors.

Technical interaction means that the fishery on one stock creates fishing mortality on other stocks because the fishery is either a multispecies fishery or because of inevitable by-catches.

This chapter presents a brief discussion of some aspects of these three kinds of interaction. It is not intended to explain the models at a level which would enable the reader to apply them in practice. One good reason for not doing this is that many aspects of the models are still not well investigated and not well understood. Therefore, most stocks also in temperate waters are still assessed by means of single species models.

Various approaches to models taking interactions into account have been suggested during the last decade. The majority of the models are extensions of the single species/single fleet models presented in the foregoing chapters, so that the theory of the simple systems is a necessary background to multispecies/multifleet theory. Multispecies/multifleet assessment with special reference to tropical fish stocks is reviewed in e.g. FAO (1978), Pope (1979, 1980), Saila and Roedel (1980) and Pauly and Murphy (1982).

10.1 SURPLUS PRODUCTION MODELS APPLIED TO MULTISPECIES/MULTIFLEET SYSTEMS

The simplest way to deal with the multispecies/multifleet system is to apply the surplus production models (cf. Chapter 9) to the total catch of all species and the total effort by all fleets. Fig. 10.1.1 shows a plot of the total yield of the Gulf of Thailand trawl fishery against total trawl effort (from Pauly, 1984, after SCSP, 1978). Applying, for example, the Schaefer model (Eqs. 9.1.2 and 9.1.5) to the yield of all species caught by all fleets would give an estimate of the total MSY for the sea area in question. This approach, however, combines so many complex interactions in such a simple model that its general applicability can be questioned. For example, the surplus production model assumes that the curve is reversible which cannot be true if each stock follows the Schaefer model. This aspect is illustrated in Fig. 10.1.2 which shows a hypothetical fishery on three species. After effort level F1 (see figure) species A is eradicated and after level F2 species B has gone. Thus, going back in effort from, say, level F3 would, according to the Schaefer model, be along the curve for species C. Further, if we assume each stock to conform to the Schaefer model, the total yield of all three stocks may not be a parabola, but the descending part of the total yield curve will resemble an exponential decay curve, as illustrated in Fig. 10.1.2.

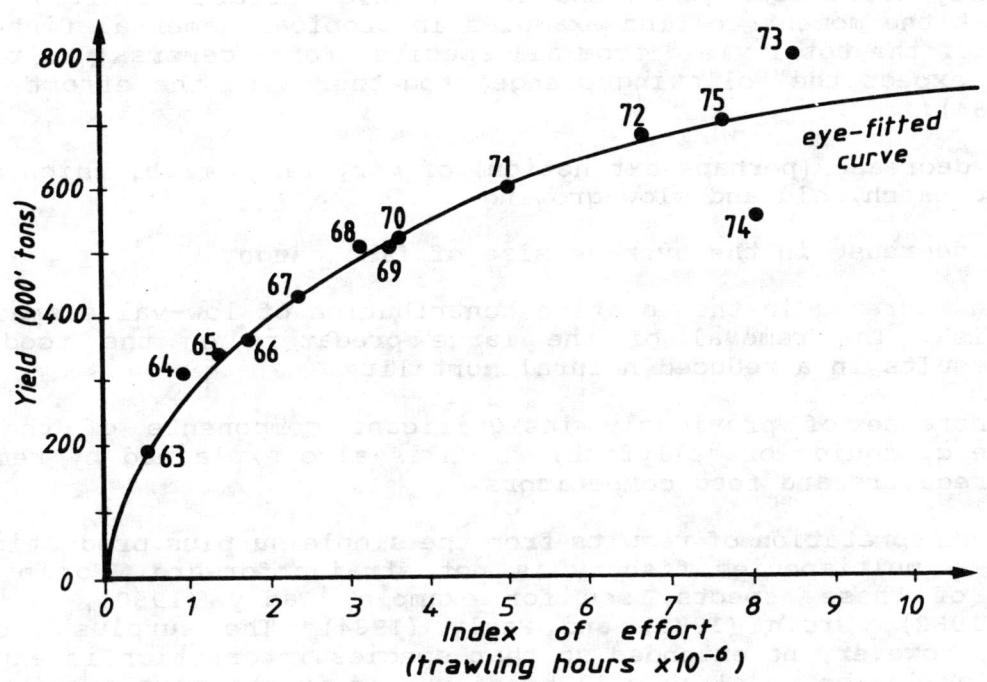

Fig. 10.1.1 Catch and effort from the Gulf of Thailand trawl fishery (from Pauly, 1984. Data derived from SCSP, 1978)

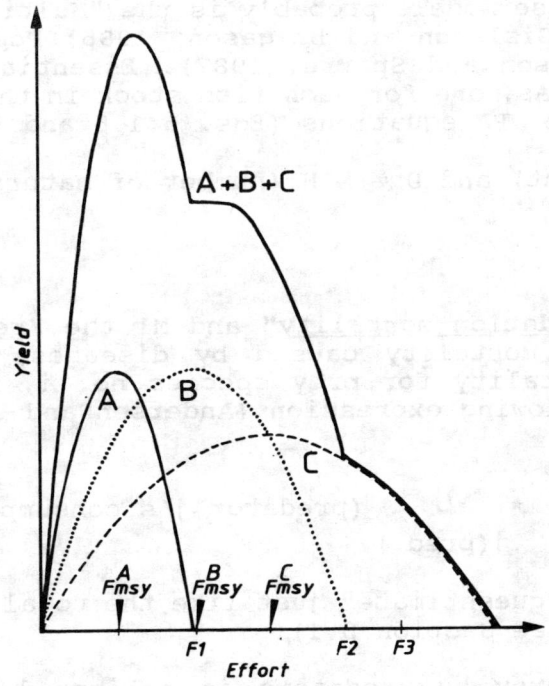

Fig. 10.1.2 Illustration of the changes of yields in a fishery on three species according to the simple Schaefer model

In practice, however, a picture is often observed, as in Fig. 10.1.1, in which a steady increase in effort does not produce a drop in total yield of all species combined. The curve appears to have a continued upwards trend. This probably would come to an end at some high effort level. Yet, it is difficult at the moment to find examples in tropical demersal fisheries of a collapse of the total yield from all species. For a demersal multispecies fishery we expect the following changes together with the effort increase (Pauly, 1984):

1. A decrease (perhaps extinction) of very large fish, which are easy to catch, old and slow-growing

2. A decrease in the average size of fish caught

3. An increase in the relative contribution of low-value small-sized fish. The removal of the large predators on the food fishes results in a reduced natural mortality

4. Increase of previously insignificant components of the system (e.g. squids or jellyfish) which is also explained by removal of predators and food competitors

Thus, the interpretation of results from the simple surplus production model applied to a multispecies fishery is not straightforward. For a further discussion of these aspects, see for example, Caddy (1980), Marten and Polovina (1982), Ursin (1982) and Pauly (1984). The surplus production models can, however, be extended so that species interaction is explicitly accounted for in the model as will be discussed in the next section.

10.2 BIOLOGICAL INTERACTION

Several authors have more or less successfully extended single species assessment models to cover biological interactions between species. This is done by introducing terms for mutual predation and food competition. The most promising of these models probably is the "Multispecies VPA" (Helgason and Gislason, 1979; Gislason and Helgason, 1985; Pope, 1979a; ICES, 1984, 1986, 1987 and Gislason and Sparre, 1987). Essentially, it consists of a number of parallel VPAs, one for each fish stock in the sea area considered. The second of the two VPA equations (Eqs. 5.1.5 and 5.1.6):

$C = F*\overline{N}$ (number caught) and $D = M*\overline{N}$ (number of natural deaths) is extended to:

$$D = (M1 + M2)*\overline{N}$$

where M2 is the "<u>predation mortality</u>" and M1 the "<u>residual natural mortality</u>" (i.e. natural mortality caused by diseases, starvation, old age, etc.). Predation mortality for prey species no. i, M2(i) is in principle derived from the following expression (Andersen and Ursin, 1977):

$$M2(i) = (1/\overline{N}(i)) * \sum_{j(pred.)} \text{(predator j's consumption of prey no. i)}$$

while M1 has to be a "guesstimate" just like the total natural mortality (M) in an ordinary VPA (see Section 5.1).

The consumption of prey by predators is estimated from data on stomach contents of predators (Sparre, 1980 and Gislason and Sparre, 1987). In addition to catch data, the multispecies VPA also requires stomach content data and data on food requirements (from feeding experiments) as input. The multispecies VPA was tested in ICES for the North Sea for the first time in 1984 (ICES, 1984; Sparre, 1984; Gislason and Sparre, 1987), but it has not yet been used as the basis for management of the fisheries.

Pope (1980a) developed a version of multispecies VPA based on length data. This method is the extension of Jones' length cohort analysis to multispecies length cohort analysis.

Pope (1979, 1980) also extended the surplus production model (the Schaefer model) to the multispecies case by introducing biological interaction parameters. In the case of two species Pope's model reads (cf. Eq. 9.1.2):

prey   : $Y_1/f_1 = a_1 + b_1*f_1 + c_1*f_2$

predator: $Y_2/f_2 = a_2 + b_2*f_2 - c_2*f_1$

If the interaction parameters, c_1 and c_2, have zero value we get two independent Schaefer models. When the interaction parameters are positive ($c_1 > 0$ and $c_2 > 0$) we can interpret the two-species model as a predator/prey system since the interaction term c_1*f_2 produces a higher yield of the prey species when effort on the predator is increased, i.e. when the predators are removed. The interaction term for the predator, c_2*f_1, has the opposite effect. If effort on the prey is increased the predator gets short of food and the stock is thus less productive, resulting in a reduction in yield from the predator stock.

10.3 ECONOMIC INTERACTION

The economic interaction of several fleets was already introduced with the age-based Thompson and Bell model (Section 8.6) to which the reader is referred. Economic interaction can also be described by the Beverton and Holt yield per recruit model (Beverton and Holt, 1957), but we shall not go into that here.

10.4 TECHNICAL INTERACTION

In fisheries the catch consists of a mixture of different species. Pauly (1984, p. 161) gives a table showing a typical trawl catch from the Java Sea. It contains over 55 species distributed over 29 different families. However, the ten most abundant species in that trawl haul constituted 70% of the catch. The most abundant single species constituted 32% of the total catch in weight. Disregarding the rare species, we often end up with, say, 5 to 15 species which are important from a commercial and/or ecological point of view.

10.4.1 A yield per recruit model for mixed fisheries

To illustrate the problem we consider a simple system of two species, A and B. A is a large slow-growing species and B is small fast-growing species. A has a low natural mortality and B a high one. The typical shapes of Y/R curves for species with the characteristics of A and B are shown in Fig. 10.4.1.1. If A is the target species of the fishery and B is the (inevitable) by-catch then a Y/R-curve for B has little practical applicability. Management measures have to be directed at the fishery for the target species, A. If there is a simple relationship between the fishing mortality of the two species, for example a linear one:

$F(A) = k*F(B)$

where $F(A)$ and $F(B)$ are the fishing mortalities on A and B respectively and k is a constant, the effects of various fishing strategies on A can easily be transformed to B.

Two Y/R-curves cannot be added since they are in different units ("per A recruit" is something different from "per B recruit").

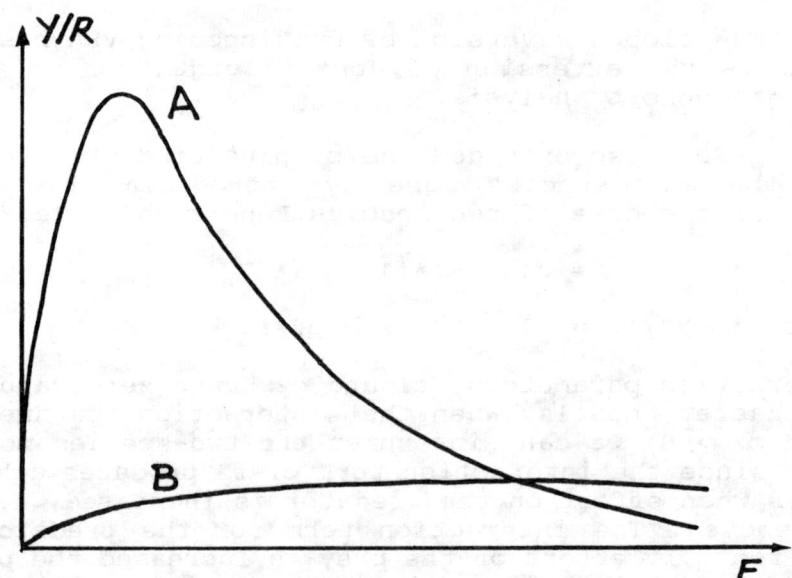

Fig. 10.4.1.1 A: Y/R-curve for a large slow-growing species with a low natural mortality

B: Y/R-curve for a small fast-growing species with a high natural mortality

If estimates of relative recruitment are available we can transform the Y/R curves into comparable units (Sparre, 1980; Murawski, 1984). For example, if the number of recruits of A (on an average) is 0.001 times that of B we can express the yield for A in "per recruit" and the yield for B in "per 1000 recruits", and we can then add the two curves. However, if the prices per kg of the two species are different it may make more sense from a management point of view to multiply by the price before adding. If A preys on B the situation becomes more complicated and simple adding of yield curves may lead to erroneous conclusions (Sparre, 1979). To handle this situation a multispecies VPA (Sparre, 1980) might be adequate.

Munro (1983) developed an alternative approach which is less data demanding than the multispecies VPA approach. His model probably represents the simplest way of extending the Beverton and Holt model to take into account species interaction. Murawski (1984) developed an extension of the Beverton and Holt Y/R analysis in which technical interaction in a mixed-species fishery is accounted for.

10.4.2 Assessment of mixed fisheries based on length-frequency data

Suppose a certain type of commercial gear catches a mixture of three major species, called A, B and C. One may think of a bottom trawl catching up to 50 different species of which three make up the bulk of the catch. It may well be so that none of these species can be considered the target species.

Thus, the catch consists of a mixture of species which is not determined by the fishing operation but by the availability of the fish. The species in question inhabit the same fishing grounds and are caught together. We assume length-frequency data from all three major species to be available, and also that the prices per kg differ between species and between size categories within species.

In this case one cannot treat each species separately and subsequently sum the results in terms of yield. Before a summation makes sense the yield must be converted into units of value. A fishery may for example catch shrimps as the target species and squid as a by-catch. From the point of view of the fishermen the catch of shrimps is far more important than the catch of squid. Moreover, even if yield is converted into value it is still not possible to sum the results of single species assessments such as the length-based Thompson and Bell analyses. It will usually be so that the effort level which for species A gives the maximum sustainable economic yield, MSE, will not be the MSE level for species B and C.

The approach suggested below combines all species in the estimation of MSE. This assessment of a mixed fishery based on length-frequency data works as follows:

Step 1: Do a single species length-based cohort analysis on each species separately. This gives estimates of the current fishing pattern for each species, e.g. if species A ranges in length from 5 to 25 cm the F-pattern can be denoted:

$$FA(5,6), FA(6,7), FA(7,8), \ldots, FA(25,\infty).$$

Thus, species A is measured in 1 cm classes. If species B ranges from 15 to 100 cm and is measured in 5 cm classes the fishing pattern estimated from cohort analysis may be denoted:

$$FB(15,20), FB(20,25), FB(25,30), \ldots, FB(100,\infty)$$

and the results for species C may be

$$FC(10,12), FC(12,14), FC(14,16), \ldots, FC(50,\infty).$$

For each species we further get the numbers in the first length group (the recruitment number) from length-based cohort analysis.

Step 2: Perform three separate length-based yield analyses of the Thompson and Bell type for the three species, as illustrated by Table 8.7.4. Use the same F-factor for each of the three fishing patterns in each prediction. Add up the values of the yields of the three species. The result of this exercise is a table similar to Table 8.7.4. This table contains the yield, biomass and value for each species together with the sum of values.

You may, for example, calculate 16 predictions as is done in Table 8.7.4, where each prediction is made as in Table 8.7.3 with the assumptions on fishing patterns listed below:

First prediction:

0,0,..0
0,0,..0
0,0,..0

Second prediction:

$0.2*FA(5,6), 0.2*FA(6,7), \ldots, 0.2*FA(25,\infty)$
$0.2*FB(15,20), 0.2*FB(20,25), \ldots, 0.2*FB(100,\infty)$
$0.2*FC(10,12), 0.2*FC(12,14), \ldots, 0.2*FC(50,\infty)$

Third prediction:

$0.4*FA(5,6), 0.4*FA(6,7), \ldots, 0.4*FA(25,\infty)$
$0.4*FB(15,20), 0.4*FB(20,25), \ldots, 0.4*FB(100,\infty)$
$0.4*FC(10,12), 0.4*FC(12,14), \ldots, 0.4*FC(50,\infty)$
....
....

Sixteenth prediction:

$3.0*FA(5,6), 3.0*FA(6,7), \ldots, 3.0*FA(25,\infty)$
$3.0*FB(15,20), 3.0*FB(20,25), \ldots, 3.0*FB(100,\infty)$
$3.0*FC(10,12), 3.0*FC(12,14), \ldots, 3.0*FC(50,\infty)$

Step 3: Use the sum of values to determine the optimum effort level. The assumption behind the method is that when you increase the fishing mortality on species A by, say, 20% the fishing mortality on species B and C will automatically be increased also by 20%.

This approach may, for example, be used to assess the combined effect of mesh size changes. If selection ogives are estimated for each species you may use the logistic model (cf. Section 6.1) to determine the fishing pattern. That is, the fishing patterns for the three species are determined by the 50% and 75% retention lengths and the fishing mortality Fm for size classes under full exploitation (cf. Fig. 6.1.1.4) multiplied by the F-factor, X. The effect of, say, a 20% increase in mesh size is estimated by calculating three new fishing patterns, using the parameters:

Species A: 1.2*L50%A, 1.2*L75%A, X*FmA
Species B: 1.2*L50%B, 1.2*L75%B, X*FmB
Species C: 1.2*L50%C, 1.2*L75%C, X*FmC

and then do the length-based Thompson and Bell analysis for a suitable range of X (F-factor) values, enabling you to determine the F-level for MSE.

Computer programs

The above suggested computational procedure involves a large number of calculations and it is recommended to use a computer. The LFSA-package contains the program "MIXFISH" which can perform this assessment of a mixed fishery. The program carries out the single species length-based Thompson and Bell analysis for each species as well as the combined assessment and it calculates the MSE in each case, using iteration techniques. A similar program has been incorporated in FiSAT.

10.4.3 Multifleet mixed fisheries

We discussed in Section 8.6 the case of two competing fleets (economic interaction) and in Section 10.4.2 the case where one type of boat caught several species (technical interaction or mixed fishery). Most fisheries have features of economic interaction and technical interaction as well as biological interaction (Section 10.2). In this section we shall ignore the biological interaction and consider only the combination of economic interaction and technical interaction.

Table 10.4.3.1 shows an example of input data for a multifleet mixed fishery. We are considering seven different pelagic species and four different gears (fleets). Here we assume that only these four gears (fleets) exploit the seven stocks considered. The basic data are numbers caught by size group (or age group):

$C(y,s,g,i)$ = number of length class (or age group) i fish of species s caught by gear (fleet) g during time period y

$s = 1,2,\ldots,7$ $g = a,b,c,d$
$i = 1,2,\ldots,n(s)$ $y = y1, y1+1, \ldots, y2$

Table 10.4.3.1 actually shows only a small fraction of the data. It is just one table in a time series, and each i-index symbolizes a whole length-frequency of $i = 1,2,\ldots,n(s)$ observations, where n(s) is the number of length classes in the length range of species s. The right-hand column of the table contains the total number caught:

$$C(y,s,i) = \sum_{g=a}^{d} C(y,s,g,i) = \text{number of length class (or age group) i fish of species s caught by all gears (fleets) during time period y}$$

Table 10.4.3.1 Example of catch input data for multifleet mixed fisheries for a single time period (e.g. a year). $C(s,g,i)$ = number of length class i fish of species s caught by gear g.
$s = 1,2,\ldots,7$; $g = a,b,c,d$; * = sum of all gears
$i = 1,2,\ldots,n(s)$;

s	g \ gear \ (fleet) \ species \	a gill net	b long lines	c purse seine	d pole and line	total catch VPA input
1	Yellowfin tuna	C(1,a,i)	C(1,b,i)	C(1,c,i)	C(1,d,i)	C(1,*,i)
2	Skipjack tuna	C(2,a,i)	C(2,b,i)	C(2,c,i)	C(2,d,i)	C(2,*,i)
3	Kawakawa	C(3,a,i)	C(3,b,i)	C(3,c,i)	C(3,d,i)	C(3,*,i)
4	Frigate tuna	C(4,a,i)	C(4,b,i)	C(4,c,i)	C(4,d,i)	C(4,*,i)
5	Seerfish	C(5,a,i)	C(5,b,i)	C(5,c,i)	C(5,d,i)	C(5,*,i)
6	Marlin	C(6,a,i)	C(6,b,i)	C(6,c,i)	C(6,d,i)	C(6,*,i)
7	Sail-fish	C(7,a,i)	C(7,b,i)	C(7,c,i)	C(7,d,i)	C(7,*,i)

Table 10.4.3.2 Output from the seven individual length-based cohort analyses

s	species	population numbers	fishing mortality
1	Yellowfin tuna	N(1)	F(1)
2	Skipjack tuna	N(2)	F(2)
3	Kawakawa	N(3)	F(3)
4	Frigate tuna	N(4)	F(4)
5	Seerfish	N(5)	F(5)
6	Marlin	N(6)	F(6)
7	Sail-fish	N(7)	F(7)

These numbers by length (or age) group and by time period form the input for individual cohort analyses or VPAs for each species.

Suppose in the following that we are working with length groups and have performed Jones' length-based cohort analyses. We have used as input the average number caught over a time period of several years and have estimated the average number in the population and the overall fishing mortalities for each species. The output of the seven analyses of the example can be summarized as shown in Table 10.4.3.2.

Each entry in Table 10.4.3.2 represents an array (a vector):

$$\underline{N(s)} = (N(s,1), N(s,2), \ldots, N(s,n(s)))$$

$$\underline{F(s)} = (F(s,1), F(s,2), \ldots, F(s,n(s)))$$

where each vector element corresponds to a length group. Each such F of a length group stems from the combined fishing mortality created by all four gears (fleets). The total F for each of the length groups can be redistributed on the four gears (fleets) using:

$$F(s,g,i) = F(s,i)*C(s,g,i)/C(s,i)$$

Table 10.4.3.3 **Fishing mortalities partitioned into gear (fleet) components estimated from cohort analysis and the numbers caught by each gear**

s	g (fleet) species	a gill net	b long lines	c purse seine	d pole and line	total fishing mort.
1	Yellowfin tuna	F(1,a)	F(1,b)	F(1,c)	F(1,d)	F(1,*)
2	Skipjack tuna	F(2,a)	F(2,b)	F(2,c)	F(2,d)	F(2,*)
3	Kawakawa	F(3,a)	F(3,b)	F(3,c)	F(3,d)	F(3,*)
4	Frigate tuna	F(4,a)	F(4,b)	F(4,c)	F(4,d)	F(4,*)
5	Seerfish	F(5,a)	F(5,b)	F(5,c)	F(5,d)	F(5,*)
6	Marlin	F(6,a)	F(6,b)	F(6,c)	F(6,d)	F(6,*)
7	Sail-fish	F(7,a)	F(7,b)	F(7,c)	F(7,d)	F(7,*)

Table 10.4.3.3 illustrates the total fishing mortality partitioned into fishing mortalities by gear (or by fleet). The type of data given in Table 10.4.3.3 together with the estimate of the stock numbers (Table 10.4.3.2) form the input for a "length-based mixed fisheries Thompson and Bell catch prediction". The first step is to make assumptions on the F-arrays for each gear (fleet), as discussed in Section 10.4.2. This could, for example, be the assumption of a ten percent increase of the gill net fishery, in which case all arrays in the gill net column should be multiplied by 1.1. Other assumptions could be made for the other gears.

The second step is to sum up the fishing mortalities for each species. The third step is to make a length-based Thompson and Bell catch prediction on each stock. The fourth step is to distribute the catches by length group between the fleets and convert the catches into values. Finally, the values of catches of different species are summed up for each fleet and for the total multifleet fishery.

Computer program

The microcomputer package BEAM 4 (Bio-Economic Analytical Model No. Four, Sparre and Willmann, 1992) consists of two sub-models. The biological/technical sub-model of BEAM 4 is the age-based multifleet mixed fisheries version of the Thompson and Bell model. Although BEAM 4 is based on age compositions, it contains options to convert input given as length compositions or commercial size categories into age composition data.

The BEAM 4 model also has options to account for migration of fish and fleets between geographical areas. The economic sub-model of BEAM 4 links the Thompson and Bell analysis to a cost and earning analysis of the harvesting sector and the processing sector of a fishery.

11 ASSESSMENT OF MIGRATORY STOCKS

Several of the methods presented so far are often insufficient when applied to migratory or schooling fish stocks. They were all based on the assumption that we can take representative random samples of the stock, for example, the whole range of length-frequencies. When stocks are not vulnerable to fishing, due to horizontal or vertical migrations to areas not covered by the fleet or normal fishing gears during parts of their life span, it is usually not possible to sample such stocks during these periods. This may simply lead to gaps in the samples for shorter or longer periods, but it is also possible that the samples taken represent different parts of the stock and in such cases it is likely that the data are misinterpreted. In particular, the length-based methods, such as modal progression analysis and catch curve analysis are difficult to apply in the case of migratory fish stocks.

The migrations of fish stocks which have been exposed to fisheries for centuries and to fisheries research for some 100 years (the herring in the North Sea, for example) are sufficiently well known to avoid bias in sampling and misinterpretation of results. The general knowledge of most tropical fish stocks, however, is often very limited and therefore sampling bias, incomplete coverage of the stock and misinterpretation of the data may easily occur.

In this chapter the problems related to migration will be illustrated. Some methods are given which may help in the interpretation of the length-frequency data obtained from migratory fish stocks. These methods should only be considered as first steps to solving complex problems.

11.1 THE CONCEPT AND STUDY OF MIGRATION

Harden Jones (1968) recognized three types of "migratory movements": drifting with the currents, random locomotory movements and oriented locomotory movements. In a later paper Harden Jones (1984) stated: "I use the word migration in the sense of coming and going with the seasons on a regular basis". Gerking (1953) defined the concept of "homing" as "the return to a place formerly occupied instead of going to other equally probable places". Homing is known to be an important feature of the migration of many fish stocks.

For fish stock assessment purposes, the explanation of why fish migrate is of little importance. The important thing is to know where the fish are at which time of the year.

In this manual we shall look at migration primarily as a source of bias. Migration is here defined as "<u>any systematic type of movement of individuals belonging to a stock</u>" (cf. the definition of bias given in Section 7.1). Random movements are not considered migrations in the present context because we are only interested in such types of migration that create bias for a length-frequency sampling programme. If a fish moves at random the movement will not change the relative probability of being sampled. The distinction between random movements and systematic movements is not obvious. The fish might encounter concentrations of food by moving randomly, and by random movement remain there until the food is nearly exhausted. If the movement of the food items is systematic (e.g., determined by current and other systematic oceanographic features) we would consider the above described feeding movement as a migration, although it was composed of random movements.

Migration is also characterized by being predictable, e.g., for some stocks we are able to predict at which time and where high concentrations can be found.

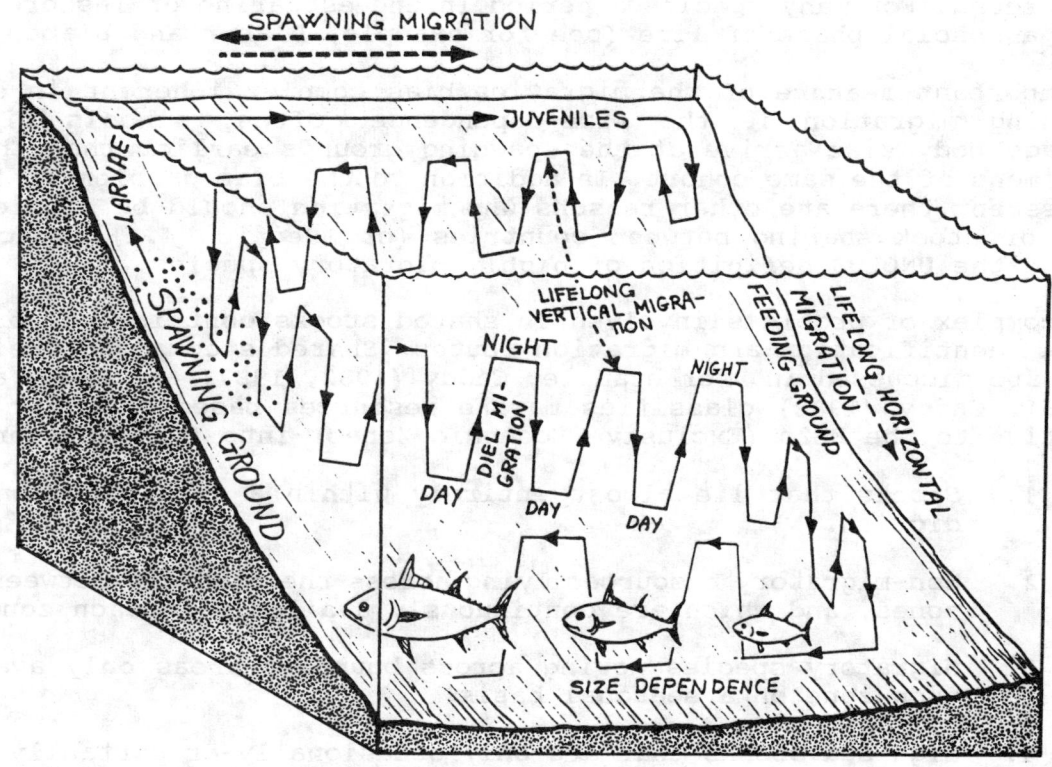

Fig. 11.1.1 Schematic illustration of five types of potential bias created by migrations

Most of the migrations causing bias are horizontal movements, along the coast, inshore/offshore or even between rivers/lagoons and marine areas. However, also relatively small vertical migrations and differences in distribution in the water column between different size groups (ages) of the same species may cause bias in our data.

Migrations creating bias may be classified into five main types, of which some are illustrated in Fig. 11.1.1.

1a. Daily vertical migration (e.g., at the bottom during day and in the water column during night).

1b. Daily horizontal migration (e.g., skipjack tuna has been observed to move away and return to a precise location each day (Yuen, 1970).

2. Spawning migration (i.e., annual return to spawning grounds followed by a movement to a feeding ground, or in the case of some cephalopods and penaeid shrimps with only one spawning during their life span, migration to the spawning grounds followed by extinction. Homing may be an important feature of this type of migration.

3. Size-dependent vertical migration of adults (e.g., skipjack tuna, the younger year classes occur in surface schools, whereas the older specimens move to greater depths).

4. Size-dependent horizontal migration (i.e., larger specimens move to deeper waters, while still undertaking the migrations mentioned above).

5. Migration of juveniles. Often juveniles undertake migrations different from those of the adults. The juveniles may remain in the upper layers day and night, whereas the adults are at the bottom during the day. They may also occupy special nursery grounds where the adults are not found. For many species a period in the estuarine or inshore waters is a special phase of life (see for example, Blaber and Blaber, 1980).

An important feature of the migration/bias complex inherent in especially spawning migration is the size-dependence. Often, schools with larger average body size arrive at the spawning grounds earlier than the smaller specimens of the same cohort. In addition to the bias problem in fish stock assessment there are other reasons why migration should be studied. One is that of stock sharing between countries (cf. Section 4.1), which is also behind the UNCLOS definition of highly migratory species.

The complex of problems involved in shared stocks contains the problems of stock identification and migration routes. Shared stocks and their management are discussed in, for example, Caddy (1982, 1987) and Caddy and Garcia (1986). Caddy (1982) classifies marine resources based on their movements relative to the EEZs (Exclusive Economic Zones) into five categories:

1. Stocks that lie almost entirely within a single national jurisdiction.

2. Non-migratory resources lying across the boundary between adjacent zones, and which are continuously available in each zone.

3. Migratory species moving across boundary areas only available in each zone on a seasonal basis.

4. High sea stocks that are only occasionally or partially available inside national zones.

5. High sea stocks which occur exclusively outside EEZs.

Especially for categories 3 and 4 the stock identification is a prerequisite for intelligent management of shared stocks. Stock identification and stock assessment has to be based on knowledge of the migration routes of the stock.

There are several ways to acquire knowledge about possible migrations of the species under study. An obvious and cheap way is to tap the knowledge of the fishermen, who surely have noticed seasonal or daily variations in the availability of the various target species.

Such fluctuations may also be reflected in the records of landings of various types of fishing boats (statistics) and a general knowledge of the movements of the fleets. Common echosounders are very useful for the detection of vertical migrations, while more sophisticated acoustic equipment can be used to map the distribution and estimate the abundance, in relatively short time, in particular of small pélagic fish.

The classical way to study movements (and growth) of fish and invertebrates is a tagging programme. Identification tags are attached externally or placed in the body cavity, the fish is measured and released at a known spot, and a reward is given for any tag returned with information on the date and place of capture. Such tagging programmes, if successful, may provide a lot of useful information on net displacements between the point of release and the point of recapture. They do not provide information on what has happened in between those moments and points.

Sophisticated acoustic and radio tags have been developed which allow the continuous observation of the movements of single fish as followed from a research vessel. The latest developments include tags that are released at a pre-set moment, pop to the surface and transmit to a satellite (applied in tuna research) and tags that record the compass bearing and tilt angle of an individual fish.

In Section 11.5 below some aspects of ordinary tagging programmes are discussed.

11.2 BIAS CAUSED BY MIGRATION

To illustrate the problem of bias caused by migration, consider a simplified hypothetical fish stock as illustrated in Fig. 11.2.1. This hypothetical fish stock spends one half of the year on the fishing grounds and the other half in an area where it is not exploited and which is also its nursery ground. The stock is composed of two components A and B (you may think of A as the fish recruited before the monsoon and B as the fish recruited after the monsoon). The two components undertake the same migrations but at different times of the year, as shown in Fig. 11.2.1. Suppose the species has a life span of only two years and suppose for simplicity that the migrations take place on the 1st of January and the 1st of July. Samples are available only from the fishing grounds.

Assume further that we do not know the migration pattern, but erroneously believe that the entire stock is on the fishing grounds all year round. In this case we are in trouble when sampling data for estimation of growth parameters because the important data for the first half year of life are not available. We may wrongly explain the absence of small fish to be caused by gear selection. We may also observe an apparent negative growth during the period from June to August if we believe that the fish on the fishing grounds in June are the same as those in August. In June component A consists of 1 and 2 year old A-fish, which in July are replaced by 0.5 and 1.5 year old B-fish.

In this (not very realistic) example one would probably not need to sample the fishery for any long period before one guessed the reason for the missing data for small fish and the apparent negative growth. In reality pelagic fish stocks show a much more complex migratory behaviour and the risk exists that we wrongly interpret phenomena caused by migration as something else. One obvious example is to misinterpret migration as mortality. If, for example, the fish at a certain body size moves to deeper waters where it cannot be caught by the fishing gears it will appear to us as if the fish had died, because the larger fish do not occur in the samples.

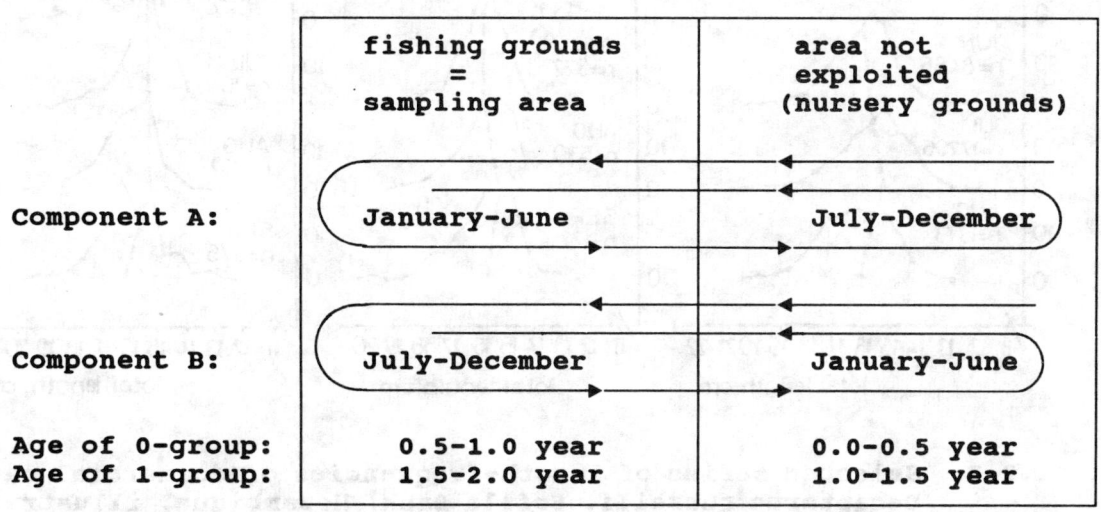

Fig. 11.2.1 Hypothetical simplified system to illustrate the main problem in obtaining random samples from a migratory stock

The usual situation is that on a given date on given fishing grounds samples can be taken only from a certain fraction of the stock. Also the time of the day and the gear used may restrict the fraction of the stock available for sampling. This section is partly based on the work by Sousa (1988) who discusses sources of bias when assessing stocks of small pelagics, with the scad Decapterus russelli in Mozambique waters as an example. The current knowledge of migration patterns of small pelagics in Mozambique waters is limited, and tagging experiments on small pelagics have not been conducted.

Severe difficulties, such as apparent negative growth were encountered when trying modal progression analysis on length-frequency samples. Fig. 11.2.2 shows an example of such a problematic time series of length-frequencies.

In the absence of exact data on migration routes, Sousa (1988) searched for plausible explanations of some of the apparent inconsistencies in the time series of length-frequencies. The first step was to suggest hypotheses for the migration pattern and the next, to test the hypotheses on the available data.

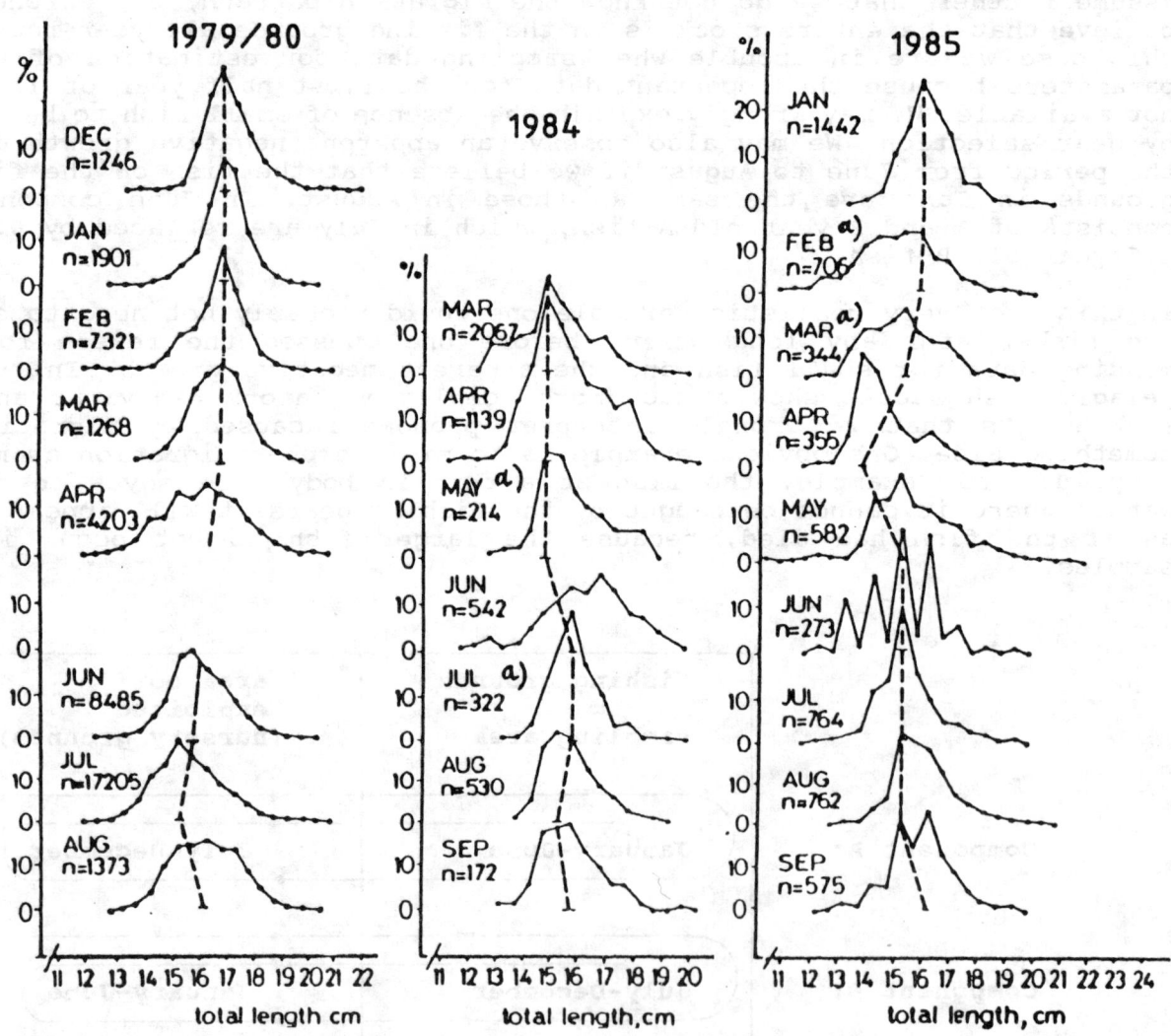

Fig. 11.2.2 Selected series of length-frequencies of commercial catches of Decapterus russelli, Sofala Bank, Mozambique, illustrating the problems with the interpretation of modal progression (from Sousa, 1988)

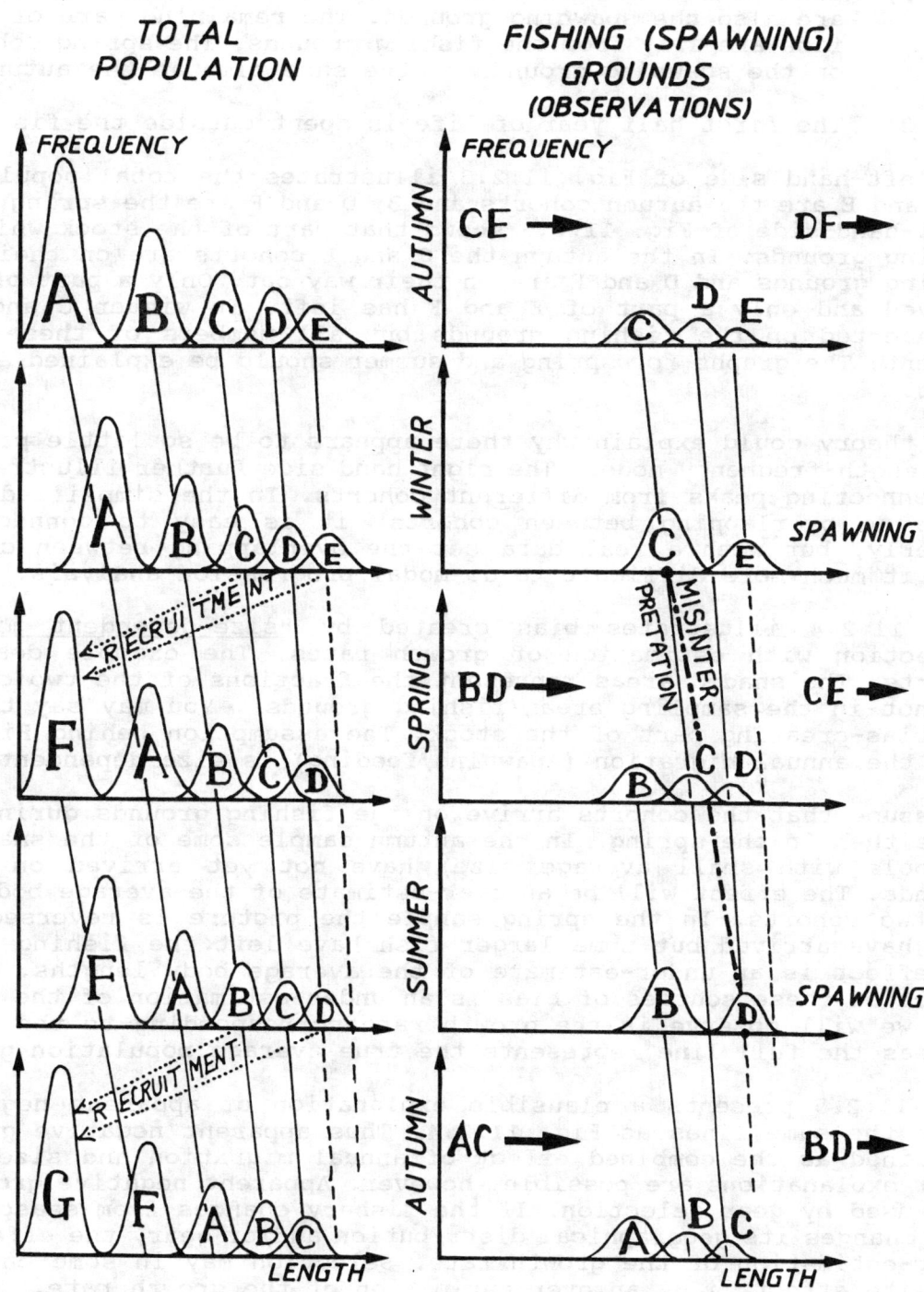

Fig. 11.2.3 Schematic illustration of the hypothesis of the migration of small pelagics. One unrealistic simplification is that the cohorts do not overlap. For further discussion see text (from Sousa, 1988)

The reasoning assumes the existence of the migration patterns illustrated in Figs. 11.1.1 and 11.2.3. The hypotheses behind Fig. 11.2.3 are:

1) There are two peak recruitments per year. They are denoted: "Autumn cohorts" and "Spring cohorts".

2) Each year a certain time is spent on the fishing grounds, which are also the spawning grounds. The remaining part of the year the fish are away from the fishing grounds. The spring cohorts are not on the spawning grounds at the same time as the autumn cohorts.

3) The first half year of life is spent outside the fishing grounds.

The left-hand side of Fig. 11.2.3 illustrates the total population, where A, C and E are the autumn cohorts and B, D and F are the spring cohorts. The right-hand side of Fig. 11.2.3 shows that part of the stock which is on the fishing grounds. In the autumn the C and E cohorts are on their way to the fishing grounds and D and F are on their way out. Only a part of C and D has arrived and only a part of E and F has left. In winter C and E only are represented on the fishing grounds but all members of these cohorts are present. The graphs for spring and summer should be explained along similar lines.

This theory could explain why there appears to be so little progression in the length-frequency modes. The right hand side further illustrates the risk of connecting peaks from different cohorts. In the simplified Fig. 11.2.3 (with no overlapping between cohorts) it is easy to connect the peaks properly, but with a real data set the overlapping between cohorts would make it much more difficult to do modal progression analysis.

Fig. 11.2.4 illustrates bias created by "size-dependent migration" in connection with estimation of growth rates. The example deals with two cohorts. The shaded areas represent the fractions of the two cohorts which are not in the sampling area (fishing grounds). You may say that they are the bias-creating part of the stock. The assumption behind Fig. 11.2.4 is that the annual migration (spawning/feeding) is size-dependent.

We assume that the cohorts arrive on the fishing grounds during autumn and leave them in the spring. In the autumn sample some of the small specimens (schools with small average size) have not yet arrived on the fishing grounds. The effect will be an over-estimate of the average body lengths of the two cohorts. In the spring sample the picture is reversed, all small fish have arrived but some larger fish have left the fishing grounds. Now the effect is an under-estimate of the average body lengths. The combined effect of these sources of bias is an under-estimation of the growth rate. What we will observe is the growth rate corresponding to the broken line, whereas the full line represents the true average population growth rate.

Fig. 11.2.5 presents a plausible explanation of apparent negative growth along the same lines as Fig. 11.2.4. Thus apparent negative growth may be explained as the combined effect of annual migration and size-dependence. Other explanations are possible, however. Apparent negative growth may also be caused by gear selection. If the fishery changes from season to season, e.g. changes its geographical distribution and/or gear, the effect may be an under-estimation of the growth rate. Selection may in some cases have the opposite effect, i.e. an over-estimation of the growth rate.

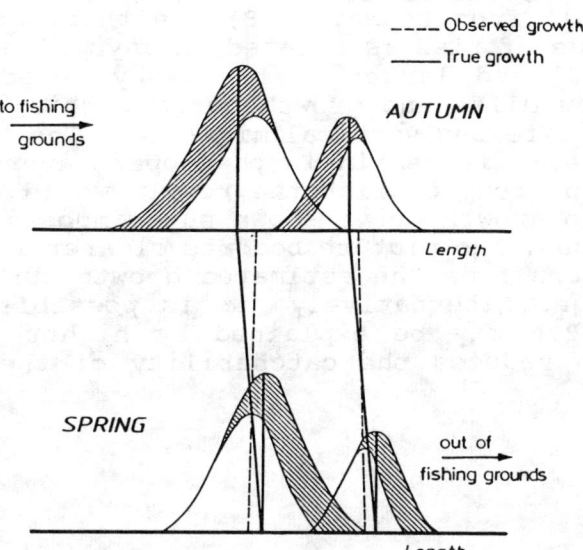

Fig. 11.2.4 **Schematic representation of the bias caused by size-dependent migration. Dashed zones indicate that part of the population that is not sampled (from Sousa, 1988)**

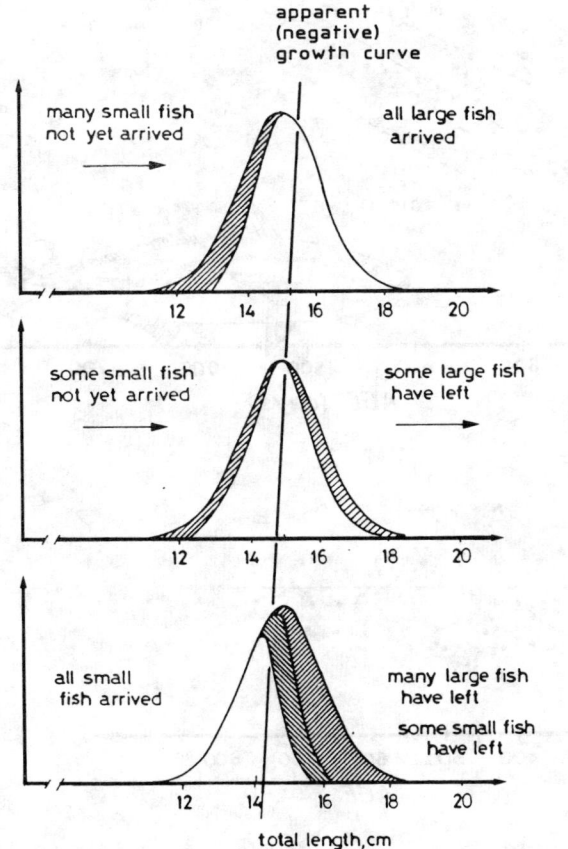

Fig. 11.2.5 **Schematic representation of possible reasons why apparent negative growth can be observed (from Sousa, 1988)**

Fig. 11.2.6 illustrates bias caused by size-dependent vertical migration and/or gear selection. Fig. 11.2.6A shows age/length data obtained from otolith readings of 118 D. russelli caught by two different gears, bottom trawl and pelagic trawl (from Sousa, 1988). In this case daily rings were observed so that the age of fish is counted in days. Smaller specimens were caught by pelagic trawl and larger specimens by demersal trawl. The two groups appear to follow different growth curves. This can be explained by size-dependence of the lifelong vertical migration. The slower the individuals grow the longer they will remain in the upper layers. If growth curves were estimated for each group of fish the result would be two quite different growth curves. The growth curve shown superimposed on Fig. 11.2.6A is based on all observations. The picture becomes clearer in Fig. 11.2.6B where the residuals (deviations from the estimated growth curve in Fig. 11.2.6A) are plotted against age. Alternatively, it is possible that the findings presented in Fig. 11.2.6 may be explained by higher swimming speeds of larger specimens which reduces the catchability of the pelagic trawl for these size groups.

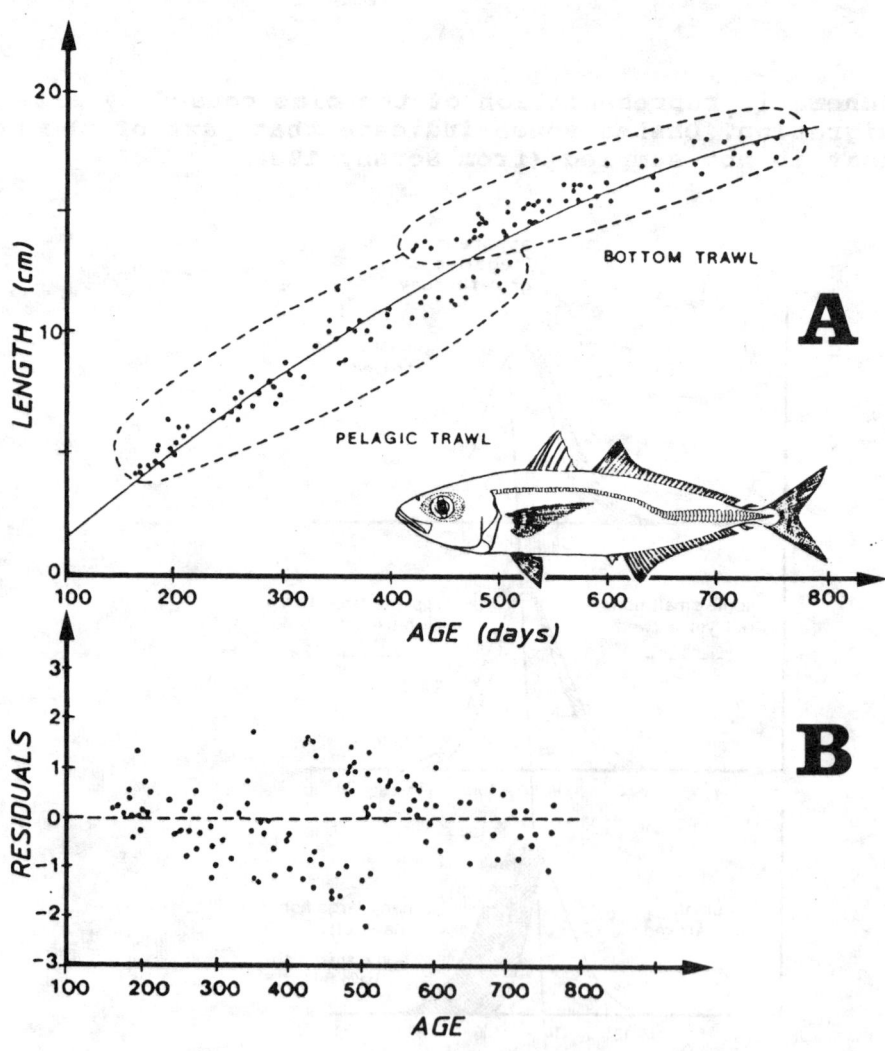

Fig. 11.2.6 Growth curve of Decapterus russelli on Sofala Bank, Mozambique, based on age readings (daily rings) of fish sampled by the commercial bottom trawl fishery and by a pelagic trawl survey, both in 1982 (from Sousa, 1988)

11.3 THE ANNUAL-RETURN MATCHED SAMPLES METHOD

The method presented in this section is a simple special case of the "general matched samples method", which will be discussed in the next section. The "matched samples method" is a simple method for the estimation of growth parameters and mortality rates of migratory stocks. As other simple methods it is based on rather strong assumptions. The question is whether these assumptions are reasonable approximations to the reality.

The method is based on the assumption that a fish stock follows a predictable migration route. If this migration route is known (e.g. from tagging experiments) in time and space we are in a position to follow the cohorts and to "match" samples so that they originate from the same cohort.

Consider a simple hypothetical model:

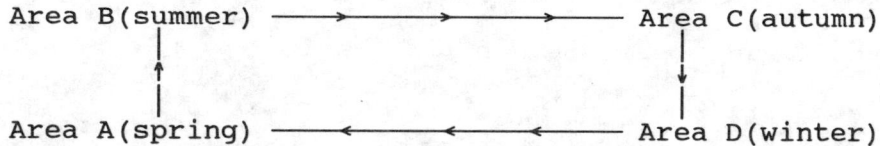

A, B, C and D symbolize geographical areas. We assume the stock to undertake the same migration each year and the timing for, say, the spring cohort is as indicated above. To "match" samples in the above model for the spring cohort means to perform the analysis based on samples taken:

In spring in area A, in summer in area B,
in autumn in area C and in winter in area D

The samples may originate from one area only, say, A, and in that case only samples collected in the spring (in different years) should be matched.

This model may fit to small pelagics like scads and Indian mackerels and it is known to fit the migration pattern of small pelagics in temperate waters (cf. Section 11.6). It can be used to estimate growth parameters, as described below.

11.3.1 Estimation of growth parameters by the annual-return matched samples method

Fig. 11.3.1 depicts a series of length-frequency samples taken every four months over a 2 years period. Note that the distributions in subsequent periods often indicate a negative growth. Following the model we can connect the modes in the distribution for the corresponding months in subsequent years (e.g. Jan.1982, Jan.1983 and Jan.1984), as in a normal modal progression.

The next step is then to use Chapman's method (Eq. 3.3.2.2), as the time interval Δt is relatively long, assumed to be one year throughout:

$$L(t+\Delta t) - L(t) = c*L_\infty - c*L(t) \text{ where } c = 1 - \exp(-\Delta t*K)$$

There are four estimates of $\Delta L/\Delta t$ with $\Delta t = 1$ year, namely: (L5-L1), (L6-L2), (L7-L3) and (L8-L4). The mean value of the increments of these four observations is 8 cm while the mean of the mean lengths of the first modes is (L1+L2+L3+L4)/4 = 4.25 cm. Then we have:

the mean of $L(t+1) - L(t) = 8$ cm and $\overline{L}(t) = 4.25$ cm

$8 = c*L_\infty - c*L(t) = c*L_\infty - c*4.25$ or

$c = 8/(L_\infty - 4.25)$ or as

$K = -(1/\Delta t)*\ln(1-c)$ (see Section 3.3.2), we obtain:

$K = -(1/1)*\ln[1 - 8/(L_\infty - 4.25)]$

If we use 18 cm as a first rough estimate of L_∞ we get:

$K = -\ln(1 - 8/(18-4.25)) = 0.87$

Using $L_\infty = 18$ cm and $K = 0.87$ per year we obtain a modal progression from 4.3. to 12.2 cm at the ages 0.31 and 1.31 years (Fig. 11.3.1):

$L(0.31) = 18*(1 - \exp(-0.87*0.31)) = 4.3$ cm
$L(1.31) = 18*(1 - \exp(-0.87*1.31)) = 12.2$ cm

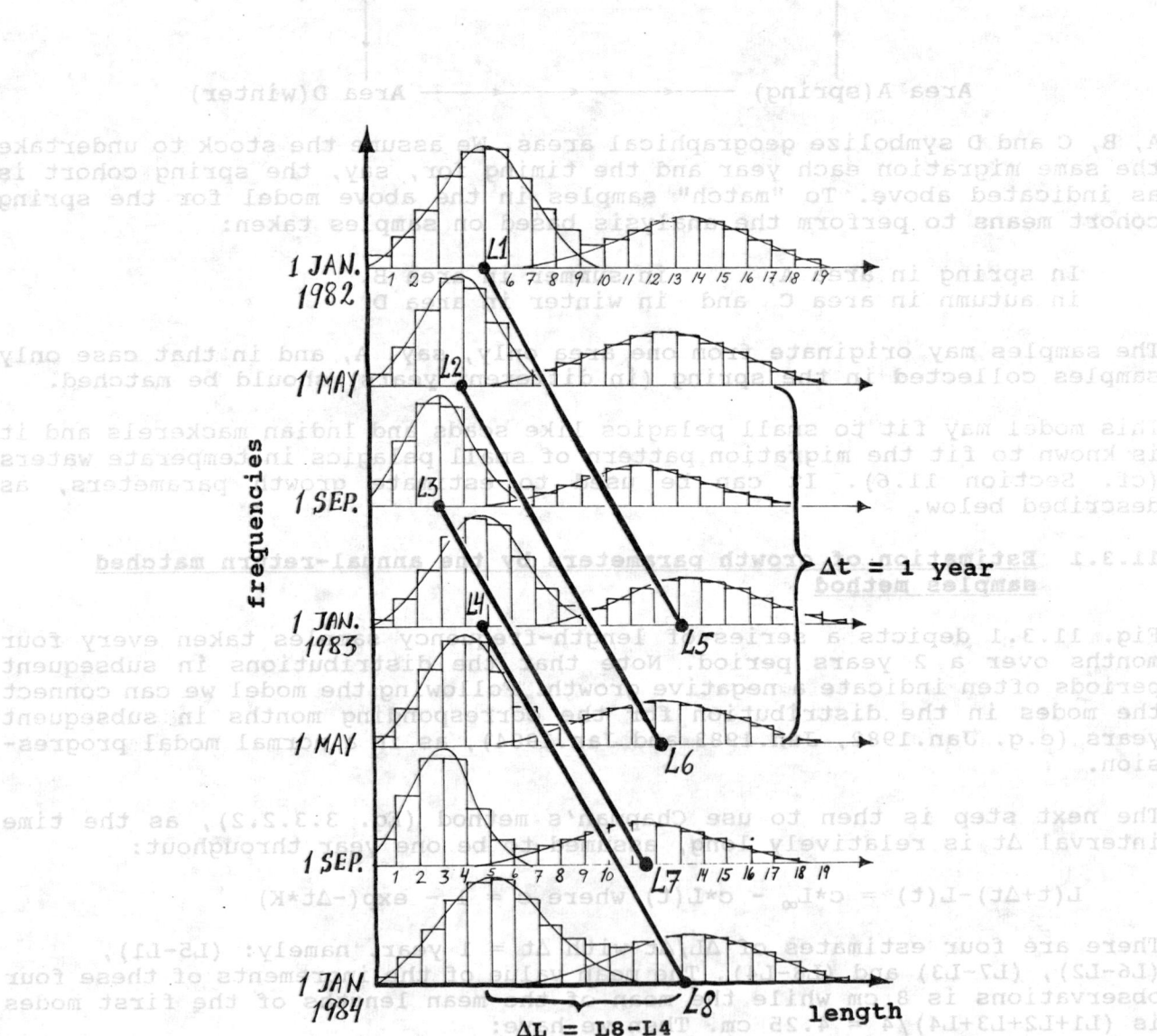

Fig. 11.3.1.1 Illustration of the estimate of growth parameters by the annual-return matched samples method. For further details, see text

Table 11.4.1 Hypothetical example to illustrate the "matched samples method"

length				Area A	(number	caught	per	unit	of effort)	
	1	2	3	4	5	6	7	8	9	10	total
JAN		8465									8465
FEB			5732								5732
MAR				1758	1153						2911
APR					629	357					986
MAY						125	42				167
JUN											0
JUL											0
AUG											0
SEP							4	15	11		30
OCT								104	123	32	259
NOV								201	355	152	708
DEC								178	424	264	866
Total	0	8465	5732	1758	1782	482	46	498	913	448	20124

length				Area B							
	1	2	3	4	5	6	7	8	9	10	total
JAN											0
FEB			1433								1433
MAR				1905	1249						3154
APR					2306	1308					3614
MAY						1870	634				2504
JUN						271	606	111			988
JUL							94	74			168
AUG							50	98	30		178
SEP							98	351	239		688
OCT								422	499	129	1050
NOV								212	375	161	748
DEC								43	103	64	210
Total	0	0	1433	1905	3555	3449	1482	1311	1246	354	14735

length				Area C							
	1	2	3	4	5	6	7	8	9	10	total
JAN											0
FEB											0
MAR											0
APR					341	193					534
MAY						1252	424				1676
JUN						738	1650	302			2690
JUL							1650	1297			2947
AUG							690	1350	418		2458
SEP							216	771	526		1513
OCT								233	276	71	580
NOV								28	49	21	98
DEC											0
Total	0	0	0	0	341	2183	4630	3981	1269	92	12496

- 294 -

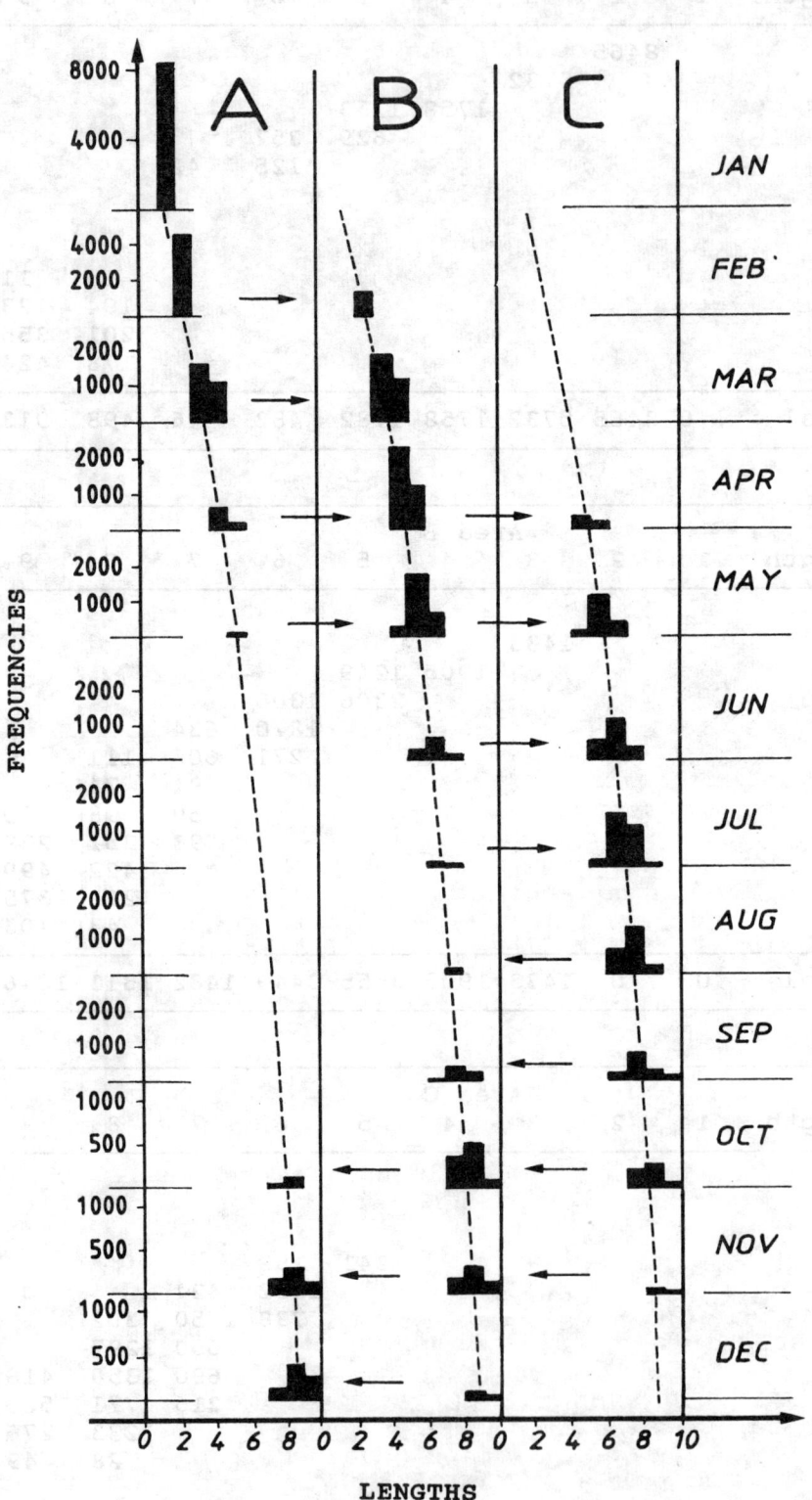

Fig. 11.4.1 Hypothetical example to illustrate the estimation of mortality rates by the "matched samples method". Note that the scales differ from one month to another

11.4 THE GENERAL MATCHED SAMPLES METHOD

This approach assumes that we have knowledge or a hypothesis of the migration route in time and space, and therefore are able to "match" samples so that they originate (or can be hypothesized to originate) from the same cohorts (cf. the beginning of the previous section).

To illustrate the features of the general matched samples method for estimation of total mortalities a simple hypothetical example was constructed. This simplified example deals with one cohort migrating through three areas, A, B and C. From A the cohort moves to B and then to C where it stays for a while; it moves back to B and ends where it started in A.

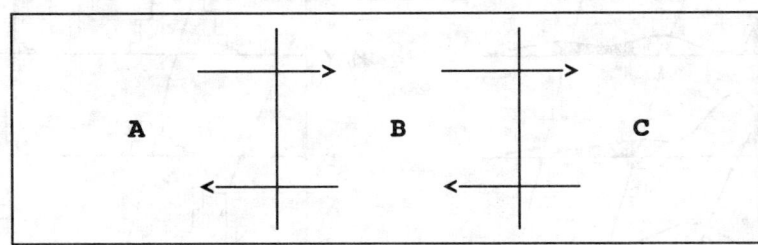

Table 11.4.1 and Fig. 11.4.1 show the example in numbers caught per unit of effort. The curves superimposed on the figure correspond to the growth parameters: L_∞ = 10 cm, K = 2.0 per year and t_o = 0. The arrows in Fig. 11.4.1 indicate the migration between areas. Ignoring migration we would interpret the first part of the length-frequencies in area A as a cohort becoming extinct in May and we would estimate the mortality rates as follows (cf. Table 11.4.1):

$Z(JAN) = \ln(8465/5732) = 0.39$ per month

$Z(FEB) = \ln(5732/2911) = 0.68$ per month

$Z(MAR) = \ln(2911/986) = 1.08$ per month

$Z(APR) = \ln(986/167) = 1.77$ per month

The real mortalities (including the entire cohort in all three areas) are (cf. Table 11.4.1):

$Z(JAN) = \ln(8465/(5732+1433))$
       $= \ln(8465/7166)$ $= 0.17$ per month
$Z(FEB) = \ln((5732+1433)/(2911+3154))$
       $= \ln(7166/6065)$ $= 0.17$ per month
$Z(MAR) = \ln((2911+3154)/(986+3614+534))$
       $= \ln(6065/5134)$ $= 0.17$ per month
$Z(APR) = \ln((986+3614+534)/(167+2504+1670))$
       $= \ln(5134/4346)$ $= 0.17$ per month

The differences between the two sets of Z-values are called "<u>migration coefficients</u>".

Fig. 11.4.2 illustrates the difficulties often encountered when trying to estimate growth curves for migratory species. Part C shows the migration route for the stock considered. In this hypothetical example we assume that the samples are obtained from the fishing grounds only. The fish are assumed to migrate to and from the fishing grounds depending on their size. During the first year of life they will be available on the fishing grounds when they are at lengths between L1 and L2 and for the second year between lengths L3 and L4.

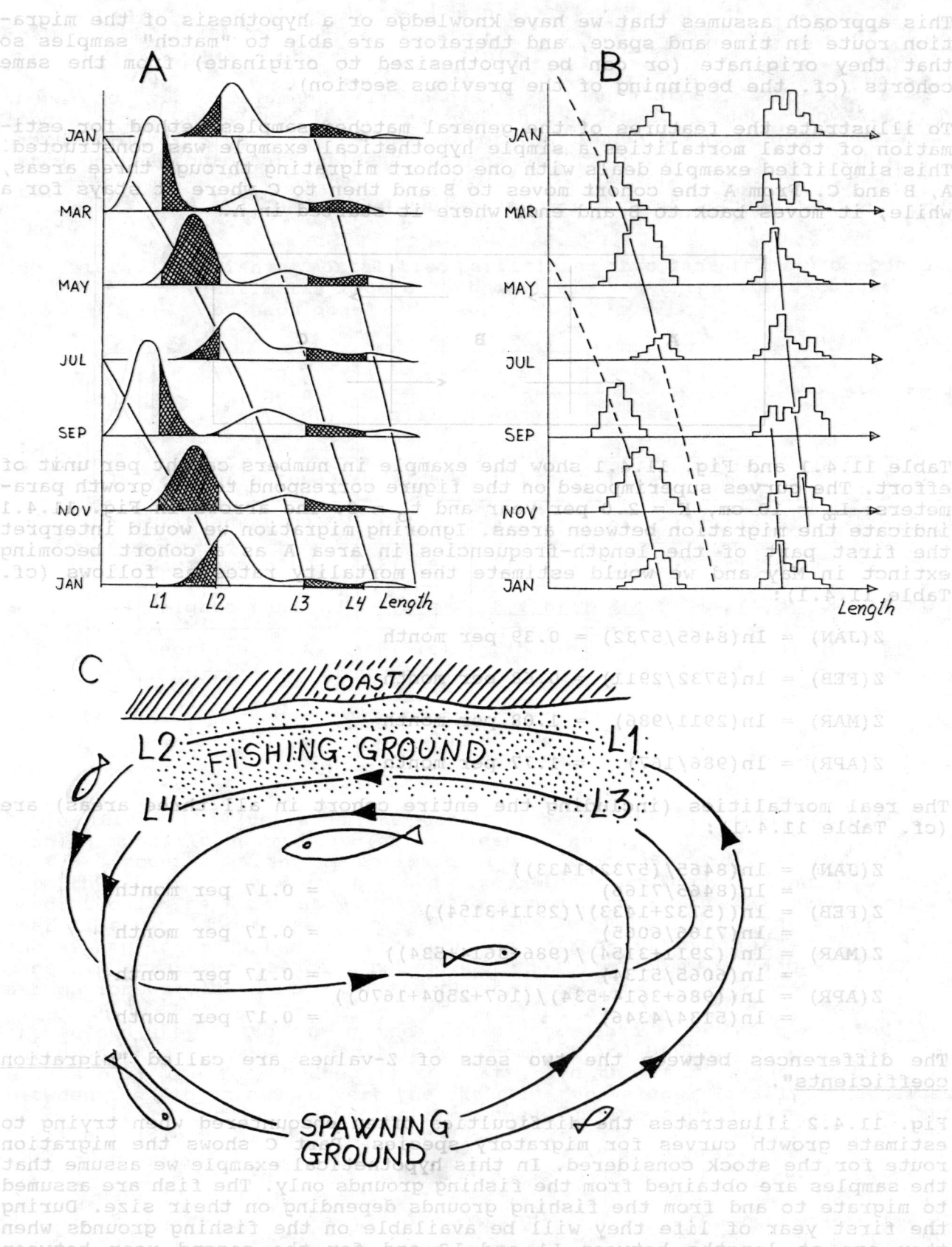

Fig. 11.4.2 Illustration of bias problems when estimating the growth curve of a migratory species from "unmatched" samples

Part A shows a time series of unbiased length-frequency samples and the growth curves we would estimate from these samples. These samples could be obtained from a sampling scheme which covers the entire distributional range in space and time. However, we would not be in a position to obtain such unbiased samples if all samples were taken on the fishing grounds only. If the fish arrived on the fishing grounds exactly at the lengths L1 and L3 and left exactly at the lengths L2 and L4 we would obtain the samples represented by the shaded parts of the length-frequency distribution on Part A. In practice we would, of course, not observe such a "knife-edge" recruitment to the fishing grounds, but something like the length-frequency samples shown in Part B which show two distinct groups of fish without any overlap in length-frequencies. The left-hand groups of small fish show some progression in the modes, whereas the length-frequencies of the larger fish are open to any interpretation. The intermediate length group is absent. This general pattern is often observed when sampling length-frequencies from a restricted area, i.e. an area which is only a small part of the distributional range of the stock in question.

The curves superimposed on Part B are the curves we would obtain from a modal progression based on those length-frequencies. These curves are steeper than the true growth curves drawn in Part A. The modes of the smallest fish of the samples in Part B show some progression, but less than the progression in the modes of Part A, which represent the true growth curve. The effect of size-dependent migration (cf. Section 11.2) is therefore an under-estimation of the growth rate of the small fish.

The apparent modes of the large fish in Part B are difficult to interpret in terms of growth. In practice, these modes may have little to do with growth but rather reflect random noise and/or bias from the sampling procedure. In this case the growth curves fitted to the large fish represent an underestimate of the growth rate, but in other cases might produce an overestimate. In general, one should be cautious when interpreting modal progression for length groups approaching the maximum length of the species.

11.5 ASSESSMENT BASED ON TAGGING DATA

The success of tagging experiments depends on the ability and willingness of the fishermen and others dealing with the catch to report on where and when the marked fish was caught. If the data are used also for estimation of growth parameters the size of the recaptured fish should be reported as well. The fishery must cover a relatively large part of the distribution in space and time of the stock to secure a reasonable number of recaptures for the estimation procedure. It is assumed that the tagged fish constitute a representative sample of the population, and thus have the same basic parameters as the untagged part. Models along these lines were suggested by Gulland (1955), Paulik (1963), Seber (1973) and Jones (1977). Seber (1973) presents a comprehensive discussion of the analysis of capture/recapture data.

Kleiber, Argue and Kearney (1983) suggested a model for assessment of Pacific skipjack tuna based on tagging data. This is the traditional catch curve model (Eq. 4.2.7) with modifications to take into account mortality due to tagging, shedding of tags and missing reports on recaptures. The basic equation of the model for estimation of population size, P, and "attrition rate A", (see definition below) reads:

$$r(t) = a*b*N_o*\exp(-t*A)*\frac{C(t)}{P*A}*[\exp(A) - 1] \qquad (11.5.1)$$

where

a = fraction of short-term survivors, while 1-a = short-term mortality, due to the trauma of being tagged

A = attrition rate (includes mortalities (F and M) and shedding of tags, while growth out of vulnerability to the gear and migration out of the area are also considered)
b = fraction of recaptured tags actually returned with usable recapture information
C(t) = catch in biomass units during time period t
N_o = number of fish tagged (at time t = 0)
P = standing stock in biomass units (assumed constant in time)
r(t) = number of tag returns during time period t
t = index of time period

To see that Eq. 11.5.1 is basically the same as the catch equation (Eq. 4.2.7) we introduce:

$$N(t) = N_o * \exp[-(t-1)*A] = \text{number of tagged fish at the beginning of time period t}$$

$$F(t) = C(t)/P, \text{ fishing mortality}$$

If we assume a = b = 1 and A = Z (no shedding or other type of attrition factors), after substitution and rearranging Eq. 11.5.1 becomes:

$$r(t) = N(t) * \frac{F}{Z} * [1 - \exp(-Z)] \quad \text{i.e. the catch equation (Eq. 4.2.7)}$$

The above description is only a short introduction of the basic equation. For a complete description of the model the reader is referred to the original work.

11.6 ESTIMATION OF THE GROWTH PARAMETERS OF A MIGRATORY STOCK: THE ATLANTIC MACKEREL

As mentioned in Section 11.2, seasonally migrating species sometimes migrate earlier in the season the older and bigger the fish are. The problem is analysed in well documented studies of a stock of the North Atlantic mackerel, *Scomber scombrus*. This stock lives north and west of the United Kingdom and Ireland. The main migration route is shown in Fig. 11.6.1. The stock undertakes an annual migration from the "over-wintering area" to the "spawning area", from the spawning area to the "feeding area" and from the feeding area back to the over-wintering area. The total distance travelled is in the order of 500-1000 nautical miles per year. Spawning takes place from March to July in the area south of Ireland (see Fig. 11.6.1).

We shall concentrate here on aspects related to size-dependent migration as discussed in Sections 11.2 and 11.3. Size-dependent migration of this stock has been demonstrated by, for example, Dawson (1986) and Eltink (1987). The following is based on these two papers, which arrive at similar conclusions from independent sets of observations.

Because it is a species of temperate waters it is relatively easy to read the age of *Scomber scombrus* from the otoliths. The findings of Dawson (1986) and Eltink (1987) were based on random samples of mackerel caught by commercial fishing vessels as well as research vessels on the spawning grounds during the spawning period of the sampled fish ages, lengths and maturity stages were recorded. Samples were taken on a monthly basis. The principal migration routes and the spawning grounds were known beforehand from other investigations (e.g. tagging experiments).

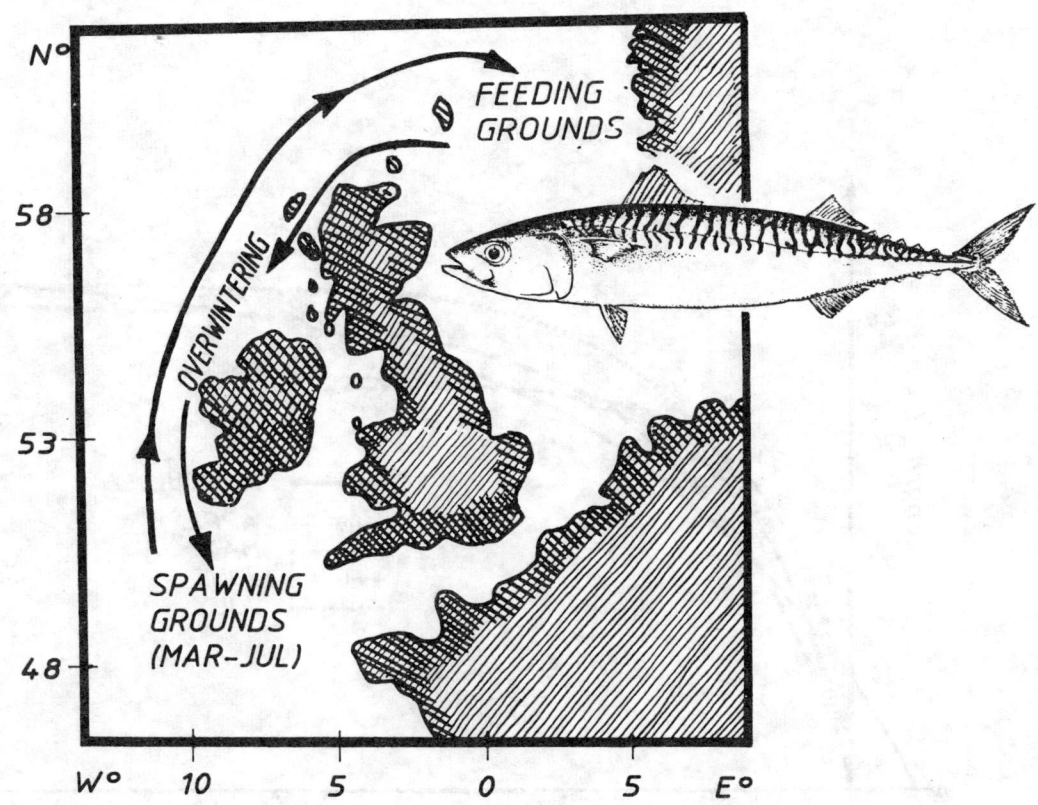

Fig. 11.6.1　The main annual migration route of a stock of Atlantic mackerel (<u>Scomber</u> <u>scombrus</u>) based on various sources including numerous tagging experiments (redrawn from Eltink, 1987)

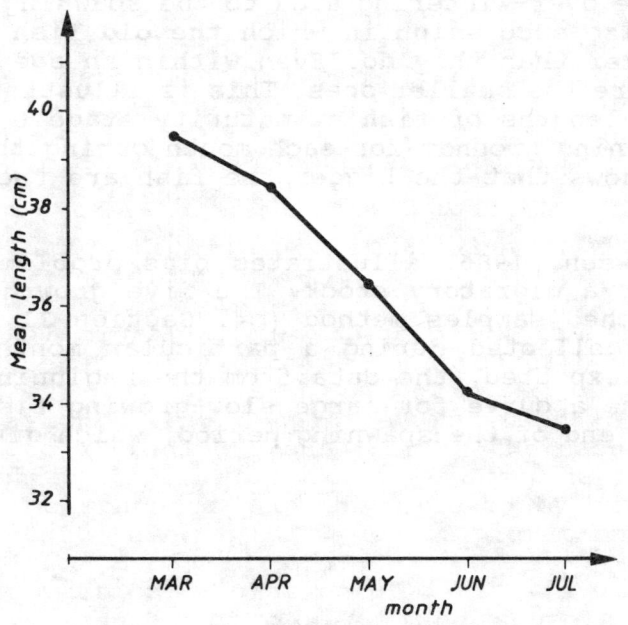

Fig. 11.6.2　Average length of spent (maturity stage 6) Atlantic mackerel (<u>Scomber</u> <u>scombrus</u>) by month. Note that the biggest fish spawn first (redrawn from Eltink, 1987)

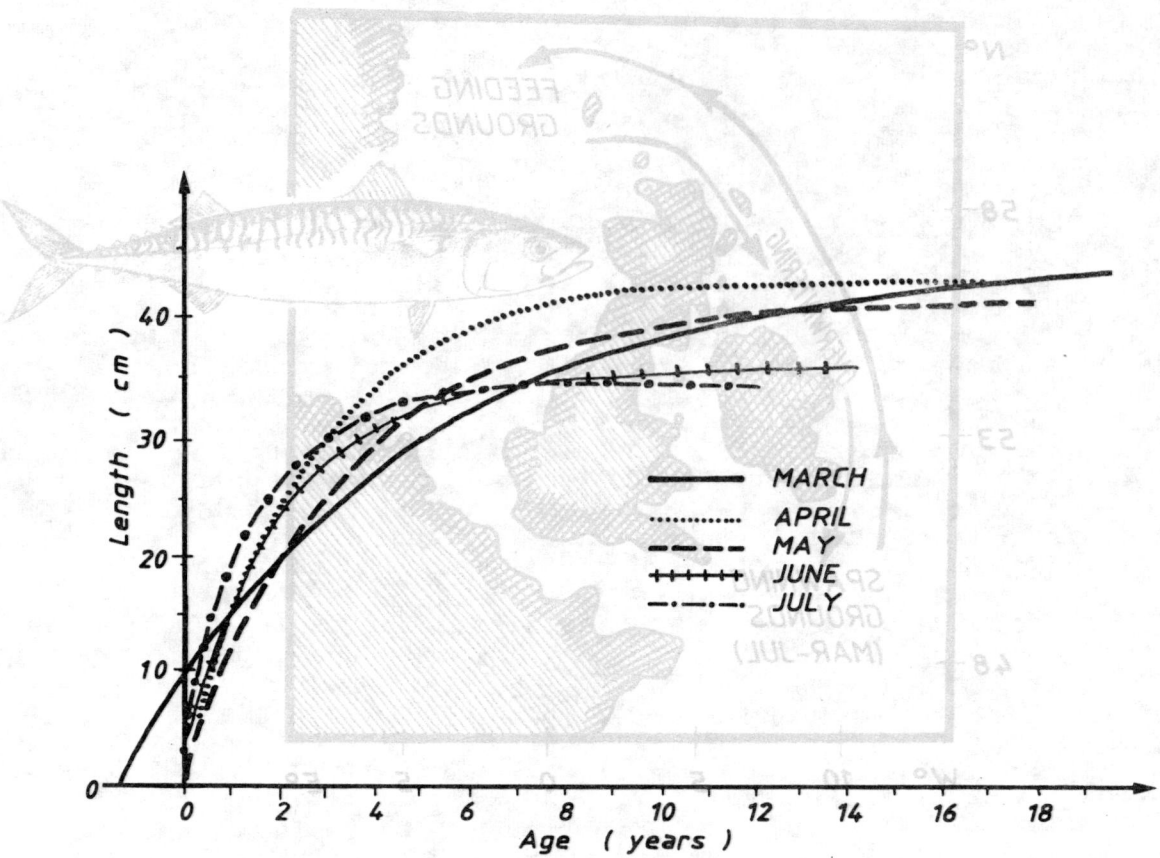

Fig. 11.6.3 Growth curves for Atlantic mackerel (<u>Scomber scombrus</u>) estimated from age/length data of specimens sampled on the spawning grounds from March to July - illustration of a bias problem (from Dawson, 1986)

The migration from the over-wintering area to the spawning grounds was found to occur in an age-size succession in which the old fish arrive before the young and leave earlier than they do. Even within an age group, the larger specimens arrive before the smaller ones. This is illustrated in Fig. 11.6.2 which shows the mean lengths of fish at maturity stage 6 of random samples collected on the spawning grounds for each month during the spawning season (Eltink, 1987). It shows that the bigger the fish are the earlier spawning occurs.

Fig. 11.6.3 (from Dawson, 1986) illustrates bias problems when estimating growth parameters for a migratory stock. The five growth curves were each estimated by the matched samples method (cf. Section 11.3). Each curve is estimated from data collected during a particular month on the spawning grounds. As could be expected, the data from the beginning of the spawning period (March) produce a curve for large slow-growing fish compared to the data collected at the end of the spawning period, which give a steeper curve with a smaller L_∞.

12 THE STOCK/RECRUITMENT RELATIONSHIP

The stock and recruitment problem may be considered as the search for the relationship between parental stock size and the subsequent recruitment in numbers or the year class strength (cf. Section 8.3). This is a central problem of fish population dynamics, since it represents nature's regulation of population size, whether or not the populations are being exploited.

This chapter presents some considerations on the stock/recruitment (S/R) relationship problem although it does not present methods to solve actual problems. The reason why the S/R relationship is given this kind of treatment is not that the subject is less important, but rather that really convincing models to handle the problem have not yet been developed.

The following essay on the S/R relationship is based on Beyer and Sparre (1983) and Pauly (1984). Discussions of the S/R relationship with special reference to tropical fish stocks are given in, among others, Murphy (1982) and Pauly (1984).

To present the S/R relationship an exceptionally clear example has been chosen (Fig. 12.0.1). It deals with the false trevally (<u>Lactarius lactarius</u>) in the Gulf of Thailand (Pauly, 1980a). In this case there appears to be a well defined relationship between recruitment and spawning stock size. However, as already mentioned in Section 8.3 (Fig. 8.3.2) this is not a typical example of an S/R plot. Fig. 12.0.2 shows the S/R plot for herring (<u>Clupea harengus</u>) in the North Sea for the years 1949-1978. The estimates of stock sizes and recruitment (numbers of one year old fish) are derived from VPA (cf. Section 5.1). This example is more representative for S/R plots in general.

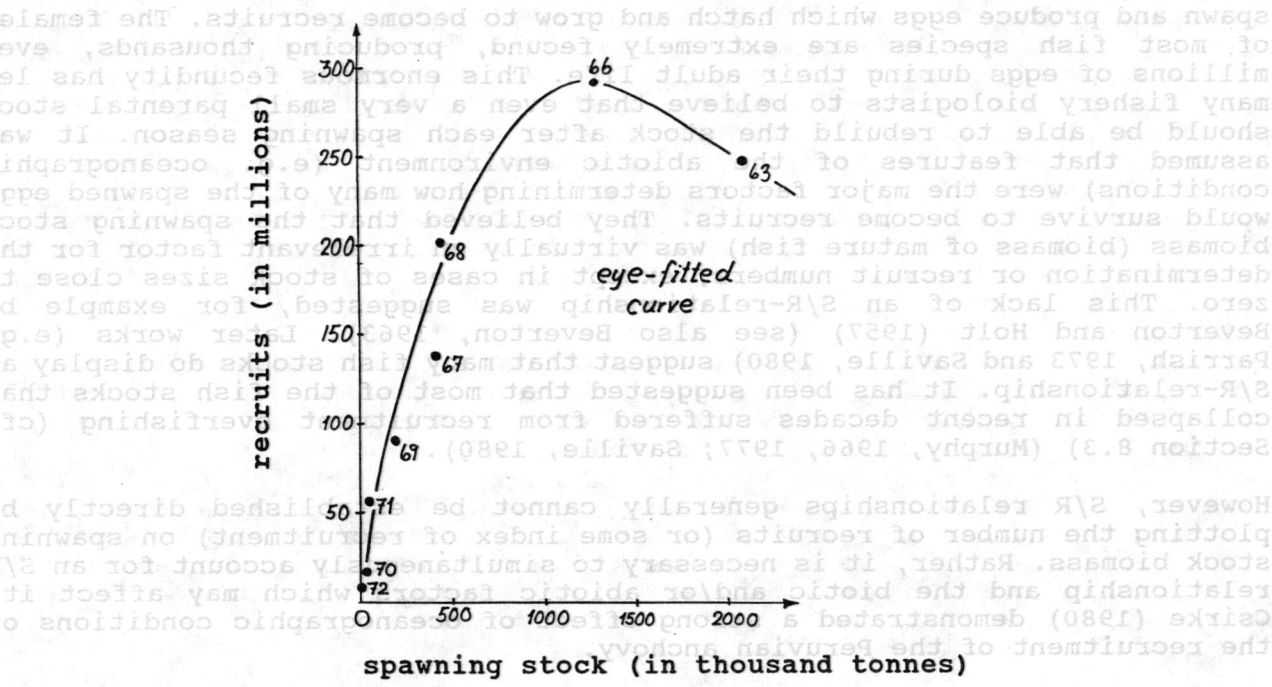

Fig. 12.0.1   Stock and recruitment plot for the false trevally (<u>Lactarius lactarius</u>) in the Gulf of Thailand (from Pauly, 1980a)

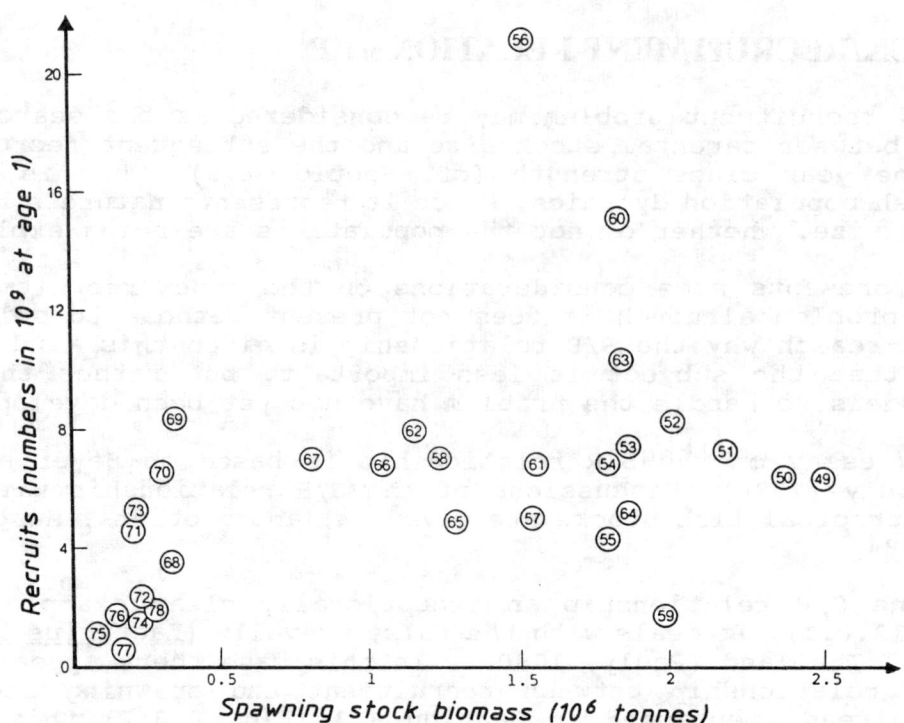

Fig. 12.0.2   Stock and recruitment plot of North Sea herring (Clupea harengus). Number of one year old herring vs spawning stock biomass. Data derived from VPA (reproduced from Beyer and Sparre, 1983)

Obviously, there can be no recruits if no adult fish are left to mature, spawn and produce eggs which hatch and grow to become recruits. The females of most fish species are extremely fecund, producing thousands, even millions of eggs during their adult life. This enormous fecundity has led many fishery biologists to believe that even a very small parental stock should be able to rebuild the stock after each spawning season. It was assumed that features of the abiotic environment (e.g. oceanographic conditions) were the major factors determining how many of the spawned eggs would survive to become recruits. They believed that the spawning stock biomass (biomass of mature fish) was virtually an irrelevant factor for the determination or recruit numbers, except in cases of stock sizes close to zero. This lack of an S/R-relationship was suggested, for example by Beverton and Holt (1957) (see also Beverton, 1963). Later works (e.g. Parrish, 1973 and Saville, 1980) suggest that many fish stocks do display an S/R-relationship. It has been suggested that most of the fish stocks that collapsed in recent decades suffered from recruitment overfishing (cf. Section 8.3) (Murphy, 1966, 1977; Saville, 1980).

However, S/R relationships generally cannot be established directly by plotting the number of recruits (or some index of recruitment) on spawning stock biomass. Rather, it is necessary to simultaneously account for an S/R relationship and the biotic and/or abiotic factors which may affect it. Csirke (1980) demonstrated a strong effect of oceanographic conditions on the recruitment of the Peruvian anchovy.

12.1   CLASSICAL S/R CONSIDERATIONS

It has been observed for many fish stocks that the mean recruitment level is almost constant in a large intermediate domain of variation of the parental stock (see Figs. 12.0.2 and 12.1.1). In the case of the North Sea herring (Fig. 12.0.2) it has been estimated that this intermediate range comprises at least an interval corresponding to 10^{12}-10^{13} eggs. The mean recruitment level is about $8*10^9$ one year old fish. Thus the probability of an egg developing into a one year old recruit is about 0.0001.

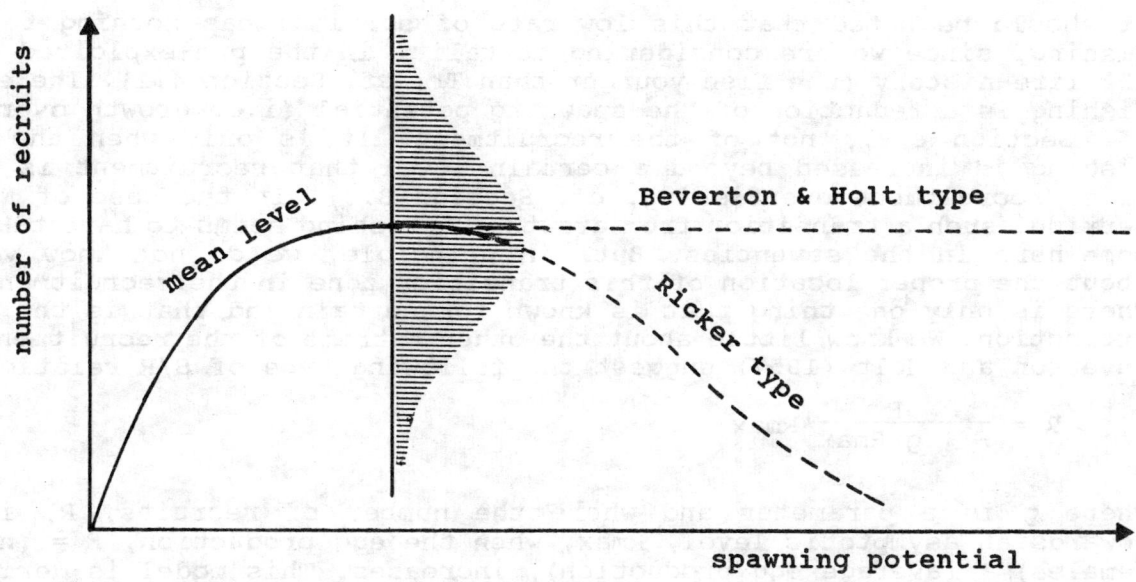

Fig. 12.1.1 The basic form of a recruitment curve. The shaded distribution indicates variation in recruitment about the mean level

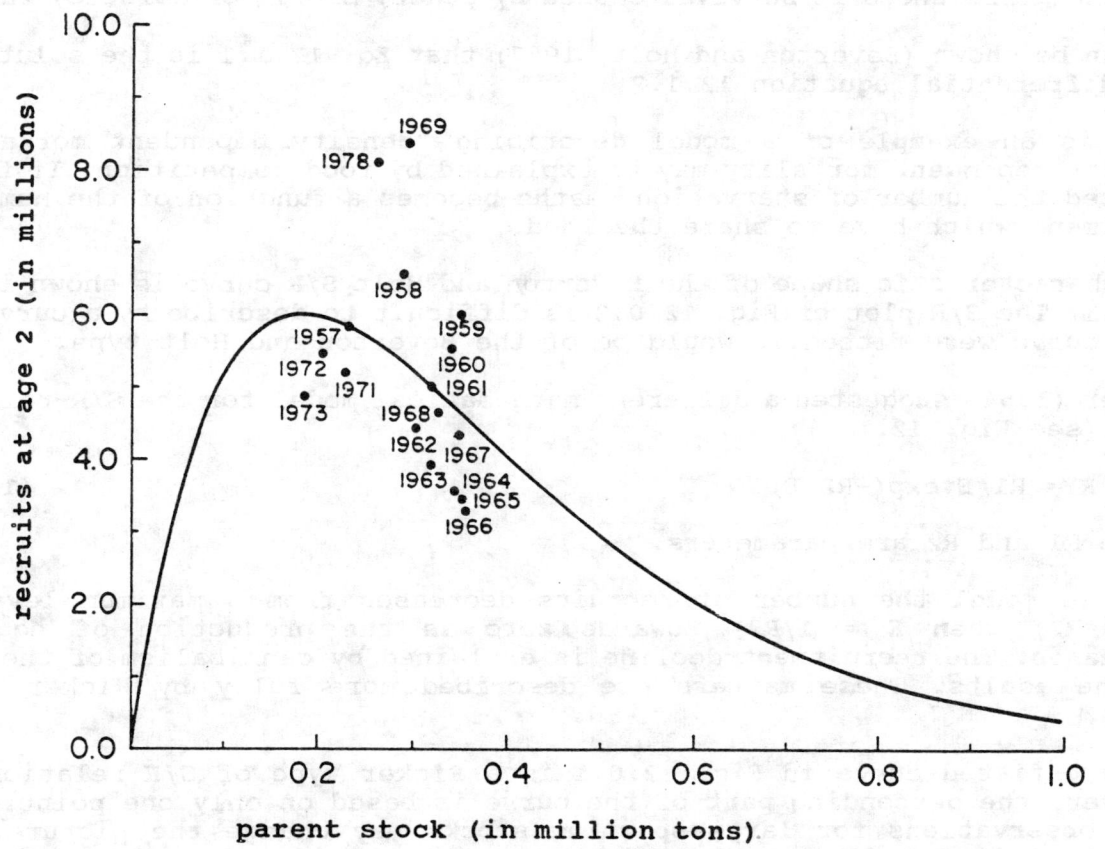

Fig. 12.1.2 Ricker curve fitted to the spawning stock biomass and recruitment of southern bluefin tuna. Data derived from cohort analysis (reproduced from Murphy, 1982, with permission)

It should be noted that this low rate of survival has nothing to do with fishing, since we are considering mortality in the pre-exploited stage of the life-history (the fish younger than Tr, cf. Section 4.1). The effect of fishing is a reduction of the spawning potential (i.e. growth overfishing, cf. Section 8.2), not of the recruitment. It is only when the rate of fishing is increased beyond a certain level that recruitment is affected (i.e. recruitment overfishing, cf. Section 8.3). In the case of North Sea herring, such a transition from growth overfishing seems to have taken place somewhere in the seventies. But, in principle, we do not know very much about the proper location of this transition zone in the recruitment graph. There is only one thing that is known for certain and that is the point of extinction. We know little about the other extreme of the recruitment curve. Beverton and Holt (1957) suggest the following type of S/R relationship:

$$R = \frac{E}{E + g*Rmax}*Rmax \qquad (12.1.1)$$

where g is a parameter and where the number of recruits, R, increases towards an asymptotic level, Rmax, when the egg production, E = (number of females) * (average egg production), increases. This model is derived from a simple density-dependent mortality model:

$$dN(t)/dt = -M(t)*N(t) \quad \text{and} \quad M(t) = m1 + m2*N(t) \qquad (12.1.2)$$

where $N(t)$ is the number of survivors at age t, $M(t)$ the rate of natural mortality of juveniles and m1 and m2 are parameters. Thus, the more survivors there are at age t, the higher their mortality. This mechanism evens out the differences in survival caused by other, biotic or abiotic, factors.

It can be shown (Beverton and Holt, 1957) that Eq. 12.1.1 is the solution to the differential equation 12.1.2.

This is an example of a model describing "density dependent mortality". Density dependent mortality may be explained by food competition. If food is limited the number of starvation deaths becomes a function of the number of specimens which have to share the food.

The characteristic shape of the Beverton and Holt S/R curve is shown in Fig. 12.1.1. The S/R plot of Fig. 12.0.2 is difficult to describe by a curve, but if a curve were fitted it would be of the Beverton and Holt type.

Ricker (1954) suggested a different mathematical model for the S/R-relationship (see Fig. 12.1.1):

$$R = R1*E*exp(-R2*E) \qquad (12.1.3)$$

where R1 and R2 are parameters.

In this model the number of recruits decreases from a maximum level (of R1/(e*R2) when E = 1/R2) towards zero as the production of eggs, E, increases. The recruitment decline is explained by cannibalism of the young by the adults. These matters are described more fully by Ricker (1954, 1975).

The eye-fitted curve in Fig. 12.0.1 is a Ricker type of S/R relationship. However, the descending part of the curve is based on only one point. A few more observations for large spawning stocks may change the picture. Fig. 12.1.2 shows a Ricker curve fitted to data for southern bluefin tuna (<u>Thunnus</u> <u>maccoyii</u>) (Murphy, 1982). This example also illustrates the difficulties often encountered when trying to fit an S/R curve. As Murphy (1982) says about this plot: "there appears to be an underlying density dependent relationship overlaid by considerable environmental variability".

Essentially, the Beverton and Holt model says that above a certain level of spawning stock there is no relationship between parent stock and recruitment, whereas the Ricker model says that this relationship exists for all sizes of the spawning stock, and that there is an optimum spawning stock size (cf. Fig. 12.1.1).

Deriso (1980) and Schnute (1985) suggested a general stock/recruitment model:

$$R = R1*E*[1 - R2*R3*E]^{1/R3}$$

where R1, R2 and R3 are parameters. For large values of the "shape" parameter R3, the above model reduces to the Ricker model as we then get

$$R = R1*E*\exp(-R2*E)$$

If R3 = -1 we get the Beverton and Holt model by redefining the parameters to R1 = 1/g and R2 = 1/(g*Rmax) as

$$R = R1*E/[1 + R2*E] = (R1/R2)/(E + 1/R2)$$

Most available data on stock and recruitment refer to the medium range of variation in the spawners. It is here that most species appear to maintain a constant mean recruitment level and this constitutes the motivation for the assumption of constant recruitment in classical fish population dynamics. The most remarkable fact, perhaps, is that recruitment shows only relatively small variations about this mean level considering the great reduction in numbers from the egg stage until maturity (Ursin, 1982)

12.2 THE STABILITY OF RECRUITMENT

Table 12.2.1 gives relative recruitment variations for eleven commercially important North Sea species in the period 1963-75 (Ursin, 1982). The table is derived from VPAs from various ICES working group reports. We see for example, that the cod year classes show a factor of six in their variation from the weakest to the strongest.

To consider this variation range in the light of the range of population reduction of cod during the first year of life we may apply the following consideration as an approximation of the situation in the seventies. The mature stock comprises about 200,000 tonnes of cod, half of which are females. Once a year each female spawns 10% of her body weight as eggs, giving a total annual production of 10,000 tonnes of eggs or $2*10^{13}$ eggs, since each egg weighs about 0.5 mg wet weight. The mean recruitment level, however, is only $2*10^8$ one year old fish. Thus on the average only one egg out of 100,000 survives and grows into a one year old cod. The factor of six in recruitment variation implies that the probability of a fish dying in the first year of life at most changes from 0.999,997 for a bad year class to 0.999,983 for a good year class, or in other words the probability of a fish surviving changes from 0.000,003 to 0.000,017.

We could also consider stability in terms of recruitment to the mature stock. During her lifetime a female cod must produce an average of one mature female and one mature male. Only extremely small deviations from this magic number 2 are feasible if the cod stock is to remain at approximately the same level, as it apparently has done for the last 150 years or more. This approximate one-for-one replacement between successive generations holds at least as long as the mature stock is within the medium range of variation as referred to above. Thus, whether we are considering a situation of heavy fishing, where the mature female cod on the average lays about two million eggs in its short lifetime, or whether we are considering an average production of 40 million eggs per female at a low fishing level, the number 2 still holds.

Table 12.2.1 Recruitment variation in North Sea fish stocks 1963-1975. Numbers adjusted to a value of 100 for the most outstanding year class of each species. "Ratio" is the ratio between the numbers in the strongest and the weakest year class (after Ursin, 1982)

year class	gadoids					flatfish		clupeids		others	
	cod	haddock	whiting	saithe	Norway pout	plaice	sole	herring	sprat	sand eel	mackerel
1963	52	1	14	17	4	100	100	100	–	–	10
1964	49	1	26	23	6	29	21	53	–	–	26
1965	70	2	30	18	0.5	28	11	47	–	–	43
1966	63	12	37	50	7	25	11	66	59	–	62
1967	20	100	100	51	–	21	18	65	62	–	10
1968	19	6	33	55	2	27	9	36	37	–	16
1969	82	2	30	29	–	32	26	78	41	–	100
1970	100	14	33	29	45	25	6	62	20	100	11
1971	18	21	68	30	7	20	14	41	19	21	17
1972	35	4	90	40	16	62	19	19	46	47	4
1973	31	21	63	100	100	40	18	47	91	28	15
1974	51	40	92	27	38	25	7	–	100	86	11
1975	27	9	37	50	18	37	22	–	79	41	4
Ratio	6	100	7	6	200	5	17	5	5	5	25

We do not know what causes this enormous reduction in numbers from the egg stage and leads to the fine adjustment of the number of recruits to the number of mature fish. Note in passing that the North Sea haddock and Norway pout show a remarkable instability in recruitment (see Table 12.2.1). It could be that stabilizing mechanisms do not exist in the North Sea for these two species.

12.3 TOWARDS MODELLING RECRUITMENT

It is difficult to attack the recruitment problem in a sensible way by means of statistical tools only because of the small amount of useful data that in general seem to be available. We only achieve one new data point per species each year. Thus, to explain recruitment variability such as the occurrence of extremely strong year-classes, we may advance a great variety of conflicting hypotheses, all of which lead to different conclusions but none of which can be rejected on the basis of the available data.

The situation is no better with respect to explanations of recruitment stability. It is not possible to discriminate between empirical recruitment models such as those given by Eqs. 12.1.1 and 12.1.3. Neither is it possible beforehand to sort out irrelevant mechanisms of mortality regulation by means of data. Apparently we need to build more biological knowledge on the causes of natural mortality in the first year of life into the recruitment models. See also Sinclair (1988).

13 DEMERSAL TRAWL SURVEYS

Bottom trawl surveys are widely used for monitoring demersal stocks when only an index of abundance is required. From unfished stocks (or stocks for which no or few data on the fishery are available) biomass and annual yield estimates may also be derived by bottom trawl surveys. The estimation of total biomass from the catch per unit of effort (or unit area), however, involves several crucial assumptions, leaving such estimates rather imprecise.

The mean catch (either in weight or in numbers) per unit of effort or per unit of area is an index of the stock abundance (i.e. assumed to be proportional to the abundance). This index may be converted into an absolute measure of biomass using the so-called "swept area method" (Section 13.5). This method falls under the so-called holistic methods (cf. Section 1.8).

Reviews of the theory are given in, for example, Gulland (1975), Saville (1977), Troadec (1980), Doubleday (1980) and Grosslein and Laurec (1982). These reviews also give guidelines for conduct of surveys (planning, design, data collection, data recording, analysis and reporting), some of which are summarized in Sections 13.2 to 13.4), see also Butler et al. (1986), ICOD (1991) and Strømme (1992).

This Chapter gives first a short description of demersal trawl survey planning and a few guidelines for data recording and for field work (Sections 13.1 to 13.4). For more detailed descriptions of these subjects the reader is referred to, among others, Alverson and Pereyra (1969), Alverson (1971), Mackett (1973), FAO/UNDP (1975), Gulland (1975), Saville (1977), Flowers (1978), Doubleday (1981), Grosslein and Laurec (1982) and Fogarty (1985). The remainder of the Chapter (Sections 13.5 to 13.8) is a short account of the theory necessary to perform a stock assessment based on trawl survey data.

13.1 THE BOTTOM TRAWL

The bottom trawl (Fig. 13.1.1) is a conical net bag with a wide mouth fitted with weights on the ground-rope and floats on the head-rope (cf. Section 6.3). When the vessel is under way the net is kept open by two otter boards, wooden or iron structures which are towed by the warps attached forward of their centre so they tend to diverge. The two otter boards are connected to the net by bridles. These may be up to 200 m long and sweep the sea bed over a wide area. They frighten the fish towards the advancing net and so increase its effectiveness. The shape of net varies depending on the kinds of fish to be caught and on the types of bottom. The ground-rope may be fitted with roller gear (bobbins) so that the trawl can be used on stony bottom without being damaged.

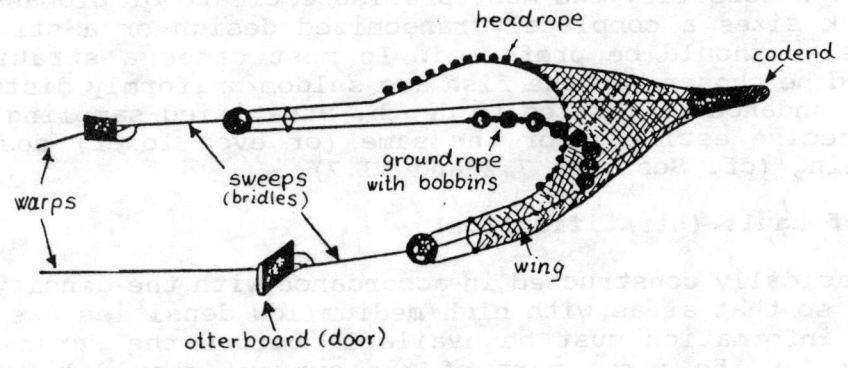

Fig. 13.1.1   Bottom trawl

The tail end of the gear from which the captured fish are removed is called the "codend". This is where most of the size selection takes place. In most cases a relatively small mesh size is required in the codend, in order to obtain a representative sample for the entire size range of the species under investigation (see Section 13.2).

13.2 PLANNING A DEMERSAL TRAWL SURVEY

Below is a list of some important items to consider before conducting a survey.

Definition of objectives

The objective(s) should be specified. Examples of objectives are:

1. Estimation of the total biomass and catch rates
2. Estimation of the biomass of selected species
3. Collection of biological data (e.g. length-frequencies) for estimation of growth and mortality parameters
4. Collection of environmental data

Information about the survey area

Information about depth and bottom conditions to point out trawlable areas and decisions on strata may be obtained from a preliminary survey with echo-sounding. Information from local fishermen may be valuable. Information on seasonal winds, currents and migration patterns of fish stocks are important as well.

Choice of gear

The design of the trawl should fit with the expected bottom conditions and the vessel used. If rough bottom is prevalent the gear should be fitted with bobbins in order to avoid damage to the gear. If semi-pelagic species are common a high-opening trawl should be used. The mesh size in the codend should be chosen so that the entire size range of the fishable part of the stock is retained by the trawl. Often 10-20 mm stretched mesh is appropriate. The mesh size used for assessment surveys is usually much smaller than the size used by the commercial fisheries, because samples of small fish are important for assessment methods based on length-frequencies.

If different trawls are used parallel or alternate hauls should be carried out in order to estimate correction factors for pooling of data.

Survey design

A procedure for the selection of stations should be decided on. A fixed grid of stations ensures maximum information on the distribution throughout the area, but not necessarily the most precise estimate of biomass. For estimation of stock sizes a completely randomized design or a stratified random sampling design should be preferred. In most cases a stratified sampling design should be chosen because fish are seldom uniformly distributed and in most cases abundance is related to depth. Stratified sampling often gives a much more precise estimate for the same (or even lower) cost than simple random sampling (cf. Sections 7.2 and 13.7).

Allocation of hauls (stratification)

Strata are basically constructed in accordance with the density distribution of the fish, so that areas with high/medium/low densities are separated. To do so, some information must be available before the survey. Perhaps the first survey (or the first part of the survey) should have a completely randomized survey design, or the trawl hauls might be evenly spaced. But in the next part of the programme some information on densities and standard deviations of density estimates will be available, and that information can be used for stratification. The standard deviation of catch per unit area (CPUA) is often found to be proportional to the CPUA and will therefore be higher in areas with high abundance.

The optimum allocation of a given number of hauls between strata will be to sample each stratum in proportion to its standard deviation multiplied by the stratum size (cf. Sections 7.2 and 13.7). The distribution of hauls within strata should preferably be at random, but often practical matters dictate the sample design. For example, obstacles on the bottom may not allow a proper random distribution of stations.

Possible number of hauls

To estimate how many hauls it is possible to make in a given period the following information should be available:

      Total number of days available .... N
      Time spent going to and from the fishing ground (hrs) t1
      Duration of one haul (hrs) t2
      Time used for shooting and hauling the trawl (hrs) t3
      Time to cover distance between stations (average, hrs) t4
      Number of hours available per day (depending on crew,
      behaviour of investigated species, navigation, etc.).......... T
      Time used for preparations: loading of ice, food, water,
      repair of gear and equipment (days) t5

Except for the first and the last day of a cruise, when T should be reduced by t1, the number of hauls per day can be calculated from:

$$(\text{number of hauls per day}) = T/(t2+t3+t4)$$

$$(\text{Total number of hauls}) = (N-t1-t5)*(\text{number of hauls per day}) + (\text{hauls made first and last day})$$

It is important to <u>standardize</u> the length of a haul throughout the survey, since the catchability of species and sizes often depends on the duration of the haul. For survey purposes hauls of 0.5 hour or 1 hour are usually the most adequate (see also Section 13.6).

13.3 DATA RECORDING

When setting up a plan for a trawl survey a crucial point is to decide on the data items to be recorded, the required precision and how often they should be recorded.

The data items to be recorded are determined by the models by which the objectives of the survey can be achieved, e.g. the swept area method (cf. Section 13.6), length-frequency analysis (cf. Chapter 3), mortality estimation (cf. Chapter 4). Such data items would usually include specifications of the gear used, and for each haul the time and position at start and at end of the haul, wire length, wingspread, bottom type and depth. The catch record should include total weight, species composition by weight and length-frequencies (for selected species).

The required precision depends on the subsequent use of the data. However, often the precision of a trawl survey is controlled by the number of hauls, which limits our ability to decide on the precision.

It must be known how the data are going to be processed before data recording takes place, particularly in cases where data reduction takes place before the recording stage. Consequently, before you can design appropriate log sheets you must have a rather precise idea of how data will subsequently be processed. When designing a log sheet, practical considerations are important, e.g. where the data are entered, on the bridge, in the laboratory or on the deck.

It is important that the data are well organized to facilitate processing, e.g. by computer. This is especially the case when data from more than one vessel have to be combined. Properly designed recording documents will make the correct recording much easier and more dependable. Several documents are usually required to record all the information collected during a survey. These are in the following categories:

1. A log which summarizes the whole cruise.

2. Details of individual stations (or hauls), the "fishing log", which will generally provide information on the vessel's position, time of start and end of haul, gear rigging etc. Summary information on the catch such as total weight and weight composition by species should also be recorded on the fishing log.

3. Detailed information on the catch. This may be in terms of length, weight, sex, maturity stage etc. for each specimen, or samples of length-frequency distributions.

A detailed description of a data processing system for demersal trawl surveys is given in Flowers (1978) who reproduces forms suitable for the work specified as 1 to 3 above.

13.4 DECK SAMPLING AND CATCH RECORDING PROCEDURES

It is important, before the survey begins, to make sure that the equipment and working conditions are such that the sampling can be carried out easily and without risk. Also, the crew must be instructed not to remove any part of the catch before the sampling has been completed.

The following steps pertain to methods for sorting the catch of a fishery research vessel such that the catch composition, both by weight and by number of each species (species group) can be established. The procedure outlined here is from Pauly (1980, adapted from Losse and Dwiponggo, 1977).

Step 1: Remove all sea snakes and other venomous or otherwise dangerous animals. Also remove turtles, and if alive, return these to sea. Record number and kind of animals removed.

Step 2: Remove inorganic debris and plant material. Record type of material removed.

Step 3: Remove the larger fish that are readily visible and place them in a box.

Step 4: Wash the remainder of the catch (of small fish) if necessary, and mix with shovels.

Step 5: Put the mixed catch in boxes, while continuing to remove larger fish and putting them into the box mentioned in Step 3. The boxes should be filled simultaneously, not one after the other, and it should be made certain that all boxes contain approximately the same weight of fish.

Step 6: Count the number of boxes with small fish and record.

Step 7: A rule of thumb is to take one box out of every five at random for subsampling. Record number of boxes taken for subsampling as B1, B2, B3, ... etc.

Step 8:   The box(es) taken for sub-sampling is (are) then treated as follows:

- Weigh the total catch in B1 and record.

- Place fish of B1 on a sorting table and sort to species level as far as food fishes and valuable crustaceans (e.g. shrimps) are concerned, and to taxonomic groupings as well-defined as possible (e.g. genus, family, etc.) for other groups (the non-edible fish and miscellaneous crustaceans).

- Repeat procedure, if appropriate, for the other boxes, B2, B3, ... etc.

Step 9:   If more than one box was sorted, compute, for each species (or higher taxonomic group) the total weight and number in all sorted boxes.

Step 10:   Multiply the numbers and weight of fish and invertebrates by species (or higher taxonomic group) by the ratio of the number of unsorted to sorted boxes.

Step 11:   Weigh and count the larger fish mentioned in Steps 3 and 5, by species (very large fish should be weighed individually and measured).

Step 12:   Add, when there is an overlap (when the fish of a certain species occurred both in the sorted boxes of small fish and in the large fish box) the weights and numbers obtained in Step 11 to the weights and numbers in Step 10.

Step 13:   Step 12 (as well as Step 11, when there is no overlap) provides estimates of total catch, both in weight and number, by species or higher taxonomic groups. Record the total, both in weight and numbers into an appropriate fishing log and convert to catch per unit if fishing time was less or more than an hour. During surveys, this step must be completed after each haul, or every evening at the latest, to preclude loss of information.

Step 14:   In addition to catch sampling, identifying and recording, the work of the fishery scientist generally includes among other things:

- Collecting length-frequency data
- Collecting miscellaneous biological information on the fish caught, e.g., concerning their weight and maturity
- Collecting and preserving specimens for further studies onshore
- Collecting oceanographic data.

It is important to sort and sample the catch to the greatest extent possible. The extra cost involved in sorting and sampling all the edible species is usually only a fraction of the total costs of the survey.

13.5 THE SWEPT AREA

From Fig. 13.5.1 it appears that the trawl sweeps a well defined path, the area of which is the length of the path times the width of the trawl, called the "swept area" or the "effective path swept". The swept area, a, can be estimated from:

$$a = D*h*X_2, \quad D = V*t \qquad (13.5.1)$$

where V is the velocity of the trawl over the ground when trawling, h is the length of the head-rope (see Fig. 13.5.1), t is the time spent trawling. X_2 is that fraction of the head-rope length, h, which is equal to the width of the path swept by the trawl, the "wing spread", $h*X_2$ (see Fig. 13.5.1).

For southeast Asian bottom trawls values of X2 from 0.4 (Shindo, 1973) to 0.66 (SCSP, 1978) are reported. Pauly (1980) suggests X2 = 0.5 as the best compromise. In the Caribbean a value of X2 = 0.6 was used by Klima (1976).

For the estimation of biomass we use the catch per unit area (CPUA), as will appear in the next section. It is estimated by dividing the catch by the swept area (in square nautical miles or square kilometers). This estimate thus depends on how accurately the swept area is estimated. The swept area as defined in Fig. 13.5.1 assumes that the bridles have no herding effect, which they are in fact known to exhibit. The wing spread is calculated as the fraction X2 of the head rope length. The wing spread varies with hauling speed, weather conditions, current velocity and direction and warp length and is therefore not well defined. It can be accurately measured by special devices, but an approximate value can be calculated from measurements of the distance between the warps at the gallows and at say, 1 m, towards the net.

When exact positions of the start and the end of the haul are available the distance covered can be estimated in units of nautical miles (nm), by:

$$D = 60 * \sqrt{(Lat1-Lat2)^2 + (Lon1-Lon2)^2 * \cos^2(0.5(Lat1+Lat2))} \qquad (13.5.2)$$

where

Lat1 = latitude at start of haul (degrees)
Lat2 = latitude at end of haul (degrees)
Lon1 = longitude at start of haul (degrees)
Lon2 = longitude at end of haul (degrees)

If exact positions are not available, but only the velocity of the vessel and its course together with direction and speed of the current, then the distance covered per hour can be calculated from:

$$D = \sqrt{VS^2 + CS^2 + 2*VS*CS*\cos(dirV-dirC)} \quad (nm) \qquad (13.5.3)$$

where

VS = velocity of vessel (knots = nm/hr)
CS = velocity of current (knots)
dirV = course of vessel (degrees)
dirC = direction of current (degrees)

Eq. 13.5.3 is an application of "vector addition" (see Fig. 13.5.2)

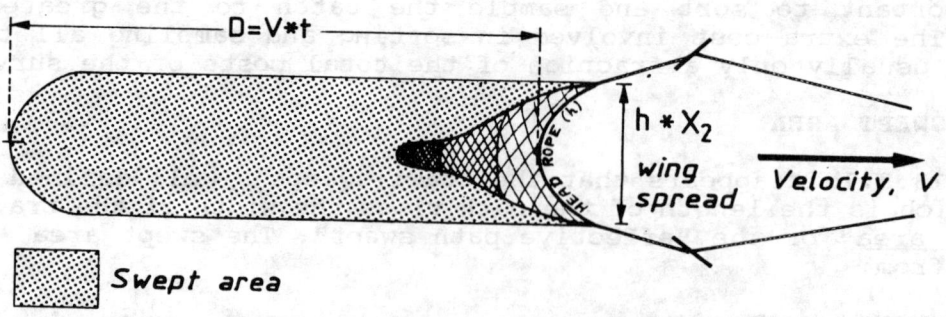

Fig. 13.5.1 Swept area (cf. Fig. 4.3.0.1)

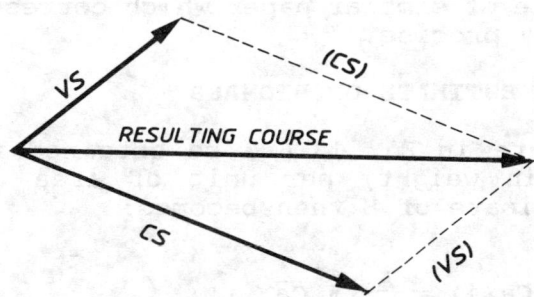

Fig. 13.5.2 Vector addition of vessel and current course and velocity

13.6 BIOMASS ESTIMATION BY THE SWEPT AREA METHOD

Let Cw be the catch in weight of a haul. Then Cw/t is the catch in weight per hour, when t is the time spent hauling (in hours). Let a be the area swept (cf. Eq. 13.5.1). Then a/t is the area swept per hour, and

$$\frac{Cw/t}{a/t} = \frac{Cw}{a} \quad kg/nm^2 \qquad (13.6.1)$$

becomes the catch in weight per unit of area. Let X1 be the fraction of the biomass in the *effective* path swept by the trawl which is actually retained in the gear and let $\overline{Cw/a}$ be the mean catch per unit area of all hauls. Then an estimate of the average biomass per unit area, b, is:

$$\overline{b} = (\overline{Cw/a})/X1 \quad kg/nm^2 \qquad (13.6.2)$$

Let A nm^2 be the total size of the area under investigation. Then an estimate of the total biomass, B, in this area, A, is obtained from:

$$B = \frac{(\overline{Cw/a})*A}{X1} \qquad (13.6.3)$$

It is difficult to estimate which proportion of the fish that is present in the area swept by the trawl gear is actually retained by the gear, in other words it is difficult to give a precise estimate of X1. Underwater television recordings show that the reaction of fish to trawls varies markedly between species. The value of X1 is usually chosen between 0.5 and 1.0. For trawlers in southeast Asia a value of X1 = 0.5 is commonly used in survey work (Isarankura, 1971, Saeger, Martosubroto and Pauly, 1980). Dickson (1974), on the other hand, suggests X1 = 1. The difference between these two values of X1 is difficult to resolve. Using X1 = 0.5 doubles the estimate of biomass compared to that obtained by using X1 = 1.0.

The duration of the haul is proportional to the distance covered so the duration should have no direct influence on the catch per unit of area. However, the catchability, X1, of different species may vary according to the duration of the haul because some species, when herded by the trawl get tired soon and get captured while other species are able to swim in front of the trawl for a long period and thus avoid being caught. It is therefore important to standardize the duration of the haul so that results from different hauls can be compared. To investigate the dependence of catchability on haul duration, parallel hauls of different duration (e.g half an hour and one hour) could be made.

The total area surveyed can be estimated from a mercator projection by means of a planimeter. Another method is to copy the area on transparent paper which then is cut out and weighed. Then the weight of the paper is compared to the weight of a piece of similar paper which corresponds to a known area. This method is not very precise.

13.7 PRECISION OF THE ESTIMATE OF BIOMASS

Let the biomass estimate in Eq. 13.6.3 be obtained from n hauls, and let $Ca(i)$ be the catch (in weight) per unit of area of haul no. i, where $i = 1, 2, \ldots, n$. The estimate of B then becomes:

$$B = \frac{A}{X1} * \frac{1}{n} * \sum_{i=1}^{n} Ca(i) = \frac{A}{X1} * \overline{Ca} \qquad (13.7.1)$$

and the variance (cf. Eq. 2.1.2):

$$VAR(B) = \left[\frac{A}{X1}\right]^2 * \frac{1}{n} * \frac{1}{n-1} * \sum_{i=1}^{n} [(Ca(i) - \overline{Ca})]^2 \qquad (13.7.2)$$

Thus, a higher precision (a smaller variance) can be obtained by increasing the number of hauls, n (cf. Section 7.1, Fig. 7.1.1). Another way of reducing the variance is to apply stratified sampling (cf. Section 7.2). Suitable stratification may reduce the variance considerably for the same number of hauls and thus improve survey efficiency for the available research vessel time. The distribution of many fish species is determined by depth and bottom type. Therefore stratification based on these factors is widely used. Fig. 13.7.1 shows an example of stratification based on depth.

The areas of the four strata in Fig. 13.7.1 are A1, A2, A3 and A4 and the total area $A = A1+A2+A3+A4$. Let $B(i)$ be the estimated biomass in stratum no. i calculated by Eq. 13.7.1. Then the estimate of the total biomass of the total area, A, becomes:

$$B = B(1) + B(2) + B(3) + B(4) \qquad (13.7.3)$$

and the variance:

$$VAR(B) = VAR(B(1)) + VAR(B(2)) + VAR(B(3)) + VAR(B(4)) \qquad (13.7.4)$$

where $VAR(B(i))$ is calculated by Eq. 13.7.2. If we work with densities, in units of biomass per square nm. (cf. Eq. 13.6.2) the parallel to Eq. 13.7.3 becomes:

$$b = [b(1)*A1 + b(2)*A2 + b(3)*A3 + b(4)*A4]/A \qquad (13.7.5)$$

and the parallel to Eq. 13.7.4 is

$$VAR(b) = VAR(b(1))*(A1/A)^2 + \ldots + VAR(b(4))*(A4/A)^2 \qquad (13.7.6)$$

where Ai is the size of stratum no. i and A is the total area of all strata. Eq. 13.7.6 is then seen to be analogous to Eq. 7.2.11 with A as N, Ai as $N(j)$. Confidence limits of B and b can be calculated as described in Section 2.3.

Stratification may be based on the total catch of a mixture of species or on the catch of a single species. It is often desirable, however, to focus on several species, or groups of species, each with its own type of distribution. In such cases a stratification procedure must be decided on for each species, or for each group of species with similar distribution patterns within the area surveyed.

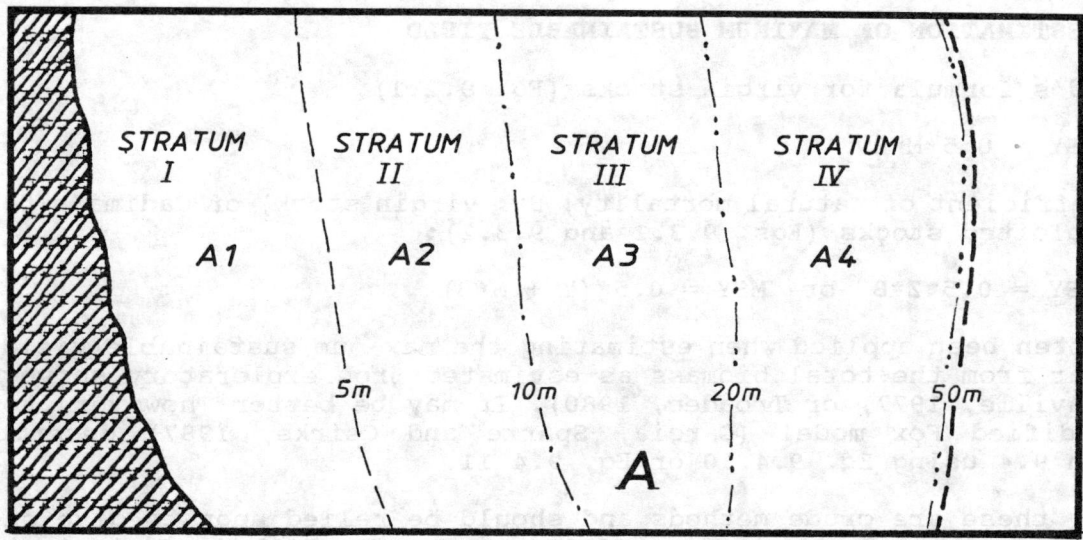

Fig. 13.7.1 Example of stratification

For a more detailed description of the statistical aspects of a trawl survey see Fogarty (1985) and Gavaris and Smith (1987). The latter discuss the effect of stratification by comparing the precision obtained from stratified sampling to that of simple random sampling, using the method of Sukhatme and Sukhatme (1970).

The above formulas all assume that the number caught per trawl hour is a normally distributed random variable. A better description of survey data is often obtained by assuming the so-called "delta-distribution" (Pennington, 1983). This distribution takes into account that survey data often contain a large proportion of zero observations and that the non-zero observations form a log-normal distribution rather than a normal distribution.

A log-normal distribution is asymmetric and skewed to the right hand side. Fig. 13.7.2 shows a schematic delta-function. Note that the point zero has a positive probability. In his paper, Pennington gives the expression for the unbiased estimators of the parameters of the delta-function.

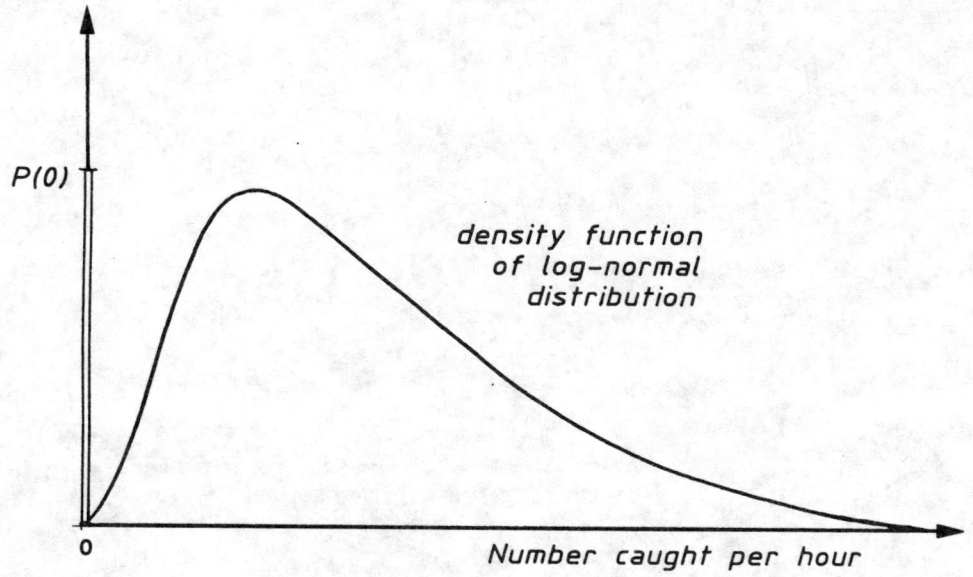

Fig. 13.7.2 Delta-function

13.8 ESTIMATION OF MAXIMUM SUSTAINABLE YIELD

Gulland's formula for virgin stocks (Eq. 9.2.1):

$$MSY = 0.5*M*Bv$$

(M: coefficient of natural mortality; Bv: virgin stock) or Cadima's formulas for exploited stocks (Eqs. 9.3.1 and 9.3.2):

$$MSY = 0.5*Z*B \quad \text{or} \quad MSY = 0.5*(Y + M*B)$$

have often been applied when estimating the maximum sustainable yield (MSY) per year from the total biomass as estimated from exploratory surveys (see e.g. Saville, 1977, or Troadec, 1980). It may be better, however, to apply the modified Fox model (Garcia, Sparre and Csirke, 1987) described in Section 9.4 using Eq. 9.4.10 or Eq. 9.4.11.

Anyway, these are crude methods and should be relied upon to give only the order of magnitude (1000 tonnes, 10000 tonnes, etc.) of the fish stock. Yet, if nothing is known in advance, even such an estimate of the order of magnitude may be valuable.

One problem is that M is usually not known before the surveys are carried out. When the stock is virgin (unexploited) an estimate of M can be obtained by analyzing the catch curves obtained during the exploratory survey (cf. Section 4.4) because when the stock is unexploited, then F = 0 and thus Z = M. Pauly's empirical formula, Eq. 4.7.2.1 (Pauly, 1980b) can be applied when growth parameters are available.

(See **Exercise(s)** in Part 2).

14 SUMMARY OF FISH STOCK ASSESSMENT

The introduction to this manual contains flow-charts for fish stock assessment (Fig. 1.3.0.1) and for the organization of the manual (Fig. 1.8.1). The beginner in fish stock assessment might have had some difficulties with the exact interpretation of parts of these figures. This chapter attempts to give a resumé of the methodology along the same lines as those of the introduction, but now assuming that the reader is familiar with the theory presented in the foregoing chapters.

The level of precision of fish stock assessments increases with the availability of data, which is usually positively correlated to the development of the fishery. In the case of unexploited stocks, assessments will have to be based on general ecological principles, or surveys with fishery research vessels. As soon as exploitation starts, the fishery itself can provide the data base for the application of more sophisticated assessments. In a very highly developed fishery a large portion of the stocks is landed and therefore accessable to sampling.

The methodology of fish stock assessment and prediction changes according to increasing availability of data. While preliminary assessments may be based on, for example, relationship between primary and secondary production or comparisons between unexploited areas and exploited areas with similar environmental characteristics, the first real assessments are usually derived from resource survey data obtained with a fishery research vessel. Assessments of the standing stocks of pelagic and demersal fish can be made by acoustic methods and by a bottom trawl. Although surveys with other types of gear/equipment, such as traps, gill nets and hooks yield data on catch rates, they can not be used directly for estimating the standing stock, because it is not known which area has been covered by the gear. The holistic model (Fig. 1.8.1B) most frequently used for assessment purposes at this stage, is the swept area method, which was discussed in Chapter 13.

An intermediate level of stock assessment can be reached when time series of catch and effort data are available from a developing fishery. Such data can be used in other holistic models, the so-called surplus production models (Fig. 1.8.1B and Chapter 9).

Once a fishery has developed and biological sampling schemes have been set up, it is possible to apply more sophisticated analytical models, which can be divided into two major categories, viz. age-based and length-based models (Fig. 1.8.1A).

In areas where it is possible to determine the age of a fish from otoliths or other hard parts, an age-length key can be established each year which can be used to assign ages to large length-frequency samples taken from the commercial fisheries. This data is then used to determine the number of fish in the sea per cohort or age group. Fairly reliable predictions can be made on that basis.

In areas where ages of fish cannot be determined on a routine basis from otoliths etc. or for resources which do not have structures that permit age-reading e.g. crustaceans, analyses have to be mainly based on length-frequencies. As long as the routine age-reading of tropical fish species is still under development most assessments in the tropical zone will depend on the interpretation of length-frequencies.

Unfortunately, length-frequency data are easily affected by biases caused by gear selectivity, migration and sampling errors, and this further complicates the assessment of tropical species. In this manual a lot of space has been dedicated to the analysis of length-frequency data. This does not mean, however, that the models used are fully length-based. In most of the methods given, the length-frequency data are used to separate presumed cohorts from each other, which are then assigned a relative age. The rest of the procedures is more age-based than length-based. It may be useful to point out here that the determination of growth parameters is only an intermediate step in assessments based on length-frequencies. The final objective is always to determine the amount of fish per cohort and the level of fishing mortalities it is subjected to. This is the basis for predictions of future yields and therefore the basis for future development or management measures.

14.1 GENERAL ASPECTS OF FISH STOCK ASSESSMENT

The main objective of fish stock assessment of exploited stocks is to predict what will happen in terms of future yields, biomass levels (sustainability) and value of the catch, if the level of fishing effort remains the same or if it is changed in one way or another.

Figure 14.1.1 is an extended version of Fig. 1.3.0.1, giving a number of examples for each box. It shows both the final output of fish stock assessment and the methodology suggested in the present manual to obtain that output. The methodology can be divided into two parts: firstly the estimation of vital parameters, and secondly, the use of these parameters to predict future catches and stock biomasses under various assumptions about the development of the fishery.

Fig. 14.1.1, however, does not contain unquantifiable features of fish stock assessment, such as subjectivity of interpretation and bias in data. The more biased data are, the more subjective is their interpretation. If data were exactly as the models assume them to be, everything would be straightforward and any two persons would independently of each other come to the same result. Essentially, the models by which historical data are analysed assume that data are random samples, that is, unbiased.

Bias may have many causes, some of which can be eliminated, and some of which are very hard to tackle. Bias caused by badly planned and/or executed sampling programmes should be easy to overcome, while bias caused by migration and gear selectivity, may be difficult to handle. Often the technical solution to such biological bias problems is to expand and adjust the sampling programme, based on the first analyses. On top of the practical problems mentioned above, there are statistical (or theoretical) problems in sampling. For example: "How many fish should be sampled to secure a successful Bhattacharya analysis". There is probably no exact solution to this problem, but it is evident that analyses based on length-frequencies need a large amount of unbiased input data.

The prerequisites for any meaningful fish stock assessment are biological data on the resources and technical data on the fisheries. There are two main types of data, which in order of priority are:

1) Data sampled from the catches of commercial fisheries

2) Data from research vessel surveys

Essentially, fish stock assessments of exploited stocks can be made with data from the commercial fisheries only. In the case of unexploited resources, research vessel data are essential. However, the latter may also be very useful as a supplement to commercial fisheries data. Samples from commercial fisheries are usually much cheaper and much easier to obtain in large quantities than research vessel survey data, a feature that should be taken into account when planning fishery research.

```
INPUT:              ┌─────────────────────────────────────────┐
                    │ FISHERIES DATA (+ ASSUMPTIONS)          │
                    │ (Data preferably by stock or species)   │
                    │ Catch/effort of total commercial catch. │
                    │ Length- or age-frequency data from com- │
                    │ mercial fisheries or surveys with       │
                    │ research vessels.                       │
                    │ Biological data (maturity stage, length/│
                    │ weight etc) from research vessel surveys│
                    │ and/or from sampling commercial fisheries│
                    │ General knowledge of the biology of the │
                    │ species (spawning season, migration,    │
                    │ etc.) and of the fisheries (area, gear, │
                    │ season)                                 │
                    └─────────────────────────────────────────┘
                                       v
PROCESS:            ┌─────────────────────────────────────────┐
                    │ Analyses of historical data:            │
                    │   Estimation of growth parameters       │
                    │   Catch curve analysis                  │
                    │   Virtual population or cohort analyses │
                    │   Surplus production models             │
                    │   Swept area method (research vessel data)│
                    └─────────────────────────────────────────┘
                                       v
INTERMEDIATE        ┌─────────────────────────────────────────┐
OUTPUT:             │ ESTIMATES OF POPULATION PARAMETERS      │
(used as input for  │ Von Bertalanffy growth parameters       │
the next process)   │ Length-weight relationship              │
                    │ Total mortality                         │
                    │ Fishing mortality and natural mortality │
                    │ Stock size (size composition of stock)  │
                    └─────────────────────────────────────────┘
                                       v
PROCESS:            ┌─────────────────────────────────────────┐
                    │ Predictions of yield:                   │
                    │   Beverton and Holt's yield/recruit model│
                    │   Thompson and Bell's model (for single │
                    │     stock or for mixed fisheries)       │
                    │   Surplus production model              │
                    │   Swept area method combined with a     │
                    │   Gulland-type of model                 │
                    └─────────────────────────────────────────┘
                                       v
FINAL OUTPUT:       ┌─────────────────────────────────────────┐
                    │ OPTIMUM FISHING LEVEL                   │
                    │ (the optimum level of fishing mortality)│
                    │ MAXIMUM SUSTAINABLE YIELD OR VALUE and  │
                    │ present position with respect to optima │
                    └─────────────────────────────────────────┘
```

Fig. 14.1.1 General flow-chart for fish stock assessment
            (compare Fig. 1.3.0.1)

The two types of data essentially contain the following elements:

> Commercial fishery data:
>
> Total catch (by species, area and type of gear)
> Effort (by area and type of gear)
> Length-frequencies (by species and sex)
> Age-frequencies (by species and sex)
> Biological data (e.g. maturity stages, length/weight relationship)
> Data on fishing gear (e.g. mesh size) and fishing operations
>
> Research survey data:
>
> Catch and effort (e.g. number caught per hour by species)
> Catch per unit area (for swept area calculation)
> Length-frequencies (by species and sex)
> Age-frequencies (by species and sex)
> Biological data (e.g. maturity stages, length/weight relationship)
> Data on fishing gear (e.g. mesh size)

Some data are easy to define and record such as the length of a fish and the number of fish. Other data are more problematical to obtain, such as the age of a fish read from daily rings. Finally, there are data which are very problematical to define, such as a measure for effort that is proportional to fishing mortality. (This type of data problem may in certain cases never be solved, and perhaps the way out is to search for alternative models, which require a different type of input data.)

There is no systematic way to overcome the problems of defining data and the bias problems. The choice of an adequate methodology is to a certain degree a matter of personal judgement combined with skill and experience. This manual may appear as a kind of cookery-book with pre-determined recipes for each case, but unfortunately, it cannot be used as such. The most important background for a fish stock assessment worker is a profound knowledge of the biology of the living resources and the fisheries exploiting them gained through continuous sampling of as many (relevant) data as possible, year after year. A long time series, say several decades, of fisheries data may contain data reflecting extreme situations, and thus define the upper and lower limits for predictions of yield and stock size.

In some cases, only a part of the catch from a stock is covered by the sampling programme. That happens, for example, when several countries exploit a shared resource and do not exchange or pool their data, or when some countries do not collect data. In that case the catches should in one way or another be raised to account for the total. The raising would be based on certain assumptions, such as country B (which does not collect samples) has a fishery which corresponds to, say, 20% of that of country A (which collects samples). In such cases it is always better to make a qualified guess than to ignore an important component of the catch taken from a stock.

Often there are doubts about the stock definition. If that is the case, remember that it is better to make the mistake of pooling two small stocks than to make assessments of only a part of a stock. Because, even when two stocks have been wrongly pooled, the conclusions for the combined stocks may be correct also for each stock individually. If on the other hand, a stock is being exploited by more than one fleet, say from different countries, and the fishing mortality caused by the other fleet(s) is not taken into account when doing the assessments and making the predictions; the effect of management measures may not be as expected. For example, a decrease in fishing effort by one fleet may be offset by an increase in the other fleet. Then the total fishing pressure may remain the same or actually increase and the predicted improvements in yields may not occur.

Fisheries data may include other types of data than those dealt with in the present manual. Such data may concern economic, sociological and environmental aspects. Although a fishery biologist is concentrating on the biological/technical aspects of fisheries, she/he should also be aware of progress in these other research fields. This applies in particular to bio-economics, which is now developing rapidly (see Chapter 8).

14.2 REVIEW OF METHODS TO BE USED ACCORDING TO THE TYPE OF DATA AVAILABLE

One of the most difficult aspects of fish stock assessment is to decide which methods should be used to analyze a particular data set. In this section a series of data sets are presented in a systematic way, grouped at different levels of availability and quality. These sets will be discussed at a case-by-case basis.

In the following five main levels of availability of data are considered, namely:

  Level A: When only survey data are available
  Level B: When only catch or catch/effort data are available
  Level C: When only length-frequencies are available
  Level D: When both catch/effort data and length-frequencies are available. (One case with limited age-frequency data is also included.)
  Level E: When all kinds of data are available, in particular time series of age-frequencies.

The review starts at the lowest level of data availability and gradually moves to the highest level. Within each main level, a number of cases is considered, which are categorized with the aid of a small table where the existence/non-existence of data from the commercial fishery and research survey data are indicated by the following:

  -                     : No data collected
  Single   : A single or a few samples collected
  Single time series   : A time series for one year has been collected
  Multiple time series : Time series for two or more years are available

Level A: When only survey data are available

Case A.1

A.1 Data	Catch/Effort	Length-frequencies	Age-frequencies	Biological data
Commercial fishery	-	-	-	-
Research survey	Single	-	-	-

This case and the following may apply to hitherto unexploited resources, or exploited resources which have never been investigated by fish stock assessment workers. Case A.1 deals with a virgin stock. Use the swept area method for analysis (see Chapter 13) and the Gulland formula for prediction of the potential yield (see Section 9.3). The result of this exercise gives only the order of magnitude of the sustainable potential yield (that is: 100 tonnes or 1000 tonnes, or 10 000 tonnes, etc.).

Case A.2

A.2 Data	Catch/Effort	Length-frequencies	Age-frequencies	Biological data
Commercial fishery	Single	-	-	-
Research survey	Single	-	-	-

In this case the resource is not a virgin stock, because there is a commercial fishery. Use the swept area method (see Chapter 13) for the historical analysis and the Garcia et al. formula for the catch prediction (see Section 9.5).

Level B: When only catch or catch/effort data are available

With this type of data only surplus production models can be used to predict the maximum sustainable yield. The models at hand are mathematically very simple, and therefore large deviations from the models can be expected.

Case B.1

B.1 Data	Catch/Effort	Length-frequencies	Age-frequencies	Biological data
Commercial fishery	Single	-	-	-
Research survey	-	-	-	-

This is the lowest possible level allowing for a sort of fish stock assessment. The method suggested is the simplest possible one - it is so simple that it has not been mentioned earlier in the manual. Prediction for future years equals the catch observed. Effort is difficult to use, as nothing is known for sure about the state of exploitation. If only the catch is known (effort not known) for a series of years, use the average catch for prediction. There is no real historical analysis involved here.

Case B.2

B.2 Data	Catch/Effort	Length-frequencies	Age-frequencies	Biological data
Commercial fishery	Multiple time series	-	-	-
Research survey	-	-	-	-

Use surplus production models (Schaefer or Fox models, see Chapter 9). If results of alternative methods deviate substantially, select the result giving the best fit to the data. If data are not by species, but by species group (family) the method may still be applied. The models are applied for the historical analysis as well as prediction. If data are by fishing grounds in the form of catch rates by area, for example in coral reef areas, the Munro and Thompson plot should be used (see Section 9.5).

Level C: When only length-frequencies are available

The methodology of length-frequency analysis is based on the assumption that recruitment is seasonal with one or at most two peaks per year (see Section 1.6). It assumes that samples cover the entire range of lengths. Often, there are problems with the smallest length classes due to gear selectivity and sometimes certain length groups are missing from the samples due to the migratory behaviour of the species in question.

The length-based methodology depends on the life span of the species under investigation. Naturally, only a short time series of data is needed for estimation of growth of a species with a short life span. On the other hand, frequent samples (say monthly) are needed for short-lived species like shrimps, while for long-lived species such as groupers one annual sample may be adequate.

As a matter of routine, try various forms of data-massage, that is, see what happens, for example, when doubling the size of the length class or when doubling the time period (see Section 3.4.2). Samples where bias problems can be demonstrated may be excluded. In cases of migration only data which are separated with a time period of one year should be used (see Section 11.3.1). There are many different ways of combining massaged data, so do not hesitate to try any sensible combination.

Be careful with computerized methods. Some computer programs give results irrespective of the degree to which the data conform to the underlying model. Choose the methods which provide a warning when data are in conflict with the model. Be critical of the results. Do not, for example, accept a growth curve fitting, unless a modal progression can already be seen in the original data.

If the Bhattacharya and similar methods do not give convincing results even after all possible massaging, then the only solution is to try to read ages from hard parts.

In the first, most data-limited case (C.1), data do not include total catches and effort of the commercial fishery. Thus, no estimate of total catch is available, and all prediction results consequently become relative. The cases under C deal with the situation where research on a resource has just been started.

Case C.1

C.1 Data	Catch/Effort	Length-frequencies	Age-frequencies	Biological data
Commercial fishery	-	Single	-	Single
Research survey	-	-	-	-

C.1 Historical analysis

<u>Growth parameters</u>: Use length-frequency analysis for example, ELEFAN I or the Bhattacharya method (see Section 3.4.1) combined with (pseudo) modal progression analysis (see Section 3.4.2) or some more sophisticated maximum likelihood method, to estimate growth parameters from a single sample under the assumption that the normally distributed components correspond to one or two recruitments per year.

<u>Mortality rates</u>: Use the length-converted catch curve to estimate Z (see Section 4.4.5), under the assumption of constant recruitment and constant mortality. Use Pauly's formula for M (Eq. 4.7.2.1) and estimate F by

subtracting M from Z. Estimate L50% from the catch curve and convert it to age (t50%) (see Section 6.5).

C.1 Catch prediction

Use Beverton and Holt's yield per recruit model (Sections 8.2 to 8.3). Use t50% as the "knife-edge-age". In cases where the length-weight relationship has an exponent significantly different from 3, or Z was found not to be constant after recruitment is completed, use the Thompson and Bell model (see Sections 8.6 to 8.7). If the fishery considered is a mixed fishery (most tropical fisheries are) then use the method for mixed fishery (see Section 10.4.2) for as many species as possible combined. Whenever possible, try to make the catch prediction for all fleets in one go (see Section 10.4.3) and calculate the value of all yields combined (value of landings in weight).

C.1 Comments

In theory this procedure is possible when only one single length-frequency sample is available. However, the results are subject to an unknown bias and uncertainty, and should always be followed up by further investigations. Length-frequency analysis may be dubious because the pseudo modal progression has to be based on the assumption that the single sample can be used as an estimate for the entire life span of the species. If the species is short-lived (for example shrimps) and therefore only shows one or at most two peaks, the method may not be applicable at all. In the latter case the sampling must be continued to obtain a time series (for example for each month of the year) before the estimation of growth parameters can be started.

Case C.2

C.2 Data	Catch/Effort	Length-frequencies	Age-frequencies	Biological data
Commercial fishery	-	Single	-	Single
Research survey	Single	Single	-	Single

C.2 Historical analysis

Growth parameters: Use length-frequency analysis on samples combined from research survey and from commercial fishery as under case C.1. The research survey data often provide a better coverage of the smaller length groups than the data from the commercial fishery.

Mortality rates: Use the length-converted catch curve based on the survey data to estimate Z, under the assumption of constant recruitment. The survey data are assumed to represent the entire stock better than the commercial fishery, which does not attempt to collect random samples. Use Pauly's formula for M and estimate F by subtracting M from Z. Do a length-based cohort analysis (see Section 5.4) and compare the average F for the fully-exploited age groups with the estimate obtained from the catch curve analysis. Estimate L50% from the catch curve and convert it to age (t50%).

C.2 Catch prediction

If estimates of stock size (or absolute recruitment) are available, use the Thompson and Bell model. The Beverton and Holt Y/R model can be used if the estimated F-array looks like knife-edged, and if the exponent in the length-weight relationship is not significantly different from 3. In all other cases it is better to use the Thompson and Bell model.

Level D: When size distributions and catch/effort data are available

In this case it is possible to raise the size-frequencies to the total catch taken from the stock in question. It is assumed that all major gear categories (fleets) have been sampled, and have been raised and summed to represent the total catch taken from the stock. If only part of the fishery is covered by the sampling programme, the estimate of total catch must be raised in one way or another to account for all catches. A skilled guess on the unknown catches will have to be used in the worst case.

Case D.1

D.1 Data	Catch/Effort	Length-frequencies	Age-frequencies	Biological data
Commercial fishery	Single	Single	-	Single
Research survey	-	-	-	-

D.1 Historical analysis

Growth parameters: Use length-frequency analysis on samples from the commercial fishery as under case C.2.

Mortality rates: Use Pauly's formula for M and length-converted cohort analysis for the estimation of F and stock size by length group. Note that the recruitment number is also estimated.

D.1 Catch prediction

Use the length-based Thompson and Bell model. Use the recruitment estimated by cohort analysis as input to predict the absolute yield and stock biomass. Use the Thompson and Bell model for assessing the effect of changing the fishing pattern. Notice that in this case, the resultant selection parameters are not explicitly estimated - they are embedded in the F-by-length-class-array.

Case D.2

D.2 Data	Catch/Effort	Length-frequencies	Age-frequencies	Biological data
Commercial fishery	Single time series	Single time series	-	Single time series
Research survey	-	-	-	-

In case D.2 there is a time series over one year, for example, length-frequency samples each month of the year, for all major gear categories. In the case of short-lived species like shrimps the time series may cover the entire life span of the species.

D.2 Historical analysis

Growth parameters: Use length-frequency analysis on samples from the commercial fishery. In this case it is possible to make a complete modal progression analysis for short-lived species, but for long-lived species (life span of two or more years) the modal progression becomes a mixture of pseudo and real cohorts.

Be critical of the results from growth curve fitting. If the time series contains parts with apparent negative growth (see Section 11.2) or periods of no growth, the possibility of bias in the data should be considered carefully, and perhaps a better way of sampling should be found. Do not ignore the parts of the time series which do not show modal progression, and do not use only those parts which show modal progression, unless there is a good rational justification for so doing. Be aware that bias caused by migration may appear as seasonality of the growth rate, and that it may be more or less impossible to separate the two phenomena.

Mortality rates: As for D.1. As input to the length-based cohort analysis, use all samples summed over the year (properly weighted). In the case of a short-lived species it is possible to make an age-based cohort analysis, if, for example, length-frequencies are available for each month and have been resolved into age-group components.

It is possible to do an age-based cohort analysis or a VPA (see Section 5.1), using the age groups estimated by the Bhattacharya analysis as input (see Fig. 7.3.5).

D.2 Catch prediction

Same as for D.1. For short-lived species the effect of a closed season can now also be assessed (see Section 8.6).

D.2 Comments

As the entire analysis is based on only one year's data, the results should be taken with reservation. Whether the year in question is an exceptional year or is close to the average year is not known. For example, there is no information on the variability in recruitment. It is of utmost importance to know if recruitment remains at a stable level or if it is highly variable.

Once length-frequency samples have been resolved it is possible to use only age-based methods, which have the advantage of being easier to work with, while the results are easier to interpret.

Case D.3

D.3 Data	Catch/Effort	Length-frequencies	Age-frequencies	Biological data
Commercial fishery	Multiple time series	Multiple time series	-	Multiple time series
Research survey	-	-	-	-

In case D.3 there is a multiple time series, for example, length-frequency samples each month of the year, during, say, ten years, for all major gear categories. In the case of short-lived species like shrimps the time series may cover the life span of the species many times, and even for long-lived species some cohorts may be represented during their entire life span.

D.3 Historical analysis

Growth parameters: Use length-frequency analysis on samples from the commercial fishery. In this case it is possible to make a complete modal progression analysis for both short-lived and long-lived species.

Mortality rates: The average annual length composition may be used as input for a length-based cohort analysis, which would give the average fishing mortalities and stock size for the period covered. An age-based cohort analysis, with age groups from the Bhattacharya analysis as input may also be attempted. In the latter case the recruitment number will be estimated for each cohort, and some understanding of the variability of recruitment will be gained.

D.3 Catch prediction

Use the length-based and/or the age-based Thompson and Bell model, depending on the type(s) of cohort analysis made.

D.3 Comment

There may still be problems with the estimation of growth parameters, because certain length groups may be lacking, for example, because of gear selection and/or migration. The commercial fishery may not cover the entire distributional area of the resources, for example, the boats may only be able to fish to certain depths or their range may be limited.

Case D.4

D.4 Data	Catch/Effort	Length-frequencies	Age-frequencies	Biological data
Commercial fishery	Multiple time series	Multiple time series	-	Multiple time series
Research survey	Single time series	Single time series	-	Single time series

D.4 Historical analysis

In this case there are two independent data collections and they may be used for mutual verification, or to disclose bias problems.

Growth parameters: It may be necessary to decide to use only the survey data for estimation of growth parameters, if the commercial data are suspected to be heavily biased. The two types of data may also be combined. Missing data from the commercial fishery may be filled in with research vessel data.

Mortality rates: Use cohort analysis or VPA. If the research vessel survey gave a swept area estimate of biomass, it may be possible to select a fishing mortality which produces a recruitment number which in the cohort analysis produces the same stock biomass as estimated from the survey. If this is not possible, then either the survey or the commercial sampling programme is biased. If both of them are biased it may not be possible to get agreement between the two methods. Z may be estimated from the catch curve analysis and be compared to the M plus the average F estimated from cohort analysis.

D.4 Catch prediction

Same as case D.3.

C.4 Comment

With the data of case D.4, there is no way to verify the estimates of growth parameters from length-frequencies.

Case D.5

D.5 Data	Catch/Effort	Length-frequencies	Age-frequencies	Biological data
Commercial fishery	Multiple time series	Multiple time series	Single	Multiple time series
Research survey	Multiple time series	Multiple time series	Single	Multiple time series

D.5 Historical analysis

In this case a few age readings are available which can be used to verify the results based on length-frequency analysis.

Growth parameters: Use the least squares method to estimate growth parameters from age/length data (see Section 3.3.4). If the results from the otolith data deviate from results from the length-frequency analysis, check the otolith readings carefully, for example, by comparison with other workers. If results still differ, use the otolith results rather than the length-based results, unless there are really good reasons to prefer the length-based methods, such as in the case where there are severe difficulties in reading the hard parts, while at the same time the length-frequency analysis shows a beautiful modal progression. For some small pelagic species it has proved to be very difficult to apply length-frequency analysis, and then the only solution seems to be age readings on hard parts.

Try to create an age/length key (see Section 3.2.1) and use it to convert length-frequencies into age-frequencies. This approach would replace the Bhattacharya method. However, as age/length keys may vary from year to year, otoliths should preferably be collected and read on a routine basis to apply this technique.

Mortality rates: Same as case D.4.

D.5 Catch prediction

Same as case D.4.

D.5 Comment

Some tropical fish deposit annual rings in otoliths or other hard parts, and for such species otoliths should be collected on a routine basis. For many tropical species, only the daily rings are useful for age determination, and then age reading becomes rather tedious and expensive. The results of age reading from daily rings may also depend on the equipment used. In some cases more rings can be observed when using a scanning electronic microscope than with a light microscope (see Morales-Nin, 1991).

Level E: When all kinds of data are available

This is the ideal case, where any kind of data you may think of is available. Forget all about von Bertalanffy growth curves, and do not use length-based methods. However, length-frequencies may be used in connection with an age/length key, and also for an assessment of gear selection. Mortality rates and stock sizes should be estimated with age-based methods.

If all these data are available several methods can be applied which are not described in the present manual, such as "tuning" of VPA and multi-species VPA. Tuning of VPA means that the VPA results are compared with independent observations which can be assumed to be proportional to the VPA results. For example, effort is supposed to be proportional to fishing mortality and catch per unit of effort (CPUE) from a research vessel survey is supposed to be proportional to the stock numbers estimated from VPA. For pelagic species estimates obtained by acoustic surveys may be useful for VPA tuning.

These are methods used in the North East Atlantic, where ICES (International Council for the Exploration of the Sea) is the scientific body that provides advice to the managers of fishery resources.

Case E.1

E.1 Data	Catch/Effort	Length-frequencies	Age-frequencies	Biological data
Commercial fishery	Multiple time series	Multiple time series	Multiple time series	Multiple time series
Research survey	Multiple time series	Multiple time series	Multiple time series	Multiple time series

E.1 Historical analysis

Growth parameters: Not really needed. The results of otolith readings provide a weight-at-age-array, which can be used as input to the Thompson and Bell model. Actually, any growth curve can be used. (The growth of some fish does not conform very well to the von Bertalanffy growth model.)

Mortality rates: Use VPA. If stomach content data are available then predation can be estimated and multispecies VPA be applied. If effort data are available then use VPA-tuning methods.

E.1 Catch prediction

Use the multispecies and multifleet Thompson and Bell model with all landings converted into value.

E.1 Comment

For some countries with large fisheries it is justifiable to organize expensive data collection programmes, while for countries with small fishery resources it may be difficult to justify large expenditures for fisheries research irrespective of the per caput income. For several reasons, the fishery researcher's dream of an ideal supply of data may never materialize, and that may not necessarily be a negative thing seen from the point of view of the entire society of the country. In that case the researcher will have to manage with a less demanding methodology.

15 MICROCOMPUTER PROGRAM PACKAGES

There are several microcomputer program packages which can be used in conjunction with the present manual in particular the Length-based Fish Stock Assessment (LFSA) package developed by FAO (Sparre, 1987) and the COMPLEAT ELEFAN package developed by ICLARM (Gayanilo, Soriano and Pauly, 1988). These two packages complement each other and files created for one package can be used in the other. The large amount of overlap and the simultaneous use of both packages in FAO/DANIDA training courses on fish stock assessment have led to the development of a new package, entitled "FAO-ICLARM Stock Assessment Tools" or FiSAT, integrating the routines contained in the above-mentioned packages and various other routines. The FiSAT package is expected to be ready for release in 1993.

15.1 THE LFSA PACKAGE

The LFSA (Length-based Fish Stock Assessment) package of microcomputer programs (Sparre, 1987), written in BASIC, was designed to match the present manual. Most of the programs follow exactly the procedures explained in the preceding chapters.

The LFSA package was originally developed for Apple II computers as indicated by the title of its manual (Sparre, 1987). Later, these programs were converted for use on IBM compatible computers. The IBM version has been updated and considerably expanded whereas the Apple II version has not.

The LFSA package runs on Apple II and IBM (PC, XT or AT) microcomputers or their compatibles with minimum memory, a monochrome screen and an 80 columns printer. The graphics of the IBM version are based on the IBM Color Graphics Monitor Adapter (High resolution graphics, 640*200).

The user is neither assumed to have knowledge of BASIC nor of the operation system used (PRODOS for Apple II and MS-DOS for IBM). The programs are executed via self-explanatory screen dialogues. The prompts often take the form of a "menu", from which the user selects a "course". In addition to the programs dealing with the methods described in this manual, the LFSA package also contains a number of utility programs for data handling, and initial data processing.

Not every method given in this manual has been incorporated. Emphasis has been placed on incorporating the time-consuming operations, such as the Bhattacharya analysis.

15.1.1 Length-frequency (LF) programs

Fig. 15.1.1 shows a flow-chart of the most important part of the LFSA package, namely the sub-package for length-frequency analysis.

LFINPUT: LF data entry/edit program.
The program is used to enter/edit time series of length-frequencies and store these on a hard disk or diskette (the type of data required for modal progression analysis, cf. Section 3.4.2). LFINPUT automatically sorts the length-frequency samples in chronological order.

LFINF: LF additional data program (information).
This program is an optional continuation of LFINPUT, which allows the attachment of a variety of additional data to each length-frequency sample. Such additional data may be effort, total catch, position, depth, gear type, bottom type, ... etc.

LFTABLE: LF table program.
Produces tables of time series of length-frequencies formatted according to the user's choice.

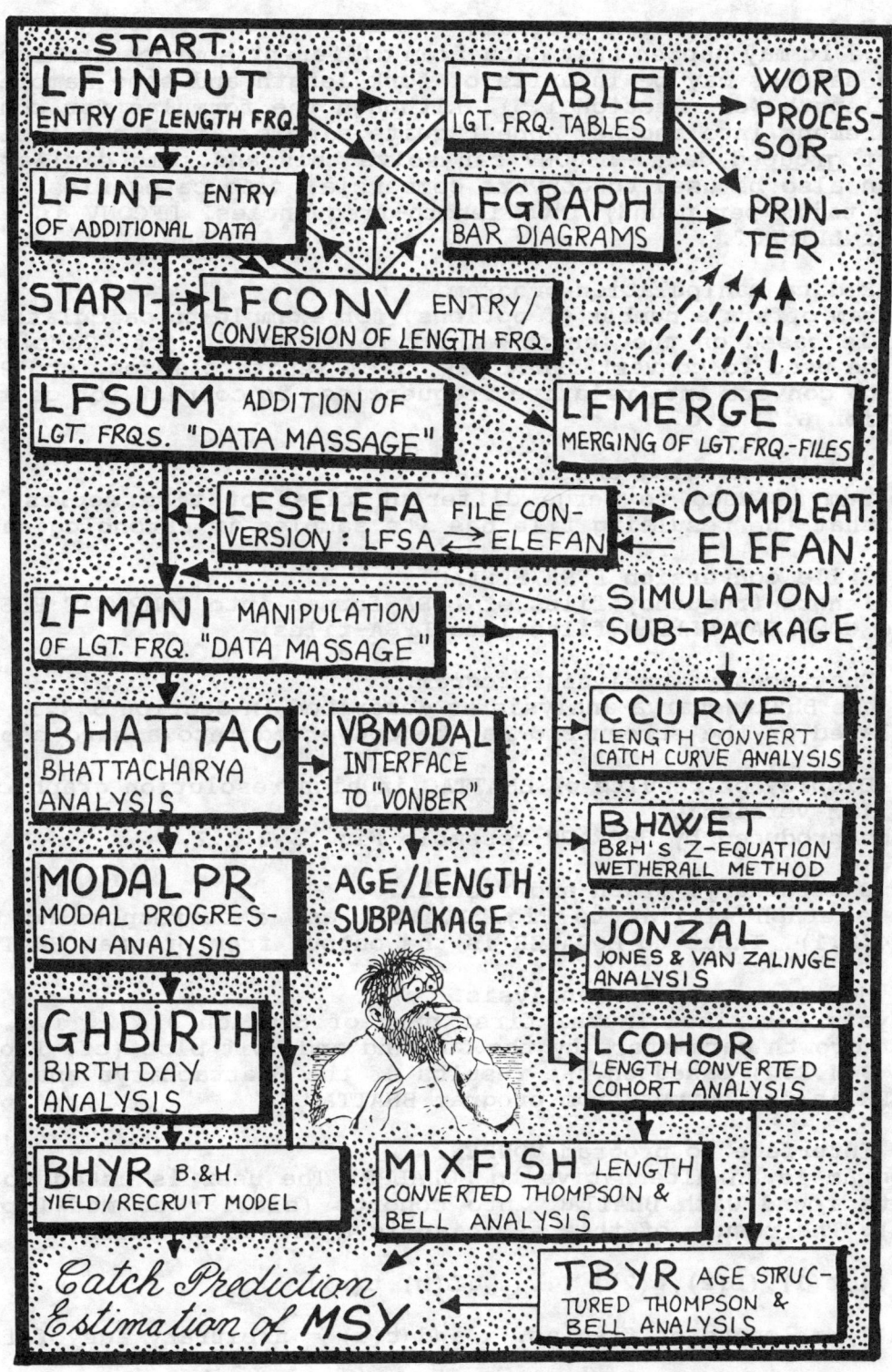

Fig. 15.1.1.1 Flow-chart for the major programs of the LFSA subpackage for length-frequency analysis (not all programs are included)

LFBAR and **LFGRPH**: LF bar diagram programs.

LFSUM: LF data summing program.
This program contains a number of options for raising and adding up length-frequency samples (cf. Chapter 7).

LFCONV: LF data conversion program.
In practice it may happen that data from different sources are incompatible, e.g. some samples may be in units of fork length and some samples in units of total length (cf. Section 1.5). Although the formulas for conversion of one measurement into another appear quite simple, it is not a simple thing to convert <u>grouped</u> data. LFCONV can solve this and some similar problems. LFCONV can also be used to convert commercial size categories (e.g. number of shrimp tails per pound) into length-frequencies. LFCONV is an extended version of LFINPUT.

LFMANI: LF data manipulation program.
This program offers a number of options, for example to calculate the weight of a sample based on the length-weight relationship (cf. Section 2.6). It may also be used to convert from, say, 1 cm length classes into 2 cm length classes, to convert into relative frequencies, to correct for gear selection (cf. Section 6.7) etc.

LFMERGE: LF file merging program.
This program is used to merge different files of time series. It merges files so that the resulting file has its samples in chronological order.

LFSELEFA: File conversion LFSA/ELEFAN.
Converts length-frequency files of LFSA-format into COMPLEAT ELEFAN-format files and COMPLEAT ELEFAN-files into LFSA-files.

BHATTAC: Bhattacharya analysis program.
Performs the Bhattacharya analysis as described in Section 3.4.1. The graphs are displayed on the screen and can be converted into a hard copy.

BHGRAPH: Display of results of BHATTAC in high resolution graphics (only in IBM version).
The graphs produced by BHGRAPH resemble Fig. 3.2.2.2.

MODALGR: Bhattacharya graph and results.
Produces a graph similar to Fig. 3.4.2.1 and a corresponding table (cf. Table 3.4.2.1). Input to MODALGR is the output from program BHATTAC.

MODALPR: Modal progression analysis.
The analysis described in the first part of Section 3.4.2, i.e. the estimation of growth parameters by the Gulland and Holt plot (cf. Table 3.4.2.3 and Fig. 3.4.2.2) based on the results of the Bhattacharya analysis. Input to MODALPR is the output from program BHATTAC.

VBMODAL: Interface to program VONBER.
This program is an alternative to MODALPR. The user is asked to group the components found with BHATTAC into cohorts (similar to modal progression analysis). The result of this exercise is a series of pairs:

$(L(1),t(1),(L(2),t(2)),\ldots,(L(n),t(n))$

where $L(i)$ = length of fish no. i and $t(i)$ = arbitrary age of fish no. i.

These data are in turn used as input to the program "VONBER", which estimates the growth parameters.

GHBIRTH: Birthday analysis.
The analysis described in the second part of Section 3.4.2, i.e. the estimation of approximate birthday by the von Bertalanffy plot (cf. Table 3.4.2.4 and Figs. 3.4.2.3 and 3.4.2.4). Input to GHBIRTH is the output from program MODALPR.

CCURVE: Length-converted catch curve analysis.
The analysis described in Sections 4.4.5 and 6.5, i.e. estimation of Z and estimation of the gear selection ogive from the catch curve.

JONZAL: Jones and van Zalinge method.
Estimation of Z as described in Section 4.4.6.

BHZWET: Beverton and Holt's Z-equation and the Powell-Wetherall analysis.
Estimation of Z as described in Section 4.5.2 and estimation of L_∞ and Z/K as described in Section 4.5.4.

LCOHOR: Jones' length-based cohort analysis.
Estimation of stock size and fishing mortality as described in Section 5.3.

TBYR: Age-based Thompson and Bell model yield calculations.
This program uses a special version of the Thompson and Bell yield and stock prediction model for the single stock single fishery situation (cf. Section 8.6). TBYR takes its starting point in the stock numbers by length group calculated by LCOHOR and converts them into age groups. The calculations are as described in Section 8.6. Because the conversion from length groups to age groups (whole years) is problematic for short-lived species, this program should be used only for long-lived species (5 years or more).

MIXFISH: Length-based Thompson and Bell model with option for analysis of a mixed fishery (only in IBM version).
This program is based on the theory explained in Sections 8.7 and 10.4.1. It does essentially the same as TBYR, but without the conversion of lengths into age groups. It may be used for long-lived as well as short-lived species. Although designed for analysis of a mixed fishery, MIXFISH contains the single species case as an option.

SIMULL: Monte Carlo simulation of length-frequency samples (only in IBM version).
The program SIMULL simulates length-frequency samples by simulating the life history of each individual fish, using Monte Carlo techniques.

SIMMIG: Monte Carlo simulation of length-frequency samples accounting for migration (only in IBM version).
The program SIMMIG simulates length-frequency samples the same way as SIMULL. It does the simulation for several areas in parallel and simulates the movements of fish between the areas.

15.1.2 Age/length analysis: estimation of growth parameters from age/length data

VONBER: Estimation of von Bertalanffy parameters.
This program estimates the growth parameters in the ordinary von Bertalanffy growth equation (Eq. 3.1.0.1) by non-linear regression analysis as described in Section 3.1. Because the computation procedure is based on iteration techniques requiring a very large number of calculations this method has not been dealt with as an example in this manual.

The input to VONBER is a series of pairs:

$(L(1),t(1)),(L(2),t(2)),\ldots,(L(n),t(n))$

where $L(i)$ = length of fish no. i and $t(i)$ = age of fish no. i.

The program VONBER determines the parameters, L_∞, K and t_0 so as to minimize the sum of squares of deviations:

$$\sum_{i=1}^{n} [L(i) - L_\infty*\{1 - \exp(-K*(t(i)-t_0))\}]^2$$

VBINPUT: Age/length (AL) data entry program.
Resembles program LFINPUT.

VBRESU: Results from VONBER.

VBGRAPH: Graph of results from VONBER.

VBMERGE: Age/length file merging program.

15.1.3 Miscellaneous programs

BHYR: Beverton and Holt's yield per recruit analysis.
The estimation of yield/recruit and biomass/recruit (Chapter 8).

PAULYM: Pauly's M-formula (Eq. 4.7.2.1) (only in IBM version).

REGRES: Ordinary regression analysis.
As described in Sections 2.4 to 2.6, contains a options for linearization.

REGRAPH: Graph for ordinary regression analysis.

15.2 THE COMPLEAT ELEFAN PACKAGE

The COMPLEAT ELEFAN package was developed around the ELEFAN (**E**lectronic **LE**ngth **F**requency **AN**alysis) program developed by Pauly and David (1981), which now appears in the package as the ELEFAN I program. This program has already been described in detail in Chapter 3. Both the ELEFAN I program and the COMPLEAT ELEFAN package built around it have undergone several modifications. The latest version is described by Gayanilo et al. (1988). The package is menu-driven and very user-friendly. It contains the following routines and sub-routines:

UTILITIES: Peripheral installation, disk copy routine, communication port installation, disk formatting routine.

ELEFAN 0: Data entry and editing, print routine, probability of capture entry and file correction, adjust class interval, delete file routine, pooling of samples, merging of files, smoothing of samples, sample weight estimation.

ELEFAN I: Restructuring of data, curve-fitting-by-eye, automatic search, response surface analysis, output routine.

ELEFAN II: Fitting selection curve, catch curve and probability of capture, recruitment pattern, Powell/Wetherall plot, relative yield per recruit analysis.

ELEFAN III: Age-based VPA, length-based VPA, combined VPA, catch file creation.

ELEFAN IV: Estimation of M.

ELEFAN V: Response surface analysis, compatability plot, growth increment analysis.

MPA: Bhattacharya's method, means/standard deviation entry, linking of means and coefficient of variance of L_∞.

15.3 THE FiSAT PACKAGE

In 1990, FAO and the International Center for Living Aquatic Resources Management (ICLARM) agreed to develop a new program package for length-based fish stock assessment, called FiSAT (FAO-ICLARM Stock Assessment Tools). The package is structured around an integration of routines incorporated in LFSA and the COMPLEAT ELEFAN package, but it also contains a number of new routines. FiSAT overcomes various problems of the two "parent packages".

Whenever possible, the models and their outputs are presented in graphical format for easy interpretation, thus keeping the user-friendliness of the COMPLEAT ELEFAN package. Examples of graphical outputs are given in Figs. 15.3.1 and 15.3.2.

Although the bulk of the routines manipulate and/or analyze length-frequency data, FiSAT also works with other types of data, such as weight-frequencies, size-at-age data, mark-recapture data, selection data (trawl and gill net) and other vector files. However, the analyses that can be performed with that kind of data are more limited.

Other key features contributing to FiSAT's user-friendly interface are:

   A spreadsheet-like "feel", with the file being processed at most with one click of a button away

   Distinctly grouped routines, for easy access through pull-down menus

   Pop-up windows, with help messages and a description of the model and its main assumptions

   Support of numerous micro-computer configurations (CGA, EGA, VGA, HERCULES, etc.).

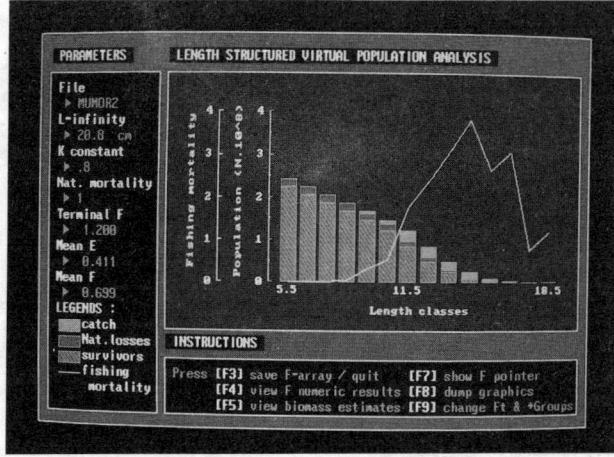
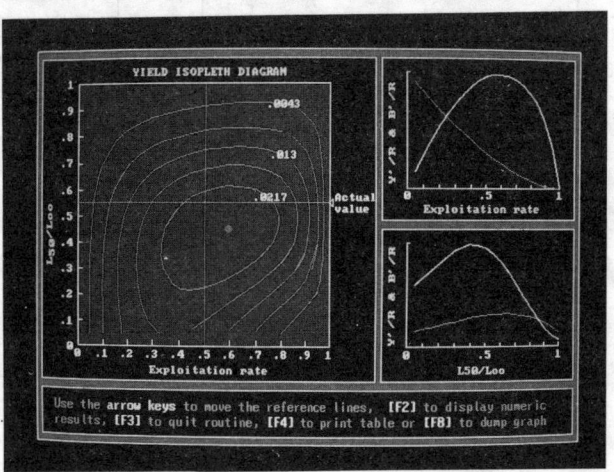

Fig.15.3.1 Facsimile representation of the results of a length-based VPA

Fig.15.3.2 Yield isopleth diagram, using the Beverton and Holt model

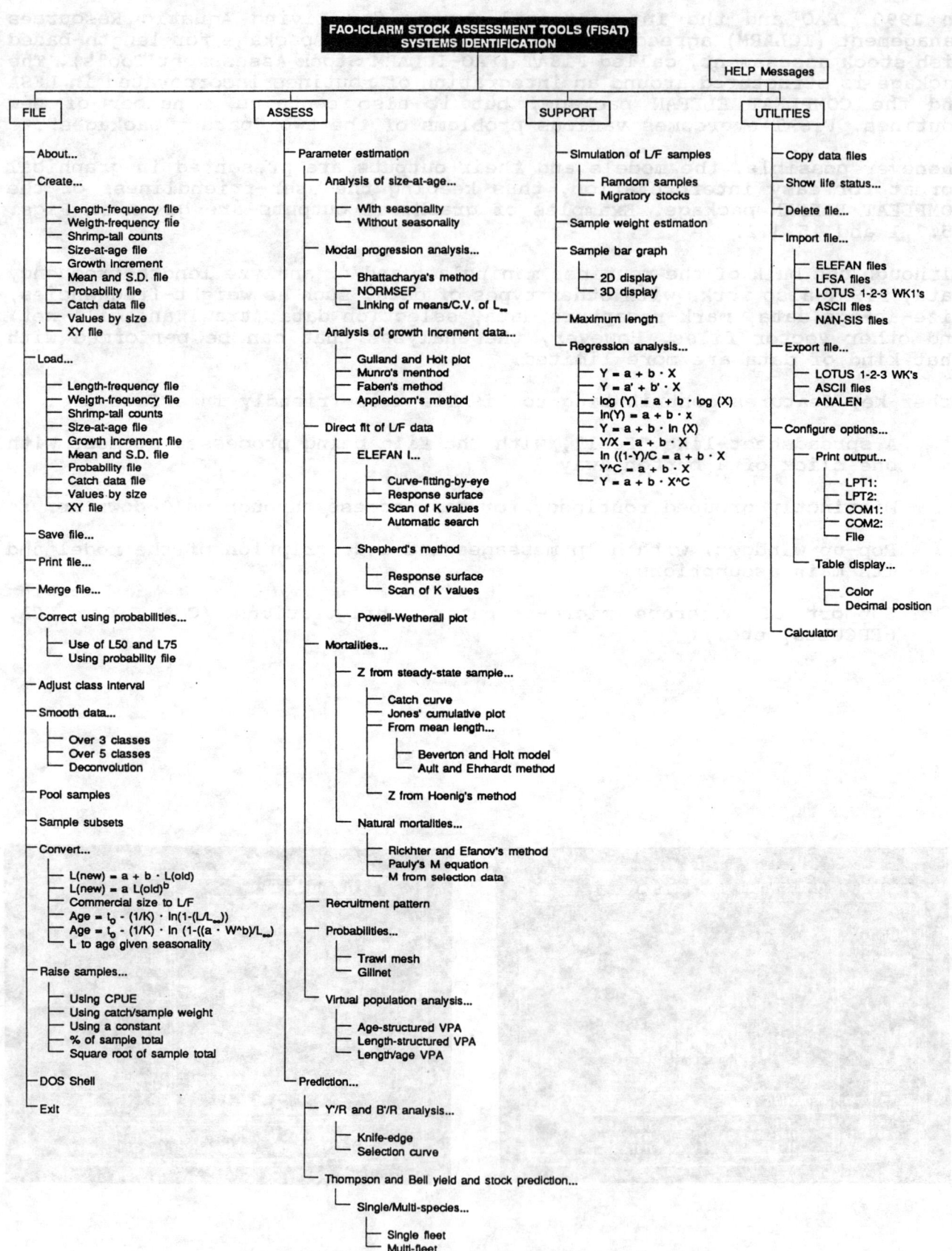

Fig. 15.3.3 Systems identification of the FiSAT package

The general layout of the package is given in Fig. 15.3.3. There are four major groups of routines:

FILE, ASSESS, SUPPORT and **UTILITIES**.

ASSESS contains the stock assessment programs, while the other three groups contain the preparatory (**FILE**) and supporting routines (**SUPPORT, UTILITIES**).

The program package contains some methods/models which have not been incorporated in this manual, except in the list of references (see Appeldoorn, 1987; Ault and Ehrhardt, 1991; Fabens, 1965; Munro, 1982 and Pauly et al., in press and Shepherd, 1987). However, the manual that will be produced for the users of the FiSAT program package will contain the necessary explanations.

15.4 OTHER FISH STOCK ASSESSMENT PROGRAMS PRODUCED BY FAO

15.4.1 The ANACO package

The ANACO package contains two sets of programs related to VPA. It is supported by a manual (issued in English, French and Spanish, Mesnil, 1989)

- **SIMUCO** is primarily designed as a training tool to explore the mechanism and properties of sequential analyses of catch-at-age data. It also includes a facility for analyses of single-year data, with the possibility to correct for departures from equilibrium conditions, and gives an introduction to sensitivity analyses.

- **VPBAS-COHORT-VPUTIL** is software for real assessments, using time series of catch-at-age data, as is common practice in many working groups and fishery commissions. **VPBAS** and **VPUTIL** are used for construction and maintenance of the data bases, and **COHORT** for the actual VPA computations.

15.4.2 The ANALEN package

The package consists of four programs, **ANALEN, ANAJON, SENJON** and **MONOJO**, which can be used for the analysis of catch-at-length data and for the simulation of multi-gear fisheries with sensitivity analysis. The package is supplemented with a manual in French (Chevaillier and Laurec, 1990).

- **ANALEN** is based on Jones' length-based cohort analysis.

- **ANAJON** calculates, the yield per recruit and the mature biomass per recruit, for different levels of exploitation. It also provides short-term and long-term forecasts of catches.

- **SENJON** is used for sensitivity analyses of the principal results.

- **MONOJO** is used to evaluate technical interactions between fishing gear.

15.4.3 BEAM 1 and BEAM 2

BEAM 1 and **BEAM 2** are programs for simple bio-economic modelling of artisanal and industrial sequential shrimp fisheries, based on an age-based Thompson and Bell yield per recruit model and a simple input-output microeconomic model. BEAM 1 simulates results by age groups, while BEAM 2 uses standard commercial categories. The manual is the first of the FAO Computerized Information Series (Fisheries) (Coppola et al., 1992).

15.4.4 BEAM 3

This program provides a more powerful analysis of complex fisheries than BEAM 1 and 2. It is a stochastic model that can handle up to four species (or both sexes of two species) and many fleets operating sequentially or simultaneously. The manual was first produced in French by Cochet and Gilly, 1990.

15.4.5 BEAM 4

This program by Sparre and Willmann (1992), is the most general and most complicated one in the BEAM series. Its general features allow for more realistic simulations of real fisheries. The data requirements are much higher than for the other programs.

All BEAM programs can, in principle, be used on any IBM PC or 100% compatible computer with minimal requirements under PC-DOS or MS-DOS.

15.4.6 The NAN-SIS package

This package was developed by Strømme (1992) to process data generated by the surveys of the R/V DR. FRIDTJOF NANSEN, in particular for the swept area method.

NAN-SIS is a Survey Information System for logging, editing and analysis of scientific trawl survey data (trawl catch and length-frequency data). It provides summaries of selected subsets of data defined by trawl station numbers, species, genera, families or user-defined species groups. Its main outputs are:

Tables of catches, mean catch rates and variances for user-defined subsets of trawl stations.

Swept-area calculations (mean densities per square nautical mile).

NAN-SIS contains routines for transferring data accessible to other program packages, such as STATGRAPHICS, LFSA, COMPLEAT ELEFAN and FiSAT. The multitude of species in tropical waters is handled through a mnemonic species code system, which converts codes into scientific or local names in the printouts.

An expanded version of NAN-SIS includes programs for logging, editing and analysis of acoustic data. These programs, however, are specific to certain models of acoustic equipment and are not covered by the manual.

15.4.7 CLIMPROD

CLIMPROD (Fréon et al., 1992) is a program that combines environmental variables with surplus production models. It requires annual data-series on catch and effort of a fishery on a single stock, and annual (or seasonal) data-series on an environmental variable known to influence the abundance or the catchability of this stock. The data-set should span at least 12 years. The results of fitting the appropriate model to the data-set may explain how the environment and fishing effort have governed the yields of the fishery in the period considered. Based on estimates of effort and different environmental factors for the next two years it is possible to make predictions.

CLIMPROD can also be used for fitting conventional surplus production models (linear, exponential, generalized) without environmental data, or for fitting the relationship between environment and production without fishing.

16 REFERENCES

Alagaraja, K., 1984. Simple methods for estimation of parameters for assessing exploited fish stocks. Indian J.Fish., 31:177-208

Allan, K.R. and G.P. Kirkwood, 1988. Marine mammals. In Fish population dynamics: the implications for management, edited by J.A. Gulland, Chichester, John Wiley and Sons Ltd., pp. 251-69

Allen, G.R., 1985. FAO species catalogue. Vol. 6. Snappers of the world. An annotated and illustrated catalogue of lutjanid species known to date. FAO Fish.Synop., (125)Vol.6:208 p.

Alverson, D.L. (ed.), 1971. Manual of methods for fisheries resource survey and appraisal. Part 1. Survey and charting of fisheries resources. FAO Fish.Tech.Pap., (102):80 p.

Alverson, D.L. and W.T. Pereyra, 1969. Demersal fish explorations in the northeastern Pacific Ocean - An evaluation of exploratory fishing methods and analytical approaches to stock size and yield forecasts. J.Fish.Res. Board.Can., 26:1985-2001

Amaratunga, T., 1983. The role of cephalopods in the marine ecosystem. FAO Fish.Tech.Pap./FAO Doc.Tech.Pêches/FAO Doc.Téc.Pesca, (231):379-415

Andersen, K.P. and E. Ursin, 1977. A multispecies extension to the Beverton and Holt theory of fishing, with accounts of phosphorus circulation and primary production. Medd.Dan.Fisk.- og Havunders.(Ny Ser.), 7:319-435

Appeldoorn, R.S., 1987. Modification of a seasonally oscillating growth function for use with mark-recapture data. J.Cons.CIEM, 43:194-8

Ault, J.S. and N.M. Ehrhardt, 1991. Correction to the Beverton and Holt Z-estimator for truncated catch length-frequency distributions. ICLARM Fishbyte, 9(1):37-9

Australian Journal of Marine and Freshwater Research, 1987. Prawn ecology and biology. Aust.J.Mar.Freshwat.Res., 38(1):190 p.

Bagenal, T.B. (ed.), 1974. Ageing of fish. Proceedings of an international symposium on the ageing of fish held at the Univ. of Reading, England 19-20 July 1973. Old Woking, Surrey, England, Unwin Bros. Ltd., 234 p.

Bakhayokho, M., 1983. Biology of the cuttlefish Sepia officinalis hierredda off the Senegal coast. FAO Fish.Tech.Pap./FAO Doc.Tech.Pêches/ FAO Doc.Téc.Pesca., (231):204-63

Bakun, A., J. Beyer, D. Pauly, J.G. Pope and G.D. Sharp, 1982. Ocean sciences in relation to living resources. Can.J.Fish.Aquat.Sci., 39: 1059- 70

Baranov, F.I., 1914. The capture of fish by gillnets. Mater.Poznaniyu Russ.Rybolov. 3(6):56-99 (Partially translated from Russian by W.E. Ricker)

Baranov, F.I., 1918. On the question of the biological basis of fisheries. Nauchn.Issled.Ikhtiol.Inst.Izv., 1:81-128 (in Russian)

Baranov, F.I., 1926. On the question of the dynamics of the fishing industry. Nauchn.Byull.Rybn.Khoz., 8(1925):7-11 (in Russian)

Baranov, F.I., 1948. Theory and assessment of fishing gear. Moscow, Pishchepromizdat. English version issued as Theory of fishing with gill nets. Transl. from Russian by Ontario Department of Land and Forests, Toronto, 45 p.

Bard, F.X., 1984. Le listao de l'Atlantique. 3. Resultats des campagnes de marquages effectuées de 1980 a 1982. Pêche Marit., 1275:319-24

Bard, F.X., P. Cayre and T. Diouf, 1988. Les migrations. FAO Doc.Tech. Pêches, (292):111-156

Bayliff, W.H. and K.N. Holland, 1986. Materials and methods for tagging tuna and billfishes, recovering the tags and handling the recapture data. FAO Fish.Tech.Pap., (279):36 p.

Beamish, R.J. and G.A. McFarlane, 1983. The forgotten requirement for age validation in fisheries biology. Trans.Am.Fish.Soc., 112(6):735-43

Beddington, J.R. and J.G. Cooke, 1983. The potential yield of fish stocks. FAO Fish.Tech.Pap., (242):47 p.

Berkeley, S.A. and E.D. Houde, 1980. Swordfish, *Xiphias gladius*, dynamics in the Straits of Florida. ICES C.M. 1980/H:59:11 p. (mimeo)

Berkeley, S.A., 1983. Atlantic swordfish stock structure data and suggestions for its interpretation. Collect.Vol.Sci.Pap.ICCAT/Recl.Doc.Sci. CICTA/Colecc.Doc.Cient.CICAA, 18(3):839-45

Bertalanffy, L. von, 1934. Untersuchungen über die Gesetzlichkeiten des Wachstums. 1. Allgemeine Grundlagen der Theorie. Roux'Arch.Entwicklungsmech.Org., 131:613-53

Beverton, R.J.H., 1963. Maturation, growth and mortality of clupeid and engraulid stocks in relation to fishing. Rapp.P.-V.Réun.CIEM, 154:44-67

Beverton, R.J.H. and S.J. Holt, 1956. A review of methods for estimating mortality rates in exploited fish populations, with special reference to sources of bias in catch sampling. Rapp.P.-V.Réun.CIEM, 140:67-83

Beverton, R.J.H. and S.J. Holt, 1957. On the dynamics of exploited fish populations. Fish.Invest.Minist.Agric.Fish.Food G.B.(2 Sea Fish.), 19: 533 p.

Beverton, R.J.H. and S.J. Holt, 1959. A review of the lifespans and mortality rates of fish in nature, and their relation to growth and other physiological characteristics. In CIBA Foundation, colloquia on ageing. Vol. 5. The lifespan of animals, edited by G.E.W. Wolstenholme and M. O'Connor. London, Churchill, Vol.5:142-80

Beverton, R.J.H. and S.J. Holt, 1966. Manual of methods for fish stock assessment. Part 2. Tables of yield functions. Manuel sur les méthodes d'évaluation des stocks ichtyologiques. Partie 2. Tables de fonctions de rendement. Manual de métodos para la evaluación de los stocks de peces. Parte 2. Tablas de funciones de rendimiento. FAO Fish.Tech.Pap./FAO Doc. Tech.Pêches/FAO Doc.Téc.Pesca, (38)Rev.1:67 p.

Beyer, J. and P. Sparre, 1983. Modelling exploited fish stocks. In Application of ecological modelling in environmental management. Part A, edited by S.E. Jørgensen. Amsterdam, Elsevier Scientific Publishing Co., pp. 485-582

Bhattacharya, C.G., 1967. A simple method of resolution of a distribution into Gaussian components. Biometrics, 23:115-35

Blaber, S.J.M. and T.G. Blaber, 1980. Factors affecting the distribution of juvenile estuarine and inshore fish. J.Fish Biol., 17:143-62

Booke, H.E., 1981. The conundrum of the stock concept - are nature and nurture definable in fishery science? Can.J.Fish.Aquat.Sci., 38:1479-80

Boonyubol, M. and V. Hongskul, 1978. Demersal resources and exploitation in the Gulf of Thailand, 1960-1975. In Report on the Workshop on the demersal resources of the Sunda Shelf, Penang, Malaysia. Manila, South China Sea Fisheries Development and Coordinating programme, SCS/GEN/77/13: 56-70

Brander, K., 1988. Multispecies fisheries of the Irish Sea. In Fish population dynamics: the implications for management, edited by J.A. Gulland, Chichester, John Wiley and Sons Ltd., pp. 303-28

Brêthes, J.-C. et R.N. O'Boyle (éditeurs), 1990. Méthodes d'évaluation des stocks halieutiques. Projet CIEO-860060, Centre international d'exploitation des océans, Halifax (Nouvelle-Écosse, Canada): 963 p. (2 Volumes)

Brody, S., 1927. Growth rates. Bull.Univ.Miss.Agric.Exp.Sta., (97)

Brothers, E.B., 1980. Age and growth studies on tropical fishes. In Stock assessment for tropical small-scale fisheries. Proceedings of an International Workshop held 19-21 September at the University of Rhode Island, Kingston, R.I., edited by S.B. Saila and P. Roedel International Center for Marine Resource Development, University of Rhode Island, pp. 119-36

Butler, M.J.A., C. Le Blanc, J.A. Belbin and J.L. MacNeill, 1986. Marine resource mapping: an introductory manual. FAO Fish.Tech.Pap., (274):256 p.

Caddy, J.F., 1980. Surplus production models. In Selected lectures from the CIDA/FAO/CECAF seminar on fishery resource evaluation. Casablanca, Morocco, 6-24 March 1978: Rome, FAO Canada Funds-in-Trust, FAO/TF/INT (c)180 Suppl.: 29-55. Issued also in French

Caddy, J.F., 1982. Some considerations relevant to the definition of shared stocks and their allocation between adjacent economic zones. FAO Fish.Circ., (749):44 p. Issued also in Spanish

Caddy, J.F. (ed.), 1983. Advances in assessment of world cephalopod resources. Progrès réalisés dans l'évaluation des ressources mondiales de céphalopodes. Progresos realizados en la evaluación mundial de cefalópodos. FAO Fish.Tech.Pap./FAO Doc.Tech.Pêche/FAO Doc.Téc.Pesca, (231):452 p.

Caddy, J.F., 1983a. The cephalopods: factors relevant to their population dynamics and to the assessment and management of stocks. FAO Fish.Tech.Pap./FAO Doc.Tech.Pêche/FAO Doc.Téc.Pesca, (231): 416-52

Caddy, J.F., 1987. Size-frequency analysis for crustacea: moult increment and frequency models for stock assessment. Kuwait Bull.Mar.Sci., (9):43-61

Caddy, J.F., 1987a. Types of resource sharing relevant to Lesser Antilles stocks. FAO Fish.Rep., (383):115-24

Caddy, J.F., 1989. Marine invertebrate fisheries: their assessment and management. New York, John Wiley and Son Inc., Interscience, 752 p.

Caddy, J.F. and J. Csirke, 1983. Approximations to sustainable yield for exploited and unexploited stocks. Océanogr.Trop., 18(1):3-15

Caddy, J.F. and S. Garcia, 1986. Fisheries thematic mapping - a pre-requisite for intelligent management and development of fisheries. Océanogr. Trop., 21(1):31-52

Carey, F.G. and R.J. Olson, 1982. Sonic tracking experiments with tunas. Collect.Vol.Sci.Pap.ICCAT/Recl.Doc.Sci.CICTA/Colecc.Doc.Cient.CICAA, 17: 458-66

Carpenter, K.E., 1988. FAO species catalogue. Vol. 8. Fusilier fishes of the world. An annotated and illustrated catalogue of caesionid species known to date. FAO Fish.Synop., (125)Vol.8:75 p.

Carpenter, K.E. and G.R. Allen, 1989. FAO species catalogue Vol. 9. Emperor fishes and large-eye breams of the world (Family Lethrinidae). An annotated and illustrated catalogue of lethrinid species known to date. FAO Fish.Synop., (125)Vol.9: 118 p., 8 colour plates

Cassie, R.M., 1954. Some uses of probability paper in the analysis of size frequency distributions. Aust.J.Mar.Freshwat.Res., 5:513-22

CECAF, 1978. Report of the Ad hoc working group on hake (Merluccius merluccius, Merluccius senegalensis, Merluccius cadenati) in the northern zone of CECAF held at the Instituto Español de Oceanografia, Santa Cruz de Tenerife, Canary Islands, Spain, 5-9 June 1978. CECAF/ECAF Ser., (48/9): 93 p. Issued also in French

Chapman, D.G., 1961. Statistical problems in dynamics of exploited fisheries populations. Proc. 4th Berkeley Symp.Math.Stat. and Probability. Cont.Biol. and Probl.Med., 4:153-68. Univ.Calif. Press

Chapman, D.G. and D.S. Robson, 1960. The analysis of a catch curve. Biometrics, 16(3):354-68

Chevailler, P. et A. Laurec, 1990. Logiciels pour l'évaluation des stocks de poisson. ANALEN: Logiciel d'analyse des données de capture par classes de taille et de simulation des pêcheries multi-engins avec analyse de sensibilité. FAO Doc.Tech.Pêches, (101) Suppl.4:124 p. (with programs on diskettes)

Chikuni, S., 1983. Cephalopod resources in the Indo-Pacific region. FAO Fish.Tech.Pap./FAO Doc.Tech.Pêches/FAO Doc.Téc.Pesca, (231):264-305

CIDA/FAO/CECAF, 1980. Selected lectures from the CIDA/FAO/CECAF seminar on fishery resource evaluation. Casablanca, Morocco, 6-24 March 1978. Rome, FAO, Canada Funds-in-Trust, FAO/TF/INT 180(c) Suppl.:166 p.

Clark, W.G., 1981. Restricted least-squares estimates of age composition from length composition. Can.J.Fish.Aquat.Sci., 38:297-307

Cloern, J.E. and F.H. Nichols, 1978. A von Bertalanffy growth model with a seasonally varying coefficient. J.Fish.Res.Board Can., 35:1479-82

Cochet, Y. et B. Gilly, 1990. Logiciels pour l'analyse bioéconomique des pêcheries. BEAM 3: Simulation bioéconomique analytique de pêcheries de crevettes tropicales avec recrutement fixe ou aléatoire. FAO Doc.Tech. Pêches, (310.2):57 p.

Cochran, W.G., 1977. Sampling techniques. New York, John Wiley and Sons, Inc., 428 p. 3rd ed.

Cohen, D.M., T. Inada, T. Iwamoto and N. Scialabba, 1990. FAO species catalogue Vol. 10. Gadiform fishes of the world (Order Gadiformes). An annotated and illustrated catalogue of cods, hakes, grenadiers and other gadiform fishes known to date. FAO Fish.Synop., (125)Vol.10:442 p.

Colette, B.B. and C.E. Nauen, 1983. FAO species catalogue. Vol. 2. Scombrids of the world. An annotated and illustrated catalogue of tunas, mackerels, bonitos and related species known to date. FAO Fish.Synop., (125)Vol.2:137 p.

Compagno, L.J.V., 1984. FAO species catalogue. Vol. 4. Sharks of the world. An annotated and illustrated catalogue of shark species known to date. Part 1. Hexanchiformes to Lamniformes. FAO Fish.Synop., (125)Vol.4, Pt.1:249 p.

Compagno, L.J.V., 1984a. FAO species catalogue. Vol. 4. Sharks of the world. An annotated and illustrated catalogue of shark species known to date. Part 2. Carcharhiniformes. FAO Fish.Synop., (125)Vol.4,Pt.2:251-655

Coppola, S.R., S.M. Garcia and R. Willmann, 1992. Software for bio-economic analysis of fisheries BEAM 1 and BEAM 2. Simple bio-economic analytical simulation models for sequential fisheries on tropical shrimp using age groups or commercial categories. FAO Computerized Information Series (Fisheries), No. 1. Rome, FAO. 61 p. (with programs on diskettes)

Csirke, J., 1980. Recruitment in the Peruvian anchovy and its dependence on the adult population. Rapp.P.-V.Réun.CIEM, 177:307-13

Csirke, J., 1980a. Introducción a la dinámica de poblaciones de peces. FAO Doc.Téc.Pesca, (192):82 p.

Csirke, J., 1988. Small shoaling pelagic fish stocks. In Fish population dynamics: the implications for management, edited by J.A. Gulland, Chichester, John Wiley and Sons Ltd., pp. 271-302

Csirke, J. and G.D. Sharp (eds), 1984. Reports of the expert consultation to examine changes in abundance and species composition of neritic fish resources, San José, Costa Rica, 18-29 April 1983. FAO Fish.Rep., (291) Vol.1:102 p.

Cushing, D.H., 1968. Fisheries biology: a study in population dynamics. Madison, University of Wisconsin Press, 200 p.

Cushing, D.H., 1988. The study of stock and recruitment. In Fish population dynamics: the implications for management, edited by J.A. Gulland, Chichester, John Wiley and Sons Ltd., pp. 105-28

Daan, N. and M.P. Sissenwine, 1991. Multispecies models relevant to management of living resources. Proceedings of a Symposium held in The Hague, 2-4 October 1989. ICES Mar.Sci.Symp., (193): 358 p.

Dall, W., B.J. Hill, P.C. Rothlisberg and D.J. Staples, 1990. The biology of the Penaeidae. Advances in marine biology Vol. 27, 489 p. London, Academic Press Ltd.

Dawson, W.A., 1986. Changes in western mackerel (Scomber scombrus) spawning stock composition during the spawning season. J.Mar.Biol.Assoc.U.K., 66:367-83

Dayaratne, P. and J. Gjøsaeter, 1986. Age and growth of four Sardinella species from Sri Lanka. Fish.Res., 4:1-33

Deriso, R.B., 1980. Harvesting strategies and parameter estimation for an age-structured model. Can.J.Fish.Aquat.Sci., 37:268-82

Derzhavin, A.N., 1922. The stellate sturgeon (Acipenser stellatus Pallas), a biological sketch. Byulleten' Bakinskoi Ikhtiologicheskoi Stantsii, 1: 1-393 (In Russian)

Dickson, W., 1974. A review of the efficiency of bottom trawls. Bergen, Norway, Institute of Fisheries Technology and Research, 44 p.

Doubleday, W.G., 1980. Survey methods for resource evaluation. In Selected lectures from the CIDA/FAO/CECAF seminar on fishery resource evaluation. Casablanca, Morocco, 6-24 March 1978. Rome, FAO, Canada Funds-in-Trust, FAO/TF/INT 180(c) Suppl:65-71

Doubleday, W.G. (ed.), 1981. Manual on groundfish surveys in the Northwest Atlantic. NAFO Sci.Counc.Stud., (2):7-55

Dwiponggo, A., 1979. Review of the demersal resources and fisheries in the Java Sea. IPFC:RRD/II/79/Inf.12. Paper presented at the SCORRAD Meeting, 1979, Hong Kong

Edser, T., 1908. Note on the number of plaice at each length in certain samples from the southern part of the North Sea, 1906. J.R.Stat.Soc., 71:686-90

Edwards, E.F. and B.A. Megrey (eds), 1989. Mathematical analysis of fish stock dynamics. American Fisheries Society Symposium, 6:210 p. + appendices

Edwards, R.R.C., 1985. Growth rates of Lutjanidea (snappers) in tropical Australian waters. J.Fish Biol., 26:1-4

Ehrhardt, N.M., P.S. Jacquemin, F. García B., G. González D., J.M. López B., J. Ortiz C., and A. Solís N., 1983. On the fishery and biology of the giant squid Dosidicus gigas in the Gulf of California, Mexico. FAO Fish. Tech.Pap./FAO Doc.Tech.Pêches/FAO Doc.Téc.Pesca, (231):306-39

Elliott, J.M., 1971. Some methods for the statistical analysis of samples of benthic invertebrates. Sci.Publ.Freshwat.Biol.Assoc.,Ambleside,U.K., (25):144 p.

Eltink, A.T.G.W., 1987. Changes in age-size distribution and sex ratio during spawning and migration of western mackerel (Scomber scombrus L.). J.Cons.CIEM, 44:10-22

Fabens, A.J., 1965. Properties and fitting of the von Bertalanffy growth curve. Growth, 29:265-89

FAO, 1978. Models for fish stock assessment. Partial translation of the annex to the report of the second FAO/CNEXO Training Centre on methods for fish stock assessment, Brest (France), 26 July-27 August 1976. Original French edition 164 p. FAO Fish.Circ., (701): 122 p.

FAO, 1978a. Some scientific problems of multispecies fisheries. Report of the expert consultation on management of multispecies fisheries, Rome, 20-23 September 1977. FAO Fish.Tech.Pap., (181):42 p.

FAO, 1978b. FAO catalogue of fishing gear designs. Catalogue FAO de plans d'engins de pêche. Catalogo de la FAO de planos de aparajos de pesca. Farnham, Surrey, Fishing News Books for FAO, 159 p.

FAO/IOP, 1979. Report of the FAO/IOP Workshop on the fishery resources of the western Indian Ocean south of the equator. Rome, FAO/UNDP, Indian Ocean Programme, IOFC/DEV/79/45:102 p.

FAO/UNDP, 1975. FAO regional fishery survey and development project, Doha (Qatar). Report of the Ad hoc working group on survey technique and strategy. Rome, FAO, FI:DP/REM/71/278/1: 45 p.

Fischer, W. (ed.), 1978. FAO species identification sheets for fishery purposes. Western Central Atlantic. (fishing area 31). Rome FAO, 7 vols: pag.var

Fischer, W. and G. Bianchi (eds), 1984. FAO species identification sheets for fishery purposes. Western Indian Ocean (fishing area 51). Prepared and printed with the support of the Danish International Development Agency (DANIDA). Rome, FAO, 6 vols. pag.var.

Fischer, W., G. Bianchi and W.B. Scott (eds), 1981. FAO species identification sheets for fishery purposes. Eastern Central Atlantic; fishing areas 34, 47(in part). Canada Funds-in-Trust. Ottawa, Department of Fisheries and Oceans Canada, by arrangement with the Food and Agriculture Organization of the United Nations, 7 vols: pag.var.

Fischer, W. and J.C. Hureau (eds), 1985. FAO species identification sheets for fishery purposes. Southern Ocean (Fishing areas 48, 58 and 88) (CCAMLR Convention Area). Prepared and published with the support of the Commission for the Conservation of Antarctic Marine Living Resources. Rome, FAO, 2 vols:470 p. Issued also in French and Spanish.

Fischer, W., M. Schneider et M.-L. Bauchot, (rédacteurs), 1987. Fiches FAO d'identification des espèces pour les besoins de la pêche. (Révision 1). Méditerranée et Mer Noire. Zone de pêche 37. Publication préparée par la FAO, résultat d'un accord entre la FAO et la Commission des Communautés Européennes (Project GCP/INT/422/EEC) financée conjointement par ces deux organisations. Rome, FAO, 2 vols:1532 p.

Fischer, W. and P.J.P. Whitehead (eds), 1974. FAO species identification sheets for fishery purposes. Eastern Indian Ocean (fishing area 57) and Western Central Pacific (fishing area 71). Rome, FAO, 4 vols:pag.var.

Flowers, J.M., 1978. A data processing and basic analysis system for demersal fisheries surveys. Regional fishery survey and development project. Rome, FAO, FI:DP/REM/71/278/4: 151 p.

Fogarty, M.J., 1985. Statistical considerations in the design of trawl surveys. FAO Fish.Circ., (786):21 p.

Fonteneau, A., 1982. Eléments pour l'aménagement des pêcheries d'albacore (Thunnus albacares) de l'Atlantique. Collect.Vol.Sci.Pap.ICCAT/Recl.Doc. Sci.CICTA/Colecc.Doc.Cient.CICAA, 17:79-163

Fonteneau, A. et J. Marcille (rédacteurs), 1988. Ressources, pêche et biologie des thonidés tropicaux de l'Atlantique centre est. FAO Doc.Tech.Pêches, (292):391 p.

Ford, E., 1933. An account of the herring investigations conducted at Plymouth during the years from 1924 to 1933. J.Mar.Biol.Assoc.U.K., 19: 305-84

Forsbergh, E.D., 1980. Synopsis of biological data on the skipjack tuna, Katsuwonus pelamis (Linnaeus, 1758), in the Pacific Ocean. Spec.Rep. I-ATTC, (2):295-360

Fox, W.W. Jr., 1970. An exponential surplus-yield model for optimizing exploited fish populations. Trans.Am.Fish.Soc., 99:80-8

Francis, R.C., 1974. Relationship of fishing mortality to natural mortality at the level of maximum sustainable yield under the logistic stock production model. J.Fish.Res.Board Can., 31:1539-42

Fréon, P., 1984. La variabilité des tailles individuelles a l'intérieur des cohortes et des bancs de poissons. 1. Observations et interprétation. Oceanol.Acta, 7(4):457-68

Fréon, P., C. Mullon and G. Pichon, 1992. CLIMPROD: Experimental interactive software for choosing and fitting surplus production models including environmental variables. FAO Computerized Information Series (Fisheries) No. 5. Rome, FAO, 82 p. (with programs on diskette)

Fry, F.E.J., 1949. Statistics of a lake trout fishery. Biometrics, 5:27-67

Garcia, S., 1985. Reproduction, stock assessment models and population parameters in exploited penaeid shrimp populations. In Second Australian national prawn seminar, edited by P.C. Rothlisberg, B.J. Hill and D. Staples. Cleveland, Queensland, Australia, NPS2:139-58

Garcia, S., 1988. Tropical penaeid prawns. In Fish population dynamics: the implications for management, edited by J.A. Gulland, Chichester, John Wiley and Sons Ltd., pp. 219-49

Garcia, S. and L. Le Reste, 1981. Life cycles, dynamics, exploitation and management of coastal penaeid shrimp stocks. FAO Fish.Tech.Pap., (203): 215 p. Issued also in French and Spanish.

Garcia, S., P. Sparre and J. Csirke, 1987. A note on rough estimators of fisheries resources potential. ICLARM Fishbyte, 5(2):11-6

Garcia, S., P. Sparre and J. Csirke, 1989. Estimating surplus production and maximum sustainable yield from biomass data when catch and effort time series are not available. Fish.Res., 8:13-23

Garcia, S. and N.P. van Zalinge, 1982. Shrimp fishing in Kuwait: methodology for a joint analysis of the artisanal and industrial fisheries. In Report on the Workshop on assessment of the shrimp stocks of the west coast of the Gulf between Iran and the Arabian Peninsula. Fisheries development in the Gulf. Rome, FAO, FI:DP/RAB/80/015/1: 119-42

Garrod, D.J., 1961. The selection characteristics of nylon gill nets for Tilapia esculenta Graham. J.Cons.CIEM, 26(2):191-203

Garrod, D.J., 1988. North Atlantic cod: fisheries and management to 1986. In Fish population dynamics: the implications for management, edited by J.A. Gulland, Chichester, John Wiley and Sons Ltd., pp. 185-218

Gavaris, S. and S.J. Smith, 1987. Effect of allocation and stratification strategies on precision of survey abundance estimates for Atlantic cod (Gadus morhua) on the eastern Scotian Shelf. J.Northwest Atl.Fish.Sci., 7:137-44

Gayanilo Jr., F.C., M. Soriano and D. Pauly, 1988. A draft guide to the COMPLEAT ELEFAN. ICLARM Software Project 2.: 65 p. and 10 diskettes (5.25 inches, 360K)

Gerking, S.D., 1953. Evidence for concepts of home range and territory in stream fishes. Ecology, 34:347-65

Gislason, H. and Th. Helgason, 1985. Species interaction in assessment of fish stocks with special application to the North Sea. Dana, 5:1-44

Gislason, H. and P. Sparre, 1987. Some theoretical aspects of the implementation of multispecies virtual population analysis in ICES. ICES C.M. 1987/G:51 (mimeo)

Gjøsæter, J.P., P. Dayaratne, O.A. Bergstad, H. Gjøsæter, M.I. Sousa and I.M. Beck, 1983. Ageing tropical fish by primary growth rings in their otoliths. Bergen, University of Bergen, 67 p. Issued also in 1984 as: FAO Fish.Circ., (776):54 p.

Goeden, G.B., 1978. A monograph of the coral trout, *Plectropomus leopardus* (Lacépède). Res.Bull.Fish.Serv.Queensl., (1):42 p.

Graham, M., 1935. Modern theory of exploiting a fishery and application to North Sea trawling. J.Cons.CIEM, 10(3):264-74

Grosslein, M.D. and A. Laurec, 1982. Bottom trawl surveys design, operation and analysis. CECAF/ECAF Ser., (81/22):25 p. Issued also in French

Gulland, J.A., 1955. On the estimation of population parameters from marked members. Biometrika, 42:269-70

Gulland, J.A., 1961. The estimation of the effect on catches of changes in gear selectivity. J.Cons.CIEM, 26(2):204-14

Gulland, J.A., 1965. Estimation of mortality rates. Annex to Arctic fisheries working group report ICES C.M. Doc. 3 (mimeo)

Gulland, J.A., 1966. Manual of sampling and statistical methods for fisheries biology. Part 1. Sampling methods. FAO Man.Fish.Sci., (3):87 p. Issued also in French and Spanish

Gulland, J.A., 1969. Manual of methods for fish stock assessment. Part 1. Fish population analysis. FAO Man.Fish.Sci., (4):154 p. Issued also in French and Spanish

Gulland, J.A. (comp.), 1971. The fish resources of the ocean. West Byfleet, Surrey, Fishing News (Books), Ltd., for FAO, 255 p. Revised edition of FAO Fish.Tech.Pap., (97):425 p. (1970)

Gulland, J.A., 1975. Manual of methods for fisheries resources survey and appraisal. Part 5. Objectives and basic methods. FAO Fish.Tech.Pap., (145):29 p.

Gulland, J.A., 1983. Fish stock assessment: a manual of basic methods. Chichester, U.K., Wiley Interscience, FAO/Wiley series on food and agriculture, Vol.1:223 p.

Gulland, J.A. (ed.), 1988. Fish population dynamics: the implications for management. Chichester, John Wiley and Sons Ltd., 422 p. 2nd ed.

Gulland, J.A., 1988a. The problems of population dynamics and contemporary fishery management. In Fish population dynamics: the implications for management, edited by J.A. Gulland, Chichester, John Wiley and Sons Ltd., pp. 383-406

Gulland, J.A. and D. Harding, 1961. The selection of *Clarias mossambicus* (Peters) by nylon gill nets. J.Cons.CIEM, 26:215-22

Gulland, J.A. and S.J. Holt, 1959. Estimation of growth parameters for data at unequal time intervals. J.Cons.CIEM, 25(1):47-9

Gulland, J.A. and A.A. Rosenberg, 1992. A review of length-based approaches to assessing fish stocks. FAO Fish.Tech.Pap., (323):100 p.

Gulland, J.A. and B.J. Rothschild, (eds), 1984. Penaeid shrimps - their biology and management. Farnham, Surrey, England, Fishing News Books Limited, 308 p.

Gunderson, D.R. and P.H. Dygert, 1988. Reproductive effort as a predictor of natural mortality rate. J.Cons.CIEM, 44:200-9

Hald, A., 1952. Statistical theory with engineering applications. New York, Wiley, 783 p.

Hamley, J.M., 1975. Review of gillnet selectivity. J.Fish.Res.Board Can., 32:1943-69

Hamley, J.M. and H.A. Regier, 1973. Direct estimates of gillnet selectivity to walleye (Stizostedion vitreum vitreum). J.Fish.Res.Board Can., 30:817-30

Hanumara, R.C. and N.A. Hoenig, 1987. An empirical comparison of a fit of linear and non-linear models for seasonal growth in fish. Fish.Res., 5:359-81

Harden Jones, F.R., 1968. Fish migration. London, Arnold, 325 p.

Harden Jones, F.R., 1984. A view from the ocean. In Mechanisms of migration in fishes, edited by J.D. McCleave et al. New York, Plenum Press, pp. 1-26

Harding, J.P., 1949. The use of probability paper for the graphical analysis of polymodal frequency distributions. J.Mar.Biol.Assoc.U.K., 28:141-53

Hasselblad, V., 1966. Estimation of parameters for a mixture of normal distributions. Technometrics, 8:431-44

Hasselblad, V. and P.K. Tomlinson, 1971. NORMSEP. Normal distribution separator. In Computer programs for fish stock assessment compiled by N.J. Abramson. FAO Fish.Tech.Pap., (101):11(1)2.1-11(1)2.10

Heincke, F., 1913. Investigations on the plaice. General report. 1. The plaice fishery and protective regulations. Part. I. Rapp.P.-V.Réun.CIEM, 17A:1-153 + Annexes

Helgason, T. and H. Gislason, 1979. VPA-analysis with species interaction due to predation. ICES C.M. 1979/G:52:10 p. (mimeo)

Hilborn, R. and C.J. Walters, 1992. Quantitative fisheries stock assessment. Choice, dynamics and uncertainty. Chapman and Hall, Inc., London, New York: 570 p. (with programs on a diskette)

Hoenig, J.M, 1983. Empirical use of longevity data to estimate mortality rates. Fish.Bull.NOAA/NMFS, 81(4):898-903

Holden, M.J. and D.F.S. Raitt (eds), 1974. Manual of fisheries science. Part 2. Methods of resource investigations and their application. FAO Fish.Tech.Pap., (115)Rev.1:214 p. Issued also in French and Spanish

Holland, K., R. Brill, S. Ferguson, R. Chang and R. Yost, 1985. A small vessel technique for tracking pelagic fish. Mar.Fish.Rev., 47(4):26-32

Holt, S.J., 1962. The application of comparative population studies to fishery biology - an exploration. In The exploitation of natural animal populations, British Ecological Society Symposium Number Two, edited by E.D. Le Cren and M.W. Holdgate. Oxford, Blackwell Scientific Publications pp:51-69

Holt, S.J., 1963. A method for determining gear selectivity and its application. ICNAF Spec.Publ., (5):106-15

Holt, S.J., 1965. A note on the relationship between mortality rate and the duration of life in an exploited fish population. ICNAF Res.Bull., (2):73-5

Holthuis, L.B., 1980. FAO species catalogue. Vol. 1. Shrimps and prawns of the world. An annotated catalogue of species of interest to fisheries. FAO Fish.Synop., (125)Vol.1:271 p.

Holthuis, L.B., 1991. FAO species catalogue. Vol. 13 Marine lobsters of the world. An annotated catalogue of species of interest to fisheries known to date. FAO Fish.Synop., (125)Vol.13:292 p.

Hoydal, K., C.J. Rørvik and P. Sparre, 1980. A method for estimating the effective mesh size and the effect of changes in gear parameters. ICES C.M. 1980/G:28:33 p. (mimeo)

Hoydal, K., C.J. Rørvik and P. Sparre, 1982. Estimation of effective mesh sizes and their utilization in assessment. Dana, 2:69-95

Hunter, J.R., A.W. Argue, W.H. Bayliff, A.E. Dizon, A. Fonteneau, D. Goodman and G.R. Seckel, 1986. The dynamics of tuna movements: an evaluation of past and future research. FAO Fish.Tech.Pap., (277):78 p.

I-ATTC (Inter-American Tropical Tuna Commission/Comisión Interamericana del Atún Tropical), 1984. Annual report of the Inter-American Tropical Tuna Commission. Informe anual de la Comisión Interamericana del Atún Tropical, 1983. Annu.Rep.I-ATTC/Inf.Anu.CIAT, (1983):272 p.

ICES, 1979. Report of the North Sea round fish working group, Charlottenlund, 7-11 May 1979. ICES, C.M. 1979/G:7 (mimeo)

ICES, 1980. Report of the North Sea round fish working group, Copenhagen, 14-18 April 1980. ICES, C.M. 1980/G:8 (mimeo)

ICES, 1981. Report of the Ad hoc working group on the use of effort data in assessment, Copenhagen, 2-6 March 1981. ICES C.M. 1981/G:5 (mimeo)

ICES, 1984. Report of the Ad hoc multispecies assessment working group, Copenhagen, 18-22 June 1984. ICES C.M. 1984/Assess:20:99 p. (mimeo)

ICES, 1986. Report of the Ad hoc multispecies assessment working group, Copenhagen, 13-19 November 1985. ICES. C.M. 1986/Assess:9:141 p. (mimeo)

ICES, 1987. Report of the Ad hoc multispecies assessment working group, Copenhagen, 12-18 November 1986. ICES C.M. 1987/Assess:9 (mimeo)

ICOD, 1991. A guide to the management and operation of marine research and survey vessels. Halifax, Nova Scotia, International Centre for Ocean Development: 18 sections, 1441 p.

Ihssen, P.E., H.E. Booke, J.M. Casselman, J.M. McGlade, N.R. Payne and F.M. Utter, 1981. Stock identification: materials and methods. Can.J.Fish.Aquat.Sci., 38:1838-55

Ingles, J. and D. Pauly, 1984. An atlas of the growth, mortality and recruitment of Philippine fishes. ICLARM Tech.Rep., (13):127 p.

Isarankura, A.P., 1971. Assessment of stocks of demersal fish off the west coast of Thailand and Malaysia. FAO/UNDP, Indian Ocean Programme, Rome, IOC/DEV/71/20:20 p.

Jamieson, G.S. and N. Bourne (eds), 1986. North Pacific Workshop on stock assessment and management of invertebrates. Can.Spec. Publ.Fish.Aquat.Sci, (92):430 p.

Jensen, J.W., 1986. Gillnet selectivity and the efficiency of alternative combinations of mesh sizes for some freshwater fish. J.Fish.Biol., 28: 637-46

Jones, R., 1961. The assessment of the long term effects of changes in gear selectivity and fishing effort. Mar.Res.Scot., 2:19 p.

Jones, R., 1963. Some theoretical observations on the escape of haddock from a codend. ICNAF Spec.Publ., (5):116-127

Jones, R., 1974. Assessing the long term effects of changes in fishing effort and mesh size from length composition data. ICES C.M. 1974/F: 33:13 p. (mimeo)

Jones, R., 1976. Mesh regulation in the demersal fisheries of the South China Sea area. Manila, South China Sea Fisheries Development and Coordinating Programme, SCS/76/WP/34: 75 p.

Jones, R., 1977. Tagging: theoretical methods and practical difficulties. In Fish population dynamics, edited by J.A. Gulland. New York, Wiley Interscience, pp. 46-66

Jones, R., 1984. Assessing the effects of changes in exploitation pattern using length composition data (with notes on VPA and cohort analysis). FAO Fish.Tech.Pap., (256):118 p.

Jones, R. and N.P. van Zalinge, 1981. Estimates of mortality rate and population size for shrimp in Kuwait waters. Kuwait Bull.Mar.Sci., 2: 273-88

Juanicó, M., 1983. Squid maturity scales for population analysis. FAO Fish.Tech.Pap./FAO Doc.Tech.Pêches/FAO Doc.Téc.Pesca, (231):341-78

Julien-Flüs, M., 1988. A study of growth parameters and mortality rates of Scomberomorus brasiliensis from coastal areas of Trinidad, West Indies. FAO Fish.Rep., (389):385-400

Karlsen, L. and B.A. Bjarnasson, 1986. Small-scale fishing with driftnets. FAO Fish.Tech.Pap., (284):64 p.

Kimura, D.K., 1977. Logistic model for estimating selection ogives from catches of codends whose ogives overlap. J.Cons.CIEM, 38(1):116-9

Kleiber, P.M., A.W. Argue and R.E. Kearney, 1983. Investigation of skipjack stock for management purposes. Collect.Vol.Sci.Pap.ICCAT/Recl.Doc. Sci.CICTA/Colecc.Doc.Cient. CICAA 18(3):794-811

Klima, E.F., 1976. An assessment of the fish stocks and fisheries of the Campeche Bank. WECAF Stud., (5):24 p.

Lablache, G. and G. Carrara, 1988. Population dynamics of emperor red snapper, Lutjanus sebae, with notes on the demersal fishery on the Mahé Plateau, Seychelles. FAO Fish.Rep., (389):171-92

Lange, A.M.T. and M.P. Sissenwine, 1983. Squid resources of the Northwest Atlantic. FAO Fish.Tech.Pap./FAO Doc.Tech.Pêches/FAO Doc.Téc.Pesca, (231):21-54

Larkin, P.A., 1972. The stock concept and management of Pacific salmon. H.R. MacMillan lectures in fisheries. Vancouver, University of British Columbia Press, 231 p.

Larkin, P.A., 1977. An epitaph for the concept of maximum sustainable yield. Trans.Am.Fish.Soc., 106(1):1-11

Larkin, P.A., 1988. Pacific salmon. In Fish population dynamics: the implications for management, edited by J.A. Gulland, Chichester, John Wiley and Sons Ltd., pp. 153-83

Laurec, A. et J.-C. LeGuen, 1981. Dynamique des populations marines exploitées. Tome I. Concepts et Modèles. CNEXO Rapports Scientifiques et Techniques (45):117 p.

Losse, G.F. and A. Dwippongo, 1977. Report on the Java Sea south east monsoon trawl survey, June-December 1976. Jakarta, Marine Fisheries Research Report (Special report). Contribution of the Demersal Fisheries Project No. 3:119 p.

Ludwig, D., 1987. Estimation of delay-difference model parameters from catch-effort data. Coop.Fish.Res.Unit.Rep.Ser.Univ.B.C.

Ludwig, D. and C.J. Walters, 1985. Are age-structured models appropriate for catch-effort data? Can.J.Fish.Aquat.Sci., 42:1066-72

Macdonald, P.D.M. and T.J. Pitcher, 1979. Age-groups from size-frequency data: a versatile and efficient method of analyzing distribution mixtures. J.Fish.Res.Board Can., 36:987-1001

Mackett, D.J., 1973. Manual of methods for fisheries resource survey and appraisal. Part 3. Standard methods and techniques for demersal fisheries resource surveys. FAO Fish.Tech.Pap., (124):39 p.

MacLean, J.A. and D.O. Evans, 1981. The stock concept, discreteness of fish stocks and fisheries management. Can.J.Fish.Aquat.Sci., 38:1889-98

MacLennan, C.N. (ed.), 1992. Fishing gear selectivity. Fisheries Research, Special issue. Fish.Res., 13(3):201-352

Mahon, R. (ed.), 1987. Report and proceedings of the expert consultation on shared fishery resources of the Lesser Antilles region. Mayaguez, Puerto Rico, 8-12 September 1986. FAO Fish.Rep., (383):278 p.

Mahon, R. and S. Mahon, 1987. Seasonality and migration of pelagic fishes in the eastern Caribbean. FAO Fish.Rep., (383):192-273

Márquez M., R., 1990. FAO species catalogue Vol. 11. Sea turtles of the world. An annotated and illustrated catalogue of sea turtle species known to date. FAO Fish.Synop., (125) Vol.11:81 p.

Marten, G.G., 1978. Calculating mortality rates and optimum yields from length samples. J.Fish.Res.Board.Can., 35(2):197-201

Marten, G.G. and J.J. Polovina, 1982. A comparative study of fish yields from various tropical ecosystems. ICLARM Conf.Proc., (9):255-85

Mather, F.J., H.L. Clark and J.M. Mason, 1975. Synopsis of the biology of the white marlin *Tetrapterus albidus* Poey (1861). NOAA Tech.Rep.NMFS (Spec.Sci.Rep.-Fish.Ser.), (675)Pt.3, Species Synopses:55-94

Mather, F.J., J.M. Mason and H.L. Clark, 1974. Results of sailfish tagging in the Western North Atlantic Ocean. NOAA Tech.Rep.NMFS SSRF (Spec.Sci. Rep.-Fish.Ser.), (675)Pt.2. Proceedings of the international billfish symposium, Kailua-Kona, Hawaii, 9-12 August 1992. Review and contributed papers:194-210

Mather, F.J., J.M. Mason and H.L. Clark, 1974a. Migrations of white marlin and blue marlin in the Western North Atlantic Ocean - tagging results since May, 1970. NOAA Tech.Rep.NMFS (Spec.Sci.Rep.-Fish.Ser.), (675) Pt.2. Proceedings of the international billfish symposium, Kailua-Kona, Hawaii, 9-12 August 1992. Review and contributed papers:211-22

Matsumoto, W.M., R.A. Skillman and A.E. Dizon, 1984. Synopsis of biological data on skipjack tuna, (<u>Katsuwonus pelamis</u>). <u>FAO Fish.Synop.</u>, (136): 92 p. Issued also as <u>NOAA Tech.Rep.NMFS Circ.</u>, (451):92 p.

McCombie, A.M. and F.E.J. Fry, 1960. Selectivity of gill nets for lake whitefish <u>Coregonus clupeaformis</u>. <u>Trans.Am.Fish.Soc.</u>, 89:176-84

McNew, R.W. and R.C. Summerfelt, 1978. Evaluation of a maximum-likelihood estimator for analysis of length-frequency distributions. <u>Trans.Am.Fish. Soc.</u>, 107(5):730-6

Megrey, B.A., 1989. Review and comparison of age-structured stock assessment models from theoretical and applied points of view. <u>American Fisheries Society Symposium</u>, 6:8-48

Mesnil, B., 1989. Computer programs for fish stock assessment. ANACO: Software for the analysis of catch data by age group on IBM PC and compatibles. <u>FAO Fish.Tech.Pap.</u>, (101)Suppl.3:73 p. (with programs on diskettes) (Also issued in French in 1988, Spanish version in press)

Mohamed, K.H., N.P. van Zalinge, R. Jones, M. El-Musa, M. Al-Hossaini and A.R. Abdul-Chaffar, 1979. Mark-recapture experiments on the Gulf shrimp, <u>Penaeus semisulcatus</u>, De Haan, in Kuwait waters. Shrimp stock evaluation and management project, Kuwait. Report TF/KUW-6/R-10:59 p. (mimeo) (Also issued in 1981 as <u>KISR Tech.Rep.</u> (401):62 p.

Morales-Nin, B., 1991. Determinación del crecimiento de peces óseos en base a la microestructura de los otolitos. [Determination of growth of bony fishes based on micro-structures in the otoliths]. <u>FAO Doc.Téc.Pesca</u>, (322):58 p. (Will also be issued in English)

Moreau, J., C. Bambino and D. Pauly, 1986. Indices of overall fish growth performance of 100 tilapia (Cichlidae) populations. <u>In</u> The first Asian fisheries forum, edited by J.L. Maclean, L.B. Dizon and L.V. Hosillos, Manila, Philippines, Asian Fisheries Society: pp. 201-6

Munro, J.L., 1974. The mode of operation of Antillean fish traps and the relationships between ingress, escapement, catch and soak. <u>J.Cons.CIEM</u>, 35(3):337-50

Munro, J.L., 1982. Estimation of the parameters of the von Bertalanffy growth equation from recapture data at variable time intervals. <u>J.Cons. CIEM</u>, 40:199-200

Munro, J.L. (ed.), 1983. Caribbean coral reef fishery resources. <u>ICLARM Stud.Rev.</u>, (7):276 p.

Munro, J.L. and D. Pauly, 1983. A simple method for comparing growth of fishes and invertebrates. <u>ICLARM Fishbyte</u>, 1(1):5-6

Munro, J.L. and R. Thompson, 1983. The Jamaican fishing industry. <u>ICLARM Stud.Rev.</u>, (7):10-4

Munro, J.L. and R. Thompson, 1983a: Areas investigated, objectives and methodology. <u>ICLARM Stud.Rev.</u>, (7):15-25

Murawski, S.A., 1984. Mixed species yield-per-recruitment analyses accounting for technological interactions. <u>Can.J.Fish.Aquat.Sci.</u>, 41:897-916

Murphy, G.I., 1966. Population biology of the pacific sardine (Sardinops caerulea). Proc.Calif.Acad.Sci., 34:1-84

Murphy, G.I., 1977. Clupeoids. In Fish population dynamics, edited by J.A. Gulland. New York, Wiley Interscience, pp. 283-308

Murphy, G.I., 1982. Recruitment of tropical fishes. ICLARM Conf.Proc., (9):141-8

Naamin, N. and S. Noer, 1980. The status of the shrimp fishery in the Arafura Sea. In Report of the Workshop on the biology and resources of penaeid shrimps in the South China Sea area. Part 1. Manila, South China Sea Fisheries Development and Coordinating Programme, SCS/GEN/80/26:13-32

Nakamura, I., 1985. FAO species catalogue. Vol. 5. Billfishes of the world. An annotated and illustrated catalogue of marlins, sailfishes, spearfishes and swordfishes known to date. FAO Fish. Synop., (125)Vol.5: 65 p.

Nédélec, C., 1982. Definition and classification of fishing gear categories. FAO Fish.Tech.Pap., (222):51 p. Issued also in French and Spanish

Nishikawa, Y., et al., 1985. Average distribution of larvae of oceanic species of Scombroid fishes, 1956-1981. Far Seas Fish.Res.Lab., S Series (12):99 p.

Osako, M. and M. Murata, 1983. Stock assessment of cephalopod resources in the Northwestern Pacific. FAO Fish.Tech.Pap./FAO Doc.Tech.Pêches/FAO Doc. Téc.Pesca, (231):55-144

Oxenford, H.A. and W. Hunte, 1986. Migration of the dolphin (Coryphaena hippurus) and its implications for fisheries management in the Western Central Atlantic. Proc.Gulf.Caribb.Fish.Inst., 37:95-111

Packard, A., 1966. Optional convergence between cephalopods and fish: an exercise in functional anatomy. Arch.Zool.Ital., 51:523-43

Paloheimo, J.E., 1958. A method of estimating natural and fishing mortalities. J.Fish.Res.Board Can., 15(4):749-58

Paloheimo, J.E., 1961. Studies on estimation of mortalities. I. Comparison of a method described by Beverton and Holt and a new linear formula. J.Fish.Res.Board Can., 18(5):645-62

Paloheimo, J.E., 1980. Estimation of mortality rates in fish populations. Trans.Am.Fish.Soc., 109:378-86

Paloheimo, J.E. and E. Cadima, 1964. On statistics of mesh selection. ICNAF Serial No. 1394/Doc.No. 98

Panella, G., 1971. Fish otoliths: daily growth layers and periodical patterns. Science, Wash., 137(4002):1124-7

Parrish, B.B. (ed.), 1973. Fish stocks and recruitment. Proceedings of a symposium held in Aarhus 7-10 July 1970. Rapp.P.-V.Réun.CIEM, (164):372 p.

Paulik, G.J., 1963. Estimates of mortality rates from tag recoveries. Biometrics, 19:28-57

Pauly, D., 1979. Theory and management of tropical multispecies stocks: a review with emphasis on the Southeast Asian demersal fisheries. ICLARM Stud.Rev., (1):35 p.

Pauly, D., 1980: A selection of simple methods for the assessment of tropical fish stocks. FAO Fish.Circ., (729):54 p. Issued also in French. Superseded by FAO Fish.Tech.Pap., (234)

Pauly, D., 1980a: A new methodology for rapidly acquiring basic information on tropical fish stocks: growth, mortality and stock-recruitment relationships. In Stock assessment for tropical small-scale fisheries. Proceedings of an International Workshop held 19-21 September 1979, edited by S.B. Saila and P. Roedel. Kingston, University of Rhode Island, International Center for Marine Resource Development, pp. 154-72

Pauly, D., 1980b: On the interrelationships between natural mortality, growth parameters, and mean environmental temperature in 175 fish stocks. J.Cons.CIEM, 39(2):175-92

Pauly, D., 1981: Tropical stock assessment package for programmable calculators and micro-computers. ICLARM Newsl., 4(3):10-3

Pauly, D., 1983. Some simple methods for the assessment of tropical fish stocks. FAO Fish.Tech.Pap., (234):52 p. Issued also in French and Spanish

Pauly, D., 1983a. Length-converted catch curves. A powerful tool for fisheries research in the tropics. (Part I). ICLARM Fishbyte, 1(2):9-13

Pauly, D., 1984. Fish population dynamics in tropical waters: a manual for use with programmable calculators. ICLARM Stud.Rev., (8):325 p.

Pauly, D., 1984a. Length-converted catch curves. A powerful tool for fisheries research in the tropics. (Part II). ICLARM Fishbyte, 2(1):17-9

Pauly, D., 1984b. Length-converted catch curves. A powerful tool for fisheries research in the tropics. (III: Conclusion). ICLARM Fishbyte, 2(3):9-10

Pauly, D., 1987. A review of the ELEFAN system for analysis of length-frequency data in fish and aquatic invertebrates. ICLARM Conf.Proc., (13):7-34

Pauly, D., 1988. Fisheries research and the demersal fisheries of Southeast Asia. In Fish population dynamics: the implications for management, edited by J.A. Gulland, Chichester, John Wiley and Sons Ltd., pp. 329-48

Pauly, D., 1990. Can we use traditional length-based fish stock assessment when growth is seasonal? Fishbyte, 8(3):29-32

Pauly, D. and J.F. Caddy, 1985. A modification of Bhattacharya's method for the analysis of mixtures of normal distributions. FAO Fish.Circ., (781):16 p.

Pauly, D. and N. David, 1981. ELEFAN I, a BASIC program for the objective extraction of growth parameters from length-frequency data. Meeresforschung, 28(4):205-11

Pauly, D. and G. Gaschütz, 1979. A simple method for fitting oscillating length growth data, with a program for pocket calculators. ICES C.M. 1979/G:24:26 p. (mimeo)

Pauly, D. and G.R. Morgan (eds), 1987. Length-based methods in fisheries research. ICLARM Conf.Proc., (13):468 p.

Pauly, D. and J.L. Munro, 1984. Once more on the comparison of growth in fish and invertebrates. ICLARM Fishbyte, 2(1):21

Pauly, D. and G.I. Murphy (eds) 1982. Theory and management of tropical fisheries. ICLARM Conf.Proc., (9):360 p.

Pauly, D. and N.A. Navaluna, 1983. Monsoon-induced seasonality in the recruitment of Philippine fishes. FAO Fish.Rep., (291)Vol.3:823-33

Pauly, D., M.Soriano-Bartz, J.Moreau and A. Jarre (in press). A new model accounting for seasonal cessation of growth in fishes. Austr.J.Mar.Freshwat.Res. (special issue on growth)

Pauly, D. and P.Sparre, 1991. A note on the development of a new software package, the FAO-ICLARM Stock Assessment Tools (FiSAT). ICLARM Fishbyte, 9(2):47-49

Pearson, K., 1894. Contribution to the mathematical theory of evolution. Philos.Trans.R.Soc.Lond.(Ser.A.), 185:71-110

Penn, J.W., 1984. The behavior and catchability of some commercially exploited penaeids and their relationship to stock and recruitment. In Penaeid shrimps - their biology and management, edited by J.A. Gulland and B.J. Rothschild. Farnham, Surrey, England, Fishing News Books Ltd., pp. 173-86

Pennington, M., 1983. Efficient estimators of abundance, for fish and plankton surveys. Biometrics, 39:281-6

Petersen, C.G.J., 1892. Fiskenes biologiske forhold i Holbaek Fjord, 1890-91. Beret.Danm.Biol.St., 1890(1)1:121-183. (in Danish)

Peterson, I. and S.J. Wroblewski, 1984. Mortality rate of fishes in the pelagic ecosystem. Can.J.Fish.Aquat.Sci., 41:1117-20

Pitcher, T.J. and P.D.M. Macdonald, 1973. Two models for seasonal growth in fishes. J.Appl.Ecol., 10:597-606

Polovina, J.J., 1984. An overview of the ECOPATH model. ICLARM Fishbyte, 2(2):5-7

Polovina, J.J., 1984a. Model of the coral reef ecosystem. Part I. The ECOPATH model and its application to French frigate shoals. Coral Reefs, 3(1):1-11

Pope, J.A., A.R. Margetts, J.M. Hamley and E.F. Akyüz, 1975. Manual of methods for fish stock assessment. Pt3. Selectivity of fishing gear. FAO Fish.Tech.Pap., (41)Rev.1:65 p.

Pope, J.G., 1972. An investigation of the accuracy of virtual population analysis using cohort analysis. Res.Bull.ICNAF, (9):65-74

Pope, J.G., 1979. Stock assessment in multispecies fisheries with special reference to the trawl fisheries in the Gulf of Thailand. Manila. South China Sea Development and Coordinating Programme, SCS/DEV/79/19:106 p.

Pope, J.G., 1979a. A modified cohort analysis in which constant natural mortality is replaced by estimates of predation levels. ICES C.M. 1979/H:16:7 p. (mimeo)

Pope, J.G., 1980. Assessment of multispecies resources. In Selected lectures from the CIDA/FAO/CECAF seminar on fishery resource evaluation. Casablanca, Morocco, 6-24 March 1978. Rome, FAO, Canada Funds-in-Trust, FAO/TF/INT 180(c) Suppl.:93-137

Pope, J.G., 1980a. Phalanx analysis: an extension on Jones' length cohort analysis to multispecies cohort analysis. ICES C.M. 1980/G:19:18 p.(mimeo)

Pope, J.G., 1988. Collecting fisheries assessment data. In Fish population dynamics: the implications for management, edited by J.A. Gulland, Chichester, John Wiley and Sons Ltd., pp. 63-82

Postel, E., 1955. Contribution a l'étude de la biologie de quelques Scombridae de l'Atlantique tropico-oriental. Ann.Stn.Oceanogr.Salammbo, (10): 167 p.

Powell, D.G., 1979. Estimation of mortality and growth parameters from the length frequency of a catch. Rapp.P.-V.Réun.CIEM, 175:167-9

Pütter, A., 1920. Studien über physiologische Ähnlichkeit. VI. Wachstumsähnlichkeiten. Pflüger Arch.Ges.Physiol., 180:298-340

Raj, D., 1968. Sampling theory. New York, McGraw-Hill.

Ralston, S., 1982. Influence of hook size in the Hawaiian deep-sea handline fishery. Can.J.Fish.Aquat.Sci., 39:1297-302

Randall, J.E., 1962. Tagging reef fishes in the Virgin Islands. Proc.Gulf Caribb.Fish.Inst., 14:201-41

Rao, C.R., 1965. Linear statistical interference and its applications. New York, John Wiley, 625 p. 2nd ed.

Regier, H.A. and D.S. Robson, 1966. Selectivity of gill nets, especially to lake whitefish. J.Fish.Res.Board Can., 23:423-54

Ricker, W.E., 1954. Stock and recruitment. J.Fish.Res.Board Can., 11: 559-623

Ricker, W.E., 1973. Linear regressions in fishery research. J.Fish.Res. Board.Can., 30:409-34

Ricker, W.E., 1975. Computation and interpretation of biological statistics of fish populations. Bull.Fish.Res.Board Can., (191):382 p.

Riedel, D., 1963. Contribution to the experimental determination of the selection parameter of gill nets. Arch.Fischereiwiss., 14:85-97

Rikhter, V.A. and V.N. Efanov, 1976. On one of the approaches to estimation of natural mortality of fish populations. ICNAF Res.Doc., 76/VI/8: 12 p.

Robson, D.S., 1966. Estimation of relative fishing power of individual ships. Res.Bull.ICNAF, (3):5-14

Robson, D.S. and D.G. Chapman, 1961. Catch curves and mortality rates. Trans.Am.Fish.Soc., 90(2):181-9

Rodríguez, A., 1977. Contribución al conocimiento de la biologia y pesca del langostino, Penaeus kerathurus (Forskål 1775) del golfo de Cádiz (Región Sudatlántica española). Invest.Pesq.Barc., 41(3):603-35

Roper, C.F.E., M.J. Sweeney and C.E. Nauen, 1984. FAO species catalogue. Vol. 3. Cephalopods of the world. An annotated and illustrated catalogue of species of interest to fisheries. FAO Fish.Synop., (125)Vol.3:277 p.

Rosenberg, A.A. and J.R. Beddington, 1988. Length-based methods of fish stock assessment. In Fish population dynamics: the implications for management, edited by J.A. Gulland, Chichester, John Wiley and Sons Ltd., pp. 83-103

Rothlisberg, P.C., B.J. Hill and D.J. Staples (eds), 1985. Second Australian national prawn seminar. Cleveland, Queensland. Australia, NPS2:361 p.

Rudstam, L.G., J.J. Magnuson and W.M. Tonn, 1984. Size selectivity of passive fishing gear: a correction for encounter probability applied to gill nets. Can.J.Fish.Aquat.Sci., 41:1252-5

Russ, G.R., 1991. Coral reef fisheries: effects and yields. In The ecology of fishes on coral reefs, edited by P.E. Sale, San Diego, Academic Press, Inc., pp. 601-35

Russell, B.C., 1990. FAO species catalogue Vol. 12. Nemipterid fishes of the world (Threadfin breams, whiptail breams, monocle breams, dwarf monocle breams and coral breams) Family Nemipteridae. An annotated and illustrated catalogue of nemipterid species known to date. FAO Fish.Synop. (125)Vol.12: 149 p + 8 colour plates

Russell, E.S., 1931. Some theoretical considerations on the "overfishing" problem. J.Cons.CIEM, 6:1-20

Saeger, J., P. Martosubroto and D. Pauly, 1976. First report of the Indonesian-German demersal fisheries project (Result of a trawl survey in the Sunda Shelf area). Jakarta, Marine Fisheries Research Report (Special report). Contribution of the Demersal Fisheries Project no. 1:46 p.

Saila, S.B. and P. Roedel (eds), 1980. Stock assessment for tropical small-scale fisheries. Proceedings of an International Workshop held 19-21 September 1979, at the University of Rhode Island, Kingston R.I., International Center for Marine Resources Development, University of Rhode Island, 198 p.

Sainsbury, K.J., 1988. The ecological basis of multispecies fisheries, and management of a demersal fishery in tropical Australia. In Fish population dynamics: the implications for management, edited by J.A. Gulland, Chichester, John Wiley and Sons Ltd., pp. 349-82

Sale, P.F. (ed.), 1991. The ecology of fishes on coral reefs. San Diego, Academic Press, Inc., 754 p.

Sato, T. and H. Hatanaka, 1983. A review of assessment of the Japanese distant-water fisheries for cephalopods. FAO Fish.Tech.Pap./FAO Doc.Tech. Pêches/FAO Doc.Téc.Pesca, (231):145-80

Saville, A. (ed.), 1977. Survey methods of appraising fisheries resources. FAO Fish.Tech.Pap., (171):76 p.

Saville, A. (ed.), 1980. The assessment and management of pelagic fish stocks. A symposium held in Aberdeen, 3-7 July 1978. Rapp.P.-V.Réun. CIEM, 177:517 p.

Schaefer, M., 1954. Some aspects of the dynamics of populations important to the management of the commercial marine fisheries. Bull.I-ATTC/Bol. CIAT, 1(2):27-56

Schaefer, M., 1957. A study of the dynamics of the fishery for yellowfin tuna in the eastern tropical Pacific Ocean. Bull.I-ATTC/Bol.CIAT, 2:247-68

Schnute, J., 1985. A general theory for analysis of catch and effort data. Can.J.Fish.Aquat.Sci., 42:414-29

Schnute, J., 1987. A general fishery model for a size-structured fish population. Can.J.Fish.Aquat.Sci., 44(5):924-40

Schnute, J. and D. Fournier, 1980. A new approach to length-frequency analysis: growth structure. Can.J.Fish.Aquat.Sci., 37:1337-51

SCSP (South China Sea Development Programme), 1978. Report on the workshop on the demersal resources of the Sunda Shelf, Part 1. Manila, South China Sea Fisheries Development and Coordinating Programme, SCS/GEN/77/12:44 p.

Seber, G.A.F., 1973. The estimation of animal abundance and related parameters. London, Griffin, 506 p.

Sharp, G.D., 1978. Behavioral and physiological properties of tunas and their effects on vulnerability to fishing gear. In The physiological ecology of tunas, edited by G.D. Sharp and A.E. Dizon. New York, Academic Press, pp. 397-449

Sharp, G.D. and A.E. Dizon (eds), 1978. The physiological ecology of tunas. New York, Academic Press, 485 p.

Sharp, G.D. and/y J. Csirke (eds), 1983. Proceedings of the expert consultation to examine changes in abundance and species composition of neritic fish resources. San José, Costa Rica, 18-29 April 1983. Actas de la consulta de expertos para examinar los cambios en la abundancia y composición por especies de recursos de peces neríticos. San José, Costa Rica, 18-29 abril 1983. FAO Fish.Rep./FAO Inf.Pesca, (291): Vol.2:1-553 and/y Vol.3: 557-1224

Shepherd, J.G., 1987. A weakly parametric method for estimating growth parameters from length composition data. In Length-based methods in fisheries research, edited by D. Pauly and G. Morgan, ICLARM Conf.Proc. 13: pp. 113-9

Shepherd, J.G., 1988. Fish stock assessments and their data requirements. In Fish population dynamics: the implications for management, edited by J.A. Gulland, Chichester, John Wiley and Sons Ltd., pp. 35-62

Shindo, S., 1973. General review of the trawl fishery and the demersal fish stocks of the South China Sea. FAO Fish.Tech.Pap., (120):49 p.

Sickle, J. van, 1977. Mortality rates from size distributions: the application of a conservation law. Oecologia, Berl., 27:311-8

Sinclair, M., 1988. Marine populations: An essay on population regulation and speciation. Books in recruitment fishery oceanography. Washington Sea Grant Program. University of Washington Press, Seattle and London: 252 p.

Sissenwine, M.P., 1978. Is MSY an adequate foundation for optimum yield? Fisheries, 3(6):22-4 and 37-42

Sissenwine, M.P., M.J. Fogarty and W.J. Overholtz, 1988. Some fisheries management implications of recruitment variability. In Fish population dynamics: the implications for management, edited by J.A. Gulland, Chichester, John Wiley and Sons Ltd., pp. 129-52

Smith, T.D., 1988. Stock assessment methods: the first fifty years. In Fish population dynamics: the implications for management, edited by J.A. Gulland, Chichester, John Wiley and Sons Ltd., pp. 1-33

Sokal, R.R. and F.J. Rohlf, 1981. Biometry. The principles and practice of statistics in biological research. San Francisco, California, Freeman and Company, 2nd ed.

Som, R.K., 1973. A manual of sampling techniques. London, Heinemann Educational Books Ltd., 384 p.

Sousa, M.I., 1988. Sources of bias in growth and mortality estimation of migratory pelagic fish stocks, with emphasis on Decapterus russelli (Carangidae) in Mozambique. FAO Fish.Rep., (389):288-307

Spain, J.D., 1982. BASIC microcomputer models in biology. Reading, Massachusetts. Addison-Wesley Publishing Company, 354 p.

Sparre, P., 1979. Some necessary adjustments for using the common methods in eel assessment. Rapp.P.-V.Réun.CIEM, 174:41-44

Sparre, P., 1979a. Some remarks on the application of yield/recruit curves in estimation of maximum sustainable yield. ICES C.M. 1979/G:41:21 p. (mimeo)

Sparre, P., 1980. A goal function of fisheries (Legion analysis). ICES C.M. 1980/G:40:81 p. (mimeo)

Sparre, P., 1983. Legion analysis game and simulation program (LAGS). NAFO SCR Doc. 83/IX/71/Ser.No. N736:49 p.

Sparre, P., 1984. A computer program for estimation of food suitability coefficients from stomach content data and multispecies VPA. ICES C.M. 1984/G:25:60 p. (mimeo)

Sparre, P., 1987. Computer programs for fish stock assessment. Length-based fish stock assessment for Apple II computers. FAO Fish.Tech.Pap., (101) Suppl.2:218 p.

Sparre, P., 1987a. A method for the estimation of growth, mortality and gear selection/recruitment parameters from length-frequency samples weighted by catch per effort ICLARM Conf.Proc., (13):75-102

Sparre, P., 1991. Introduction to multispecies virtual population analysis. ICES Mar.Sci.Symp. (193):12-21

Sparre, P. and R. Willmann, (in press). Computer programs for bio-economic analysis of fisheries. BEAM4 Manual. Analytical bio-economic simulation of space-structured multi-species and multi-fleet fisheries. FAO Computerized Information Series (Fisheries), No. 3:xxx p. (with programs on diskettes)

SPC (South Pacific Commission), 1984. An assessment of the skipjack and baitfish resources of northern Mariana Islands, Guam, Palau, Federated States of Micronesia and Marshall Islands. Final Ctry Rep.Skipjack Surv. Assess. Programme S.Pac.Comm.,(18):111 p.

Stéquert, B. and F. Marsac, 1986. La pêche de surface de thonidés tropicaux dans l'océan Indien. FAO Doc.Tech.Pêches, (282):213 p.

Strømme, T., 1992. NAN-SIS: Software for fishery survey data logging and analysis. User's manual. FAO Computerized Information Series (Fisheries), No. 4. Rome, FAO. 103 p.

Sturm, M.G. de L., 1974. Aspects of the biology of the Spanish mackerel, Scomberomorus maculatus (Mitchell) in Trinidad, West Indies. Trinidad, University of the West Indies, Ph.D. Thesis

Sukhatme, R.V. and B.V. Sukhatme, 1970. Sampling theory of surveys with applications. Ames, Iowa, Iowa State University Press, 452 p.

Sund, P.N., M. Blackburn and F. Williams, 1981. Tunas and their environment in the Pacific Ocean: A review. Oceanogr.Mar.Biol., 19:443-512

Tanaka, S., 1953. Precision of age-composition of fish estimated by double sampling method using the length for stratification. Bull.Jap.Soc.Sci.Fish., 19:657-70

Tanaka, S., 1960. Studies on the dynamics and the management of fish populations. Bull.Tokai.Reg.Fish.Res.Lab., (28):1-200 (In Japanese)

Taylor, C.C., 1960. Temperature, growth and mortality - the Pacific cockle. J.Cons.CIEM, 26:117-24

Thompson, D.W., 1910. The works of Aristotele. Vol. 4. Oxford, Clarendon Press.

Thompson, W.F. and F.H. Bell, 1934. Biological statistics of the Pacific halibut fishery. 2. Effect of changes in intensity upon total yield and yield per unit of gear. Rep.Int.Fish. (Pacific Halibut) Comm., (8):49 p.

Tiews, K. and P. Caces-Borja, 1965. On the availability of fish of the family Leiognathidae Lacépedè in Manila Bay and San Miguel Bay and on their accessibility to controversial fishing gears. Philipp.J.Fish., 7(1):59-85

Tiurin, P.V., 1962. The natural mortality factor and its importance in regulating fisheries. Vopr.Ikhtiol., (2):403-27 (in Russian)

Troadec, J.-P., 1977. Méthodes semi-quantitatives d'évaluation. FAO Circ.Pêches, (701):131-141

Troadec, J.-P., 1980. Utilization of resource survey results in stock assessment. In Selected lectures from the CIDA/FAO/CECAF seminar on fishery resource evaluation. Casablanca, Morocco, 6-24 March 1978. Rome, FAO, Canada Funds-in-Trust, FAO/TF/INT 180(c) Suppl.:139-52

Ursin, E., 1968. A mathematical model of some aspects of fish growth, respiration and mortality. J.Fish.Res.Board Can., 24:2355-453

Ursin, E., 1982. Stability and variability in the marine ecosytem. Dana, 2:51-67

Ursin, E., 1984. The tropical, the temperate and the arctic seas as media for fish production. Dana, 3:43-60

Vakily, J.M., M.L. Palomares and D. Pauly, 1986. Computer programs for fish stock assessment: Applications for the HP 41 CV calculator. Produced with the cooperation of the International Center for Living Aquatic Resources Management (ICLARM). FAO Fish.Tech.Pap., (101)Suppl.1:255 p.

Venema, S.C., J.M. Christensen and D. Pauly (eds), 1988. Contributions to tropical fisheries biology. Papers prepared by the participants at the FAO/DANIDA Follow-up Training Courses on fish stock assessment in the tropics. Hirtshals, Denmark, 5-30 May 1986 and Manila, Philippines, 12 January-6 February 1987. FAO Fish.Rep., (389):519 p.

Venema, S.C., J.M. Christensen and D. Pauly, 1988a. Training in tropical fish stock assessment: a narrative of experience. FAO Fish.Tech.Pap., (389):1-15

Walford, L.A., 1946. A new graphic method of describing the growth of animals. Biol.Bull.Mar.Biol.Lab.Woods Hole, 90:141-7

Weber, W. and A.A. Jothy, 1977. Observations on the fish Nemipterus spp. (Family: Nemipteridae) in the coastal waters of East Malaysia. Arch. Fischereiwiss., 28(2/3):109-22

Wetherall, J.A., J.J. Polovina and S. Ralston, 1987. Estimating growth and mortality in steady-state fish stocks from length-frequency data. ICLARM Conf.Proc., (13):53-74

Whitehead, P.J.P., 1985. FAO species catalogue. Vol. 7. Clupeoid fishes of the world (Suborder Clupeoidae). An annotated and illustrated catalogue of the herrings, sardines, pilchards, sprats, shads, anchovies and wolf-herrings. Part 1. Chirocentridae, Clupeidae and Pristigasteridae. FAO Fish.Synop., (125)Vol.7, Pt.1:303 p.

Whitehead, P.J.P., G.J. Nelson and T. Wongratana, 1988. FAO species catalogue. Vol. 7. Clupeoid fishes of the world (Suborder Clupeoidei). An annotated and illustrated catalogue of the herrings, sardines, pilchards, sprats, shads, anchovies and wolf-herrings. Part 2. Engraulididae. FAO Fish. Synop., (125)Vol.7.Pt.2.:305-579

Williams, T.P., 1986. Ageing manual for Kuwait fish. KISR 1915. MB-44. Kuwait Institute for Scientific Research.

Willmann, R. and S.M. Garcia, 1985. A bio-economic model for the analysis of sequential artisanal and industrial fisheries for tropical shrimp (with a case study of Suriname shrimp fisheries). FAO Fish.Tech.Pap., (270): 49 p. Issued also in French and Spanish

Worms, J., 1983. World fisheries for cephalopods: a synoptic overview. FAO Fish.Tech.Pap./FAO Doc.Tech.Pêches/FAO Doc.Téc.Pesca, (231):1-20

Wyatt, J.R., 1983. The biology, ecology and bionomics of the squirrelfishes, Holocentridae. ICLARM Stud.Rev., (7):50-8

Yates, G.T., 1983. Hydromechanics of body and caudal fin propulsion. In Fish biomechanics, edited by P.W. Webb and D. Weihs. New York, Praeger Publishers, pp. 177-213

Yonimori, T., 1982. Study of tuna behaviour, particularly their swimming depths by use of sonic tags. Newsl.Enyo (Far Seas) Fish.Res.Lab.Shimizu, (44):1-5 Transl. from Japanese

Yuen, H.S.H., 1970. Behavior of skipjack tuna, Katsowonus pelamis, as determined by tracking with ultrasonic devices. J.Fish.Res.Board Can., 27(11):2071-9

Zalinge, N. van and P. Sparre, 1986. Pakistan. Statistical systems and proposed sampling scheme. Rome, FAO, FI:DP/PAK/77/003, Field Document 6

Ziegler, B., 1979. Growth and mortality rates of some fishes of Manila Bay, Philippines, as estimated from analysis of length-frequencies. Thesis. Kiel University, 115 p.

SUBJECT INDEX

Subject: Section(s)

Age-at-first-capture, T_c 4.1, 8.1
Age-at-recruitment, T_r 4.1
Age-based Thompson and Bell model 8.6
Age determination or ageing 1.4
Age/length key 3.2.1
Age of massive maturation, $T_{m50\%}$ 4.7.3
Age-structured models 1.3.1
Age-transformed selection ogive, exact 6.4.3
Amplitude (ELEFAN), C 3.5.2
ANACO, computer program 15.4
ANALEN, computer program 15.4
Analytical models 1.3
Annual return matched samples method 11.3
Arbitrary age 3.4.2
Assessment of mixed fisheries 10.4.2
Asymptotic length, L_∞ 3.1
Asymptotic weight, W_∞ 3.1.2
Attrition rate, A 11.5
Available sum of peaks (ELEFAN), ASP 3.5.1
Average biomass, B 8.6
Average length of the entire catch, $\overline{L_c}$ 4.5.3
Average long-term catches 8.7
Average number of survivors (of a cohort), $\overline{N}$ 4.2
Average price, $\overline{v}$ 8.6

Baranov's equation 4.2
Beam 1 to 4, computer programs 8.6, 8.7, 15.4
Bell-shaped selection curve 6.2
Bertalanffy, von, growth equation 3.1
Bertalanffy, von, growth parameters 3.1
Bertalanffy, von, inverse growth equation 3.3.3
Bertalanffy, von, plot 3.3.3
Bertalanffy, von, seasonalized growth equation 3.5.2
Bertalanffy, von, weight-based growth equation 3.1.2
Beverton and Holt's biomass per recruit model, B/R 8.3
Beverton and Holt's stock/recruitment model 12.1
Beverton and Holt's relative yield per recruit model, (Y/R)' 8.4
Beverton and Holt's yield per recruit model, Y/R 8.2
Beverton and Holt's Z-equations 4.5.1, 4.5.2, 4.5.3
Bhattacharya's method 3.4.1
Bhattacharya plot 2.6
Bias 1.7, 2.2, 3.4, 6.7, 7.1
Biological interaction 10.2
Biomass estimation 5.1, 8.6, 13.6
Biomass per recruit (Beverton and Holt), B/R 8.3
Birthday, birth date 1.6, 3.2.1
Birthday, estimation of approximative 3.4.2
Body depth 2.4, 6.1
Bottom trawl 6.1, 13.1
Bottom trawl survey 13.2
By-catch 7.6

Cadima's formula 9.3
Carapace length 1.5
Cassie's method 3.4.3
Catch curve 4.4.2, 6.5
Catch curve, based on age composition (constant time interval) . 4.4.3
Catch curve, based on age composition (variable time intervals) . 4.4.4
Catch curve, based on length composition data 4.4.5
Catch curve, cumulated 4.4.4, 4.4.6

Subject:	Section(s)
Catch equation (see Baranov's equation)	4.2
Catch per unit of area, CPUA	13.6
Catch per unit of effort, CPUE	4.3, 9.0, 9.5
Catchability coefficient, q	4.3, 4.6, 9.2
Chapman's method	3.3.2
Chi-squared criterion	3.5.3
Closed season	8.6
Codend	6.1, 13.1
Coefficient of variation, $s/\bar{x}$	2.1
Cohort	1.3.1, 4.1
Cohort analysis, age-based (Pope's)	5.2
Cohort analysis, length-based (Jones')	5.4
COMPLEAT ELEFAN package	14.
Condition factor, q	3.1.2
Confidence interval	2.3
Confidence limits	2.3
Confidence limits of correlation coefficient	2.5
Confidence limits of intercept (a) and slope (b)	2.4
Constant parameter system	4.4.1, 8.1
Correction for mesh selection	6.7
Correlation coefficient, r	2.5
Covariance, sxy	2.4
Covered codend method	6.1
Cumulated catch curve	4.4.4
Cumulated catch curve based on length composition data	4.4.6
Curvature parameter, K	3.1
Daily-ring	1.4
Data massage	3.4.2
Data recording	13.3
Deck sampling	13.4
Degrees of freedom, f	2.3
Delta function or delta distribution	2.2, 13.7
Density dependent mortality	12.1
Dependent variable	2.4
Deriso and Schnute's delay/difference model	9.7
Discards	7.6
Distribution, stock	1.1
Dynamics of a cohort	4.2
Economic interaction	10.3
Effective mesh size	6.6.2
Effort (standardization of effort)	9.8
ELEFAN I	3.5.1
Equilibrium situation	9.2
Explained sum of peaks (ELEFAN I), ESP	3.5.1
Exploitation rate, $E = F/Z$	4.2
Exploited phase (of the cohort's life)	4.1
Exploratory survey	13.0
Exponential decay model	4.2, 8.1
F-at-age-array	5.1
F-factor, X	8.6
Finite population correction factor	7.1
FiSAT package, computer programs	15.3
Fishing mortality coefficient, F	4.2
Fishing pattern	5.1, 6.6.2
Fishing power	9.8
Fleet	7.0
Ford-Walford plot	3.3.2
Fork length	1.5
Fox model	9.1

Subject:	Section(s)
F-pattern	8.8
Fractiles of t-distribution (Student's distribution), t(f)	2.3
Frequency, observed, F	2.2
Frequency table	2.1
Frequency, theoretical, Fc	2.2
Functional regression analysis	2.5
Garcia et al., MSY estimators	9.4
Gear selection curve	4.5.3, 6.1, 6.4.2
Gear selectivity	6.0
General matched samples method	11.4
Gill net selection	6.2.1
Gonadal maturity	3.4.2
Growth curve	3.1
Growth overfishing	8.0, 8.2
Growth parameters	1.1, 3.1
Growth rate	3.1
Gulland and Holt plot	3.3.1
Gulland's formula	9.2
Hanging ratio or hanging coefficient	6.2.1
Head rope	6.2.1, 13.5
Heincke's method or Heincke's formula	4.3.1
Historic or retrospective models	8.6
Holistic models	1.3
Homing	11.1
Hook selection	6.3
Hyperbolic tangent, tanh	2.5
Independent variable, x	2.4
Initial condition parameter (t_o, t-zero)	3.1
Initial guess	3.5.3
Instantaneous rate of fishing mortality, F	4.2
Instantaneous rate of natural mortality, M	4.6, 4.7
Instantaneous rate of total mortality, Z	4.2
Intercept, a	2.4
Interval midpoint, $\bar{L}(j)$	2.1
Interval size, dL	2.1
Inverse regression analysis	2.5
Inverse von Bertalanffy growth equation	3.3.3
Iterative process	3.5.3
Jones and van Zalinge method	4.4.6
Jones' length-based cohort analysis	5.3
Knife-edge recruitment	8.1
Knife-edge selection	6.4.1, 8.1
Least squares method	(2.4), (3.3), 3.3.4
Length-at-age data	3.3
Length-at-first capture, Lc	4.5.3
Length-based Thompson and Bell model	8.7
Length-based cohort analysis (Jones')	5.3
Length, body, carapace, fork standard, total	1.5
Length composition of the catch	7.6
Length-converted catch curve	4.4.5
Length-weight relationship	2.6, 3.1.2
LFSA package	15.1
Linear regression	2.4
Linearized catch curve equation	4.4.2
Linearized catch curve equation with constant time interval	4.4.3

Subject:	Section(s)

Linearized catch curve equation with variable
  time interval 4.4.4
Linearized length-converted catch curve 4.4.5
Linearization of a normal distribution 2.6
Linear transformation 2.6
L-infinity, L_∞, asymptotic length 3.1
Logistic curve 6.1, 6.2.2
Log-normal distribution 13.7
Log sheets 13.3
Longevity 4.7.1

Matched samples method 11.3, 11.4
Maximum fishing mortality, Fm 6.6.1
Maximum likelihood methods 3.5.3
Maximum relative error, ϵ 7.1
Maximum sustainable economic yield, MSE 8.7
Maximum sustainable yield (MSY) 1.1, 4.5, 8.2, 9.1-9.7, 13.7
Maximum sustainable yield per recruit, MSY/R 8.2
Mean value, $\bar{x}$ 2.1
Mean value from frequency sample 2.1
Mesh selection experiments 6.1, 6.2
Mesh size 6.1
Mesh size change 8.8
Migration 1.7, 11.1
Migration coefficient 11.4
Mixed fisheries, assessment of 10.4.2
Modal progression analysis 3.4.2
Models 1.3
Monte Carlo simulation technique 3.5.4
Mortality parameters 1.1
Mortality rates 1.1, 4.2, 4.7
Moving average 3.5.1
Multifleet mixed fisheries 10.4.3
Multispecies VPA 10.2
Munro and Thompson plot 9.5

NAN-SIS, computer program 15.4
Natural mortality, M 4.1
Natural mortality coefficient, M 4.7
Neyman allocation (optimum stratified sampling equation ... 7.2
Normal distribution 2.2
NORMSEP, computer program 3.5.3
Number of survivors (from a cohort), N 4.2
Number of recruits, R, N(Tr) 4.1

Optimum length for being caught, Lm 6.2
Optimum stratified random sampling equation (Neyman allocation) ... 7.2
Overfishing, growth 1.3.1, 8.0
Overfishing, recruitment 1.3.1, 8.0, 12.0

Passive gear 6.2
Parabola method 3.4.3
Parameter 1.2, 1.3, 2.4
Pauly's empirical formula for M 4.7.2
Phi prime test (ϕ') 3.4
Plot of Z on effort 4.6
Powell-Wetherall method .... (3.3.2), 4.5.4
Predation mortality 10.2
Prediction models 8.0, 8.6
Pre-recruitment phase 4.1
Probability 2.2
Probability paper method (Cassie method) 3.4.3

Subject:	Section(s)
Proportional sampling	7.3
Pseudo-cohort	4.4.1
Raising factors	7.1, 7.4, 7.6
Random sampling	1.7, 7.1
Random variability	3.5.4
Recruit	1.6
Recruitment, $R = N(Tr)$	1.6, 4.1
Recruitment, age of, Tr	4.1
Recruitment curve	6.4.2
Recruitment intensity	1.6
Recruitment overfishing	8.0, 8.3
Recruitment pattern	1.6
Recruitment season	1.6
Reference F-at-age array	8.6
Regression	2.4
Regression line	2.4
Relative CPUE, effort, fishing power	9.6
Relative standard deviation, $s/\bar{x}$	2.1
Relative yield per recruit (Beverton and Holt), $(Y/R)'$	8.4
Residual natural mortality	10.2
Resultant curve	6.4.2
Retrospective or historic models	8.0, 8.6
Reversed logistic curve, SR	6.2.2
Ricker's stock/recruitment model (Ricker curve)	12.1
Rikhter and Efanov's formula	4.7.3
Robson and Chapman's method	4.3.2
Sampling	7.0, 13.0
Sampling commercial catches	7.4
Scatter diagram	2.4
Schaefer model	9.1
Schaefer model, multispecies	10.1
Selection curve	6.4
Selection factor	6.1
Selection range	6.1
Separation index, I	3.5.4
Shared stock	1.1
Simple random sampling	7.1
Size-dependent migration	11.2
Slope, b	2.4
Standard deviation, s	2.1
Standard error, $s/\sqrt{n}$	2.3
Standardization of effort	9.6
Standard length	1.5
Steady-state model	8.1
Step function (gear selectivity)	6.6.1
Stock concept	1.1
Stock/recruitment relationship	8.0, 12.0
Stratified random sampling	7.2, 13.1, 13.7
Stratified random sampling, optimum	7.2
Stratum	7.2, 13.7
Student's distribution or t-distribution	2.3
Summer-point (ELEFAN), ts	3.5.2
Surplus production models	1.3.2, 9.0
Surplus production models, multispecies	10.1, 10.2
Survival rate, s	4.2
Swept area method	(4.3), 13.5
Tagging data	11.5
Technical interaction	10.4
Terminal F	5.2

Subject:	Section(s)
Thompson and Bell model, age-based	8.6
Thompson and Bell, length-based	8.7
Total length	1.5
Total mortality coefficient, Z	4.2
Trap selection	6.3
Trawl selection	6.1
t-zero, t_o (initial condition parameter)	3.1
Unexploited biomass, Bv	9.1
Unit stock	1.1
Variable parameter system	4.4.1
Variance or VAR, s^2	2.1
Variance about the regression line	2.4
Variance of estimates of intercept (sa^2) and slope (sb^2)	2.4
Virgin stock biomass, Bv	8.3, 9.1
Virtual population Analysis (VPA)	5.1
Virtual population Analysis, multispecies	10.2
Weight-at-age array	8.6
Wing spread	13.5
Winter-point (ELEFAN), tw	3.5.2
Year-ring	8.6
Yield	8.6
Yield per recruit, Beverton and Holt's, Y/R	8.2
Yield per recruit, Beverton and Holt's relative, (Y/R)'	8.4
Yield per recruit model for mixed fisheries	10.4.1
Yield per recruit, Thompson and Bell's	5.3, 10.3

APPENDIX TABLES

A1 LIST OF IMPORTANT FORMULAS

A2 METHODS BASED ON LINEAR TRANSFORMATIONS AND ORDINARY
    LINEAR REGRESSION ANALYSIS: y = a + b*x

A3 IMPORTANT DATES EXPRESSED AS FRACTIONS OF A YEAR FROM 1 JANUARY

A4 FRACTILES OF THE t-DISTRIBUTION (STUDENT'S DISTRIBUTION)

TABLE A1 LIST OF IMPORTANT FORMULAS

Length-weight relationship

$$W(i) = q*L(i)^b \qquad (2.6.1)$$

Von Bertalanffy growth equation

$$L(t) = L_\infty * [1 - \exp(-K*(t-t_o))] \qquad (3.1.0.1)$$

Growth rate, von Bertalanffy growth equation

$$\frac{\Delta L}{\Delta t} = K*(L_\infty - L(t)) \quad \text{cm/year} \qquad (3.1.0.4)$$

Weight-based von Bertalanffy growth equation

$$W(t) = W_\infty * [1 - \exp(-K*(t-t_o))]^3 \qquad (3.1.2.1)$$

Inverse von Bertalanffy growth equation

$$t(L) = t_o - \frac{1}{K}*\ln(1 - L/L_\infty) \qquad (3.3.3.2)$$

Exponential decay model (1)

$$\frac{\Delta N(t)}{\Delta t} = -Z*N(t) \qquad (4.2.1)$$

Exponential decay model (2)

$$N(t) = N(Tr)*\exp[-Z*(t-Tr)] \qquad (4.2.2)$$

Baranov's equation or catch equation (1)

$$C(t1,t2) = \frac{F}{Z}*[N(t1)-N(t2)] \qquad (4.2.4)$$

Catch equation (2)

$$C(t1,t2) = N(t1) * \frac{F}{Z} * [1 - \exp(-Z*(t2-t1))] \qquad (4.2.7)$$

Average number of survivors (from t1 to t2)

$$\overline{N}(t1,t2) = N(t1) * \frac{1 - \exp(-Z*(t2-t1))}{Z*(t2-t1)} \qquad (4.2.9)$$

Total mortality from CPUE data

$$Z = \frac{1}{t2-t1} * \ln \frac{CPUE(t1)}{CPUE(t2)} \qquad (4.3.0.3)$$

Linearized length-converted catch curve

$$\ln \frac{C(L1,L2)}{\Delta t(L1,L2)} = c - Z*t(\frac{L1+L2}{2}) \qquad (4.4.5.3)$$

Pauly's formula for M (ln-based)

$$\ln M = -0.0152 - 0.279*\ln L_\infty + 0.6543*\ln K + 0.463*\ln T \qquad (4.7.2.1)$$

Logistic curve for gear selectivity (length-based)

$$S_L = \frac{1}{1 + \exp(S1 - S2*L)} \qquad (6.1.1)$$

Schaefer model

$$Y(i)/f(i) = a + b*f(i) \quad \text{if } f(i) \leq -a/b \qquad (9.1.2)$$

Fox model

$$\ln (Y(i)/f(i)) = c + d*f(i) \qquad (9.1.3)$$

VPA equations (age-based)

$$\frac{C(y,t,t+1)}{N(y+1,t+1)} = \frac{F(y,t,t+1)}{M+F(y,t,t+1)} * \left[\exp[F(y,t,t+1)+M] - 1\right] \quad (5.1.3)$$

$$N(y,t) = N(y+1,t+1) * \exp[F(y,t,t+1)+M] \quad (5.1.4)$$

Pope's age-based cohort analysis (numbers)

$$N(t) = \left[N(t+\Delta t)*\exp(M*\Delta t/2) + C(t,t+\Delta t)\right]*\exp(M*\Delta t/2) \quad (5.2.3)$$

Pope's age-based cohort analysis (fishing mortalities)

$$F(t,t+\Delta t) = \frac{1}{\Delta t}*\ln\left[\frac{N(t)}{N(t+\Delta t)}\right] - M \quad (5.2.4)$$

Jones' length-based cohort analysis (numbers)

$$N(L1) = \left[N(L2)*H(L1,L2) + C(L1,L2)\right]*H(L1,L2) \quad (5.3.4)$$

Length-based catch equation

$$C(L1,L2) = N(L1)*\frac{F}{Z}*\left[1 - \exp(-Z*\Delta t)\right] \quad (5.3.5)$$

Jones' length-based cohort analysis (fishing mortalities)

$$F(L1,L2) = M*\frac{F(L1,L2)/Z(L1,L2)}{1-[F(L1,L2)/Z(L1,L2)]} \quad (5.3.6)$$

where the exploitation rate F/Z is derived from:

$$F(L1,L2)/Z(L1,L2) = \frac{C(L1,L2)}{N(L1) - N(L2)} \quad (5.3.7)$$

Set of formulas for age-based Thompson and Bell analysis

age interval: $i = (t_i, t_i + \Delta t)$

$$Z_i = M + X*F_i$$
$$N(t_i + \Delta t) = N(t_i)*\exp(-Z_i*\Delta t)$$
$$C_i = [N(t_i) - N(t_i + \Delta t)]*X*F_i/Z_i$$
$$\overline{w}_i = w(t_i + \Delta t/2)$$
$$Y_i = C_i*\overline{w}_i$$
$$\overline{B}_i = Y_i/[F_i*\Delta t*X]$$
$$V_i = Y_i*\overline{v}_i$$

(8.6.1)

Set of formulas for length-based Thompson and Bell analysis

length interval: $i = (L_i, L_{i+1})$

$$Z_i = M + X*F_i$$

$$N(L_{i+1}) = N(L_i) * \frac{1/H_i - X*F_i/Z_i}{H_i - X*F_i/Z_i} \quad \text{where}$$

$$H_i = \left[\frac{L_\infty - L_i}{L_\infty - L_{i+1}}\right]^{M/2K}$$

$$C_i = [N(L_i) - N(L_{i+1})]*X*F_i/Z_i$$

$$\overline{w}_i = q*[(L_i + L_{i+1})/2]^b$$

$$Y_i = C_i*\overline{w}_i$$

$$V_i = Y_i*\overline{v}_i$$

$$\overline{N}_i*\Delta t_i = [N(L_i) - N(L_{i+1})]/Z_i$$

$$\overline{B}_i*\Delta t_i = \overline{N}_i*\Delta t_i*\overline{w}_i$$

(8.7.7)

TABLE A2 METHODS BASED ON LINEAR TRANSFORMATIONS AND ORDINARY LINEAR REGRESSION ANALYSIS: y = a + b*x

Method	Assumption	Independent (x)	Dependent (y)	Slope (b)	Intercept (a)	Eq.
Length-weight relationship	$W_i = q*L_i^b$	$\ln L_i$	$\ln W_i$	b	$\ln q$	2.6.1
Bhattacharya plot	lengths-at-age are $N(\bar{x}, s^2)$	$x + dL/2$	$\Delta \ln N(x+dL/2)$	$-dL/s^2$	$\bar{x}*dL/s^2$	2.6.5
GROWTH						
Gulland and Holt plot	Δt small but not constant	$\bar{L}(t)$	$\Delta L / \Delta t$	$-K$	$K*L_\infty$	3.3.1.1
Ford-Walford plot	Δt constant	$L(t)$	$L(t+\Delta t)$	$\exp(-K*\Delta t)$	$[1 - \exp(-K*\Delta t)]*L_\infty$	3.3.2.1
Chapman's method	Δt constant	$L(t)$	$L(t+\Delta t) - L(t)$	$\exp(-K*\Delta t) - 1$	$[1 - \exp(-K*\Delta t)]*L_\infty$	3.3.2.2
Von Bertalanffy plot	L_∞ known	t	$-\ln(1 - L(t)/L_\infty)$	K	$-K*t_0$	3.3.3.1
MORTALITY	**steady state**					
Age-based	$t \geq T'$					
Catch curve	Δt small and variable	$t + \Delta t / 2$	$\ln[C(t, t+\Delta t)/\Delta t]$	$-Z$	not used	4.4.4.2
Catch curve	Δt constant	t	$\ln C(t, t+\Delta t)$	$-Z$	not used	4.4.3.1
Cumulated catch curve	$\Delta t = \infty$	t	$\ln C(t, \infty)$	$-Z$	not used	4.4.4.1
Length-based	$L \geq L1 = L'$					
Catch curve	$L \geq L'$	$t\left(\dfrac{L1+L2}{2}\right)$	$\ln \dfrac{C(L1,L2)}{\Delta t(L1,L2)}$	$-Z$	not used	4.4.5.3
Jones-van Zalinge	L_∞ known	$\ln(L_\infty - L1)$	$\ln C(L1, L_\infty)$	Z/K	not used	4.4.6.1
Powell-Wetherall	$L \geq L1 = L'$	$L1$	$\bar{L} - L1$	$-1/(1+Z/K)$	$L_\infty / (1+Z/K)$	4.5.4.1

TABLE A2 (CONTINUED) METHODS BASED ON LINEAR TRANSFORMATIONS AND ORDINARY LINEAR REGRESSION ANALYSIS: $y = a + b*x$

Method	Assumption	Independent (x)	Dependent (y)	Slope (b)	Intercept (a)	Eq.
Plot of Z on effort (f)	constant catchability (q)	f_i	z_i	q	M	4.6.1
GEAR SELECTIVITY						
Covered codend	logistic curve	L	$\ln(1/S_L - 1)$	$-S2$	$S1$	6.1.2.1
Resultant curve (from catch curve)	steady state $M_t = Z - Fm*S_t$	t	$\ln(1/S_t - 1)$	$-T2$	$T1$	6.4.3.3
Gill net selection (two net experiment)	bell-shaped curve (Lm, s^2): $Lm = SF*m$	$(L1+L2)/2$	$\ln\left[\dfrac{Cb(L1,L2)}{Ca(L1,L2)}\right]$	$SF*(m_b - m_a)/s^2$ (not given in Chapter 6)	$-(SF)^2*(m_b^2 - m_a^2)/(2s)^2$	6.2.1.2
SURPLUS PRODUCTION MODELS	**STEADY STATE**					
Schaefer model	$f < -a/b$	f_i	Y_i/f_i	$-MSY/f_{MSY}^2$	$2*MSY/f_{MSY}$	9.1.2
Fox model	constant catchability (q)	f_i	$\ln(Y_i/f_i)$	$-1/f_{MSY}$	$1 + \ln(MSY/f_{MSY})$	9.1.3
STATISTICS						
Mean	n data: (x_i, y_i)	$\bar{x} = \dfrac{1}{n}*\Sigma x_i$	$\bar{y} = \dfrac{1}{n}*\Sigma y_i$	$b = sxy/sx^2$	$a = \bar{y} - \bar{x}*b$	2.1.1 2.4.5/6
Standard deviation	the y's are $N(ax+b, s^2)$ with constant s	sx	sy	sb	sa	
Variance		$sx^2 = \dfrac{\Sigma[x(i)-\bar{x}]^2}{n-1}$	$sy^2 = \dfrac{\Sigma[y(i)-\bar{y}]^2}{n-1}$	$sb^2 = \dfrac{\left[\dfrac{sy}{sx}\right]^2 - b^2}{n-2}$	$sa^2 = sb^2*[\bar{x}^2 + \dfrac{n-1}{n}*sx^2]$	2.1.2 2.4.11/12
Covariance		$sxy = \dfrac{1}{n-1}[\Sigma x(i)y(i) - \dfrac{1}{n}\Sigma x(i)\Sigma y(i)]$				
Confidence limits	t-values in Table 2.3.1			$b \pm sb*t_{n-2}$	$a \pm sa*t_{n-2}$	

TABLE A3   IMPORTANT DATES EXPRESSED AS FRACTIONS OF A YEAR FROM 1 JANUARY

date	no. of days (cumulated)	fraction of year	date	no. of days (cumulated)	fraction of year
1 Jan	0	0.00	15 Jan	14	0.04
1 Febr	31	0.08	15 Febr	45	0.12
1 March	59	0.16	15 March	74	0.20
1 April	90	0.25	15 April	104	0.28
1 May	120	0.33	15 May	135	0.37
1 June	151	0.41	15 June	165	0.45
1 July	181	0.50	15 July	196	0.54
1 Aug	212	0.58	15 Aug	227	0.62
1 Sept	243	0.67	15 Sept	257	0.70
1 Oct	273	0.75	15 Oct	288	0.79
1 Nov	304	0.83	15 Nov	318	0.87
1 Dec	334	0.92	15 Dec	349	0.96
1 Jan	365	1.00	15 Jan	365	1.00
1 Febr	396	1.08	15 Febr	379	1.04
1 March	424	1.16	15 March	407	1.12
etc.	etc.	etc.	etc.	etc.	etc.

TABLE A4   FRACTILES OF THE t-DISTRIBUTION (STUDENT'S DISTRIBUTION)

degrees of freedom f	fractiles 90% t(f)	95% t(f)	99% t(f)	degrees of freedom f	fractiles 90% t(f)	95% t(f)	99% t(f)
1	6.31	12.71	63.66	15	1.75	2.13	2.95
2	2.92	4.30	9.93	16	1.75	2.12	2.92
3	2.35	3.18	5.84	17	1.74	2.11	2.90
4	2.13	2.78	4.60	18	1.73	2.10	2.88
5	2.02	2.57	4.03	19	1.73	2.09	2.86
6	1.94	2.45	3.71	20	1.73	2.09	2.85
7	1.90	2.37	3.50	25	1.71	2.06	2.79
8	1.86	2.31	3.36	30	1.70	2.04	2.75
9	1.83	2.26	3.25	40	1.68	2.02	2.70
10	1.81	2.23	3.17	50	1.67	2.01	2.68
11	1.80	2.20	3.11	60	1.67	2.00	2.66
12	1.78	2.18	3.06	80	1.67	1.99	2.64
13	1.77	2.16	3.01	100	1.66	1.98	2.63
14	1.76	2.15	2.98	∞	1.65	1.96	2.58